Introduction to the
Engineering Profession

Second Edition

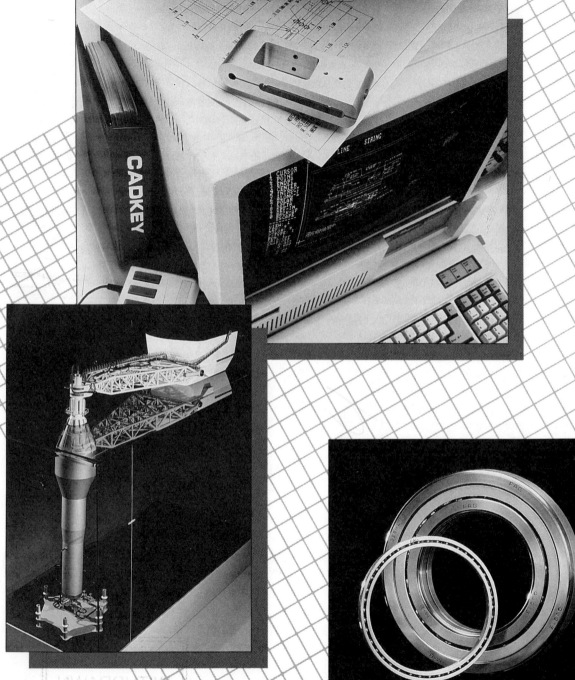

MENSURATION: w, r, b, h, R denote lengths; A denotes area; V denotes volume

Rectangular Prism

$$V = wbh$$

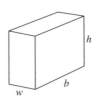

Pyramid (with base of any shape)

$$V = \frac{1}{3} hA$$

A = area of base

Cone

$$V = \frac{1}{3} hA$$

s = slant height

A = area of base

$$\text{lateral area} = \frac{1}{2} (\text{perimeter of base}) \bullet s$$

Cylinder

$$V = \pi r^2 h$$
$$\text{surface area} = 2\pi rh$$

Truncated Cylinder

$$V = \frac{\pi}{2} r^2 h$$
$$\text{surface area} = \pi rh$$

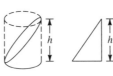

Sphere

r = radius

$$V = \frac{4}{3} \pi r^3$$

$$\text{surface area} = 4\pi r^2$$

Spherical Segment

$$V = \frac{\pi h^2}{3} (3r - h)$$

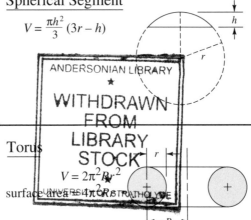

Torus

$$V = 2\pi^2 R r^2$$

$$\text{surface area} = 4\pi^2 Rr$$

Physical Quantities and Constants

Acceleration of gravity (at Earth's surface)	$= 9.807 \text{ m/s}^2$ (32.18 ft/sec^2)
Avogadro's number	$= 6.022 \times 10^{23}$ particles/mol
Density of air (15°C, 760 mm)	$= 1.225 \text{ kg/m}^3$
Density of water (4°C)	$= 1000 \text{ kg/m}^3$ (62.43 lb/ft^3)
Fundamental charge (of an electron, e)	$= 1.602 \times 10^{-19} \text{ C}$
Gravitational constant	$= 6.672 \times 10^{-11} \text{ N} \bullet \text{m}^2/\text{kg}^2$
Heat of fusion of water	$= 3.338 \times 10^5 \text{ J/kg} (143.5 \text{ Btu/lb})$
Heat of vaporization of water	$= 2.256 \times 10^6 \text{ J/kg} (970 \text{ Btu/lb})$
Mass of Earth	$= 5.976 \times 10^{24} \text{ kg}$
Mass of electron	$= 9.110 \times 10^{-31} \text{ kg}$
Mass of proton	$= 1.673 \times 10^{-27} \text{ kg}$
Planck's constant	$= 6.626 \times 10^{-34} \text{ J/Hz}$
Radius of Earth at equator	$= 6378 \text{ km}$
Speed of light in a vacuum	$= 2.998 \times 10^8 \text{ m/s}$
Speed of sound in dry air (0°C, 760 mm)	$= 331.5 \text{ m/s} (1088 \text{ ft/sec})$

Introduction to the
Engineering Profession

Second Edition

John Dustin Kemper
College of Engineering
University of California, Davis

Saunders College Publishing
Harcourt Brace Jovanovich College Publishers

Fort Worth   Philadelphia   San Diego   New York   Orlando
San Antonio   Toronto   Montreal   London   Sydney   Tokyo

Text Typeface: Times Roman
Compositor: SYNTAX, International
Senior Acquisitions Editor: Emily Barrosse
Assistant Editor: Laura Shur
Managing Editor: Carol Field
Project Editor: Anne Gibby
Copy Editor: Mary Patton
Manager of Art and Design: Carol Bleistine
Associate Art Director: Doris Bruey
Text Designer: Anne O'Donnell
Cover Designer: Lawrence R. Didona
Text Artwork: GRAFACON
Director of EDP: Tim Frelick
Production Manager: Jay Lichty
Marketing Manager: Marjorie Waldron

Cover Credit: Photos courtesy of Gabe Palmer/The Stock Market (top, center left and right), Brownie Harris/The Stock Market (bottom)

Frontispiece: Photos courtesy of Cadkey (top), FAG Kugelfischer (center and bottom right)

Printed in the United States of America

Library of Congress Catalog Card Number: 92-054104

INTRODUCTION TO THE ENGINEERING
PROFESSION, second edition

ISBN: 0-03-092858-3

2345 042 987654321

D
620
KEM

Preface

■ OBJECTIVES OF THE TEXT

Over the years, engineering faculty members have observed that freshman students are frequently unclear about what lies ahead for them. Often, students are not sure if they have made the right choices, or indeed, if they should be in engineering at all. Their concerns stem from a lack of information, and to help solve this problem, many engineering schools have developed various kinds of courses for their freshman students—orientation and motivational courses, courses in problem-solving skills, introduction to design, and various combinations of these. In all of these cases, the objective has been to give the student a preliminary view of what the activities of professional engineers are like. This text is intended for courses of this type.

It is the purpose of this book to attempt to communicate to students a "feel" for what engineers do. In carrying out this objective, I believe it is not enough to rely only on descriptive terms. It is also important to engage students in some of the technical aspects of the various fields, through personal experience with engineering problem solving. For that reason, this book contains several chapters on technical subjects. Numerous problems are provided, which permit students to sample the nature of some of the branches of engineering. Obviously, not every branch of engineering can be included. That would be impossible. But there should be enough representation from some of the major fields—mechanics, electric circuits, computers, and energy—to give a broad perspective. Even though the problems may be relatively simple ones—necessarily so, because the scientific and mathematical repertoires of freshman students are still relatively limited—this is the only way to communicate an actual feel for the various disciplines.

Engineering students spend much of their time in the first two years studying mathematics and science, and it is not surprising that they find it difficult to discover the nature of "real engineering" behind these somewhat abstract topics. A related problem is that it is often not clear to students at this stage what the purpose is in studying so much mathematics. Clearly, students will do a more effective job of planning their own educational goals if they have a broad view, right at the beginning, of the terrain which lies ahead, than if they wait two or three years for that view gradually to emerge.

■ ORGANIZATION

This text is organized in such a way that it can be adapted to different kinds of freshman orientation courses. Chapters 1 through 6 give an overview of the engineering profession and its challenges. They carry an underlying theme that engineering is immensely broad and diverse, and that there is something in engineering for almost every kind of taste. These chapters also carry the message that engineering is an "action-oriented" profession, and is of central importance to our modern society.

Chapters 7 through 10 cover some of the basic skills needed by engineers— presentation of results, statistics, units, and communication—while chapters 11 through 13 deal with the creative side of engineering.

The final six chapters deal with technical subjects, most of which will be met again, in depth, by the student at later stages of the educational program. Topics may be selected by the instructor from among these, in accordance with the objectives of the course.

■ NEW TO THIS EDITION

Major revisions have been incorporated in the second edition of this text, many of them in response to suggestions offered by users of the first edition and by reviewers.

A special chapter entitled "Getting Started in Engineering," has been included with newly added material on budgeting time, taking notes, and problem solving.

A new chapter entitled "Modern Engineering Challenges," contains discussions of major topics such as acid rain, global warming, international competitiveness, rebuilding the infrastructure, water and air pollution, and energy.

Another new chapter, entitled "Safety and Public Responsibility," incorporates case histories of the *Challenger* disaster, the DC-10, and Chernobyl. Also, material on the important topic of product liability has been included in this chapter.

The chapter on engineering design has been expanded and realistic cases drawn from actual engineering practice have been included which illustrate the design process. In discussing engineering design, the centrality of manufacturing is emphasized.

The material on statics, dynamics, and mechanics of materials has been increased to include vector algebra, simple trusses, elementary beam analysis, and additional material on dynamics.

Most of the material on graphical communication has been shifted to other chapters. In particular, the material on layouts and working sketches has been incorporated in the chapter on engineering design, where it fits properly with the "implementation" step of the design process.

Other changes include the addition of the subject of correlation to the chapter on statistics, and more examples of statistical applications. In the chapter on computers, the use of BASIC has been supplanted by the use of FORTRAN. More explanation of units systems has been included where appropriate, as well as greater emphasis on the use of free-body diagrams and upon applications. In addition, many new problems have been included throughout.

■ ADDITIONAL FEATURES

A lengthy appendix is included for the use of the student, as an aid to problem solving. In particular, it is recognized that most students will come to this book with no exposure to calculus. Yet, they will soon be studying calculus, and calculus is a powerful and essential mathematical tool for the understanding of the physical world. Therefore, calculus is encountered several places in the text, but no prior understanding of calculus on the part of the student is expected or needed. In some cases, straightforward graphical approximations to calculus are included, as in finding the area under a curve, or in describing slopes of curves. In a couple of other cases, differential equations arise, as in the case of $v = L\,di/dt$, but the solutions are given without derivation. The complete solutions are given in the appendix for those who want them, but are not essential to an understanding of the text. Even if the student does no more than read the technical chapters, without fully understanding the mathematics, something of value will have been gained, because the student will acquire some appreciation of the scope of engineering, and the role that mathematics plays in problem solving.

The appendix also includes reference sections on logarithmic functions, determinants, and summation conventions, because these topics frequently are troublesome to students even though they may have been formally exposed to them. An extensive section on graphics is included for reference, as well as several sections on units, such as approved and non-approved SI units, and a unit conversion table.

Numerous exercises have been included throughout the book, so that students may test their own skills at problem solving. These have been kept at a relatively straightforward level, recognizing that the brief treatment given here of the various scientific principles has not been sufficiently deep to permit tackling more challenging problems. Nevertheless, the problems have been constructed to have varying levels of difficulty, and some of them may indeed be perceived to be challenging to the students' ingenuity. Many of the exercises have complete solutions attached, for students to follow as models. Answers have been provided in the back of the book for those problems which have numerical answers. In the case of chapters where problem-type exercises are not appropriate, topics for "study and discussion" have been suggested.

■ ACKNOWLEDGMENTS

In preparing this book I have drawn upon my experience both in industry and in education—15 years in scientific instruments and computer systems as a design engineer and engineering manager, and 25 years in education as a professor and dean of engineering. I am deeply grateful to all my engineering colleagues who have taught me those things I value today. Among those in industry whom I believe were especially influential in this regard are Wyche D. Caldwell, William L. Martin, Trude C. Taylor, John H. Weaver, and Arthur J. Winter. In education, I am especially grateful to the following, all of whom either are or have been deans or associate deans of engineering at the various campuses at the University of California: Don O. Brush, Martin C. Duke, Clyne F. Garland, Warren H. Giedt, Ray B. Krone, Zuhair A. Munir, John B. Powers, and John R. Whinnery. Special debts of gratitude go to Roy Bainer, the founding dean of engineering at U. C. Davis, and Llewellyn Boelter, the founding dean of engineering at U.C.L.A., both now deceased.

I would also like to give my thanks to Professors James W. Baughn, Gary E. Ford, Jerald M. Henderson, and Edward D. Schroeder, all at the University of California, Davis, who graciously read portions of the manuscript and made constructive comments. I am also deeply grateful to the following reviewers, who helped immeasurably to improve the text for this new edition: John Barrett Crittenden, Virginia Polytechnic Institute and State University; John W. Hakola, Hofstra University; Robert L. Skaggs, University of Nevada, Las Vegas; Wendell C. Bean, Lamar University; B. Peter Daay, Columbus College; Thomas E. Glenn, University of Illinois at Urbana-Champaign; Bettie Stansilao, North Dakota State University; and R. E. Boughner, Western Michigan University. My thanks go, too, to the staff at Saunders College Publishing who shepherded the manuscript through to a completed text: Emily Barrosse, Senior Acquisitions Editor; Laura Shur, Assistant Editor; and Anne Gibby, Project Editor.

My deepest and most special feeling of gratitude goes to my wife, Barbara, for her patience and forbearance, who knows it is like entering a tunnel whenever a textbook-writing project comes into view.

John Dustin Kemper
Davis, California
May 1992

Contents

Introduction to the
Engineering Profession

Second Edition

Drawings from Appleby, *Appleby's Illustrated Handbook of Machinery*, London, 1877 (top), and Agricola, *De Re Metallica*, 1556 (Dover Publications, 1950) (center). Photograph at bottom right courtesy of General Electric.

1

The Historical Development of Engineering

■ 1.1 THE ORIGINS OF ENGINEERING

Engineering began when the human race did. From the moment when a rock was shaped into a sharp edge and became a tool, or when energy was consciously directed to human use in the form of a campfire, engineering existed. As civilization developed, so did engineering.

The origins of many of the devices and methods we take for granted today are lost in antiquity. No one knows, for example, when human beings first learned to create fire on demand, yet such an invention must have required intellect of a high order. Just because fire is occasionally created spontaneously in nature, it would by no means be obvious that man could do the same. Yet today almost all our methods of modern energy conversion are based on fire. Likewise, the lever and the wheel are essential elements of modern technology, though it may be easier to visualize the invention of these devices than the invention of fire. The lever could have arisen from the use of a pry stick, and the wheel from log rollers, for example. At any rate, these were all in use thousands of years ago, as were such things as the bow drill (forerunner of the lathe), the balance beam, the pulley, the crank, the bellows (complete with leather valves), and methods for smelting metals from their ores for making porcelain, for weaving, and for constructing irrigation works. Stone working was developed to a high degree, as is demonstrated by the gigantic structures of Mesopotamia, Egypt, and Central America that still exist today.

People commonly look upon the pyramids of Egypt as one of the major examples of engi-

1

neering achievement. And the pyramids do deserve our respect; after all, the largest of them, the Great Pyramid of Cheops, as originally constructed, was 481 feet high. That is about the same as a 48-story building. Furthermore, the ratio of its height to the perimeter of its base is almost exactly the same as the ratio of the radius of a circle to the circumference. It is not likely that this relationship occurred accidentally; it must have been intentional—for what reason, we can only guess. Even without knowing the reason, we must respect the knowledge that made it possible. The Cheops pyramid was built sometime between 4235 and 2450 B.C., and there it stands today, a major monument to the engineering capability that existed some 6000 years ago.

Other engineering achievements of that long-ago period are not as visible as the pyramids but actually may have been more significant in the long run. One of these had to do with the construction of canals and aqueducts, which made the creation of cities possible and expanded the use of agriculture. Long before 3000 B.C., the Sumerians had drained the marshes near the Persian Gulf and had constructed canals for irrigation. Humans have continued this kind of development ever since. Without it, civilization as we know it might be impossible. Today the fields of water supply and drainage are some of the most basic elements of civil engineering and are based upon thousands of years of human activity. It is only in this century that civilization has produced such affluence that we can afford to leave certain marshes undrained and certain water supplies undeveloped in order to provide habitats for wildlife. Only the wealthy, highly developed portion of our world can afford that kind of restraint.

Another development that in the long run has proved more significant than the pyramids is the replacement of human power with other kinds of power. Humans started very early to use oxen to haul loads, but the invention of a simple device— the horse collar—made a new power source available. Horses are faster and more efficient than oxen, but early attempts to use them had failed because their harnesses strangled them and prevented them from working effectively. The horse collar, imported from China, solved the strangulation problem. The result was a major spurt in human development. An even bigger spurt occurred when horse power was replaced by mechanical power, giving rise to the period we know as the Industrial Revolution, but that story comes later.

Other inventions imported from China were the crank, paper, and gunpowder. The crank is important because it converts rotary motion into linear motion and vice versa. Even in our modern automobiles, the crank is vital for converting the linear motion of a piston into the rotary motion of the drive shaft. Paper, of course, is fundamental to all our ideas of civilization. (Imagine what would happen to society if every form of paper suddenly disappeared.) The Egyptians had papyrus, of course, but papyrus was made by a laborious process of cutting the pith of a papyrus plant into slices and building a layered structure of these slices into a usable material. What the Chinese contributed was a way of making paper directly from vegetable fibers. As for gunpowder, there are probably plenty of people who wish it had never been invented. Even if the Chinese had not developed it, however, others would have. The Chinese are believed to have come up with the idea in about A.D. 850, but it was discovered (apparently independently) by the English monk Roger Bacon in the thirteenth century and again by a German monk, Berthold Schwartz, in the fourteenth century. And even though we may deplore the use of gunpowder to kill people, we must remember it was the forerunner of dynamite, without which we would find it impossible to build many of the structures we need today. (Think of trying to build a modern freeway without using dynamite, for example. Of course, there are people who believe they could get along very nicely without freeways, too, but the great majority of voters are in favor of them and often vote to increase their taxes only on the condition that the money goes to freeways.)

Speaking of freeways leads us to the Romans. The Romans could hardly have held their empire together as long as they did without the excellent road system they built, and making roads is one of the principal things engineers do. To help them in this and their other construction activities, the Romans had something new—a good cement. They did something else important: They extended the use of the arch to new purposes, such as bridges. The Romans did not invent the arch; it had long been used by the Chinese, the Babylonians, and the Etruscans. The Egyptians, too, were familiar with the arch, but had made little use of it. The Romans extended the arch to new applications, using it, for example, in the construction of their remarkable aqueducts. (The famous aqueduct at Pont du Gard, France, is 160 feet high and contains three tiers of arches.) The arch also enabled the Romans to build the Pantheon, a domed structure 140 feet high that still exists today.

The most enormous structure in human history was begun in the third century B.C. and took approximately 400 years to finish. This is the Great Wall of China, 25 feet high and 20 feet wide, with large watchtowers every few hundred yards. It is "officially" 3900 miles long, counting offshoots and parallel walls, although most references give its length as 1700 miles. It could accommodate five mounted horsemen riding abreast and has been referred to as almost a "superhighway" across the rugged terrain. It has also been called the world's longest graveyard because of the countless slaves who died constructing the wall and were buried in place. It is one of the most remarkable achievements of all time, and one of the few marks of human activity on the Earth's surface that are visible from outer space.

There was not much science in the ancient world, even though there was plenty of engineering. The word *science* comes from the Latin word *scientia* (knowledge), whereas the word *engineer* comes from the Latin *ingeniare* (to design, devise) and has the same root as the word *ingenious*. Even today, the word for "engineer" in both French and German is *ingenieur*. Thus, from the very beginning, engineering was based upon creating devices and structures. The main thing separating the engineering of today from that of yesterday is that we have a greater base of scientific knowledge for the things we do.

We should not, however, underestimate the capabilities of engineers who worked almost entirely from an empirical base rather than a scientific one. For example, some of the magnificent Gothic cathedrals of Europe were constructed in the tenth to thirteenth centuries, long before any firm basis for structural theory had been developed. Engineers learned from experience of the enormous load-carrying ability of the arch and used arches in inspired ways to create the lofty, soaring naves we admire today. And, because these arches imposed outward-thrusting forces on the cathedral walls, the engineers of those days devised the exterior flying buttresses to hold the walls in place.

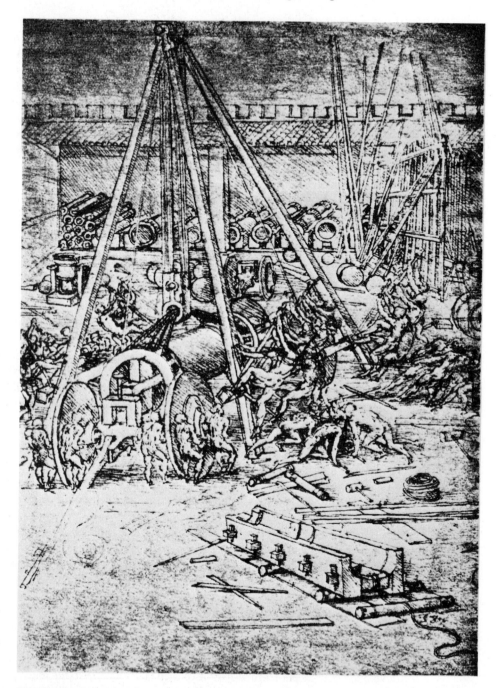

Figure 1.1
Sketch of a gun foundry by Leonardo da Vinci (1452–1519). (Osvald Sirēn, *Leonardo da Vinci, the Artist and the Man*, Yale University Press, New Haven, Conn., 1916)

⬛ Famous Engineers
Leonardo: Incredible Engineer

Leonardo da Vinci, the painter of *The Last Supper* and the *Mona Lisa*, was one of the greatest artists of all time. He was also a versatile and creative engineer and conceived ideas that were hundreds of years ahead of his time.

He was interested in everything, and apparently regarded his work with science and engineering to be a natural part of his life as an artist. In 1482 he offered his services to the duke of Milan in a letter that said, in part,

> I have a sort of extremely light and strong bridge, so contrived as to admit of being carried with the greatest ease, and with the aid of which one may pursue, and at anytime flee from, the enemy.
>
> I have also a kind of bombards, extremely convenient and easy to transport; with these it is possible to hurl forth showers of small stones, a perfect whirlwind.
>
> Further, I can make armored wagons, secure and indestructible, which, when they are introduced among the enemy, manned with shooters, can break through even the largest army of armed men.
>
> In time of peace I believe myself able to vie successfully with any in the designing of public and private buildings and in conducting water from one place to another.
>
> Further, I can execute in sculpture, and likewise also in painting, anything that is at all possible of accomplishment and against any competitor whatsoever.

At the time Leonardo wrote his letter, the city-states of Italy were engaged in recurrent warfare against each other and against France. Thus, it must have seemed natural to Leonardo to offer his services as a military engineer. In the preceding excerpt the anticipation of the modern armored tank is clearly evident. It also was while Leonardo was in Milan, presumably as a military engineer, that he painted his magnificent *Last Supper*.

Thousands of pages of Leonardo's notes and sketches have survived, and it is from them that we learn of the incredible scope of his inventive genius. His notes show that he invented canal-cutting machines, a steam-driven cannon, the submarine, and the parachute. The invention of the submarine, however, he kept secret, because he feared that evil men would "practice assassinations at the bottom of the sea."

One of Leonardo's most persistent visions was of mechanical flight. He studied the flight of birds carefully and sketched numerous machines that he thought might fly. None did, of course, because an engine of sufficient lightness and power was unavailable. Most of his machines used flapping wings, as in a bird's flight, but one of his conceptions shows clearly how original his mind was. This was a machine with a cloth-covered helix that, if turned rapidly, would carry the machine into the air. In other words, he had invented the helicopter.

References
S. Rapport and H. Wright (eds.), *Engineering* (New York: New York University Press, 1963).
O. Siren, *Leonardo da Vinci, the Artist and the Man* (New Haven: Yale University Press, 1916).

Very early, engineering was divided into two kinds: *military* engineering, dealing with engines of destruction; and *civil* engineering, meaning everything that was not military. The earliest works of civil engineers were buildings, roads, bridges, canals, and dams, and the name *civil engineer* continues to this day to be associated with the general field of public works. But, beginning in about the seventeenth century, civil, or nonmilitary, engineering underwent the technological explosion that has made engineering such an influential force in our world today and that progressively gave birth to mechanical engineering, electrical engineering, and all the rest.

■ 1.2 INFLUENTIAL INVENTIONS OF THE MIDDLE AGES

It is common to think of the Middle Ages (fifth through fifteenth centuries) as a period of stagnation characterized by lack of social progress. Yet, as we have just seen, some of the grandest architectural creations of the human race arose in that period. Also, two machines were invented during this supposed stagnant period that had enormous impact upon subsequent progress; they were the weight-driven clock and the printing press.[1]

The trouble with a weight-driven clock is that the weight moves faster the farther it falls. If the weight could be made to fall a short, fixed distance and then stopped and made to fall the same fixed distance again, the time elapsed during each interval would be identical. If a device could be invented that would make the weight fall the same short distance again and again, an accurate method for measuring a succession of identical time intervals would have been found. The mechanical device that does this is called the *escapement*, and it dates from the thirteenth century—not only from the Middle Ages, but from that portion sometimes included in the Dark Ages. The impact of an accurate time-keeping mechanism was twofold: It started a social revolution in ways of doing business, and it also laid the foundation for an accurate instrument-making trade without which later scientific advances would have been impossible.

The story of the invention by Johann Gutenberg (1394–1467) of printing from movable type is a familiar one. (See boxed item.) However, certain aspects are worth emphasizing:

1. This invention arose from the Middle Ages and undoubtedly played a large part in ushering in the Renaissance.
2. It is believed that more books were published in the 50 years following Gutenberg's invention than in the previous thousand years.
3. Gutenberg, in developing his movable type system, had to solve the problem of casting large numbers of pieces of type with closely matching dimensions, so they could be set up in a press with the type surfaces evenly aligned. Otherwise, some pieces would project higher than others and would produce an uneven impression.

As a consequence of his solutions to these problems, some have declared that Gutenberg should be considered the first production engineer.

---

[1] D. S. L. Cardwell, *Turning Points in Western Technology* (New York: Neale Watson Academic Publcations, 1972).

GUTENBERG'S FIRST PROOF.

Figure 1.2
Gutenberg's invention of printing with movable type in about 1450 was
one of the great triumphs of the Middle Ages, helping to bring on the
Renaissance. (Arthur S. Bolles, *Industrial History of the United States*,
Henry Bill Publishing Co., Norwich, Conn., 1892)

◼ 1.3 THE GROWTH OF SCIENCE IN ENGINEERING

The beginnings of a scientific basis for engineering can be traced to Georgius Agricola
(1494–1555) and Galileo Galilei (1564–1642). Agricola lived in Saxony, now a part
of East Germany, and is considered to be the father of mineralogy. His monumental
treatise on the "metallic arts," *De Re Metallica*, published posthumously in 1556,
brought together and systematically organized all known knowledge on mining and
metallurgy. It remained the principal authority for nearly 200 years—an achievement
few books can claim. As his translators Herbert and Lou Hoover have said, "His
statements are sometimes much confused, but we reiterate that their clarity is as
crystal to mud in comparison with those of his predecessors—and of most of his
successors for over two hundred years."[2]

---

[2] Herbert C. Hoover and Lou H. Hoover (translators), *Georgius Agricola, De Re Metallica* (New York:
Dover Publications, 1950; originally published in 1912), pp. xii–xiii. Herbert Hoover, of course, became
the thirty-first President of the United States in 1929.

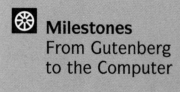

Milestones
From Gutenberg
to the Computer

History tells us that Johann Gutenberg invented movable type in the fifteenth century. Actually, movable type was in use by the Chinese as early as the eleventh century. But the Chinese language required 40,000 characters, so there was no economical advantage to the process. It is also believed by some that a Dutch printer was using movable metal type 30 years before Gutenberg, but the evidence is not clear. The important thing is that Gutenberg solved the problem in such a way that his methods took hold and spread rapidly through Europe, creating thereby a "passion for books." Gutenberg started a revolution in printing that may not be over yet.

Gutenberg and his colleagues made their type by striking an impression of each letter into a piece of soft copper with a steel punch. Each impression was the reverse of the final piece of type and formed part of a mold. All pieces were cast in the same basic mold, with one of the struck copper pieces as its bottom. Gutenberg had solved the problem of making large quantities of metal type characters whose dimensions were so precise that they were interchangeable.

The trouble with Gutenberg's method was that each piece of type had to be set by hand, a method that was not to change substantially for the next 400 years. When the next major advance came, it was in the form of a machine with a keyboard somewhat like that of a typewriter, which, in response to the operator's depressing of the keys, would automatically select and line up a row of type molds. Then it would automatically inject molten metal into the molds, casting a whole line of type at once. The machine, called a Linotype, was a mechanical engineering marvel.

But the biggest advance has occurred just since the 1950s and essentially departs from Gutenberg's invention by dispensing with metal type altogether. The "typesetting" is done photographically, under computer control. The input can be by typewriter keyboard, magnetic tape, or "floppy disk." (See Chapter 18.) In a typical machine of this kind, all the characters are contained on a transparent glass disk. As the disk rotates, a 2-microsecond burst of light flashes just at the instant the desired character passes the exposure aperture, causing that character to be caught on photographic film. If the input is from a source such as magnetic tape, as many as 1500 lines of type per minute can be set, with the computer handling everything automatically, including hyphenation, pagination, and selection of type size and style.

After exposure, the images of the pages are transferred—again photographically—onto printing plates. The photosensitive materials on the plates allow the exposed portions to retain ink, while the unexposed portions do not. During the actual printing, which can occur at very high speeds, the exposed portions of the plates pick up ink and transfer it to the paper, producing the desired images. Nowhere in the system is there any metal type, movable or otherwise.

References
Will Durant, *The Story of Civilization*: *Part VI, The Reformation* (New York: Simon and Schuster, 1957).
W. P. Spence and D. G. Vequist, *Graphic Reproduction* (Peoria, Ill.: Chas. A. Bennett Co., Inc., 1980).

Galileo, who lived most of his life in Florence, Italy, is best known for his astronomical observations and his declaration that objects of different masses fall at the same rate. He also attempted to develop a theory of stress in structural members. For example, he visualized a cantilever beam of rectangular cross section as a pair of

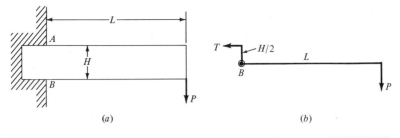

Figure 1.3
Schematic representation of Galileo's attempt to analyze the stress in a cantilever beam.

levers (see Figure 1.3). He deduced that the cantilever beam, set into a wall as shown in Figure 1.3(a), would tend to rotate around the fulcrum point B when the load P was applied at the end of the beam. In Figure 1.3(b), the load P acting with the lever arm L would be resisted by the internal force T within the beam, acting with a lever arm of half the depth of the beam. Thus, he reasoned,

$$T\frac{H}{2} = PL \tag{1.1}$$

Galileo's reasoning happened to be wrong. But the basic concept embodied in Equation (1.1) is very important because it is the Principle of Equilibrium, applied to moments. (See Chapter 14; a *moment* is the product of a force times a distance, in this case the force P acting at right angles to the lever arm, or *moment arm*, multiplied by the distance L. Likewise, the force T multiplied by its lever arm, or moment arm, $H/2$, is also a moment.) Where Galileo went wrong was in assuming that all the "fibers" in the beam were in tension above the fulcrum point. Actually, the "fulcrum," or *neutral axis*, as we now call it, is at the midpoint of the beam, so the fibers above the neutral axis are in tension and the fibers below are in compression.

Galileo apparently did not include any allowance for the elasticity, or "stretch," of materials, but those who followed him soon did. One of these was Robert Hooke, who published the first paper on elasticity in 1678. His paper included the concept now known as Hooke's law, which stated that the elongation of a material would be proportional to the load, or "As the pull, so the stretch." Upon this concept our modern *theory of elasticity* is based. Actually, the idea of Hooke's law was developed simultaneously in France by E. Mariotte, and it was also Mariotte who corrected Galileo's cantilever beam theory by noting that the fibers located below the neutral axis are in compression.

Occasionally in history there have appeared individuals who exerted an enormous influence on the subsequent development of science and technology. Galileo certainly was one of these, and so was Isaac Newton. Newton's principal legacy to engineers is his famous three laws of motion:

1. Bodies at rest tend to remain at rest, and those moving in a straight line continue to so move, unless acted upon by an external force.

Figure 1.4
An overshot water wheel from Agricola, 1556. [Georgius
Agricola, *De Re Metallica* (translated by H. C. and L. H.
Hoover and published in *The Mining Magazine*, 1912;
1950 edition by Dover Publications, New York)]

2. Force and acceleration are proportional to each other, and the constant of proportionality is called *mass*.
3. For every action, there is an equal and opposite reaction.

These laws are dealt with in more detail in Chapter 14 on statics and dynamics. The first law is an expression of what is called the *law of equilibrium,* and the second is given by the famous formula $F = ma$. Newton worked out many of the solutions to the problems of the motions of the planets in his extraordinary *Principia Mathematica* (1687) and along the way invented calculus. But calculus was simultaneously and independently invented in Germany by Gottfried Leibniz. Leibniz published his work before Newton did, but Newton claimed he was first and that he had simply neglected to publish his work promptly. Naturally, a squabble over priority ensued, and later historians have often settled the matter by considering Newton and Leibniz as co-discoverers.

■ 1.4 CONVERTING FIRE INTO MECHANICAL WORK

Galileo, Newton, and Leibniz all lived in the seventeenth century, which has some-times been called "the century of genius." Toward the end of that century there oc-curred an event that has had an incalculable effect upon all of us, because man finally learned how to convert fire into mechanical work. One wonders how anyone living then could ever have conceived that such a thing was possible. Fire was fire, and mechanical force was mechanical force, and that was that. What was there to make anyone suppose one could be converted to the other?

It came about this way. First, the atmosphere had to be "discovered." That is, it had to be recognized that we live in the bottom of an ocean of air and that air has weight. Galileo first had an inkling of this. His pupils Torricelli and Viviani for-mulated the idea of an "ocean of air"; they also invented the mercury barometer (1645). Later Blaise Pascal designed an experiment with a barometer showing that the air pressure decreases as one ascends a mountain.

Virtually everyone has seen the schoolbook picture of 16 horses attempting to separate the two halves of a sphere from which the air had been evacuated. To con-duct such an experiment required that the air pump first be invented, which is credited to Otto von Guericke, in the 1650s. It was also von Guericke who set up the demon-stration of the horses and the evacuated sphere, and then he took the crucial next step by evacuating the air from a cylinder with a movable piston, causing a weight attached to the piston to be lifted.

The invention of a cylinder with a movable piston was crucial to the development of the "fire-engine," as it later came to be called. It is believed the idea originated from the syringe used by the Egyptians in mummification, and Hero used pistons in cylinders for pumping water in the second century B.C. Von Guericke's experiment is diagrammed in Figure 1.5. When air was pumped out of the cylinder, the atmosphere pressing down on the piston caused the weight to be lifted.

Von Guericke published his work in 1672, but it took almost 20 years more be-fore the link to fire was found. This was accomplished by the Frenchman Denis Papin, who suggested in 1691 that a pool of water should be placed in the bottom of the cylinder. The cylinder would be heated by placing a fire under it, and, as the water turned to steam, the piston would be forced to the top of the cylinder. At this point, the piston would be attached to the weight to be lifted, and the fire removed. As the steam cooled it would condense, creating a vacuum. The air pressure would force the piston down, and the weight would be lifted.

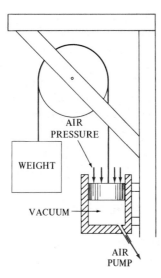

AIR
PRESSURE

WEIGHT

VACUUM

AIR
PUMP

Figure 1.5
Diagram of von Guericke's experiment to show how air pressure can lift a weight.

The steam engine had been invented, but hardly anyone knew it. It was to be another 40 years before it came into general use, and then only for pumping water from deep mines. It was almost a hundred years after Papin that James Watt devised ways to make the steam engine fast enough to be used for tasks other than pumping water.

A steam engine using a somewhat different principle from Papin's was developed by an Englishman, Thomas Savery, about 1700. Savery's device consisted of a pair of chambers that worked alternately, each equipped with inlet and exit valves. There were no moving parts other than the valves. Steam under pressure would be admitted to one of the chambers, forcing out any water that was there. Cold water would then be poured on the outside of the chamber, causing the steam to condense and create a vacuum. The vacuum would suck up water from below into the chamber, and the water would be held there by a one-way valve. Then steam would be admitted to the chamber again, pushing the water out through another one-way valve and to a higher level. Savery's machine was called "the miner's friend." Mines were always troubled by the infiltration of water; thus, it was not surprising that the early application of steam engines was to the pumping of water from mines.

Beginning about 1705, another Englishman, Thomas Newcomen, put the principle of the Papin steam engine into practical application. However, instead of boiling the steam inside the cylinder, he admitted the steam to the cylinder whenever it was wanted by operating a valve. Then someone had to pour cold water on the cylinder to get the steam inside to condense and create a vacuum. The piston, acting under atmospheric pressure, pulled down on the end of a rocker arm, and the other end of the rocker pulled up on a cable that operated a piston-and-cylinder water pump deep in the mine. By 1712 Newcomen had contrived to make his machine completely automatic, so that it was no longer necessary to operate the valves manually or throw cold water on the cylinder by hand.

Savery's engine used steam under pressure, whereas Newcomen's engine generated and used steam at atmospheric pressure. Steam under pressure was dangerous (and still is), and the pressure vessels of the time were unreliable. Thus, Newcomen's engine immediately offered much safer operation than did Savery's. Also, Newcomen's device could be placed at the surface of a mine and pump water from deep levels, whereas Savery's, dependent in part upon suction, could not lift water very far and had to be located down in the mine. Word of Newcomen's engine spread quite rapidly and within a few years the engine was in use in North and South America as well as in Europe.

Newcomen's engine was practical and useful, but slow acting and inefficient. It was a Scotsman, James Watt (1736–1819), who recognized this inefficiency and set out to do something about it. After he had struggled with the problem for some time, the solution came to him suddenly one Sunday as he was strolling on Glasgow Green. The Newcomen engine required that the cylinder be alternately heated and cooled: first heated to allow the steam to enter without being condensed prematurely, then cooled to condense the steam at the right time, then heated again, and so on. Such a process produced an inherently slow engine, and one that wasted much energy. Watt's inspiration was to have a condensing chamber separate from the piston and cylinder. The cylinder would always be hot, and the condensing chamber always cold. At the beginning of a cycle, a valve would open, admitting steam (at atmospheric pressure) to the upper side of the piston, while another valve connected the under side of the piston to the near-vacuum in the condensing chamber. Thus, with atmospheric pressure above and a near vacuum below, the piston would be driven down, doing useful work. At the end of the stroke, the inlet and outlet valves would close and an "equilibrating" valve would open, connecting the spaces in the cylinder above and below the piston so that the pressures on both sides were equal. The piston was connected to one end of a rocker arm (see Figure 1.6), and the pump linkage extending down into the mine was connected to the other. The weight of the pump linkage would cause the piston to be drawn to the top of its stroke, whereupon the equilibrating valve would close and the cycle would begin anew.

It took Watt nearly ten years from the time he first had his inspiration until he produced a successful engine in 1774. Subsequently, he added a means for converting the to-and-fro pumping action to a smooth rotary motion and invented a governor that would automatically adjust the steam supply as a function of the engine speed. The Industrial Revolution was ready to be born.

■ 1.5 THE INDUSTRIAL REVOLUTION

The Industrial Revolution is generally thought of as embracing the century from 1750 to 1850, but it was the central part of this period that saw the highest rate of change. Watt's engines, which were far more efficient than Newcomen's, began coming into general use about 1790 and by 1825 had been adapted to self-propelling locomotives.

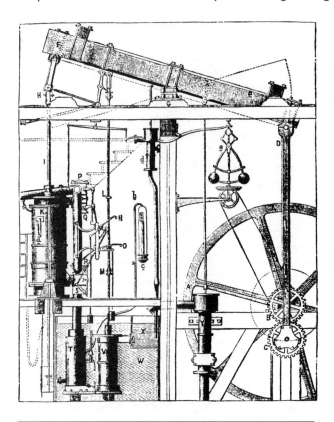

Figure 1.6
Watt's double-acting steam engine, about 1780. (R. H.
Thurston, *Histoire de la Machine à Vapeur*, Librairie
Germer Baillière, Paris, 1880)

Interestingly, before the locomotive could exist, Watt's basic patent on his steam engine had to expire. We should recall that Watt used steam at atmospheric pressure, even though he well knew that his machine could be made more powerful if he used steam at high pressure to drive the piston. But he apparently felt that high-pressure steam was too dangerous and refused to consider it. Others did not agree, and after Watt's patent expired in 1800, they hastened to use steam at high pressure and thus produced light, powerful engines with the specific view of achieving self-propelling locomotion. The first to do so was Richard Trevithick, who in 1804 demonstrated a locomotive hauling a set of cars loaded with coal. The locomotive operator did not ride aboard, but walked alongside as if he were tending a horse.

The principal effect of the Industrial Revolution was to bring the factory system into being. Before this time, water-powered wheels were in widespread use. As early as the eleventh century there were 5000 watermills in the southern part of England alone. Countries without sufficient elevations to produce an adequate fall of water,

Figure 1.7
Trevithick's locomotive, 1808. (R. H. Thurston, *Histoire de la Machine à Vapeur*, Librairie Germer Baillière, Paris, 1880)

PITTSBURGH.

Figure 1.8
Artist's impression of Pittsburgh in about 1870, with clouds of polluting smoke every-where. (Arther S. Bolles, *Industrial History of the United States*, Henry Bill Publishing Co., Norwich, Conn., 1892)

such as the Low Countries, had to rely on windmills. But even in countries with a sufficient fall of water there was a natural limit, as every feasible mill site soon became occupied. The advent of the steam engine changed all this. Factories could now be concentrated in selected localities, giving rise to factory towns, polluted and dreary.

Concurrent with the spread of the factory system came an increasing demand for fuel. Coal, of course, was the fuel, and England was abundantly supplied with it. By the seventeenth century so much coal was being burned in London that foreign

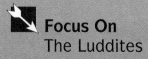

Focus On
The Luddites

In 1811–12, in the northern industrial areas of England, there occurred the most serious outcry against technology on record. The country was in a general condition of misery and distress brought on by a number of factors: the protracted Napoleonic Wars; the loss of trade caused by the closure of the American market, leading to the War of 1812; and four successive years of crop failures. The spark that caused resentment to turn into open rebellion was the growing use of automatic labor-saving equipment in the textile industry, which was occurring in spite of the fact that unemployment was at record heights.

The sound of marching men was heard nightly for months. Terror was widespread, as the demonstrators demanded money and arms from the populace. Robbers took advantage of the circumstances for their own purposes. In the end, more than 1000 of the offending labor-saving machines were broken by the mobs, and 12,000 troops had been required to quell the disorders. Several of the rioters were killed, and one of the factory owners was waylaid and murdered. After order was restored, more than 40 persons were tried and executed.

The demonstrators were called *Luddites*. The name was derived from that of an apprentice weaver named Ned Ludlow, who had allegedly been disciplined for not having his frames straight and had beaten his machine to pieces in a rage. Subsequently, anyone who broke his frames was called "Ned Ludd." Letters containing demands and warnings to the populace were frequently signed "Ned Ludd," and there were even references to "General Ludd" being in charge of the mobs.

The Luddite disorders occurred at the height of the Industrial Revolution and were the most serious of many demonstrations against the ever more efficient machines being introduced in England. In the long run, the ultimate irony of the Luddite disorders is that the machines to which the Luddites objected were the means by which their objections became irrelevant. By this is meant that as the Industrial Revolution continued its course, the general welfare of the populace gradually rose as a result, so the misery which caused the Luddite disorders receded.

Periodically in our own time there have been outcries against technology because of the fear that improved machines would cause unemployment. For example, in the 1950s there was widespread fear that the advent of computers would displace large numbers of office workers and render middle management positions obsolete. The reverse actually happened, but now the same fears are being expressed once more. Especially alarming is the worry that factory jobs will be destroyed by the intrusion of robotics. No one can say that these things will not take place, because labor displacements

visitors to London were shocked by the smoky and polluted atmosphere. Actually, the conversion to coal had been going on for some time because of the increasing pressure on wood fuel and the rapid denuding of forests. A critical metallurgical problem of the time was how to smelt iron without using charcoal, which was made, of course, from increasingly scarce wood. Finally, in the early eighteenth century, iron makers solved the tricky problems of using coal instead of charcoal, relieving the pressure on Britain's forests.

have occurred in the past. Clearly, the automobile eliminated livery stables and the carriage industry, and the hand-held calculator wiped out the mechanical calculator business. In recent years demonstrations have taken place against the automatic tomato harvester, developed by the University of California. A lawsuit (unsuccessful) was even filed against the university on the grounds that the tomato harvester enriched big growers and deprived farm workers of jobs.

The basic issue is that of productivity. The word is anathema to some groups, who claim that "labor-intensiveness," rather than productivity, is an attribute to be sought. On the opposite side, many scoff at the notion of labor-intensiveness, because it implies that "make-work" is inherently desirable, even when it is obvious that much of the activity involves useless and wasted motions. The issue is not a trivial one, however, because the relative costs of labor and capital always must be compared in order to make an investment in automation. Sometimes it turns out to be economically more desirable to use human labor for a given task than to mechanize. It should also be held in mind that human beings prefer their daily tasks to be personally more stimulating and rewarding than the tasks typical of factory assembly lines or agricultural stoop labor. Thus, some have held that any replacement of humans by machines at tasks that are boring or physically exhausting is inherently good.

A century ago, Friedrich Engels declared that the perfecting of machinery would render human labor superfluous. The result would be an idle worker population at the mercy of the bosses. The solution, said Engels, was to be found in the seizure of the means of production by the proletariat. Engels was wrong, of course. Instead of causing widespread suffering in the industrial countries, increases in productivity have been accompanied by major improvements in human welfare. But who can say that things won't reverse themselves and go downhill in the future? Many sincere people think they will. But we cannot foretell the future; we can only read the lessons of the past and assume that the same rules will continue to apply. One of the rules seems to be that the well-being of each individual in our modern industrial society is tied to the well-being of all. Put another way, in order for the producers of a commodity to prosper, there must be others who have the means to purchase the commodity. Also, the commodity must be inexpensive enough to fit the purses of the buyers. Improvements in productivity have been the driving engine of this economic cycle, causing the availability of an endless variety of affordable material goods, along with better diets, improved health care, and greater leisure.

References

J. G. Burke and M. C. Eakin (eds.), *Technology and Change* (San Francisco: Boyd and Fraser, 1979).

F. O. Darvall, *Popular Disturbances and Public Order in Regency England* (London: Oxford University Press, 1934).

T. Forester, *The Microelectronics Revolution* (Oxford: Basil Blackwell, 1980).

The emergence of the factory system with its factory towns led to the impersonal exploitation of labor and to social unrest that persisted through the nineteenth century and into the twentieth. As a consequence, much criticism has been directed toward the Industrial Revolution. But it is also true that the evolution of the factory system led to mass manufacturing techniques and to improvements in industrial productivity, which in turn has produced an enormous improvement in the standard of living in industrialized countries. Some writers have even asserted that it was the growing use of mechanical power that doomed slavery by causing human power to be more costly than machine power. (Others insist it is the other way around: that the huge death rate resulting from the Black Death in the fourteenth century had caused such a shortage of workers that a powerful stimulus was thereby given to the development of mechanical power.)

Toward the end of the Industrial Revolution, innovative leadership began to pass from Britain to the United States. A couple of explanations have been advanced

Figure 1.9
In the days before power machinery came into widespread use, horses and mules provided much of the motive power needed for underground mining. Here a horse is being lowered into a mine, probably to spend the remainder of its life underground. (Arthur S. Bolles, *Industrial History of the United States*, Henry Bill Publishing Co., Norwich, Conn., 1892)

for this. One is that, by the nineteenth century, Britain had a surplus of labor while the United States had a persistent shortage. Hence, the stimulus toward mechanization was stronger in the United States than in Britain. The other explanation is that there seems to have been a recurring pattern throughout history. We note, for example, that the earliest inventions came from India and China and that countries such as Italy and France imported and copied those ideas. Then, after a period of imitation, the imitative countries were stimulated into periods of inventiveness of their own, becoming for a time the principal sources of science and intellectual activity. Britain followed suit, first copying the inventions of France and Italy and then emerging into a burst of creative activity of its own. Then came the United States, copying Britain, then surpassing it. And some say the same thing is now happening with Japan and the United States.

Through much of the nineteenth century, the matter of railroading dominated engineering consciousness. One of the greatest achievements of American engineering

Figure 1.10
In the latter half of the nineteenth century, coal was the fuel that powered the country's growing industrial might. But as coal mining expanded, so did accidents, reaching shocking proportions. Mine safety laws improved the situation, but mining continues to be one of the most hazardous occupations in the United States. (Arthur S. Bolles, *Industrial History of the United States*, Henry Bill Publishing Co., Norwich, Conn., 1892)

was the construction of a railroad from coast to coast, begun in 1862 and completed in 1869. One man—an engineer—conceived the idea, developed the detailed plan that made it possible, sold it to a reluctant Congress, and organized a group of financial backers to carry it out. He was only 34 years old at the time. But friction soon developed between him and those who had put up the money, principally because he was interested in building a railroad and they were interested in making a financial killing. The others offered to buy him out for $100,000—a huge sum in 1863—but he died of yellow fever before he could collect it. His name was Theodore Judah; few have ever heard of him. His financial backers, however, not only became multimillionaires, but their names are famous in American history: Stanford, Huntington, Crocker, and Hopkins, usually referred to simply as the "Big Four." Judah, the engineer who made it all possible, is virtually unknown.

One more nineteenth-century engine development must be mentioned, and that is the "air-engine," which we know today as the internal combustion engine. Early experiments had been made in this direction by attempting to create small explosions inside a cylinder containing a piston, in a manner probably inspired by the example of the gun. In 1860 a Frenchman, Etienne Lenoir, invented the first successful internal combustion engine, using a mixture of air and gas made from coal. A big advance was made two years later when another Frenchman, Beau de Rochas, proposed the four-stroke cycle that we use today. This cycle was incorporated into the spectacularly successful Otto gas engine, introduced by a German, Nikolas Otto, in 1876. As a result, the four-stroke cycle is known today as the Otto cycle and is used in most of our modern automobiles.

■ 1.6 THE ELECTRICAL REVOLUTION

Historians have not as yet formally declared the existence of a period called the Electrical Revolution, but they might well do so. If they do, its beginning date would probably be established as 1831, and its ending date hasn't arrived yet.

Again it was an Englishman, Michael Faraday, who made the crucial discoveries. Oersted and Ampère had earlier demonstrated that an electric current produced a magnetic field, by showing that a magnetic compass needle was deflected when placed near a current-carrying wire. But Faraday set out deliberately to do the reverse: produce electricity from magnetism. After many failures he discovered, in an experiment that included two insulated coils wound on a common iron core, that an electric current could be induced in the circuit of the second coil when a switch was closed in the circuit of the first. From this he learned the principle upon which our entire electrical generation industry is based: it is a *change* in a magnetic field that induces an electric current, and not just the existence of a magnetic field *per se*. He went on to conduct other experiments, showing that a current could be induced in a coil by thrusting a bar magnet into the coil and showing that just moving a loop of wire through a magnetic field would induce a current in it. He even built a simple model of an electric generator, but then lost interest in the topic.

As is often the case with a new concept, Faraday's experiments seemed to offer only tiny amounts of power and were therefore unlikely to be of much use. Furthermore, there did not seem to be any particular need for more electricity than could readily be satisfied by galvanic cells, or batteries. However, some slightly improved versions of Faraday's generator did find use in medical therapy and in electroplating.

A truly significant application of electricity occurred when Samuel F. B. Morse, an American, developed a practical telegraph system, beginning in 1835. As is often the case with inventions, the telegraph was also developed simultaneously by others, Charles Wheatstone and William Cooke in England. The first use of the telegraph was for signaling along railway systems. But by the 1850s telegraph cables were being laid under the sea from Britain to France, and in 1858 the first undersea cable was laid across the Atlantic, a distance of 2200 miles.

The first electric motor was made in 1830 by Professor dal Negro of Padua University in Italy. It was a tiny affair, consisting of nothing more than a pivoted magnet that oscillated under the influence of current flowing in a coil. Upon each oscillation, the magnet would contact a one-way cog wheel, causing the wheel to rotate under the influence of the series of strokes from the magnet. Other investigators also experimented with crude electric motors, gradually improving them. By

Figure 1.11
An early electric generator called a *dynamo*, invented by C. F. Brush, about 1878. (Arthur S. Bolles, *Industrial History of the United States*, Henry Bill Publishing Co., Norwich, Conn., 1892)

1854 an electric motor powerful enough to drive a 12-ton locomotive was tested, but its batteries were bulky and would not last very long. As a result it was another 40 years before electrical distribution systems came into general use and made street railway systems practical.

In the 1870s electric lighting in the form of arc lamps made an appearance, and the demand for electricity accelerated. Many experimenters began to offer improved versions of electric generators. In 1879 Joseph Swan in England and Thomas Edison in America both introduced practical versions of incandescent bulbs, which further stimulated the demand for electricity. (See boxed item in Chapter 12: "Edison Did Not Invent the Electric Light.") By 1890 generators in essentially their modern form were in use, and the stage was set for the virtual takeover of industry by electricity.

Two observations of note can be made here. First, virtually all of our electricity today is made by first creating steam and running a steam engine (in this case a turbine) connected to a generator, so we owe an enormous debt to James Watt as well as to Michael Faraday. Second, our powerful electrical generation industry, which generates billions of watts, owes its existence to the puny force exerted by a compass needle when it deflects in the presence of a current-carrying wire. The tiny force of the compass needle seemed trivial to the investigators of the time, yet it is the same force that is responsible for those billions of watts. This observation may cause us to reflect on the existence of other tiny forces in our physical universe whose potential we may be overlooking.

■ 1.7 THE SNOWBALL OF PROGRESS

The accomplishments of two individuals in the mid-nineteenth century stand out as major milestones in putting a firm scientific base under engineering. These were the initiation of the theory of thermodynamics by the Frenchman Sadi Carnot in 1824 and of electromagnetic field theory by the Englishman James Clerk Maxwell in 1865. Both theories continue to provide inspiration for practicing engineers (and occasionally create despair for engineering students trying to understand them).

Carnot, with superb scientific insight, saw the principles upon which the efficient use of heat must be based and published his theories in a treatise called *Reflections on the Motive Power of Fire*. His theories stand unchallenged today, more than a hundred years later.

Maxwell's theories were developed in a manner similar to Carnot's: an intellectual tour de force involving superb insight. Maxwell proposed a set of mathematical equations which said, essentially, that in a steady-state condition a magnetic field can exist without causing an electric field. But if one field is changing, it causes the other to come into existence, and the two are always at right angles to each other. Maxwell's equations require an understanding of differential equations and vector analysis, so no attempt will be made to describe them here. Suffice it to say that they provided a general theory of electromagnetism. They also asserted that electro-

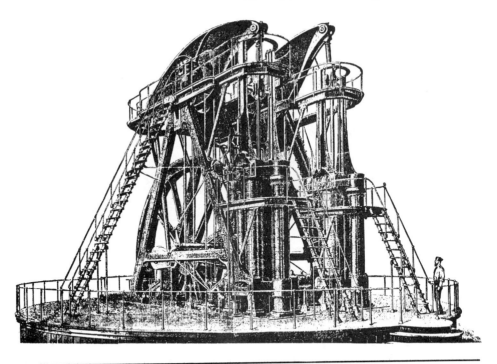

Figure 1.12
A giant Corliss steam engine, which was the wonder of the 1876 Centennial Exposition in Philadelphia. The machine was used to drive all the machinery exhibited at the exposition. (Charles T. Porter, *Engineering Reminiscences*, John Wiley & Sons, New York, 1908)

magnetic fields would travel through space with the velocity of light, although no experimental evidence existed to show that Maxwell was right. This confirmation did not come for more than 20 years, when a German, Heinrich Hertz, devised an experiment that proved Maxwell correct. Electromagnetic waves subsequently came to be called Hertzian waves, and the work of Maxwell and Hertz laid the foundation for radio and, later, radar.

At this point, there does not seem to be much need to chronicle in detail the further development of modern engineering. Most of the major events are too well known: the invention of the telephone, the advent of airplanes, and the achievement of space travel, for instance. Further, most twentieth-century achievements are based upon developments from earlier centuries. In the case of the airplane, for example, people had been imagining flying machines for centuries but lacked a sufficiently compact power source until Lenoir and Otto came up with their internal combustion engine. In like fashion, our present level of space travel is based upon Newtonian mechanics, most of which Newton worked out 300 years ago when he solved the equations of motion of the planets.

SUSPENSION BRIDGE, MOUTH OF MONONGAHELA RIVER.

Figure 1.13

An early suspension bridge across the Monongahela River. (Arthur S. Bolles, *Industrial History of the United States*, Henry Bill Publishing Co., Norwich, Conn., 1892)

Some important contributions to the development of engineering are described in boxed items distributed throughout the text. The works of Nikola Tesla, Thomas Edison, and Stephen Timoshenko, for example, are part of the history of engineering and are described in boxes. So, too, are the microprocessor, the jet engine, and photocopying.

Two major twentieth-century developments have profoundly affected engineering. The first was the rise of relativistic mechanics and quantum mechanics, the creations of Albert Einstein and others. Newtonian mechanics serves most engineers' purposes quite well, but the scientific insights of the new mechanics have led to such modern developments as lasers, nuclear reactors (both fission and fusion), transistors, and the creation of new materials.

The second development belonging wholly (or almost wholly) to the twentieth century is that of electronics. Faraday had experimented with the passage of electricity through evacuated spaces, and in the 1880s Edison had found that an electric current would pass from one element to another in a vacuum if one element were heated and the two bore suitable polarities. But it was not until the first decade of the twentieth century that Ambrose Fleming, an Englishman, and Lee De Forest, an American, found that a vacuum tube with *three* elements could amplify an electric current. Thus was electronics born.

Ironically, the vacuum tube, which started the technological explosion of electronics, may look to future historians like a brief interlude in the history of technology. The reason is that in the late 1940s Bardeen, Brattain, and Shockley invented the transistor, and it was found that everything (or almost everything) the electron tube could do in a vacuum could be done better by devices working in the solid state.

■ 1.8 CONTRIBUTIONS OF ENGINEERING TO SOCIETY

It has become commonplace to observe that there is practically nothing we use in our daily lives that has not been influenced by engineers. The connection is obvious when we drive a car, cross a bridge, fly in an airplane, use a computer, or watch television. The connection is less obvious when we buy a paperback book in a drugstore, pick up a package of frozen vegetables at the supermarket, or purchase a home.

In the case of the inexpensive paperback book, we should consider the miracle that such a thing exists at all. Paperbacks must be produced by the tens and hundreds of thousands in order to be as cheap as they are, so high-speed presses are obviously essential to the process. So are cheap paper and fast-drying inks. And just as essential is a distribution system—usually trucks or railroads—that brings the raw materials together at the printing plant and sends them out again all over the country to bookstores and supermarkets. The efficient functioning of all the foregoing systems has been brought about through the work of engineers, with the result that we can buy a book that otherwise would cost ten or perhaps a hundred times as much as it does.

With regard to the frozen vegetables, countless unseen engineers have created the system that brings the food into the freezer bin. The role of the engineers who designed the refrigeration systems is obvious, but the engineers who designed the harvesters, the processing equipment, and the packaging machinery have been just as influential.

And what of the homes we live in? Engineers have designed the equipment that harvests the logs, cuts them into lumber, and sorts and stacks the pieces. They also designed the machines that cultivate the tree seedlings and plant them for the next cycle of lumber production. All the new materials that have slowly revolutionized

Figure 1.14

The Flatiron Building, constructed in 1901 in New York City, represents a landmark in the history of structural engineering. It was the first building to use a structural steel framework, dispensing with load-bearing walls, and to have high-speed elevators. (Courtesy of Parsons Brinckerhoff)

Figure 1.15
People have largely forgotten the kind of visual pollution that once was common in oil production areas, as in this 1920s photo. The area shown is Huntington Beach, California, a popular recreational area then and now. (Courtesy of Standard Oil Company of California)

home construction over the years—insulation, wallboards, paneling, carpeting, paints—have been the result of the work of engineers.

During the nineteenth and twentieth centuries, as engineering became more important, societal concerns began to grow regarding the safety of the engineers' structures and machines. Lessons were learned the hard way. As steam engines became more common in the nineteenth century, so did boiler explosions, especially as it was learned that greater efficiencies came from using higher pressures. Then, as high-voltage electricity began to make its appearance, so did the incidence of electrocutions because of haphazard insulation practices. These accidents, coupled with occasional structural failures, brought growing demands for guarantees of public safety.

Today boiler explosions and structural failures are rare, but they do still happen. Some of the boxed items in this book deal with engineering disasters. Unfortunately, accidental electrocutions are still distressingly common, as are cases of fires started by unsafe devices, and we cannot help but be shocked by the astonishing annual death toll resulting from automobile accidents. Great societal pressure has been brought to bear upon engineers to improve these matters, and progress has been made. For example, electrocutions are fewer than they once were because of the use of double insulation; autos are safer than they were 20 years ago because of the incorporation of safety belts, padded interiors, better tires, double braking systems, and better structural integrity. But the pressure for improved safety continues, and in recent years new concerns have appeared: not only should the creations of engineers be safe, but their impact on the environment should be considered also. And then, in addition to safety and environmental impact, engineers are expected to consider the overall impact of new technological devices and whether the net public good is enhanced in each case.

The foregoing matters are obviously controversial, and the very fact that they are so controversial shows that they are important to the public. This point is worth emphasizing: *Engineering is very important to the public.* And this is why engineering as a profession is so attractive and challenging, because engineering is where the action is. If you are a person who likes to make things happen, then engineering is a good field for you. True, each engineer's acts represent only a few drops in the bucket, but the collective result is enormous. One of the most dramatic benefits of technology is the raised purchasing power of the average person. Ever since 1900, spendable income in the United States has continued to rise so that it has been possible for most Americans to satisfy more of their wants while working fewer hours. In 1909, workers in manufacturing enterprises earned an average of 19 cents per hour, with no fringe benefits. (The average work week was 51 hours.) When this figure is corrected for inflation, such a manufacturing worker existing in 1988 would have a purchasing power corresponding to a wage of $2.52 per hour ($129 weekly). The actual wage in 1988 was $10.71 per hour ($428 weekly), not counting fringe benefits. This increase in actual purchasing power is an improvement of 325 percent during a period when the work week was cut by 22 percent.[3]

The basic driving force behind this advance has been improved productivity, mostly achieved through ever more efficient manufacturing methods. The result of an increase in productivity is that more goods and services are available to more people with less human labor input than before. Thus, people are not only better fed and better housed than before and can afford better health care, but they also have a whole new range of options open to them for personal fulfillment. Rather than being forced into a life of bare subsistence, which was virtually the only course open to most people before the twentieth century, persons in industrialized nations have options available to them in education, a variety of careers, the arts, travel, sports, and entertainment that would have been beyond the dreams of those who lived only a few generations ago.

Technology has faced enormous challenges in this century and has had startling successes. The achievement of flight—first into the air and then into space—is clearly one of them. Instant communication around the world is another. The freeing of human beings from brutish labor is probably the finest of all to date. Now technology faces its greatest challenge: to move from a depletable energy base—that is, from dependency on fossil fuels—to a perpetual energy base. We do not know when, or if, this will be achieved. But if we expect to maintain the gains of past centuries, it must be done. And it will be the generations of students now in college—perhaps some of those reading this book—who will do it.

---

[3] *Statistical Abstract of the United States, 1990* (Washington, D.C.: U.S. Bureau of the Census, January 1990, p. 402; also *The Economic Almanac: 1964* (New York: National Industrial Conference Board, 1964), pp. 54, 55, 63, 76, 77, 103.

References

W. H. G. Armytage, *A Social History of Engineering* (London: Faber and Faber, 1970).

A. F. Burstall, *A History of Mechanical Engineering* (Cambridge, Mass.: M.I.T. Press, 1965).

D. S. L. Cardwell, *Turning Points in Western Technology* (New York: Neale Watson Publications, 1972).

P. Dunsheath, *A History of Electrical Engineering* (London: Faber and Faber, 1962).

J. L. Morse (ed.), *Funk & Wagnalls Standard Reference Encyclopedia* (New York: Standard Reference Works Publishing Co., 1963).

J. P. M. Pannell, *An Illustrated History of Civil Engineering* (New York: Frederick Ungar Publishing Co., 1965).

S. Rapport and H. Wright (eds.), *Engineering* (New York: New York University Press, 1963).

Exercises

1.1 Obtain a reference book on Leonardo da Vinci from the library, and list some of the inventions he made, other than those mentioned in the text.

1.2 Investigate a reference work from the library on the history of technology (some of those listed in the reference section of this chapter would help you), and prepare a diagram showing in detail how Watt's double-acting steam engine operated.

1.3 Sir Charles Parsons developed the first practical steam turbine. Make an investigation, and write a short report on this landmark event.

1.4 Write a report on the development of sky-scrapers and on the important development that occurred when the function of the "skin," i.e., the walls, was separated from that of the load-bearing skeletal structure.

1.5 John Roebling and Washington Roebling—father and son—were responsible for the design and construction of the Brooklyn Bridge (1867–1883). Write a brief report on the construction of the bridge, with particular attention to the safety problems involved.

1.6 Write a report on the development of submarine telegraph cables, leading to the laying of the first transatlantic cable in 1857–1858.

1.7 Guglielmo Marconi is generally given credit for the development of practical radio telegraphy (1895–1901), applying the principles of James Maxwell and Heinrich Hertz. Investigate library references, and write a brief report on Marconi's accomplishment.

1.8 Prepare a report on Ambrose Fleming's invention of the diode in 1904 and Lee De Forest's invention of the triode in 1906.

1.9 Prepare a brief report on the invention of the transistor (1948), starting with the early investigations into the properties of semiconductors by persons such as Roschenschold (1835), Lossev (1924), and Wilson (1931).

1.10 Write a brief report on the famous Eddystone lighthouse, which was constructed in 1698, rebuilt in 1699, destroyed by storm in 1703, rebuilt in 1709, destroyed by fire in 1755, rebuilt in 1759, and rebuilt again in 1882.

1.11 Investigate and report on the 1779 construction of the world's first iron bridge, which crosses the river Severn at Coalbrookdale, Shropshire.

1.12 The development of materials has been a vital factor in the advancement of engineering. Select a particular material and see what you can learn about the part it has played in the history of engineering. Some examples: steel, concrete, copper wire, pure silicon, gasoline, aluminum, glass, plastics, and ceramics.

1.13 Not too many years ago, epidemic diseases stemming from polluted water supplies caused thousands of deaths. See what you can learn about the techniques that were developed in the past for combating such situations, leading to modern water treatment practices.

1.14 List what you believe to be the most significant technological innovations from 1800 until now. Then list what you believe to be the most significant technological achievements from the time the human race began until 1800.

1.15 Consider, in some detail, ways in which your own life might be altered if you gave higher priorities to such things as pollution abatement, resource conservation, and reduced impact on nature. Evaluate these possible changes in your life to determine their importance to you. In each case, examine the trade-offs to see if the loss in your life style would be balanced by benefits that might be produced, either directly or indirectly. If the loss is unacceptable, see if you can think of other approaches that would be acceptable and would produce a net gain for you.

1.16 It is well known that the advent of the automobile caused a decline in the need for harness makers, but it also created a large new industry offering far more jobs than were lost with the harness industry. Discuss the positive and negative aspects of other technological innovations, such as the introduction of computers, the large-scale entry of industrial robots in manufacturing, the introduction of hand-held electronic calculators costing $10 in place of mechanical desk calculators costing $600, and the general displacement of hand-harvesting methods by machine harvesting.

1.17 What are some of the ways in which the automobile has altered our social structure? What about the computer? Television? The telephone? Inexpensive printing processes?

Photographs courtesy of Chrysler Corporation (top), American Consulting Engineers Council (center), and Cincinnati-Milacron (bottom right).

2

Modern Engineering Challenges

To many engineers in the 1950s, it seemed as if there was nothing exciting on the horizon to provide a stimulus to engineering. Television was the last big product innovation, and there appeared to be nothing new that would have the same impact.

Then suddenly electronic computers burst on the scene. But their promise soon faded, because the official conventional wisdom was that there was probably a market for only 100 or so of the huge electronic machines. Only the government and perhaps a few big research labs could afford them.

In the late 1950s, however, President Eisenhower announced the commencement of the huge interstate highway construction program, and for civil engineers, at least, matters really hummed.

In 1962, President Kennedy announced the Apollo program. By the end of the 1960s, he said, we would put a man on the moon. In 1969, we did put a man on the moon; then, because the program had achieved its purpose, the government began cutting back on the aerospace program. Once more, many engineers complained that nothing new was on the horizon. As if to underscore this prediction, the massive interstate highway construction program began to wind down, causing the construction industry to fall on hard times.

Then suddenly the computer industry arose from the early death that had been predicted for it and burst all bounds. Microchips were invented, making it possible for computers to increase spectacularly in power and decrease spectacularly in price. Instead of being saturated at 100 computers, the market seemed to be able to absorb millions of them. Then the computer industry sagged, and some people asked: What next?

The moral is clear: As each challenge arises, grows to a peak, and then subsides, a new challenge takes its place. Usually the new challenge is unforeseen and catches everyone by surprise. In the last decade of the twentieth century, however, matters

Engineering and the Environment
Oil Spills: Santa Barbara and Alaska

It is generally agreed that the 1969 Santa Barbara oil spill was the event that galvanized the environmental movement in the United States into a major political force. Yet it was neither the first great spill nor the largest. The spill at Santa Barbara involved perhaps 70,000 to 80,000 barrels. The wreck of the oil tanker *Torrey Canyon* off Great Britain involved nearly ten times as much, and the 1979 spill in the Bay of Campeche, off Mexico, spewed forth crude oil at the rate of 10,000 to 30,000 barrels *per day*. In 1989 the *Exxon Valdez* spilled 250,000 barrels into Alaskan waters.*

The Santa Barbara spill was the result of what is called a "blowout," and it appeared to have been caused because the crew was pulling out the drill string too fast. This upset the hydrostatic pressure balance at the bottom of the hole, which caused gas and oil to come surging out the top. In the past, blowouts were called "gushers" and were usually regarded as signs of good luck and prosperity. A huge gusher occurred in 1910 near Taft, California, and spewed out 9 million barrels of oil—130 times as much as at Santa Barbara, and 45 times as much as in Alaska—before it could be stopped.

Emergency blowout preventers have been developed, and one was used at Santa Barbara. The preventer shut off the blowout at the top,

* The Alaska spill has often been reported as 11 million gallons. One barrel contains 42 gallons.

but the pressure forced the oil out through weak rock strata lying just beneath the ocean floor, because the well had been protected with steel casing to a depth of only 239 feet. Since drilling at the time had extended to 3479 feet, usual practice would have been to protect the well with casing to a much deeper level than 239 feet. A waiver of the rules had been granted by the U.S. Geological Survey because it was felt the subsurface structure was too weak below the 239-foot level to cement the casing safely in place.

The residents of Santa Barbara had been fighting the encroachment of oil for years. It was bad enough to look out on their scenic channel and see it sprouting 20-story oil derricks. But to have the oil break loose and surge ashore to foul their beaches and harbors was the last straw. The Union Oil Company had to spend more than $10 million to clean up the spill, involving 1000 people and 10,000 truckloads of straw to absorb the oil. Hundreds of volunteers flocked to the scene in an attempt to save the lives of oil-soaked birds. Widespread claims of ecological damage were made following the spill. Some dead sea lions covered with oil were assumed to be casualties of the spill, and so were six dead whales that had washed ashore. But no definite proof of this could be established, and in each case where an autopsy was performed, the death was found to be from causes other than oil. Even when careful research studies were published, showing that the ecological damage was slight (except for the tragic loss of seabirds), some voices cried out that we were creating a dead sea off Southern California.

There were three principal reasons why the ecological damage was small.

1. Crude oil is relatively benign in the environment unless it washes ashore, fouls the beaches, and coats the feathers of birds.

have become somewhat different: A whole series of major challenges confronts engineers and scientists—challenges that demand solutions. Engineers no longer need wonder about what the future may bring, because there are many excruciatingly visible tasks that need to be done. There is also the nagging worry that some addi-

Crude oil degrades naturally in the environment within a few months. In the case of Santa Barbara, the actual final cleansing of the beaches occurred through natural wave-scouring. Conversely, refined products such as diesel oil are much more lethal.

2. Natural oil slicks have occurred in the Santa Barbara channel throughout history. Deposits of oil lie close to the surface, and in some places the oil oozes directly into the sea. As a result, the species in the channel were accustomed to some oil.

3. The oil spill coincided with unusually heavy winter storms, which had caused 14,000 gallons of silt to enter the channel for each gallon of oil. Since fresh water is deadly to many kinds of marine life, it was difficult to separate the damage done by the flooding from that done by the oil.

The Alaska spill was different from that at Santa Barbara in a number of ways. First, there was much more oil, and it coated a much greater area of shore. Second, access to most of the shoreline was extremely difficult, and even though 11,000 people were employed in the cleanup at its height, the cleaning was spotty. Third, Alaska in the summer is home to millions of waterbirds, and it was feared there would be an enormous die-off. Finally, many people in Alaska depend on the fishing industry, and in some areas their livelihood was destroyed for the year.

But what about overall ecological damage to Alaskan waters? Many dire predictions were made, but a year after the spill, observers could not say whether there was any detectable diminution in the numbers of birds, partly because they are hard to count and partly because there are so many of them. Nearly 30,000 dead birds were recovered after the spill (probably only a fraction of the total die-off), yet it is estimated

that Alaskan waters contain 30 million of the so-called *alcids*—the puffins, murres, and auklets—which are at most risk from oil spills. Many sea otters were also killed by oil, yet no one could say for sure what the impact on their total population might be. And, in spite of the massive cleanup efforts, it was finally generally agreed that the only effective cleanup would come from the scouring of wave action, just as at Santa Barbara.

In spite of the fact that the damage apparently was not nearly so great as reported by the news media, there are no grounds for complacency when it comes to oil spills. The damage to wildlife is shocking, and the economic loss to the oil industry (and to the country) is too great to shrug off. Furthermore, there is a pervasive feeling that most of these spills are preventable. In the case of three of the spills mentioned above—the *Torrey Canyon*, the *Exxon Valdez*, and Santa Barbara—lapses of professional judgment were involved. In the aftermath of the *Exxon Valdez* spill, everyone from the Coast Guard to the captain of the tanker was blamed. Clearly, double hulls on tankers should be mandatory. And so should stricter navigational rules, backed up by potentially massive fines and legal judgments. Even when all reasonable precautions are taken, accidents will still occur as long as humans depend on oil. The best we can do is to minimize them.

References

R. Easton, *Black Tide: The Santa Barbara Oil Spill and Its Consequences* (New York: Delacorte, 1972).

P. Spencer, *White Silk and Black Tar: A Journal of the Alaska Oil Spill* (Minneapolis, Minn.: Bergamot Books, 1990).

C. Steinhart and J. Steinhart, *Blowout: A Case Study of the Santa Barbara Oil Spill* (Belmont, CALIF.: Duxbury, 1972).

tional problems may appear out of nowhere and surprise everyone. In fact, some have said that the only thing we know for sure about the future is that it will contain surprises.

Some of the challenges confronting us are actually old challenges, stemming from the 1960s and 1970s. Originally these were called "crises," but then they settled down to become long-term, insistent problems: the environmental crisis, the energy crisis, the defense crisis. Others became apparent in the 1980s and also were called crises by some groups: the international competitiveness crisis, the infrastructure crisis.[1] In this chapter we will look at the latter two, as well as a couple of the older ones, in some detail.

■ 2.1 INTERNATIONAL COMPETITIVENESS

In 1973 sociologist Daniel Bell published a book called *The Coming of Post-Industrial Society.* In such a society, he said, we would shift from a goods-producing economy to a "service economy." This would bring with it an emphasis upon innovation, research, and specialized knowledge and would create a new pre-eminent professional and technical class in our society. To some, the significance of the postindustrial society was that the manufacturing of goods would become "trivially easy," and they greeted this new state of affairs with enthusiasm. Engineers and scientists took all of this pretty calmly, assuming that they would be at the center of the "pre-eminent," specialized technical class that was foreseen.

Then the United States was hit with a series of economic blows as other countries in the world began to produce products that functioned better, lasted longer, and cost less than those made in America. The United States developed huge trade deficits with virtually every one of its trading partners—especially with Japan. The gains in "real" wages, which Americans had come to take for granted, stopped and actually turned into losses. The production of goods was not "trivially easy" after all. The United States apparently was the only country in the world that had accepted the notion of the service economy; other countries were busily manufacturing goods and were beating America at a game in which it had supposed itself to be supreme.

Numerous "quick fixes" were proposed: The United States should impose protectionist tariffs and quotas; the United States should subsidize certain industries that were threatened; the United States should cut the inflated wages and benefits of factory workers; the United States should allow the value of the dollar to fall so that its products would be cheaper abroad. None of these solutions went to the core of the problem, which was that the United States had lost track of the central importance of manufacturing.

There is a terrible irony to all this, because the man who taught the Japanese how to be so good at manufacturing is an American, W. Edwards Deming. Deming

---

[1] The name *infrastructure* is often given to the entire system of constructed works that support our life-style: highways, railroads, water systems, airports, docks, utilities, and sewer systems.

is thought of by many as simply a statistical quality control expert, but he brought a great deal more to Japan than just statistical quality control: He brought an entire philosophy of management. In 1950 he was invited to Japan by the Japanese Union of Science and Engineering to tell them about quality control. Instead, he told them about his new ideas of management. In its simplest form, his philosophy says this: The route to success is through high quality; most quality defects are the fault of the manufacturing system, not the people; the job of management is to improve the system; the people who can help the most to improve the system are the ones who have to work with it. The Japanese tried Deming's ideas and found they worked, and the rest is history. He is revered today in Japan, and the Deming Prize is the highest industrial award in that country.

It may appear to some that this does not have much to do with engineering, but such a point of view underscores one of the faults of industry in the United States: The view of what constitutes "engineering" has been much too narrow. There has also been a tendency to separate engineering from manufacturing and to relegate the latter to second-class status in the corporate hierarchy. John Young, president of Hewlett-Packard Co., puts the problem thus:

> Many companies accord manufacturing a low status, and this is a key contributor to our nation's sagging productivity in comparison with Japan and other major industrial nations. Our culture has been quick to celebrate the dramatic breakthrough, the bold idea, and the brilliant concept. It has been far less inclined to praise incremental improvements and painstaking execution.
>
> ... Proper attention to manufacturing confers more than cost advantage. Often it leads to systematic continuing improvements in products and processes that improve the quality of a product, make it easier to service or use, and enhance its competitiveness in other ways. Such refinements often have more impact on a product's commercial success than research breakthroughs or radical innovations.[2]

A part of the problem has been the classic view of the way engineering and manufacturing relate to each other (Figure 2.1), which has been embedded in corporate organization charts: A new product originates in the Research and Development (R&D) Department, and a functional prototype is constructed; the prototype and working drawings are transmitted to the Production Engineering Department; Production Engineering redesigns the product so that it can be manufactured; then the working drawings are released to the Manufacturing Department, which does the tooling and produces the product.

That is the classic view, but more and more it is being heaved overboard. John Young says, "Our manufacturing engineers used to play a somewhat passive role in the innovation process. They assumed that whatever the design engineers threw over the fence, manufacturing would build. Today, manufacturing engineers are part of the product design team from day one of a project. Product and process design go on in parallel."

---

[2] J. A. Young, "Technology and Competitiveness: A Key to the Economic Future of the United States," *Science*, 15 July 1988, pp. 313–316.

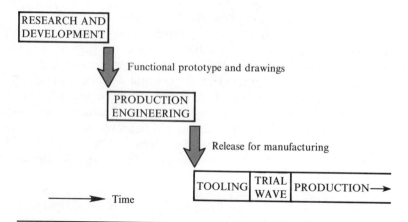

Figure 2.1

The "classic" view of the sequence from research and development through production engineering to manufacturing. This scheme is being superseded by the one shown in Figure 2.2.

The new style of doing business has been dubbed "simultaneous engineering" and is a familiar process in Japan. Two Japanese professors of management, Hirotake Takeuchi and Ikujiro Nonaka, investigated the methods used by several Japanese manufacturers (Canon, Brother, Epson, Honda, NEC, and Fuji-Xerox) to develop new products. They described the results in an article entitled "The *New* New Product Development Game." At these companies, the "classic" way of doing everything in series, like a relay race, had long since been discarded. Instead, all the phases had become highly overlapped like *sashimi*, they said, referring to the traditional Japanese delicacy of slices of raw fish overlapping each other on a plate. This new view, they asserted, could also be called the "rugby" approach: A hand-picked, multidisciplinary team moves down the field together, passing the ball back and forth, working together from start to finish. Their visualization of this approach is shown in Figure 2.2, where the six functions of conception, testing, design, development, pilot production, and final production are all very much overlapped.

To be sure, product development and engineering are not the only elements that must play a part in reviving the ability of the United States to compete. Some other elements follow:

Emphasis on Short-Term Profits   Some have blamed the way our business schools teach their students for the fact that U.S. firms tend to focus on the short term instead of the long. The result is that businesses are often reluctant to undertake long R&D programs and thus fall behind in the new product race. Deming says, "Pursuit of the quarterly dividend and short-term profit defeat constancy of purpose."

Lack of a Product Focus   Many executives proudly make the claim that the purpose of their businesses is "to make a profit." Profits are important, of course; without them a company will soon cease to exist. But if a company is managed by those who

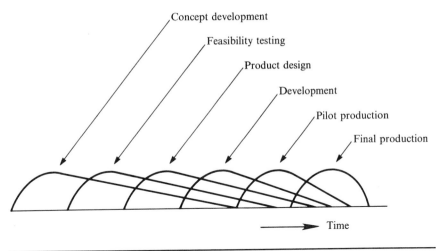

Figure 2.2
The "overlapping" model of the product development sequence. (Source: H. Takeuchi and I. Nonaka, "The *New* New Product Development Game," *Harvard Business Review*, January–February 1986, pp. 137–146)

know or care little about the products being produced and who focus only on financial performance, the company is asking for trouble. Deming quotes a foreign observer, Dr. Yoshi Tsurumi:

> Part of America's industrial problem is the aim of its corporate managers. Most American executives think they are in the business to make money, rather than products and services. . . . The Japanese corporate credo, on the other hand, is that a company should become the world's most efficient provider of whatever product and service it offers. Once it becomes the world leader and continues to offer good products, profits follow.

Capital Investment   A critical factor, emphasized strongly by some economists, is that the United States is unable to make the necessary capital investments to become fully competitive and that the fundamental cause is an insufficient rate of savings. When comparisons are made with Japan, we find the following: (1) Our net annual capital investment is a small fraction of Japan's; (2) our interest rates are much higher, causing our overall cost of investment to be three times as high; (3) our productivity growth is only one-fourth as great. These matters sound esoteric to many engineers, but since capital investment is the fulcrum of new product development, they are actually an indispensable part of engineering life.

Automation   It became apparent in the early 1980s that Japan had many times more robots in its manufacturing plants than did the United States (in 1982, for example, Japan had 32,000 and the United States had 6000). This led to the conclusion that the answer to America's industrial competitiveness problem was to introduce more robots, perhaps in fully automated factories that would not even require any lighting

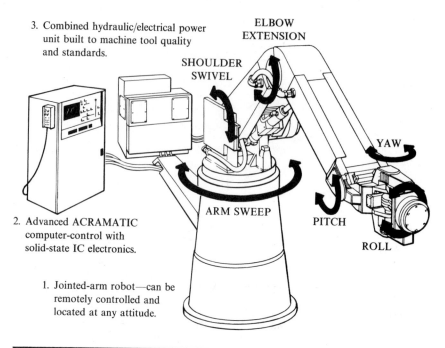

3. Combined hydraulic/electrical power unit built to machine tool quality and standards.

ELBOW EXTENSION

SHOULDER SWIVEL

YAW

ARM SWEEP

PITCH

ROLL

2. Advanced ACRAMATIC computer-control with solid-state IC electronics.

1. Jointed-arm robot—can be remotely controlled and located at any attitude.

Figure 2.3

The Cincinnati-Milacron T³ Robot System has variable six-axis positioning, permitting location of a part or a tool at any attitude. It can lift 100 lb (45 kg) and has a position repeatability of 0.05 in. (1.27 mm). (Courtesy of Cincinnati-Milacron)

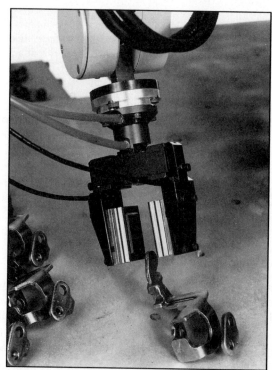

Figure 2.4

A major element of a robotics device is the gripper. This unit is equipped with a sensitive tactile sensor that exerts a carefully controlled amount of force on the object to be lifted. (Courtesy of Lord Corporation)

because there would be no human workers. But these images, like so many others, turned out to be too simplistic. Automation was important but was only one element among many.

For one thing, just introducing more robots would not solve the problem. Product design and process design usually depend upon each other, so they have to go hand in hand. The introduction of a brand-new product may also require the introduction of a brand-new process. For another thing, without a solution to the problem of a shortage in capital, automation would be an unattainable dream.

■ 2.2 REBUILDING THE INFRASTRUCTURE

Anyone who drives a car knows first-hand that the nation's highways are in terrible shape. What is less well known is that virtually all of America's constructed works need major repairs. Not only that, even if they are repaired, they have fallen seriously behind in their ability to meet our expanding needs. Some estimates indicate that the required level of expenditure is greater than that now committed to defense. The big question is whether, given other demands, government will be able to come up with the money. Here are some of the dimensions of the problem:

Highways and bridges are in bad shape and are unable to meet the demands placed on them. Nearly half of the nation's bridges have been judged deficient in one way or another. The federal Department of Transportation has estimated that by the year 2005 American motorists will be subjected to nearly 4 billion hours of traffic delays per year if no improvements are made. Another study has claimed that more than a thousand lives per year would be saved by a $4 billion investment in our bridges and highways.

Clean water is a critical problem. Expenditures for sewers and water systems were in excess of $10 billion per year during the 1980s but could still not keep up with the deterioration of such systems in our older cities. In 1987, the federal government provided $20 billion through the Clean Water Act, primarily intended for new sewage treatment facilities, but an even greater amount of money would have to come from state and local governments. An especially complex issue, which will persist well into the twenty-first century, is the disposal of toxic wastes. The method of the past—burying them—almost invariably allowed them to show up later as pollution of the groundwater. A related general issue is the disposal of all kinds of solid wastes. By the 1970s and 1980s, it was becoming difficult to find sites for new landfills, and more innovative methods of solid waste disposal were being considered, including the recovery of some kinds of wastes for the resources they contain and the burning of others for their energy content.

Utilities have done little to add to their capacity since the 1970s, primarily because gains in energy conservation made new plants unnecessary. Some observers have declared that utilities will soon have to start building again.

Airports are clogged, and so are the highways leading to them. Airport expansion is not easy, partly because of citizens' lawsuits regarding aircraft noise, but it is needed. Probably the most difficult problem is that of access to airports. At some airports, the addition of more highway lanes and more parking is simply not feasible. More reliance on public transport will be necessary, as well as new solutions such

Figure 2.5

Railroads are a major part of our transportation infrastructure. The increased use of continuous welded rail (CWR) has resulted in a large number of accidents attributable to train derailments induced by thermal buckling of railroad track. If tracks are laid when the temperature is relatively low, a normal climatic temperature increase can create compressive forces sufficient to lead to track buckling. These "before-and-after" photographs show the effects of a deliberately induced static track buckling test on a major eastern railroad. Research is under way to improve the buckling safety of CWR track. (Courtesy of Federal Railroad Administration, U.S. Department of Transportation)

as the use of remote passenger check-in at satellite terminals, coupled with frequent limousine service.

In 1981 the U.S. Army Corps of Engineers announced the results of a three-year study of the safety of dams in the United States. More than 3000 dams were declared to be unsafe. Some states simply threw up their hands at the magnitude of the problem, since corrective action would require billions of dollars. But one state that had many unsafe dams, Pennsylvania, quickly authorized $300 million for dam repair and in four years reduced the number of unsafe dams by half. (In 26 cases, unsafe dams were simply destroyed.) More remains to be done, and if all states followed Pennsylvania's example, the cost would be enormous.

In view of all these needs, some have said that the decades from 1990 to 2010 will be among the most challenging of all times, especially for civil engineers. All kinds of engineers will be needed to meet the infrastructure crisis, but the needs we face are mostly of the type that civil engineers are trained to deal with. It has been said that even though computers and aerospace may grab the headlines, "civil engineering will have far greater impact on life quality for most of the world's population."

■ 2.3 AIR POLLUTION

Air pollution has many components; the most publicized are sulfur oxides and nitrogen oxides. Sulfur oxides arise principally from the burning of fossil fuels such as coal and petroleum, although many other kinds of industrial activity also produce them. SO_2 is a gas that reacts with oxygen in the atmosphere to form SO_3, which then combines immediately with water to yield H_2SO_4 in the form of droplets. Within the United States, the highest SO_2 values have been reported in the Northeast, where large quantities of high-sulfur fossil fuels have been burned. In most large U.S. cities, pollution from SO_2 has diminished in recent years, mostly because of the shift from high-sulfur coal to low-sulfur natural gas.

Even in the absence of sulfur, combustion of fossil fuels causes serious air pollution in the urban atmosphere. Exhaust gases usually contain unburned hydrocarbons (HC), carbon monoxide (CO), nitrogen oxides (NO_x), and the "normal" combustion products such as CO_2 and H_2O. In the atmosphere many of these products react chemically to produce new contaminants. The reaction processes are stimulated by sunlight, and the products are thus generally termed *photochemical oxidants*. Two of the principal photochemical oxidants are ozone and peroxyacetyl nitrate (PAN). (Ozone is also continually being created in the atmosphere by natural processes, but not to a degree great enough to constitute a pollution hazard.) This type of polluted air is often referred to as photochemical smog. Although all large cities in the world are afflicted with smog, the highest concentrations of photochemical oxidants in the United States are found in the Los Angeles area. In 1989, however, the California Air Resources Board was able to report that Los Angeles was showing slow improvement in its smog problem. Concentrations of nitrogen dioxide (NO_2), carbon monoxide (CO), and oxidant in Los Angeles' atmosphere declined during the 1970s and 1980s, principally because of control of motor vehicle exhausts. (See Figure 2.6.)

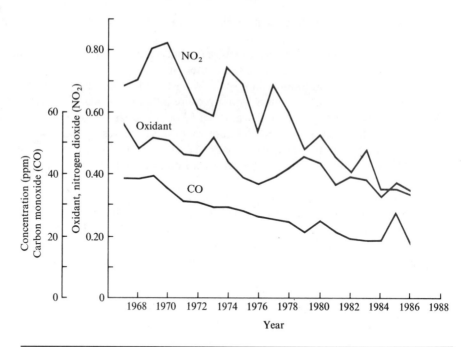

Figure 2.6
Trend lines of three major air pollution concentrations in the Los Angeles area in parts per million (ppm). The pollutants are oxidant (mainly ozone), carbon monoxide (CO), and nitrogen dioxide (NO_2). Data are given for maximum 1-hour concentration (NO_2), second highest 1-hour average concentration (oxidant), and second highest 8-hour average concentration (CO). (Source: California Air Resources Board, *Annual Summaries*, 1979–1989)

■ 2.4 WATER POLLUTION

In a historical sense, our rivers are less polluted now than they were in the nineteenth century. In that era the world had a genuine water pollution crisis. For example, in the mid-nineteenth century, 20,000 people in London were killed by cholera. As Donald Carr says, "In the Western world this was the greatest pollution disaster of history." Typhoid and cholera epidemics, stemming from water contaminated by sewage, were widespread. During the great wagon-train emigration of 1849 in the United States, more deaths occurred from cholera than from fights with Indians, attacks by wild beasts, or shootings. The greatest danger to emigrants came as they waited alongside the polluted rivers of Missouri to begin their journey west.

In the United States today, threats of cholera and typhoid are relatively remote as a result of widespread sewage treatment plants and the use of chlorine to kill bacteria in drinking water. Nevertheless, a comprehensive survey of the nation's rivers in the early 1980s showed that there had been more losses than gains in overall water quality. Widespread improvements had occurred in fecal bacteria and

Figure 2.7
In our modern, environmentally conscious era, oil companies go to great lengths to conceal their unsightly equipment. This office-building look-alike in West Los Angeles is a camouflage for 64 producing oil wells. (Courtesy of Standard Oil Company of California)

lead concentrations, but deterioration had taken place with respect to nitrate, sodium, chloride, arsenic, and cadmium. It was believed that the nitrate came mostly from the enormous increase in fertilizer use that had occurred during the previous decade and that the sodium and chloride came from the even greater increase in the winter use of salt on the nation's highways. The arsenic and cadmium apparently came from fossil-fuel combustion products and manufacturing of primary metals. The increased presence of nitrate and sodium in the nation's surface waters was especially troubling, because these come mostly from so-called nonpoint sources (i.e., from all over the place) instead of from point sources such as sewage treatment plants, power plants, and factories. Many of the nation's strategies had been developed to deal with point sources; reducing pollution from nonpoint sources would require new approaches.

It is not commonly realized, but agriculture is one of our biggest polluters. (It is also one of our biggest *users* of water. For example, it accounts for 85 percent of total water use in the west, where irrigation is vital.) As water passes across farmlands, it picks up dissolved salts, fertilizers, nitrogen from animal manure, and pesticide residues. When it drains off the land, the pollutants wind up in our surface waters. Excess nitrogen is believed to be one of the major factors behind the precipitous decline in the numbers of fish caught in Chesapeake Bay. Nutrients such as nitrogen and phosphorus promote the growth of algae. The algae then consume the dissolved oxygen in the water, rendering it unlivable for most other aquatic life.

One success story has been that of Lake Erie. In the 1970s, it became popular with the news media to say that "Lake Erie is dead" because of heavy algae growths and major declines in the fish populations. It was not dead, of course, and in fact has been slowly recovering as a consequence of pollution control programs. Full recovery cannot be expected, however. In fact, "full recovery" may not even be desirable. As two scientific investigators of the lake's ecology stated, "If [full recovery] meant the return of mosquitoes in their original numbers, with attendant epidemics of malaria, few of us would choose it."

Another success story is that of Delaware Bay, which is responding to intensive cleanup efforts. In contrast with neighboring Chesapeake Bay, the level of dissolved oxygen is on the rise, even in the areas that were most polluted. This has allowed the fish population to stabilize. The major difference between the two bays appears to be that Delaware Bay, since it is shallow, flushes itself four times faster than Chesapeake Bay.

It is commonly assumed that the pollution situation can be improved only through greater expenditures. Initially this may be so, but we should not overlook the fact that pollution *causes* enormous costs through poorer human health, crop damage, fish kills, materials decay, and the like. We should also not overlook one of the fundamental facts of our economic life ever since the industrial revolution: The foremost effects of the activities of engineers through the years have been reduced costs and increased benefits. This is what engineering is all about. There is no reason to expect that engineers will suddenly fail to be able to do in the future what they have done so well in the past.

A few examples can be cited in which lower costs and less pollution have been achieved simultaneously. In the 1970s, five paper companies in the United States and Canada disclosed a new papermaking process that could save millions of dollars per year, reduce water pollution, and produce a higher-quality product. The heart of the process was a new method for grinding wood into pulp, increasing the strength of the paper, and reducing waste and the amounts of chemicals needed. In other fields also, the amount of wastes per unit output has been lessened: Twenty years ago canneries were producing 40 cases per ton of fresh peaches, whereas today they produce 55 cases; from 1943 to 1963 the wastes generated by the container board industry dropped from 0.45 per ton of final product to 0.21 per ton. In the energy-intensive aluminum industry, Alcoa in 1973 announced plans for an experimental plant that would reduce the energy required to produce aluminum by one-third. Not only would this process have a favorable impact on resources and cut Alcoa's costs,

Figure 2.8

Insects are among the most persistent of mankind's enemies in competing for the world's food supplies. There is a constant search for chemicals which will kill only the target species and leave others unaffected. One new type of chemical, for example, harms only chewing insects, is biodegradable, and is harmless to birds, butterflies, bees, and human beings. (Courtesy of Standard Oil Company of California)

but it would also eliminate undesirable fluoride emissions and create a better working environment for employees.

The argument has sometimes been advanced that engineers should aim for no less than 100 percent safety and zero pollution and that society should allow no new technological undertakings unless it can be guaranteed that there will be no undesirable side effects. These conditions are unattainable, of course. Life cannot be made 100 percent safe, nor can any human being guarantee the future. Furthermore, zero pollution can be obtained only through zero activity. We are always going to be faced with the presence of a certain amount of risk and of undesirable side effects, although engineers, with their special technical knowledge of materials and the like, are in a better position to reduce risks and side effects than almost any other group. Furthermore, an engineer who is constantly thinking about matters such as safety and pollution control is much more likely to produce designs that excel in these categories than is one who does not think about them.

■ 2.5 ACID RAIN

In the 1980s, the phenomenon of acid rain captured the attention of the nation. Both sulfur and nitrogen oxides are implicated, because they form sulfuric and nitric acids in the atmosphere and cause rainfall to become more acidic than otherwise would be the case. Neutral water ideally has a pH of 7.0, but "normal" rainfall in remote areas that are unpolluted has a pH of about 5.0 because of the presence of small amounts of acid of natural origin. The acid rain problem arises because the rainfall in many areas of the northeastern United States has pH values lower than this, ranging down as low as 4.1. Since these same areas lie downwind from the states that produce most of the sulfur and nitrogen oxides, the conclusion is that there is a cause-and-effect connection. There is a tremendous amount of tension over this conclusion, because the downwind people obviously feel that the upwind people should take corrective action, which is enormously costly.

Originally the acid rain problem appeared to be primarily limited to lakes that had lost their fish populations as their acidity increased. But then the problem developed even broader dimensions as evidence of direct forest damage began to appear. The first alarm came in the 1970s from Europe, where spruce and fir trees were dying, presumably the victims of air pollution and acid rain. In the United States, an investigation of trees at the higher elevations of the Appalachian Mountains at first showed no signs of decline; but by 1985 another investigation showed that 39 percent of the trees were half defoliated and 7 percent were dead.

Some of the tension has come about because of variability in the data and uncertainty about the mechanisms at work. For example, lakes in the Adirondacks were shown to have become more acidic over time, but those in New Hampshire had not done so. Some of the lakes in Wisconsin had actually become less acidic. This state of affairs exists because the acidity of precipitation is not the only factor. The ability of the soil surrounding a lake to neutralize acid is also a factor, and so is the nearby land use pattern. It is known, for example, that primeval forest floors are more acidic than cut-over lands, and lakes in areas with dairy farms are more likely to have fish than those without such nearby farms.

Engineering and the Environment
Three Mile Island

The accident at Three Mile Island (TMI) on March 28, 1979, is the most serious that has occurred at a nuclear reactor in the United States. Some have said that the 1957 Windscale accident in England was far more serious. In each case the accident ruined the reactor, but in the case of Windscale radioactive contamination was widespread, whereas TMI released comparatively little radioactivity and injured no one.

To understand the accident, reference must be made to the following diagram.

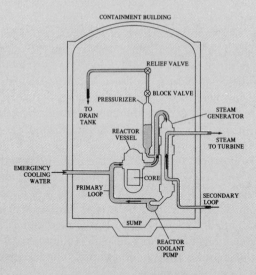

Water at high temperature and pressure circulates in the *primary loop*. This loop has a *pressurizer* connected to it, and the only place where a bubble of steam is supposed to exist in this loop is in the upper part of the pressurizer. The bubble can expand or contract as needed to maintain the pressure in the system. Heat is generated by nuclear activity in the *core* in the *re-*

actor vessel, and the primary loop carries the heated water to a heat exchanger in the *steam generator*, where the heat is transferred to water circulating in the *secondary loop*. The water in this loop is thereby converted to steam and is used to run the turbine connected to the electric generator.

Early in the morning on March 28, a blockage caused by a maintenance operation occurred in the secondary loop. This would not have been serious in itself, because an emergency feedwater pump immediately started up automatically. However, the valves in the emergency feedwater lines had accidentally been left closed. Therefore, circulation in the secondary loop stopped. Since the secondary loop was no longer removing heat from the primary loop, the pressure in the primary loop began to rise, forcing the *relief valve* at the top of the pressurizer to open. This caused the reactor to shut down automatically, 8 seconds after the accident's initiation.

As soon as the reactor shut down, it started to cool. The relief valve should then have reclosed, but it stuck open, allowing the water level to rise in the pressurizer and causing the pressure to drop. The dropping pressure automatically engaged the *emergency cooling water*, but an operator, seeing that the instruments showed the water level in the pressurizer was rising, shut the emergency cooling water off. If the water rose all the way to the top of the pressurizer, there would be no cushioning bubble in the pressurizer. The system would then have "gone solid," and operators were taught to avoid such a condition at all costs.

The reactor continued to generate "decay" heat. This was only a fraction of its normal output, but quite enough to melt the core, if left uncooled. With the emergency cooling cut off, the reactor vessel was losing its water through the stuck valve on the pressurizer, but no one knew this because the indicator light for that valve showed that it had been shut. This was because an electrical signal had indeed gone to the valve to close it, and the light on the control board was wired to respond to the presence of this signal. But no mechanical sensor existed to show whether the valve had actually closed.

Then, 8 minutes after the accident began, it was discovered that the valves to the emergency pumps in the secondary loop were closed. They were immediately opened, but the stuck valve on the pressurizer was now controlling events. The instruments in the control room gave confusing and inconsistent readings, and it was not until nearly $2\frac{1}{2}$ hours had elapsed that someone deduced that the relief valve must be stuck open and closed another valve (the *block valve*) in the same line. By this time, so much water had been forced out through the relief valve that the reactor core had been left partly uncovered for much of that time, and core melting had occurred. Also the abnormal amount of water being delivered to the drain tank through the stuck relief valve had caused a rupture disk to let go, and several feet of radioactive water had collected in the basement of the containment building. Some of this had been pumped into an auxiliary building, causing radiation to escape to the atmosphere.

Melting of the fuel also caused hydrogen to be generated because of the reaction at high temperature of zirconium with water. This hydrogen escaped into the containment building and ignited, causing a hydrogen burn that lasted several minutes. A pressure "spike" on chart recorders showed that the pressure rose to 28 psi, which was well within the 60-psi design strength of the containment building. It was not until later that the significance of the spike on the chart became clear. Then fear arose that another hydrogen bubble was forming in the primary loop and that further explosions might occur. After many alarms and demands for evacuation of the population, it was finally shown that the amount of free oxygen that could exist under the pressures inside the primary system was insufficient to produce an explosion. Finally, after six days of careful manipulation of the plant, the reactor was brought to a cold shutdown.

No one had been injured. The radioactive releases, though considerable, were finally put in perspective by the following: If an individual had been standing across the river from the plant, on the side with maximum exposure, 24 hours a day for six days, in the open and with no clothes on, he would have received about the same dosage as the average person gets annually from medical diagnostic activities. But a great deal of panic had been caused, partly by some who, playing it safe, had urged evacuations, and partly by saturation media coverage. Charges of cover-ups were made, resulting from incomplete information, confusion, and overlapping jurisdiction of various agencies. Because of the closed valves in the emergency feedwater lines, there was even a charge of sabotage. But no plausible evidence emerged that sabotage had occurred. In fact, there is not to this day any explanation as to why the valves were closed. They had been inspected only two days before the accident, and several witnesses swore they had been left open, as required.

Nuclear critics charged that TMI had proved nuclear power was too dangerous to tolerate. Advocates replied that, on the contrary, TMI proved just the opposite, because an incredible number of things had gone wrong and still no one had been injured. One thing was sure: The plant was ruined. Later, after the containment shell had been entered and inspected, much damage was found—mostly resulting from rust and prolonged exposure to water. The radioactive water in the basement, standing several feet deep, was slowly removed and decontaminated, the radioactive portions being converted to solid waste for shipment to disposal sites.

References

"Good Progress Being Made in Three Mile Island Recovery," *Nuclear Engineering International*, February 1982, pp. 21–25.

Daniel Martin, *Three Mile Island: Prologue or Epilogue* (Cambridge, Mass.: Ballinger Publishing Co., 1980).

M. Rogovin (ed.), *Three Mile Island—A Report to the Commissioners and to the Public*, Vol. I (Washington, D.C.: Nuclear Regulatory Commission, January 1980).

Further complexities arise when it comes to the die-back of trees in the Appalachians. Acid rain is believed by some to be the culprit, because the most severely damaged trees are those at higher elevations, where they are often surrounded by acidic clouds. But there are other possibilities. Ozone is one, and it has already been shown that trees are damaged by high ozone concentrations, such as those that occur in the Los Angeles basin. Another possibility is excess nitrogen in the form of nitrogen fertilizers that evaporate into the clouds. Trees can be stressed by excess nitrogen. Toxic metals are also suspect, as are various synthetic chemicals that have made their way into the atmosphere. Clearly, all of these collectively constitute air pollution, so we can say with some confidence that air pollution is the cause of tree die-back. But what is not so clear is exactly what corrective action to take; it would do little good to take massive action on one pollutant and then discover that little improvement had occurred. It seems clear, however, that the public will not long tolerate a major die-back in its forests. In Germany, where the people have a strong attachment to their forests, a 1983 poll showed that citizens were more concerned about the loss of forests than about the deployment of nuclear missiles.

In 1986 in this country, the National Research Council (NRC) declared that sufficient evidence exists to show that there is a link between the burning of fossil fuels and the acidification of lakes in the Northeast. (The NRC did not make a similar definitive statement about a connection between acid rain and forest die-back, because the evidence is inconclusive.) U.S. environmental groups had been pressing for action for some time, and so had Canada, which felt it was bearing much of the damage caused by U.S. industry. Finally, in spite of the uncertainties, and in spite of the protests by utility users that they would end up paying the billions of dollars involved, in 1990 Congress passed legislation requiring industry to reduce its contributions to atmospheric pollution.

■ 2.6 THE GREENHOUSE EFFECT

Since the end of the nineteenth century, the amount of CO_2 in the atmosphere has increased by about 23 percent, to 340 parts per million (ppm). The increase is attributed to two factors: the burning of fossil fuels and deforestation in the tropics. Carbon dioxide is vital to the earth's heat balance, because it causes the trapping of a certain amount of heat in the atmosphere—a "greenhouse" effect. If there were substantially less CO_2 in the atmosphere than there is, so little heat would be retained that the surface of the earth would be coated with ice. Conversely, if the amount of CO_2 were to increase to double the recent historical average, the Earth's average temperature might increase by 3 or 4°C, with uncertain but probably adverse consequences. The "doubling time" depends upon the rate at which we burn fossil fuel, but estimates have ranged from 88 to 220 years.

One of the consequences of global warming of which we can be confident is that the sea level will rise. In fact, the sea level *has* risen 15 cm or so in the last 100 years because of global warming. (See Figure 2.9.) A fact not commonly known is that the twentieth century is substantially warmer than the previous 500 years. In fact, those five centuries are known as the "little ice age." (See Figure 2.10.) It is the climate of

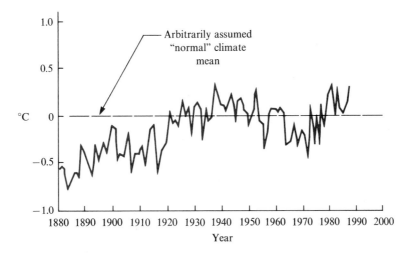

Figure 2.9

Temperature variations in the Northern Hemisphere, 1880–1987. The temperatures throughout most of the twentieth century have been higher than in the latter part of the nineteenth century. (Source: before 1980, J. Gribbin, *Future Weather and the Greenhouse Effect* [New York: Dell, 1982]; after 1980, R. A. Kerr, "The Weather in the Wake of El Niño," *Science*, 13 May 1988, p. 883)

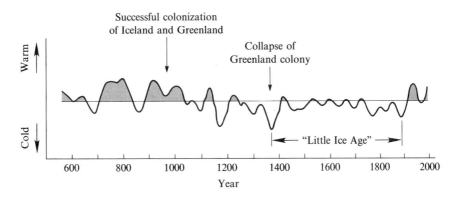

Figure 2.10

Temperature variations in the North Atlantic Ocean from A.D. 600 to the present, based upon Greenland ice cores. The temperatures in the twentieth century have been higher than in the 500 to 700 years preceding. (Source: J. Gribbin, *Future Weather and the Greenhouse Effect* [New York: Dell, 1982])

the middle part of the twentieth century that most people regard as "normal," because it is the period with which we are familiar. Yet one has to go back nearly a thousand years to find a period that was as warm. On the other hand, prior to 2 million years ago the earth was much warmer that it is now; the last 2 million years have been referred to as an "ice epoch" marked by alternations between glacial and interglacial periods. We live in a warm interglacial period now, of course, but some scientists believe it is time for us to start back into another glacial period.

The global warming of the twentieth century, and especially of the 1980s, has caused some people to declare that the greenhouse effect is already upon us. They may well be correct, but short-term (meaning a few decades') temperature variations cannot be relied upon as proof. For example, some scientists took the cooling trend from 1940 to 1970 (see Figure 2.9) as an indication that we were entering a new ice age and made public pronouncements to that effect. When the trend reversed itself and unprecedented high temperatures occurred in the 1980s, they had to eat their words. The director of the English Meteorological Office, B. J. Mason, said, "The atmosphere is wont to make fools of those who do not show proper respect for its complexity and resilience."

The aspect of the greenhouse effect that has attracted the most attention is the rise in sea level. The rise of 15 cm in this century has already caused some beaches on the Atlantic and Gulf coasts to erode at the rate of 3 to 5 feet per year. It has been predicted that in the next 100 years the sea level could rise another 8 to 25 cm if the average global temperature increases by 1.5 to 5.4°C, mostly because of the melting of glaciers. The huge glaciers in Antarctica, which make up 85 percent of the total ice on earth, are not involved in this calculation, because it is believed that they are currently subtracting water from the Earth's oceans and thus might remain in balance. The arctic ice cap is not involved at all, since it is largely floating; its melting would not substantially change the overall sea level. If, on the other hand, the antarctic ice cap (most of which is grounded on land, rather than floating) should break up, a change in sea level as great as 5 meters could be expected, which would be disastrous. The National Academy of Sciences, reporting in 1982, concluded that such a breakup is unlikely and, if it did occur, would take several centuries.

A much more worrisome scenario emerges from the computerized climate models that have been developed. Some of them have suggested that a global warming would drastically shift existing climatic patterns. The overall hydrological cycle might be intensified, with higher temperatures and greater rainfall at northern latitudes; simultaneously, rainfall at mid-latitudes might decrease. Thus, Siberia and Canada might prosper, but the great "breadbaskets" of both America and Russia might turn into dust bowls. Those who work with the models hasten to remind everyone that their models are not yet very good. Overall warming is something we can predict with reasonable confidence, but its accompanying regional effects are uncertain. The behavior of the Earth's atmosphere and oceans is so complex that the current models cannot give reliable regional projections. For example, when the models are run with known historical data, they give results that diverge widely from what actually took place. Nevertheless, the implication is that not all nations would necessarily regard the greenhouse effect as bad. Some nations might prosper under different climatic conditions.

Carbon dioxide is not the only greenhouse gas. Others are methane (CH_4), nitrogen dioxide (NO_2), ozone (O_3), the chlorofluorocarbons ($CFCl_3$ and CF_2Cl_2, called the CFCs), and various other synthetic chemicals. The CFCs are the gases that are believed to be destroying the ozone layer in the stratosphere, and they are much more potent greenhouse gases than CO_2.[3] One molecule of one of the CFCs has the same greenhouse effect as 10,000 molecules of CO_2. Major international strategies are being developed to phase out the use of CFCs, which not only will help the stratospheric ozone but will also help lessen the greenhouse effect.

Methane, too, is a more potent greenhouse gas than CO_2. Each molecule of CH_4 is about 20 times as effective as each molecule of CO_2. Most CH_4 comes from swamps, rice paddies, the rumina of cattle, and other biological sources, some of them influenced by man's activities and some not.

It is believed that about 85 percent of the CO_2 released into the atmosphere comes from fossil fuel combustion and the rest results from tropical deforestation. (Nontropical deforestation appears to be a negligible factor.) However, less than half of the released CO_2 appears to wind up in the atmosphere. Most of the rest is assumed to be absorbed by the oceans, but nearly 20 percent is unaccounted for. This is another example of the uncertainties that bedevil the makers of computer models of the world's climate.

The big question is: What should we do about all this? There is the possibility, of course, that any greenhouse warming will be completely overshadowed by "normal" variations in the Earth's climate. Such variations have occurred many times in the past. In the early seventeenth century, glaciers in the western Alps advanced to the point where they overran villages that had been inhabited for centuries. In the twelfth century there was a major climate change in North America that lasted for 200 years: Rainfall declined dramatically, and thousands of villages in the northern plains that depended upon the growing of maize completely disappeared by the end of the century. The drought of the 1970s and 1980s in the sub-Sahara, encompassing an area twice the size of the United States, was blamed by some on the Africans themselves, because they had stripped away the vegetation by grazing. Others pointed out that there had been a drought in the same region in the middle of the nineteenth century that was just as long and just as intense, and the geological record shows that the sub-Sahara has been plagued by a cyclical pattern of severe droughts.

Even with this historical perspective, we can mitigate the consequences of the greenhouse effect, assuming that we are not "rescued" by a new little ice age. We can, for example, reduce our consumption of fossil fuels. Some reduction already took place beginning about 1973, prompted by energy shortages and an emphasis upon energy conservation; it is claimed that carbon emissions were reduced by 30 percent. Some have urged a greater reliance on nuclear power, and a worldwide increase in nuclear power capacity has indeed slowly taken place. Others have urged a shift to solar energy, but a massive shift to conventional solar energy would cause a temporary *increase* in the use of fossil fuels because of the energy needed to fabricate

---

[3] There is "good" ozone and "bad" ozone. The "good" ozone is in the stratosphere, where it helps to screen out damaging ultraviolet rays. The "bad" ozone is at ground level, where it becomes directly toxic to plants and animals.

the enormous amount of solar equipment that would be required. (A successful development of photovoltaic systems would be a different matter, however, and would have a significant positive effect; see the next section, on energy.) A massive planting of trees to "sweep" the atmosphere of CO_2 has also been proposed. However, such a solution would require the planting of 3 to 7 million hectares to absorb as much CO_2 as is produced annually by fossil fuel burning.

The National Academy of Sciences, in its landmark 1982 report on CO_2, said that if the greenhouse effect does produce the predicted adverse effects, then our primary response should be mostly one of adaptation. We would have to engage in a slow retreat from low-lying coastal areas and develop new agricultural methods and plant varieties to adapt to drought conditions. It has been pointed out that the former is an activity in which we have already been engaged through much of this century because of the rise in sea level that has already occurred, and that the latter is being done all the time. What we will need most is enough time for the adaptations to take place in an orderly way, instead of under crisis conditions.

■ 2.7 ENERGY

During the 1970s, two oil shortages caused long lines at gas stations and prompted the federal government to embark upon ambitious programs to provide alternative energy sources. By the 1980s there was an oil "glut." Gasoline became cheaper, and Americans lost interest in energy. But as the last decade of the twentieth century came on, analysts warned that a new energy crisis was impending.

The energy flow of the United States in 1990 is shown in Figure 2.11, in the form of a so-called spaghetti diagram. Ninety percent of the country's energy was supplied by fossil fuels, and it was expected that this situation could not change very fast. The Department of Energy has estimated that it requires 20 to 50 years to make the transition from a new energy concept to commercial practicality. (By 1990, solar photovoltaic energy was approaching the end of that process.) Even then, after a new technology has been proved commercially viable, it is estimated that it would take another 30 to 50 years before enough capacity would exist to produce a quad of energy per year. (A "quad" equals one quadrillion, or 10^{15}, BTU. The United States used 81 quads in 1990.)

Both electrical energy and transportation energy are critical items. Transportation depends almost entirely upon oil, because oil is liquid and easily transportable. Therefore, any disruption in the country's oil supply is felt almost immediately in the form of long lines at gas stations. Electrical energy is critical because a gap between demand and supply has been predicted to occur by 2000. Utilities invested heavily in new plants in the 1960s and 1970s, based upon their predictions of future demand. After the crises of the 1970s, so much energy conservation took place that the utilities found they had too much capacity, so they stopped building. Simultaneously, the projected costs of big new plants, whether nuclear or fossil-fuel based, became so great that some utilities found themselves unable to finance the projects. The public, of course, will not long tolerate electrical "brownouts," so if a capacity gap does occur, a public outcry for the government to "do something" is likely.

U.S. Energy Flow – 1990
Net Primary Resource Consumption 81 Quads

Net imports 0.02

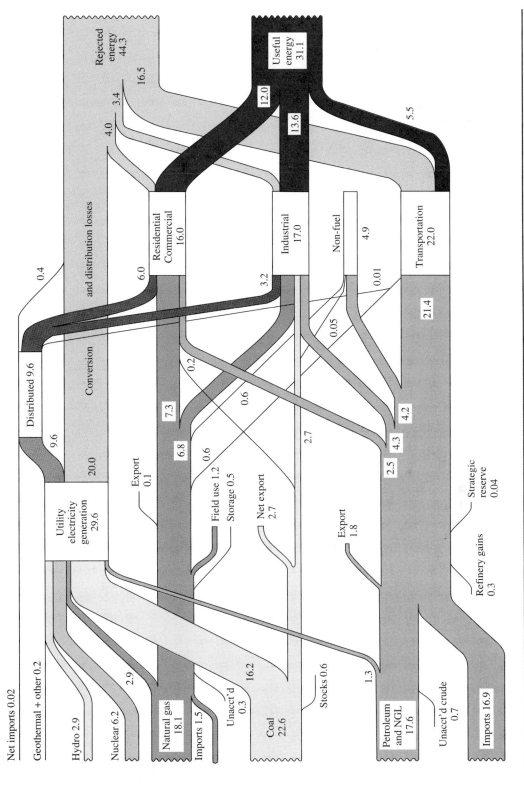

Figure 2.11

U.S. energy flow, 1990. Values are in quads. (1 quad = 10¹⁵ BTU.) Total usage in 1990 = 81 quads. (Source: I. Y. Borg and C. K. Briggs, *U.S. Energy Flow—1990*, Lawrence Livermore National Laboratory, June 1991)

Engineering and the Environment
The Fermi Breeder Reactor

It is commonly supposed that the accident at Three Mile Island in 1979 was the first of its kind. But this is not true, because an accident possessing many similarities occurred at the Fermi-I breeder reactor in 1966.

The Fermi plant was conceived in 1955 in response to an invitation from the U.S. Atomic Energy Commission (AEC) to construct an industry-sponsored demonstration breeder reactor. The size of the reactor was to be 200 MW (thermal rating; the electric output was 65.9 MW), and its cost was estimated at $60 million. Even though such a plant is small compared to more recent nuclear plants, it was far larger than the only other operating breeder reactor in existence at that time—the EBR-I, at 0.2 MW (electric output).

The plant was constructed about midway between Detroit, Michigan, and Toledo, Ohio, on the western shore of Lake Erie. A license for low-power operation (1 MW) was granted in 1963, and after successful low-power tests had been conducted, a license to operate at the full power level of 200 MW (thermal) was given in 1965. During the summer and fall of 1966, successful operation at the 100 MW (thermal) level was achieved, although it was noted that some temperature readings at several locations in the core ran as much as 35°F higher than expected. However, the temperatures were far below the boiling point of the "coolant" being used—molten sodium—and so were well within design limits. Before the tests were continued, the fuel elements in the overheated spots were shifted to new locations to see if the troubles could be attributed to the elements themselves or to something else.

When the reactor was restarted, it became apparent at about the 30-MW level that there was a discrepancy between the actual power level and what it should have been, according to the position of the control rods. As the discrepancy was being investigated, radiation alarms suddenly went off in the reactor containment building. The control rods were reinserted under normal control, and the reactor was shut down seconds after the alarm. A radiation survey of the plant site was taken, and it was determined that external radiation releases were within normal limits, so no public hazard had occurred. Samples taken from the reactor, however, allowed only one possible conclusion: Some melting of the fuel had occurred. But no one knew why or how much.

The fear of a core meltdown had plagued nuclear engineers from the beginning of the industry. If a reactor core were to lose its coolant, it could easily generate enough heat—even after being shut down—to melt its core. This led to a physicist's joking remark that a melting core might settle itself down into the earth, melting everything ahead of it, and keep going until it emerged on the other side of the world in China. He named this the *China syndrome* (the phrase was used later as the title of a movie). In the case of a breeder reactor, with its highly enriched fuel, there was a further fear: a melting core might cause the fuel to be rearranged into an unplanned geometry that could produce a so-called secondary criticality and a nuclear explosion—not like that in an atom bomb, but more like a 500- to 1000-pound TNT bomb. The reactor containment shell at Fermi had been designed to contain just such an explosion. As an additional safeguard, to prevent the plug from puncturing the containment shell an energy-absorbing structure made of crushable aluminum was constructed over the reactor in case an explosion hurled the 140-ton "plug" on the top of the reactor into the air.

None of these things happened at Fermi, although the subsequent investigative actions were taken with much physical caution in case the reactor was hanging on the verge of criticality. As things developed, it took *two years* before anyone knew what had gone wrong, because all of the examination of the reactor core had to be done remotely with a periscope-like device called a "boroscope." In order to see inside the reactor vessel, a number of the fuel elements had to be removed. These were examined (by remote control, of course), and it was determined that two of them had undergone melting. The liquid sodium was drained from around the core, and the interior of the reactor was examined by means of the boroscope. The investigators discovered a loose piece of metal lying on the bottom that at first was tentatively identified as a beer can. With infinite pains, and working remotely with a flexible device inserted through the 14-inch-diameter sodium coolant pipe and around two 90° bends, they finally fished out the offending piece. It proved to be a piece of zirconium metal that had been fastened, along with five others, at the bottom end of the reactor, where the sodium coolant entered. Obviously, the heavy flow of the coolant had torn the piece loose and plastered it against the bottom of the fuel element assembly, preventing the coolant from reaching some of the fuel elements. Deprived of coolant, those elements had overheated and melted.

It was decided that the zirconium plates actually were not needed, so, with infinite patience, all of the remaining plates were removed remotely, working through the coolant tube. As one of the participants said, "It was like trying to remove a patient's appendix through his nose." After some design modifications to prevent any other stray object from getting in a position to stop the flow of coolant, the plant was refueled and brought back into operation—almost four years after the accident. It finally was operated successfully at the full power level of 200 MW, generating a total of 32 million KWh, and then was shut down forever. The reason: lack of money.

The total cost of the reactor had been $143 million. Additional financing of $50 million was needed for a six-year program involving replacement fuel elements and a new core design. Private industry came up with most of the money, but not enough. Federal assistance was sought from the AEC, but the response was that any available federal money should be spent on the projected Clinch River breeder reactor demonstration plant, which was of a more advanced design than Fermi. The decision was then made to decommission the Fermi plant. Decommission became a three-year process, costing nearly $7 million. The highly radioactive parts had to be disassembled, packed in protective casings, and shipped to disposal sites. Most of the structure, much of it radioactive, was left in place. The containment building was sealed and the site fenced off. Surveillance and monitoring of the decommissioned plant was assigned to Detroit Edison, which had, in the meantime, constructed a new nuclear plant of conventional design adjacent to Fermi-I and named it Fermi-II.

Reference
E. P. Alexanderson (ed.), *Fermi-I, New Age for Nuclear Power* (LaGrange Park, Ill.: The American Nuclear Society, 1979).

Of all the new energy technologies that aroused excitement in the 1970s, only a few appeared to have any sustained ability to help solve our energy problems. Nuclear energy is the most highly developed, but it has aroused much uneasiness. (A special section, which follows, has been devoted to the topic.) One problem with nuclear energy is that it produces its energy only in electrical form and thus leaves most of our energy needs unanswered.

Coal is used both to produce direct heat (mostly in industry) and to make electricity. It is not very transportable and as a result has not been used much for transportation purposes in recent decades. However, we have a great deal of it, so during the crises of the 1970s there was intense interest in developing processes that would turn coal into a liquid or gaseous form. Unfortunately these processes turned out to be expensive. When the oil "glut" arrived, the enthusiasm tapered off, although research with pilot plants continued. In recent years, the research has paid off with prospective cost reductions as great as 60 percent. Thus, if oil were to become expensive again, there would be a strong incentive to invest in coal liquefaction or gasification plants. A special benefit of such plants is that they remove sulfur from the coal as a part of the process.

Until the 1990s, solar energy had produced mostly disappointment. At one time it was heralded as the sweeping answer to our energy problems, but matters did not turn out that way. The federal government offered generous tax subsidies to people who installed solar energy units, and so did many state governments. With the subsidies, solar energy units represented attractive investments. Without subsidies they did not. It was estimated that it would take 20 years or more for a typical solar hot water system to pay for itself, depending on the particular region and type of fuel being displaced.[4] Wind energy (considered by the Department of Energy to be a form of solar energy) also enjoyed a period of popularity, aided by tax subsidies and preferential treatment whereby utilities were required to purchase power from "wind farms" at generous prices. But as long as the cost of oil remained low, wind energy could not compete cost-effectively.

A more promising form of solar energy comes from photovoltaic cells. Such cells are in everyday use but are much too expensive for general purposes. Massive investments have been made in photovoltaic cell research over the years, and the cells have steadily improved in efficiency and decreased in cost. By 1988, efficiencies as high as 30 percent and costs as low as 30 to 40 cents per kilowatt-hour had been achieved. However, the costs needed to come down even more in order for solar cells to be competitive as a general source of power.

---

[4] In Los Angeles, it would take 24 years for a solar hot water system to save enough fuel to pay back its cost of installation and maintenance, if compared with the cost of natural gas, the fuel commonly employed in that city. In Boston, it would take 23 years to pay back the cost, if solar energy were compared with fuel oil. If electricity were the source of heating energy being replaced by a solar system, it would take 11 years to pay back the cost in Los Angeles and 15 years in Boston. (Electricity is much more expensive than natural gas or fuel oil when used for space heating purposes.) A basic but unanswered question is whether such systems would last for 11 to 15 years, let alone 23 to 24 years, without major maintenance or replacement costs.

Engineers have always been vitally interested in efficiency, but the energy shocks of the 1970s lent new urgency to their work. They were spectacularly successful in their efforts to improve energy efficiency, but their success received little publicity because it was accomplished in many small ways in many places. Here are a few examples:

1. Industrial plants tightened up on their energy-using systems, eliminating things like steam leaks and uncontrolled losses. In one case, a proposed new million-dollar boiler was found to be unnecessary after energy use was minimized.
2. Major improvements in lighting efficiency have been achieved over the years, going back as far as the 1940s, mostly by shifting away from incandescent bulbs to more efficient types, such as fluorescent lamps. Newer kinds of fluorescent lamps, which use high-frequency solid-state ballasts, are still more efficient and can reduce energy consumption by 25 percent.
3. Electric motors consume most of our electrical energy and account for more total energy use than do passenger automobiles. Electric motor design was neglected for many years but came back into its own with the new emphasis upon energy efficiency.
4. Large improvements in efficiency have been achieved through the adoption of new processes. Even before the energy crisis of the 1970s, major energy reductions had been achieved in the steel industry, primarily because of new blast furnace designs. Subsequently, companies found they could reduce costs, pollution, and energy usage simultaneously by developing new processes. In the food processing industry, which is a big polluter and a big energy user, it was found that the wastage from such processes as peeling and blanching could be reduced substantially. Even though the principal motivation in most of these cases was to reduce pollution, companies found they also benefited by saving the financial and energy costs of treating the wastes.

■ 2.8 NUCLEAR ENERGY

In the 1960s, nuclear energy seemed to be the wave of the future. Capacity in the United States grew quickly, and for a time electricity from nuclear sources was cheaper than that from coal. At about the same time, however, opposition to nuclear power began to develop, partly because of a report published by the Atomic Energy Commission (later renamed the Nuclear Regulatory Commission) that estimated the number of deaths and injuries that might result from a major nuclear accident. The report—known by its numerical designation, WASH-740—said that under the worst possible set of conditions, there might be as many as 3400 "prompt" deaths (meaning within a few days or weeks), and 43,000 people might develop severe radiation sickness. People who supported nuclear power scoffed at these figures as being unrealistically high; people in opposition scoffed at them as being unrealistically low. Thus began a bitter argument that has gone on ever since and that probably has a negligible chance of ever being settled.

By the last decade of the twentieth century, about 95,000 megawatts (MW) of nuclear generating capacity was in place in the United States. (A single large nuclear plant can produce about 900 to 1000 MW.) Because nuclear plant construction in the United States had come to a halt in the 1980s, it was expected that the nation's nuclear capacity would remain at this plateau until about 2010, when most plants would reach the ends of their useful lives. Nuclear supporters hoped that after 2010 new plants of improved design would be constructed to replace the old. Nuclear opponents hoped that the aging units would simply be shut down and that nuclear power would go away.

Worldwide nuclear generating capacity in 1985 was about 250,000 MW. Some countries have essentially staked almost everything on nuclear power. France, for example, got 69 percent of its electricity from nuclear sources in 1988 and planned to go to 80 percent, with most of the remaining 20 percent coming from hydropower. In other words, the French intended to free themselves entirely from dependence upon imported oil, at least as far as electricity is concerned. The percentages of electricity coming from nuclear sources in other countries in 1988 were: Belgium, 66 percent; South Korea, 53 percent; Sweden, 45 percent; Japan, 31 percent; West Germany, 31 percent; Spain, 31 percent; Bulgaria, 30 percent; Czechoslovakia, 26 percent; United States, 18 percent; Soviet Union, 12 percent.

The accidents at Three Mile Island (1979) and Chernobyl (1986) were terribly frightening to millions of people and caused confidence in nuclear power to sag. (See boxed items.) The Three Mile Island accident began when a routine maintenance operation caused a blockage in one of the main feedwater lines. By the time the event was over, more than a third of the reactor core had melted and fallen to the bottom of the reactor vessel.

The Three Mile Island event was the worst possible nuclear accident come true: The core had melted. In the scenario for a core-melt accident, the molten mass was supposed to burn its way through the bottom of the reactor vessel, then burn through the thick concrete lining at the bottom of the containment shell, and finally penetrate the groundwater table. Eventually the containment would be breached and massive amounts of radiation would spread across the countryside. It was just such a scenario that the authors of WASH-740 had in mind when they estimated that a "worst possible accident" might result in 3400 prompt deaths, with more deaths to follow later from radiation-induced cancer.

However, in the case of Three Mile Island, except for the fact that the core did indeed melt, the scenario did not take place. The molten mass did not even penetrate the shell of the reactor vessel but instead acted as an insulating layer as it cooled. The amount of radiation that escaped into the environment turned out to be negligible. The authors of WASH-740 assumed that an enormous amount of radioiodine would be released in gaseous form; in the actual event, very little radioiodine escaped. Instead, it was converted into an iodide that readily dissolved in water and thus remained inside the reactor containment. Virtually all the other radioactive materials also remained inside the containment shell, except for inert gases such as krypton and xenon, which pose little hazard.

In the aftermath of Chernobyl, a bitter argument ensued concerning whether coal power or nuclear power is more dangerous. One writer asserted that their dangers

Figure 2.12

The top of a nuclear reactor containment shell being lifted into place near Perry, Ohio. (Courtesy of Raymond Kaiser Engineers, Inc.)

are similar, even if one accepts the high estimate of 39,000 future cancer deaths from Chernobyl. (See boxed item.) He estimated that the death toll from the use of coal in the U.S.S.R. is between 5000 and 50,000 per year. Many of these deaths result from mining and transporting the coal, because the use of coal requires the handling of 100 times as much material as does the use of uranium for an equivalent energy output. In the United States, 100 or more coal miners die each year, and nearly 600 of the 1900 deaths in railroad accidents each year are the result of the shipping of coal. But the big killer is air pollution. Although it is impossible to say with certainty that any given death is caused by coal burning, it has been estimated that 50,000 people each year in the United States experience early deaths because of air pollution, mostly from coal. It is by extrapolation to similar populations and similar pollution conditions that the estimate of 5000 to 50,000 early deaths from coal burning in the U.S.S.R. is derived. This estimate suffers from the same uncertainties as do the estimates for early cancer deaths from radiation. When they were published, the coal estimates incurred the wrath of opponents of nuclear power, who were outraged that such a comparison should be made.

A valid question arises concerning Chernobyl: Why were there only 31 "prompt" deaths and 200 cases of severe radiation sickness instead of the 3400 deaths and 43,000 cases of sickness calculated by the authors of WASH-740? The amount of radiation released at Chernobyl was similar to that assumed by WASH-740, and it had certainly been spread over a very wide area. The answer lies partly in the conservatism of the WASH-740 authors, who apparently did not want to understate the possible consequences of an accident. But it lies even more in the nature of the accident itself. Instead of the radiation being dispersed around the world, the authors of WASH-740 had assumed that (1) there would be a very large city nearby; (2) the radiation cloud would be blown by the winds directly onto the city; and (3) once over the city, the cloud would be held there for a long period of time by a temperature inversion. These conditions certainly could occur, and they would be very much more damaging than the relatively thin dispersion of the Chernobyl radiation. Thus, the

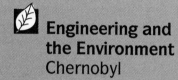

Engineering and the Environment
Chernobyl

The 1986 accident at the Chernobyl nuclear plant in the U.S.S.R. scared almost everyone in the Northern Hemisphere, and with good cause. Not only did the core melt, but the graphite blocks used for moderation purposes (almost all U.S. nuclear power plants use water, not graphite, for moderation) caught fire and burned fiercely for days, spreading radioactivity around the world.

The accident occurred during a supposedly routine low-power test. To prevent the reactor from automatically shutting itself off, the automatic trip safety system had been disconnected. Much publicity has been given to the improper procedures used by the reactor operators, which contributed to the accident, but the real problem lay with the design. At low power levels, the Chernobyl-type reactor is inherently unstable because it has what is called a "positive void co-efficient." In broad terms, this means that as the

available water in the core decreases (i.e., when steam bubbles form), the reactivity *increases*. In a "light water reactor" (LWR), just the opposite happens. As steam bubbles form, or if the water decreases for any other reason, the reactivity goes down, not up. Thus, an LWR is inherently stable. (Almost all commercial power reactors in the United States are LWRs.)

At Chernobyl, the operators apparently thought they could manage the reactivity of the plant by manipulating the controls. But the instability of the reactor at low power was so extreme that there was no chance of doing so. When the accident began, the plant was being run at about 6 percent of its full power rating. As the operators started the planned test, the power level began to rise. Because of the positive void coefficient, the rise became exponential. Within 2.5 seconds the power level had gone from 6 percent to 120 percent of full load. In just 1.5 seconds more, it was 100 times the normal full load. The reactor exploded.

The Russians had apparently designed their plants in the belief that an accident could never occur. As a result, they had provided for little containment. (U.S. plants have enormous containment shells. Because the Three Mile Island plant had such a shell, very little radioactivity

possibility of a high death toll from a major accident must remain a plausible concern. We should not acquire a false sense of security from the fact that the toll at Chernobyl was as low as it was.

Another major issue of nuclear power is the long-term disposal of highly radioactive wastes. We have not only the wastes from commercial power plants to be concerned about, but also the wastes from the production of nuclear weapons. It will be necessary to keep the wastes out of circulation for thousands of years because of the extremely long half-lives of some of the elements, such as plutonium. At present, the plan is to store such wastes in stable geological underground layers from which water has been absent for millions of years. Like most matters related to nuclear energy, the disposal of wastes is surrounded by bitter debate. One of the issues being debated is the level of certainty regarding the future. The supporters of nuclear energy admit that no absolute guarantee can be made that the wastes will remain out of human contact for thousands of years. However, they say that the risk of future expo-

escaped.) As a result, the radioactive products at Chernobyl—mostly iodine-131 and cesium-137—went directly into the atmosphere and spread around the world.

Thirty-one people died in the Chernobyl accident. Twenty-nine of them were firemen and two were reactor operators. Of the latter two, one was on top of the reactor when it exploded and was killed instantly; the other ran into the reactor building to try to find out what had happened. About 200 people were exposed to high levels of radiation and developed acute radiation sickness. Some of them were expected to die early deaths as a result of their exposure.

One year after the accident, a report by the U.S. Department of Energy (DOE) estimated that the number of cancer deaths among the exposed population worldwide might be 14,000 to 39,000 greater than would normally occur. The "normal" number of cancer deaths among the same population is expected to be 630 million, so the estimated increase in deaths because of Chernobyl lies between 0.002 and 0.006 percent. The authors of the DOE report warned that, because of the uncertainty in the statistical methods being used, they could not rule out the possibility that the future cancer increase might actually be zero. But even if the highest number

of estimated future deaths occurs, the authors of the report said it will be impossible ever to identify them as such, among the 630 million "normal" cancer deaths. A further problem, said the report, is that our current knowledge of radiation cancer risk is based upon high doses, usually delivered at high rates. There seems to be a substantial amelioration of latent health effects for those exposed to *low* doses, based upon animal experiments. Thus, they said, the actual health effects of Chernobyl might turn out to be two to ten times less severe than those predicted in their report.

References

J. F. Ahearne, "Nuclear Power after Chernobyl," *Science*, 8 May 1987, pp. 673–679.

M. Goldman, "Chernobyl: A Radiobiological Perspective," *Science*, 30 October 1987, pp. 622–623.

"Letters: Chernobyl Public Health Effects," *Science*, 2 October 1987, pp. 10–11.

A. P. Malinauskas et al., "Calamity at Chernobyl," *Mechanical Engineering*, February 1987, pp. 50–53.

E. Marshall, "Recalculating the Cost of Chernobyl," *Science*, 8 May 1987, pp. 658–659.

R. Wilson, "A Visit to Chernobyl," *Science*, 26 June 1987, pp. 1636–1640.

sure is very small. The opponents of nuclear energy, on the other hand, have insisted on a guaranteed method for keeping nuclear wastes permanently out of contact. Since they readily agree that no such guarantee can be made, they say this is reason enough to phase nuclear energy out of existence.

■ 2.9 A LOOK AT THE FUTURE

We are so regularly bombarded with varied predictions of the future that we can almost select one to suit our individual tastes. Some are bound to be correct, because almost every conceivable future has been predicted by someone. However, by the same token, most predictions will be proved wrong.

In 1937, a presidential committee of prominent scientists and engineers attempted to predict the kinds of inventions that would come in the next 10 to 25 years and the social impact they would have. They did a fair job of predicting the impacts of television, the growth of air travel, and the construction of a national system of super-highways. But they completely missed the following, all of which arrived in 25 years or less: atomic energy, radar, the jet engine, the transistor, electronic computers, xerography, and earth-orbiting satellites.

More recently, the staff of the *Wall Street Journal* prepared their version of the future, called *Here Comes Tomorrow* (1966). By 2000, they said, the following would have come to pass: Productivity would increase $2\frac{1}{2}$ times, and the average work week would shrink to 31 hours; color video would be used worldwide; nuclear energy would be cheap and plentiful; electric cars would be commonplace; virtually all homes would be heated by electricity; supersonic transports would be in wide use by the 1970s, and hypersonic (4000-mile-per-hour) transports would arrive by 1980; people would land on Mars by 1985, and manned flights around Jupiter and Saturn would occur in the 1990s; satellites would be used for communications and weather forecasting; slum areas in cities would disappear, and the crime rate would drop;

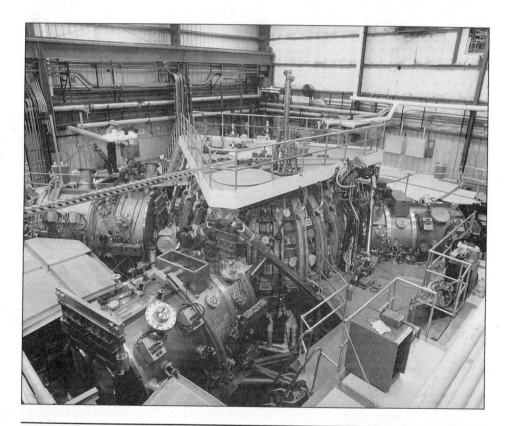

Figure 2.13

Controlled fusion represents a major possibility for unlimited energy in the future. The experimental unit shown here is a type of fusion device called a Tokamak. (Courtesy of General Atomics)

homes would be virtually free of onerous household chores; by 1975, new colleges would be founded at the rate of one per week; heart disease, arthritis, and smallpox would all but disappear, and the cure rate for cancer would increase. With sadness, the authors also predicted that war and the common cold would still be with us.

The authors of *Here Comes Tomorrow* hit the mark on color video, communications satellites, and the elimination of smallpox. They missed on almost everything else, although we would have to admit they are probably still right about war and the common cold. (Ten years later, the authors admitted they had been overly optimistic.)

Optimism about the future is a troubling matter. As one looks backward, all sorts of surprising breakthroughs of technology—which have had profound effects upon our lives—are perceivable: breakthroughs in medicine, in travel, in communication, and in food production. But, looking ahead, one dare not predict a future controlled by breakthroughs, because we have no idea what they will be. We are forced to make predictions only on the basis of what we know at present; to do otherwise is to invite dismissal as being merely a visionary. Yet we know that breakthroughs will occur, because they have always occurred. The least intelligent thing we could do would be to state, categorically, that the last breakthrough has already occurred and there will never be another one.

Some groups have made very negative predictions for the human race. In 1972 a book called *The Limits to Growth*, based on the results of a computer model, was published and achieved instant popularity. The computer results showed that disaster could be expected to overtake the world before 2100, taking the form of either unlimited pollution, catastrophic population decline, or both. Only under circumstances of population stabilization at a fairly low level, coupled with resource recycling and strict pollution controls, did the computer predict a stable future for the human race.

Most people would agree with the view that there *are* limits to growth, but many would not agree with the catastrophic outlook of the computer models. After all, they point out, the output of a computer program is fixed as soon as the input assumptions are chosen. Pessimistic assumptions produce pessimistic outcomes; optimistic assumptions produce optimistic outcomes.

Some of the most pessimistic assumptions in *The Limits to Growth* had to do with population growth. But since the publication of that book, some remarkable developments in the rate of population growth have occurred, at least in the more industrialized countries. In 1985, according to the Worldwatch Institute, all of the following countries were reported to be at or near zero population growth: Austria, Belgium, Denmark, Germany, Hungary, Italy, Sweden, and the United Kingdom. Five more, Bulgaria, Czechoslovakia, France, Netherlands, and Romania, were approaching zero population growth. Unfortunately, most of the world was still experiencing unchecked growth, especially countries like Bangladesh, Brazil, Ethiopia, India, Mexico, Nigeria, and Pakistan. (According to the Worldwatch Institute, the United States will continue to grow, partly because it has a population skewed toward the childbearing ages and partly because of its high rate of immigration, much of it illegal.)

Probably the most sensible statement that can be made about the future is that it will contain surprises. Also, there are certain to be crises, some of which will be truly major and some of which will soon fade in public importance. We know this because our history has always contained crises, and there is no reason why the future should be different from the past in this regard.

It appears clear that science and engineering will continue to increase in importance in the future and that the result will be a continuation of past trends toward better health, longer life, and increased personal freedom and quality of life, at least in industrialized countries. Computers, particularly in networking systems, will continue to grow in importance. Factory automation will boost productivity, and this in turn will lead to higher living standards.

Food will continue to be a problem for most of the world. Energy resources will be a problem for everyone. There will be a shift toward coal and oil shale, as oil and gas become more scarce. Probably the increased use of coal will entail some environmental damage, but it will continue to be a major objective of society to mitigate this damage. Renewable energy resources, such as solar photovoltaic and wind, will gradually grow in importance. Engineers will continue to find clever ways to use energy and other resources more effectively. Water will become a major problem, especially in the arid southwestern portion of the United States. The ground transportation system of the United States, especially the railroads, will probably undergo a rejuvenation as our population grows denser and fuel costs increase.

Most people appear to feel that the world of today is, by and large, a better and more interesting world than the one of 30 or 40 years ago. Perhaps that is only the result of the natural optimism of humans, but that very optimism has always been the most valuable asset of the human race. Assuming we can avoid nuclear holocaust, people 30 years hence will no doubt feel that their era is an improvement over ours. Engineers will have a lot to do with causing that more attractive future to become a reality.

References

"Back to the Energy Crisis," *Science*, 6 February 1987, pp. 626–627.

B. P. Beckwith, *The Next 500 Years* (New York: Exposition Press, 1967).

R. H. Bezdek, A. S. Hirshberg, and W. H. Babcock, "Economic Feasibility of Solar Water and Space Heating," *Science*, 23 March 1979, pp. 1214–1220.

W. Booth, "Postmortem on Three Mile Island," *Science*, 4 December 1987, pp. 1342–1345.

L. R. Brown, et al., *State of the World, 1985* (New York: W. W. Norton, 1985).

California Air Quality Data (Annual Summaries, 1979–1989) (Sacramento: Air Resources Board, State of California).

E. Cornish, *The Study of the Future* (Washington, D.C.: World Future Society, 1977).

W. E. Deming, *Out of the Crisis* (Cambridge, Mass.: Center for Advanced Engineering Study, Massachusetts Institute of Technology, 1986).

J. Gribbin, *Future Weather and the Greenhouse Effect* (New York: Dell, 1982).

Here Comes Tomorrow! Living and Working in the Year 2000 (Princeton, N.J.: Dow Jones Books, 1966).

"Infrastructure Crisis: What the Media are Saying," *Civil Engineering*, April 1987, pp. 48–50.

C. Komanoff, "Increased Energy Efficiency, 1978–1986," *Science*, 8 January 1988, p. 239.

J. R. Luoma, "Forests are Dying, But Is Acid Rain to Blame?" *Audubon*, March 1987, pp. 37–51.

D. H. Meadows, et al., *The Limits to Growth* (New York: American Library, 1972).

V. Ramanathan, "The Greenhouse Theory of Climate Change: A Test by an Inadvertent Global Experiment," *Science*, 15 April 1988, pp. 293–299.

R. A. Smith, R. B. Alexander, and M. G. Wolman, "Water-Quality Trends in the Nation's Rivers," *Science*, 27 March 1987, pp. 1607–1615.

Technological Trends and National Policy (Washington, D.C.: National Resources Committee, 1937).

J. A. Young, "Technology and Competitiveness: A Key to the Economic Future of the United States," *Science*, 15 July 1988, pp. 313–316.

Exercises

2.1 What are some of the ways in which energy and environmental matters are fundamentally associated?

2.2 Make some assumptions about the average energy efficiency of gasoline-powered automobiles and compare to the probable overall energy efficiencies of electric automobiles, including the efficiency of power plants that make the electricity. If the average efficiencies of automobiles and power plants are both about 30 percent, based on the initial heat content of their respective fuels, does this mean the two systems are about the same in energy effectiveness? Are there any other important energy losses in these two systems?

2.3 Investigate and compare the environmental risks of nuclear power plants and coal-fired power plants.

2.4 List the positive and negative aspects of the following energy sources, considering such matters as risk to life, environmental damage, wilderness values, recreation, food supply, resource exhaustion, and creation of jobs: nuclear energy, coal, large dams, offshore oil, geothermal energy, solar energy, wind energy, conservation. Are there any health or environmental risks connected with solar energy? With energy conservation?

2.5 Examine the utility bills for your household from a full year and try to estimate the portion of your utility expense that goes for hot water, making reasonable assumptions if necessary. Estimate what portion of this expense you might be able to save if you had a solar hot water heater, remembering that your solar heater will not work in cold, cloudy weather. If you believe your annual saving should be large enough to permit you to recover the cost of a solar hot water system over a ten-year period, how much could you afford to invest in such a system?

2.6 What are some of the ways in which the automobile has altered our social structure? What about the computer? Television? The telephone? Inexpensive printing processes?

2.7 What examples can you discover, in addition to those in the text, that result in a lessening of environmental damage while simultaneously preserving or even improving other benefits?

2.8 What are the trade-offs involved in recycling newspapers into new paper products versus burning them for their energy content?

2.9 Examine some of the world-growth computer models that have been developed, as in the book *The Limits to Growth*. Discuss some of the ways in which the results of the models might be significantly altered by variations in the assumptions and inputs.

2.10 Construct for yourself a simplified population growth model in which the initial population distribution is heavily weighted toward the young side. Observe the way the total population increases, even though each couple produces only the theoretical "no growth" number of two children. To keep the model simple, assume the initial distribution by age groups to be as follows:

Age Group	Number in Group
0–10	2,000,000
10–20	2,000,000
20–30	2,000,000
30–40	1,000,000
40–50	1,000,000
50–60	1,000,000
60–70	1,000,000
70–80	1,000,000
	11,000,000

Assume that everyone in the 20–30 age group produces exactly two children per couple but that no children are produced by any other age group. Also, assume that no one dies until age 80, whereupon the entire age group dies simultaneously. Construct a chart showing the number in each age group at the end of every ten years. Observe how the total population grows. What is the eventual steady-state population? (Admittedly, the above example is simplified beyond the bounds of reason, but it nevertheless demonstrates why a youthfully skewed population, perhaps occurring as a result of an earlier "baby boom," continues to grow even though each couple produces only two children.)

2.11 In some proposals for hybrid electric-gasoline engines it is assumed that the gasoline engine runs only intermittently, but when it does run, it does so at full speed to charge a set of batteries, which in turn supply current to electric motors that drive the wheels. Why would such a system be more attractive than an all-electric car running on storage batteries? What is there about the load-versus-efficiency characteristics of gasoline engines that enhances the environmental attractiveness of the above scheme?

2.12 College students are known to be great book buyers, going well beyond the purchase of required textbooks. Also, as well-informed citizens, they tend to be deeply concerned about environmental pollution. Consider and discuss the conflicts of interest for members of this group if they oppose the construction and operation of paper mills because these are big polluters, but simultaneously do not wish to give up books, even though they are responsible for the consumption of an enormous volume of paper.

2.13 From newspapers and magazines, see if you can find examples of environmental issues that go beyond the usual concerns of clean air and water, toxic wastes, and health risks and that might include other factors such as economic opportunity, jobs, and racial justice.

2.14 Examine recent newspaper articles on environmental topics to see if the writers have consciously or unconsciously stated matters in such a way as to favor one side or the other. Look for statements including phrases such as "could result in" or "might cause . . ." but presented so that the reader gains the impression that an actual fact, rather than a possibility, has been stated. Look to see if sources on only one side of an issue have been extensively quoted,

especially when the quotations contain inflammatory or unsubstantiated statements. In such a case, itemize for yourself possible alternative ways in which the writer could have prepared the article to present a more objective view of the topic.

2.15 Write a brief report on what you think the twenty-first century will be like and how developments in technology will influence our future.

2.16 List ways in which you believe that knowledge derived from our space program will benefit us.

2.17 Do you believe space exploration will expand in the future or diminish? Why?

2.18 What do you believe future attitudes in the United States toward science and technology will be? Do you believe engineers will become more important or less important? Write a brief essay describing the reasoning which supports your views.

2.19 Most current projections are that gas and oil reserves will be in substantial decline by the early part of the twenty-first century. At the same time, there is strong environmental resistance to the use of coal, oil shale, and nuclear power. What do *you* think will be the outcome of all this? Why?

2.20 What do you believe the views of college students in 2090 toward the problems and the achievements of the twentieth century will be? In answering this question, consider your own views regarding the nineteenth century. Take stock of your knowledge and opinions of things like: the Mexican War; the Civil War; the Spanish-American War; Indian Wars; malaria, cholera, and yellow fever epidemics; emigration by wagon train to the West; child labor; the industrial revolution; the construction of transcontinental railroads; Thomas Edison; Jules Verne.

2.21 The use of steel has declined in this country for many reasons, but in part because of the trend toward substitute materials such as concrete, aluminum, and plastics. What do you think will be the major trends in materials during the next 20 years? Justify your views.

2.22 Improved productivity is believed to be vital to our future. Investigate the subject of productivity by consulting recent books and periodicals in your library, and write a brief report on the topic.

2.23 Write a brief environmental impact report on the possible consequences of improved productivity. To what degree do you think your conclusions are influenced by views you held before you began looking into the topic?

2.24 Where do you believe modern life and technology are taking us? Do you believe your activities as an engineer will help lead us toward an improved world? What might the improvements be? Do you think quality of life for you and those you know is likely to improve or decline in your lifetime? Do you think matters have generally improved or worsened during most of the twentieth century for the United States and other developed countries? What, in your opinion, might be the most basic factors affecting any improvements (or losses) in quality of life of which you are aware?

Photographs courtesy of University of California, (top and bottom), and Cadkey (center).

3

Getting Started in Engineering

Most engineering education is at the under-graduate level, meaning that it is a nominal four-year program culminating in a bachelor's degree. This is very unusual for educational programs leading to professional careers. Law schools and medical schools, for example, usually take students *after* they have already received bachelor's degrees and give them three or four more years before awarding so-called professional degrees, such as the J.D. (Juris Doctor) or M.D. (Doctor of Medicine). Most engineers enter their professional careers with the B.S. (Bachelor of Science), and engineering technologists enter their careers with a degree of B.E.T. (Bachelor of Engineering Technology) or similarly designated degree.

Graduate study for engineers has steadily grown in importance over the years. Many engineers obtain master's degrees, variously designated as M.S. (Master of Science) or M.Engr. (Master of Engineering). Engineers who go into research or teaching typically earn doctor's degrees, either the Ph.D. (Doctor of Philosophy) or the D.Engr. (Doctor of Engineering). Master's degrees usually take one or two years beyond the bachelor's degree, and doctor's degrees frequently take five years or more beyond the bachelor's degree.

During the freshman year, most engineering programs look pretty much alike. Students usually take chemistry, calculus, physics, English composition, and a certain number of humanities or social sciences electives. They are also very likely to have an introductory course

69

Figure 3.1

An artist's conception of the X-29A, an unusual forward-swept-wing aircraft that is expected to be more efficient and less expensive than today's fighter planes. The wing is also expected to increase maneuverability and will be made of lightweight composite materials. (Courtesy of Grumman Aerospace Corporation)

in engineering, although this requirement is not universal. An introductory course in computing is almost certain to turn up in either the freshman or sophomore year. A course in engineering graphics (usually meaning the making and interpretation of engineering drawings and/or sketches) is generally required for civil and mechanical engineering and for related fields.

The humanities or social sciences electives have a definite purpose in engineering programs. They are *not* there, as some would have it, to provide engineers with a veneer of culture or to "make engineers acceptable in polite society." The purpose of the humanities and social sciences electives is to open engineers' eyes to the larger society in which they live, to give them some feeling for the great ideas and movements that have shaped our culture, and to show them that human beings will not always make choices on the rational bases engineers prefer, and that it is essential for engineers to understand these things. In blunter terms, the purpose of these courses is to get engineers out of their technical shells to see what is going on around them.

Even during the sophomore year, the requirements among engineering fields are usually very similar, with the major exception of chemical engineering, which emphasizes chemistry. In the junior and senior years, however, the curricula diverge along specialized paths. The result of all the foregoing is that transfer from one curriculum to another is usually pretty easy in the first year, is still relatively easy in the second year, and is difficult after that, involving loss of time.

Students who do exceptionally well in their undergraduate studies are usually invited to join one of the honorary societies listed in Table 3.1. Tau Beta Pi is the oldest and most widespread of the honorary societies. It covers all engineering disciplines, whereas the last four societies listed in Table 3.1 are restricted to specific fields. Sigma Xi requires achievement in research; hence, individuals generally do not join Sigma Xi until they are in graduate school or later. In the case of Tau Beta Pi, engineering students in the top 10 percent of their class are usually invited to join as juniors, and those in the top 20 percent are invited to join as seniors. Tau Beta Pi is more or less the engineering equivalent of Phi Beta Kappa. (Engineering students, even those who are outstanding scholars, usually do not take enough humanities courses to qualify for Phi Beta Kappa.) Tau Beta Pi not only is an honorary society, but is also a service organization. Typical service activities in which Tau Beta Pi members may engage are free tutoring for engineering freshmen who want it and student evaluation of the teaching effectiveness of engineering faculty members.

Table 3.1 Engineering Honorary Societies and Related Organizations

Name of Society	Location	Members	Year Founded	Remarks
Tau Beta Pi	Knoxville, Tenn. 37916	210,000	1885	National engineering honorary society
Sigma Xi	New Haven, Conn. 06511	125,000	1886	Recognition of noteworthy achievement in research
Eta Kappa Nu	Urbana, Ill. 61801	146,000	1904	National electrical engineering honorary society
Pi Tau Sigma	Cookeville, Tenn. 38501	65,000	1915	National mechanical engineering honorary society
Chi Epsilon	Knoxville, Tenn. 37916	46,000	1922	National civil engineering honorary society
Alpha Pi Mu	Morgantown, W.V. 26506	15,500	1949	National industrial engineering honorary society

Source: *Directory of Engineering Societies and Related Organizations* (New York: American Association of Engineering Societies, Inc., 1982).

It sometimes seems to students that they spend most of their time in the first two years of an engineering program studying mathematics (especially calculus), physics, and chemistry. They ask: "Where is the engineering?" Sometimes they even get the notion that calculus does not have much to do with engineering. But calculus and the closely related topic of differential equations are basic to our understanding of physical systems and make possible our predictions about those systems. And *making predictions* is exactly what engineers do: They predict that the structure they have designed will indeed hold up under a certain load, or that an electronic system will behave in such-and-such a way, or that a digital computer will solve problems at a predicted speed and of a specified complexity, or that a multibillion-dollar power

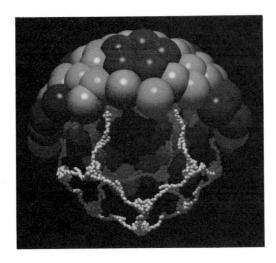

Figure 3.2
A computer-generated photograph showing a hemisphere of the 180 amino acid subunits contained in the protein coat of a particular virus. The large spheres represent amino acid chains, and the small spheres represent single amino acids, which are connected to each other and link the whole structure together. (Courtesy of the University of California Lawrence Livermore National Laboratory and the U.S. Department of Energy)

plant will produce so much electricity at a certain efficiency. These predictions are made before the systems are built. After the systems have been constructed and placed in operation, the results will be plain for all to see. The engineer must predict these things in advance, and mathematics—especially calculus—is the major tool by which these predictions can be made quantitative and correct. At least they are correct most of the time; engineers are human and do make mistakes sometimes. When they do, the results may be catastrophic and make front-page news. More will be said in Chapter 5 about catastrophes and the responsibilities engineers have for public safety and welfare.

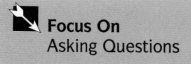

Focus On
Asking Questions

There are two periods of unusually accelerated learning in a young engineer's life. One is college, and the other is the first year on the job. Too often, young graduates worry that they have not learned enough to be good engineers. On the other hand, a few act as if they know everything worth knowing and create much unnecessary trouble for themselves.

Both extremes are bad, and young engineers should realize there is much to learn that does not come in textbooks. It is exceedingly important for them to ask questions when they do not know something, and to keep on asking. The people with the answers are almost always glad to help, because they remember their own early engineering days.

A mechanical engineer tells of his early experiences:

> On my first job, I found myself with the task of designing portions of a precision optical measuring instrument, and soon began having trouble, because I didn't know very much about manufacturing processes. I knew more or less what I wanted to accomplish, but I didn't know if it would be possible to make the parts I was thinking about, or what kind of precision could be achieved. Not knowing any better, I decided to go out in the shop and ask the shop foreman. The foreman fell all over himself in giving me

helpful advice, and I quickly found myself learning a lot about manufacturing, which helped me to become a better design engineer.

Later, when I had come to know the shop foreman better, he told me that far too few engineers were in the habit of seeking shop advice and that a lot of designs could be improved if the engineers sought shop input more. I remembered this lesson, and it served me well in subsequent years. There was one occasion, especially, where it made a sort of mini-hero of me.

I had moved to a new company, and one of my tasks was to look into a problem which had been plaguing the assembly line. However, I had also been warned that the manufacturing personnel were heavily unionized, and that if engineers went poking around in the factory, there might be a strike. I was kind of skeptical about this warning, though. I remembered my previous favorable experiences regarding shop input, and decided to try that route.

When I was ready, I approached the assembly foreman, who introduced me to the union steward. I told them what I wanted to do, which was to work in the part of the assembly line which dealt with the problem area and do all the things the assembly workers did. I also wanted to be able to discuss the assembly operations with the workers on the line. I got an immediate green light and worked for a week on the line.

There was no strike. On the contrary, the workers were eager to help. I shortly found the problem, too. A particular adjustment routine had been reorganized just prior to the time the troubles began, in a cost-cutting move, and the reorganization had gone too far. I made some minor changes, and the problem was solved. I was a hero, and all because I was willing to ask questions.

■ 3.2 STUDY HABITS

It has been noted that new college students frequently disregard advice concerning study habits, to their later regret. The problem is that they generally assume that the study habits they learned in high school will continue to serve them well in college. This assumption turns out to be wrong, because in college the student spends a smaller proportion of time in the classroom than in high school, and a much greater proportion of time must be spent on study outside the classroom.

The usual planning basis for college workloads is that one unit (or one semester hour, or one quarter hour) requires about 3 hours of work per week including class time. (Note: Semesters last approximately 15 weeks, plus examinations, and quarters last 10 weeks, plus examinations. The amount of work *per week*, however, is the same in either case.) Thus, an average 16-unit load nominally requires about 48 hours of work per week, which is already 20 percent more than the standard 40-hour work week. Furthermore, it is a fact of life that some courses—especially lab courses—require more than the standard 3 hours per unit, so the work week for an engineering student can easily go to 50 or 55 hours per week, and perhaps more.

Figure 3.3
An experimental model of a roadway-powered vehicle. The power is supplied from electrical cables laid in the roadway, and transferred to the vehicle through a power pickup that maintains a constant air gap with the road surface. (Courtesy of Lawrence Livermore National Laboratory)

According to custom, the 3 hours per week per unit is split, for a lecture course, into 1 hour in class and 2 hours spent in doing homework. Thus, a typical three-unit class may require three 1-hour (usually with 50 minutes = 1 "hour") lectures per week, and 6 hours per week in homework. Engineering classes, particularly, are much given to the assignment of daily problem sets, and the student, with some consternation, may find that some of those sets may take more than 2 hours of homework to complete. This inevitably leads to the question: What results should be expected from putting in 3 hours per week? An "A"? A "C"? Barely passing? No one can answer such a question, because human beings are all so different. There are some very brilliant people, of course, who can do their lessons in much less than the nominal time, but most people probably should be prepared to spend more than the nominal amount of time. The author has an especially vivid memory of times when it took up to 6 hours to finish calculus problem sets that were supposed to take only 2 hours.

Students frequently discover they have gotten into difficulty in the following way: They find that their instructors do an exceptionally good job of explaining things in class, and they follow and understand everything with ease. Naturally, the students are elated that things are going so well; hence, they feel that the drudgery of homework is probably unnecessary and treat it lightly, perhaps not doing it at all. Then, at examination time, they discover they cannot recall some of the things they understood so well in class and thus do poorly on the exams. The problem is a simple one, well understood by teachers, but accepted by students often only after an unpleasant experience. Here is the explanation: Students retain knowledge only on the basis of *what they do themselves* and hardly at all on the basis of what someone else does, such as a teacher's explanation in a classroom. A good teacher guides, stimulates, opens mental doors, and gives insight, but it is the student, and the student only, who learns. Does this sound like an exhortation to do the homework—all of it—every day? That's exactly what it is.

Taking examinations is an art. Some people "freeze" during an exam, and the usual reason is a lack of confidence brought on by inadequate preparation. The best way to build confidence is to be "on top" of a course at all times, doing the homework on time, attending class regularly, and seeking help from the instructor the minute something is unclear. (If instructors sometimes seem unreceptive to students seeking help, the students should persist anyway, because it is a part of an instructor's responsibility to give assistance outside class to those who did not understand the lecture or cannot do the homework.) Then, even though a student has kept on top of things throughout, a review of the material the evening before the exam is vital. Reviewing is a part of the learning process, because it is through the mental reinforcement process called repetition that we really remember and learn things. However, the sometimes-popular "cram" session just before an exam, used as a substitute for regular study, usually produces the opposite results from those wanted. In particular, if the cram session results in a lack of sleep, then mental preparation for the exam is reduced rather than enhanced.

Assuming the student is alert, well rested, prepared, and confident, the next step is to not do something stupid during the examination itself. One thing students sometimes do (and which they subsequently declare, in their own choice of words, to have been "stupid") is to fail to read the instructions carefully. In their hurry to get to the exam itself, they skim over the instructions hastily and sometimes misinterpret.

Look over the entire exam to get an idea of its scope. No doubt you will discover some problems that you can handle easily. Do these problems first, unless the instructions tell you to do otherwise. Not only do you build up your test score rapidly by this procedure, but you also develop confidence and momentum. If you run into a roadblock on any problem, do not remain hung up there through the remainder of the test period. Go to a different problem, one that may yield a solution, and come back to the tough one later. Show all your work. Many instructors give partial credit for what you did, even though your final answer is incorrect or missing. Above all, make your work clear, or the instructor will not have any idea what you are doing and will not be able to give partial credit.

If you finish an exam early, check your work, proceeding slowly and in a relaxed frame of mind. Recheck the instructions. Reread each problem or question carefully to be sure your interpretation was correct. Reconsider your procedures to be sure you were applying the right principles. Finally, recheck your calculations to be sure you did not enter some incorrect figures into your hand-held calculator (it is frustrating how often this occurs), and recheck your decimal point location. Then consider your answer for a moment to see if it violates your "commonsense" idea of the approximate value it should have. For example, if you are working with a small electrical circuit, and you instinctively feel the current in the circuit might be expected to be 10 amperes or less, and you get an answer in the neighborhood of 1000 amperes, you should consider checking your work yet again.

Some factors can interfere with a successful educational experience. One of these is motivation. If you feel you lack motivation, then perhaps what is bothering you is that you are wondering what being an engineer is really like and whether it is right for you. In a case like this, a conversation with a faculty member or a practicing engineer could help to clear your thoughts. But if your lack of motivation is simply due to the fact that college does not interest you, for whatever reason, then perhaps the solution is to take a year to do something else—work, perhaps. Of course, the kind of work you can get without a college education is likely to offer a limited future, and students have often reported that they acquired a strong sense of motivation by working for a year at a job they discovered they hated.

Another problem frequently encountered by students is the failure to develop a good time budget. Such students may have plenty of motivation but have never seriously considered developing time budgets. Here is a way to approach the matter.

Figure 3.4
The lecture method of classroom teaching has remained the most effective and popular mode of teaching because of the personal contact with the instructor and the opportunity to ask questions of an experienced professional. Here a professor is teaching a class in structural design. (Courtesy of University of California, Davis)

First, if you multiply 24 hours by 7, you discover the week has 168 hours in it. The very first thing you should put in your budget is sleep—8 hours a night for a total of 56 hours. (Many people believe they can get by on less than 8 hours of sleep, but it turns out there is little medical support for this belief. If a person attempts to get by with less, the result usually is a lessening of mental alertness and an increasing susceptibility to illness.) Next, you have to allow for getting up and getting to class. Let us say, in your case, you know that it takes you 2 hours in the morning to get up, shower, get dressed, eat, and get to class, including a small margin of safety. (Some people shorten this process by skipping breakfast, but the price of this decision is usually a lessened ability to cope with classes and lessened vitality.) You have no classes on Saturday or Sunday, but you know yourself well enough to realize you will probably just move more slowly on Saturdays and Sundays, so you allow 2 hours on these days, too. This gives 14 hours for the week. You allow 14 hours a week for lunch and dinner, and 10 or 15 minutes each day for travel from campus to your dwelling. Total for the week: 29 hours. Adding up all these numbers shows that 85 hours of your week have already been consumed, and you have not even allowed for classes yet.

Figuring the time spent in class is the easy part. Let us suppose you are carrying 16 units, with 13 of them in lecture classes, and three units of lab. The lecture classes require 13 hours, and a three-unit lab typically would require 9 hours in the lab. Now for homework. The 13 units of lecture will call for 26 hours of homework if the usual formula is applied. (It may turn out that you will actually need more than this.) The lab really is not supposed to require any time beyond that actually spent in lab, which is 9 hours for the three-unit class. But you have been forewarned that professors usually assign outside homework in labs anyway, so you provide for an extra 4 hours of homework, for a total of 30 hours. Now we can add everything up:

Weekly Time Budget	
Sleep	56 hours
Get up, shower, dress, eat breakfast, travel to and from class	29
Lecture	13
Lab	9
Homework	30
Free time	31
Total	168 hours

Miraculous! There are 31 hours of free time in this budget! Unfortunately, a considerable fraction of this free time occurs in small unusable patches throughout the week—time between classes, time after dinner and before starting on homework, and so forth. Nevertheless, a little calculation shows that all day Saturday can be used in recreational pursuits, even including staying out until midnight Saturday night and getting up late on Sunday morning. But another little calculation shows that every night except Saturday is going to have to be spent in doing homework. If you spend 3 hours a night for six nights, including Sunday, this adds up to 18 hours; you still have to find 12 more hours someplace. On the two afternoons a week

Figure 3.5
Laboratory classes are important in the education of engineering students. (Courtesy of University of California, Davis)

when you do not have lab, you could put in 4 hours each for a total of 8 hours. (There go your free afternoons.) And you still need 4 hours. (There go your Sunday afternoons.) But at least you have all day Saturday free unless, of course, you let things slide during the week, in which case your Saturdays are gone, too.

One further item may prove useful, and that has to do with the taking of notes. Most of the classes that engineering students take are mathematical in nature. Hence, the note-taking process often consists of transcribing the professor's mathematics from the blackboard to the note paper. Sometimes such a literal transcription is ridiculed ("Taking notes is the process of transferring the professor's notes from the blackboard into the student's notebook, without the information going through the brain of either one"), but the process actually has value. The principal value is that it keeps you, the student, directly involved in what the professor is doing. (It also helps keep you awake.) Maybe you are even able to see ahead to what comes next. Later, when you review the notes, you will have a much better chance of remembering what it was all about if you were intensely involved in the process at the time.

In contrast with the primarily mathematical lecture is the kind that is primarily descriptive. In a mathematical lecture, you try to record everything the professor does. In a descriptive lecture, writing down every word the professor says is impossible. You must try to understand the important points and get these into your notes. Most professors provide an outline of the lecture, either written on the board or in the form of a syllabus. In the latter case, you can spend your note-taking time elaborating on the points already written in the syllabus, with less frantic effort spent in just trying to keep up. It is a good idea to raise your hand at the very first class meeting and ask if the course has a syllabus. Sometimes just the asking of such a question may spur a professor into providing a syllabus.

Finally, a word about asking questions. Do not sit there in silence when you really do not understand something. Raise your hand and ask. It is your right to do

Figure 3.6
Engineering students often have a chance to try their hand at designing and building real devices. The unit shown here was designed and constructed by classes of senior students as a vertical takeoff and landing (VTOL) aircraft. (It actually flew—two feet off the ground.) (Courtesy of University of California, Davis)

so, and a good professor expects students to ask questions. (If you are unfortunate enough to find yourself in a class where the professor does not seem to tolerate questions, maybe you can transfer into a class with a different professor.) If the professor goes too fast for you to keep up, ask him or her to slow down. Professors are so familiar with the material and may be so enthusiastic about what they are doing that they are often not aware they are going too fast. But for you everything is unfamiliar, and it is by no means out of line for you to ask that the pace be a little slower.

■ 3.3 PROBLEM SOLVING

Chapter 11, Engineering Design, will present information on the *design process*. You will learn that there are five distinct phases to this process: problem definition, invention, analysis, decision, and implementation. A somewhat similar sequence of phases appears in any *problem-solving* approach:

1. Problem definition
2. Collection of applicable information
3. Selection of mathematical model (the "theory" to be used)
4. Carrying out the solution
5. Checking the solution

In college, the first step is usually complete before the problem is presented to the student. Textbook problems, for example, are usually stated very precisely because they are drawn up in such a way as to demonstrate certain principles. In real life, on the other hand, no professor or textbook author is there to define the problem. *You* have to do this. Sometimes it is not as easy as it might seem. At any rate, it is absolutely necessary that the problem to be solved be stated unambiguously.

Step 2, for most textbook problems, is not very mysterious. You have just finished reading a section of the textbook, so you have a pretty good idea that the problems at the end of that section require the use of the information you just read. However, professors sometimes (on a final exam, perhaps) give you more information in a problem statement than is applicable to the problem. This means you have to decide what is relevant. Students sometimes find this exasperating because they are used to cases where every bit of given information is used in the solution. If they are given extraneous information (and especially if they are given no hint in the problem statement that some of the information is indeed extraneous), they become upset when they get to the end of a problem and find they have some information left over. But engineering school is supposed to prepare its graduates for real-life situations, and real-life situations invariably come burdened with all sorts of information, some of it applicable, some of it not.

Step 3—selection of a mathematical model—often turns out to be "grabbing a formula." Again, students expect that problems at the end of a particular section of a textbook will require the application of the formulas in that section—usually a correct assumption. But the problems in a final exam draw on many textbook sections and many formulas. No exhortation in a textbook (like the one you are reading now) can really equip a student to face this problem. Only experience can do that. College gives you four years of experience at doing exactly what we are talking about here—selecting the right theory and the right mathematics for the right problem. (After graduation you will make an interesting discovery when you apply to your state licensing board for registration as an engineer. Licensing boards require a certain number of years of experience—usually eight—and they will give you four years of credit toward that experience if you have graduated from an accredited program.) For the present, the most we can say in this book is that you should be constantly aware that what you are doing in your educational program is learning to judge which theories and which formulas are appropriate to the problem at hand. "Grabbing a formula" is almost always wrong. (More will be said about mathematical models in Chapter 11.)

Step 4—carrying out the solution—probably requires little elaboration. Students know they are required to do this. Doing this step *correctly* is, of course, vital, and it is easy to make procedural errors—leaving something out, transcribing a number incorrectly, making a calculation error, omitting a step. Students are so familiar with the solution step that it sometimes dominates their thoughts and obscures the other steps. After moving into professional practice, they are sometimes disturbed by the fact that the solution step looks a lot less important than it did during college, when it seemed to be everything worth thinking about. It is still vital, in professional practice, that solutions be carried out correctly. But it is just as vital that the problem be defined correctly, that the right information be gathered, and that the mathematical model be appropriate for the problem at hand.

Step 5—checking the solution—should really be renamed "checking everything you have done." Probably the first task is to check the reasonableness of the answer. This was mentioned above in the context of taking exams. In the pressure-cooker conditions of an exam, making calculation errors is understandable, even partially forgivable. But when you are doing homework—or, even more to the point, when you are working on the job as a graduate engineer—calculation errors are totally unforgivable. If you are a structural engineer, for example, your structure could collapse because of a decimal point error just as easily as for any other reason. Not only calculations, but every step of the process should be checked—including your problem definition, the information you collected, and the theory you chose to apply.

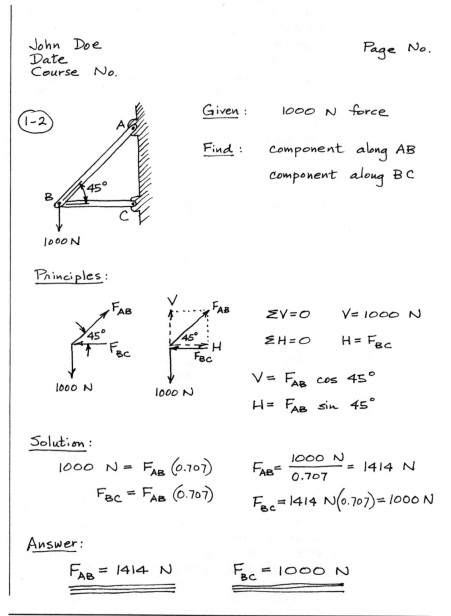

John Doe
Date
Course No.

Page No.

(1-2)

Given :   1000 N force

Find :   component along AB
          component along BC

Principles :

$\Sigma V = 0$ $V = 1000$ N

$\Sigma H = 0$ $H = F_{BC}$

$V = F_{AB} \cos 45°$

$H = F_{AB} \sin 45°$

Solution :

1000 N $= F_{AB} (0.707)$

$F_{BC} = F_{AB} (0.707)$

$F_{AB} = \dfrac{1000 \text{ N}}{0.707} = 1414$ N

$F_{BC} = 1414$ N$(0.707) = 1000$ N

Answer :

$F_{AB} = 1414$ N $F_{BC} = 1000$ N

Figure 3.7
A sample problem format.

■ 3.4 PROBLEM SETS

Engineering students spend a great deal of their time preparing problem sets. In some
courses, a problem set will be due at every meeting of the class, because instructors
know that students learn only on the basis of what they do for themselves.

Different instructors prefer different formats for problem sets, and students should observe the procedures specified by the instructor in each class. There is no one "official" format, but the one shown in Figure 3.7 has often been used. (In case you have not yet encountered metric units, the label "1000 N" stands for 1000 newtons. The newton is a unit of force. See Chapter 9.) The format shown in Figure 3.8 is also popular. It is extremely important that every step be shown neatly and clearly. Instructors often give partial credit for work done correctly even though the final answer itself is wrong; however, if the instructor cannot figure out what you have done, partial credit is hard to give.

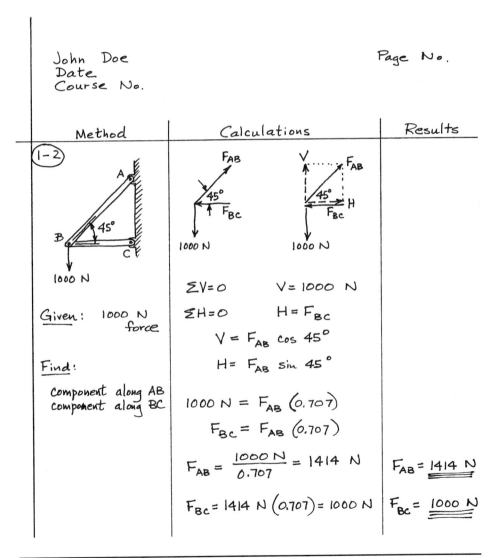

Figure 3.8
An alternative form of the sample problem format.

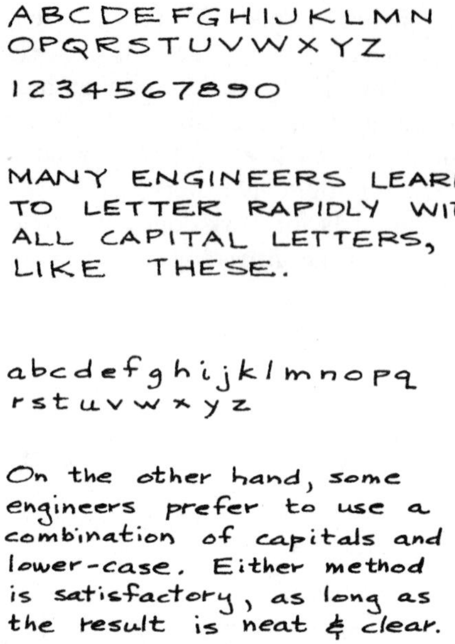

ABCDEFGHIJKLMN
OPQRSTUVWXYZ

1234567890

MANY ENGINEERS LEARN
TO LETTER RAPIDLY WITH
ALL CAPITAL LETTERS,
LIKE THESE.

abcdefghijklmnopq
rstuvwxyz

On the other hand, some
engineers prefer to use a
combination of capitals and
lower-case. Either method
is satisfactory, as long as
the result is neat & clear.

Figure 3.9

Examples of two styles of lettering used by engineers.

A few rules to remember:

Unless the problem is purely mathematical, start off with a clear sketch showing the given conditions. At each point along the way, make a clear sketch of what you are doing. Clarity is the keynote here. A messy sketch may not communicate your intentions accurately to the person who grades your paper. A messy sketch may even mislead *you* at some point along the way.

Print everything clearly. *Do not write in longhand.* Longhand is notoriously difficult for other people to decipher. An engineer must cultivate the ability to print letters rapidly. Either of the styles shown in Figure 3.9 is satisfactory, but much practice is necessary before they become second nature and stop looking like hen scratches.

Always show the units of your answer. An answer without units is wrong unless it is fundamentally a unitless quantity. (For example, if the answer to a problem is the value of sin 45°, which is a pure number, 0.707, then of course the answer has no units.) You should carry your units along through each of your calculations.

Clearly identify your answer, such as by underlining it two or three times.

Show all your calculations, and do so as clearly as possible, so the instructor (and you) will know what you are doing.

Do not start a second problem on a page unless it can be finished on the same page.

When you get your answer, stop to consider it for reasonableness. Finally, check all your work, your assumptions, your sketches, and your calculations.

Spell all words correctly, even on problem sets.

If erasures are necessary, make sure they are neat and complete. Do not leave smudges.

A popular method for handling problem sets is to fold them vertically as shown in Figure 3.10, and put your name, date, course number, and problem numbers on the outside. However, instructors' preferences on such things may differ, so you should follow the directions given.

Figure 3.10
When a problem set is complete, it should be folded vertically down the center with the student's name, the date, the course number, and the problem number(s) written on the outside, as shown.

■ 3.5 HAND-HELD CALCULATORS

The two basic types of hand-held calculators are the algebraic-entry type and the RPN type. (RPN stands for *Reverse Polish Notation.*) The instinctively natural mode of calculator use is algebraic entry. In this mode, entries for a mathematical operation are made exactly as they would be written. If we wish to add 2 and 3, we enter it as

$$2 + 3 =$$

and the answer is displayed immediately as 5. On the other hand, in an RPN calculator we would enter the same operation as

2 ENTER↑ 3 +

which also gives us 5 as the answer as soon as we press the + key.[1]

A bewildering variety of hand-held calculators are available on the market. The most basic machine performs primarily the four elementary functions—addition, subtraction, multiplication, and division—with perhaps a few other special keys provided, such as % and $\sqrt{}$. A unit with these capabilities is known as a *four-function*

---

[1] When function keys such as ENTER, SIN, COS, and y^x are shown in a diagram of a key-in sequence, they will be enclosed in a box. When numerical values, parentheses, decimal points, and operation signs such as +, −, ×, ÷, and = are shown in such a diagram, they will be shown without a box in order to preserve the familiar appearance of algebraic equations wherever possible.

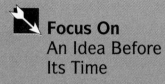

Focus On
An Idea Before
Its Time

Everyone knows that the advent of hand-held calculators destroyed the mechanical calculator industry. Some have said: "Why couldn't they see what was coming and do something about it?" The answer is that they did.

In the mid-1950s a group of engineers was organized to work on a secret project at one of the major mechanical calculator companies. Their assignment was to develop an electronic desk calculator. The transistor had recently been invented. Electronic computers existed, although they were still big and expensive. Nevertheless it was obvious even to laypersons that the mechanical calculator was doomed to extinction.

Within a few months, the engineers had developed a preliminary design for their miniature computer. But they had two major problems. First, transistors of computer quality were almost impossible to procure and cost about $20 apiece. This would not do; there were to be hundreds of transistors in the new device, and even the future projected price of $2 per transistor was too high. The engineers solved this problem by arbitrarily deciding that the future price of transistors would decline to 25 cents each. Some of their associates were inclined to award them the lunacy prize of the decade for such rashness, but they did it anyway. (Twenty years later, the individual cost of a transistor was to be measured in a tiny fraction of a cent.)

The other major problem was not to be resolved so easily. What were they going to do about random-access memory? A memory using transistors could be designed, but its cost would be outlandish. Magnetic ferrite cores could be used, but they were also out of the question because of high cost. The engineers finally settled on a magnetic drum as the least costly. Drums were widely used in existing computers, but their cost continued to be one of the biggest problems in the project. (In later years, with the advent of integrated circuits, the use of transistors for memory became very cheap. Furthermore, the descendants of the magnetic drum exist today in the form of hard disk files and floppy disk inputs.)

The project team finally came up with an estimate and produced a shocking result: Their machine would be four times as big as the mechanical machine it was to replace and would have to sell for ten times the price, or about $8000. The big stumbling blocks in the estimates were the costs of the transistors and the magnetic drum. There was no way around them—not at that time—and the project was canceled. (Today, of course, because of the invention of integrated circuits, such machines sell for less than $10.)

The company later made two more attempts to enter the hand-held calculator market. Both attempts were swept away in a competitive avalanche in the early 1970s, when about 20 companies came out with cheap calculators. All three of the companies making mechanical machines quietly phased out their production facilities. Thousands of people were thrown out of work. In the meantime, the muscular new microelectronics industry that rose over the ashes of the mechanical calculator became many times larger than the industry it displaced. Did those thousands of misplaced workers find jobs in the new industry? Possibly, but no one will ever know for sure.

Figure 3.11
An algebraic-entry hand-held calculator.

calculator and can be purchased for less than $10. Engineering students usually purchase a *scientific calculator*, having functions such as SIN, COS, TAN, LOG, π, x^2, $\sqrt{\ }$, e^x, y^x, and $\sqrt[x]{y}$. Also, many calculators have the ability to accept calculations containing expressions within parentheses, and some are programmable. There are variations among models and manufacturers, and no book such as this can serve as an instruction manual for even a few of the models, let alone all of them. This discussion should be regarded as giving a general appreciation for the capabilities of such machines. You can learn the specifics of your own calculator from the instruction manual that comes with it.

Generally, a calculator intended for use by scientists and engineers is equipped with keys that give trigonometric and exponential functions with a single keystroke. For a trigonometric function, the argument is entered first, and then the function key is pressed. To get the value of sin 30°, for example, we would depress the keys in the following order:

Enter   30

Press $\boxed{\text{SIN}}$

In this book we will diagram such a sequence as

30 $\boxed{\text{SIN}}$

In this case, the value of the argument, 30°, is entered in degrees. Some machines have the ability to switch over so that arguments are entered in radians, and some even make provision for entering in "grads," where one grad is equal to one-hundredth of a 90° angle.

To get 27^4, we would typically press keys as follows:

27 $\boxed{y^x}$ 4 =

and get the answer, 531 441.

We should note, however, that some machines have keys labeled x^y, rather than y^x. In such a case (using an RPN machine as an example), we would have to enter

Figure 3.12
The hand calculator shown here is a powerful problem solver with full programming capability and 21 memory registers. In case you accidentally go away and leave it on, it will turn itself off after 10 or 15 minutes in order to save battery power. Even if the power is turned off, data keyed into the calculator are not lost, but are preserved by a "continuous memory" feature. (Courtesy of Hewlett-Packard Company)

27^4 in the reverse order, as

4 $\boxed{\text{ENTER}\uparrow}$ 27 $\boxed{x^y}$

Since machines differ in specifics like these, particularly with regard to which term is to be regarded as "x" and which as "y," it is necessary to find out which procedure applies to your own machine by using the manual and experimenting.

Some calculators provide for entry of numbers in so-called scientific notation. As a typical example, let us suppose we are to enter a number such as 1.357×10^{-6}. Calculators vary in the manner in which their keys are marked, but one common method is to provide a key marked $\boxed{\text{EXP}}$. Then, to enter the above number, one performs the following sequence:

1.357 $\boxed{\text{EXP}}$ 6 $\boxed{+/-}$

Entry of the number 1.357 causes it to appear in the display in the normal fashion. Pressing the $\boxed{\text{EXP}}$ key causes the display to change in a manner that depends upon the particular calculator. One of the following displays will probably appear:

1.357×10^{00}

1.357   E00

1.357   00

These are all equivalent. The first one should require no explanation. In the second, the "E" stands for "exponent," meaning the exponent of 10. The third display is similar to the second, except that the "E" is omitted. When the number 6 is entered, these displays become

1.357×10^{06}

1.357   E06

1.357   06

These, of course, are all positive exponents. Pressing the $\boxed{+/-}$ key changes the signs of the exponents, and we get

1.357×10^{-06}

$1.357 \quad E - 06$

$1.357 \quad -06$

Expressions structured like the following are frequently encountered in engineering:

$$\frac{3.2 \times 64}{60 \times 214} = \tag{3.1}$$

The correct answer for this (to four significant figures) is 0.01595. However, if we blindly went through and pressed the keys in the following order, we would get the wrong answer of 730.4533:

WRONG: $3.2 \times 64 \div 60 \times 214 =$

The reason the answer is wrong is that we are telling the machine that we should first compute the quotient $(3.2 \times 64) \div 60$ and then multiply the quotient by 214. Actually, we want to divide by 214, not multiply. A good rule in such cases is to take the figures in pairs and calculate successive quotients of these pairs, as follows:

$3.2 \div 60 \times 64 \div 214 =$

In case you are unsure of such procedures, the expression can always be written in parentheses, as follows:

$(3.2 \times 64) \div (60 \times 214) =$

Then, if your machine has parenthesis keys, you simply key in exactly what you see in the foregoing expression and get the right answer.

Errors

A good rule in calculating any expression is to make a mental estimate of the order of magnitude of the answer. We can mentally rewrite expression (3.1), for example, as

$$\frac{3.2}{60} \times \frac{64}{214}$$

The first term we know is about 1/20, or 0.05. The second is about 1/3 or, let us say, 0.3. Therefore, the answer will be on the order of 0.05(0.3), which is 0.015. If an answer like 730.4533 came up, as it would in the case of our wrongful-key-entry example, we would know immediately that something was the matter. In dealing with a hand-held calculator, which does not produce a printed record of its entries, it is distressingly easy to make an entry error and not know it. One of the easiest errors to make is the omission of the decimal point. A mental estimate of the magnitude of the answer will guard against many of the errors you might make, including most of those involving decimal-point error.

Memory

Most calculators provide at least one memory register. One typical use of such a register is to store a constant for subsequent repetitive use. Suppose, for example, that the following series of calculations is to be made:

$$37 \times 0.735 =$$

$$42 \times 0.735 =$$

$$22 \times 0.735 = \tag{3.2}$$

$$\vdots$$

Rather than keying in all factors each time, the constant multiplier 0.735 can be placed in memory by entering the number in the visible register, pressing the "enter memory" key, and subsequently pressing the "recall" key each time the factor is needed. Calculators have different ways of labeling such keys. Two examples of "enter memory" key labels are:

$$\boxed{x \to M} \qquad \text{and} \qquad \boxed{\text{STO}}$$

where the first label implies that the value x displayed in the visible register goes into "M" (for memory) and the second label is an abbreviation for "store," meaning that the value is thereby stored in the memory.

Two examples of "recall" keys are

$$\boxed{\text{RM}} \qquad \text{and} \qquad \boxed{\text{RCL}}$$

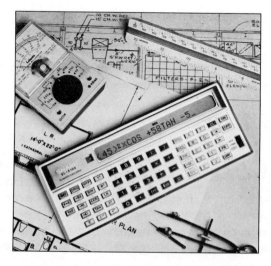

Figure 3.13
Mathematical equations can be entered into the scientific calculator shown here in much the same way they are written on a piece of paper. (Courtesy of Sharp Corporation)

where the first label stands for "recall memory" and the second is an abbreviation for "recall."

A typical sequence computing the expressions in (3.2) on an algebraic calculator would then be

$$0.735 \; \boxed{x \rightarrow M}$$
$$37 \times \boxed{RM} =$$
$$42 \times \boxed{RM} =$$
$$22 \times \boxed{RM} =$$
$$\vdots$$

RPN Calculators

As already mentioned, RPN stands for *Reverse Polish Notation*. The original Polish notation was devised by a Polish mathematician named Jan Lukasiewicz (woo-ka-SHAY-vich) as a method of mathematical notation that avoids the use of parentheses. In this notation, two factors are listed with an operational sign preceding them. Thus, if we write

$$-AB$$

we mean that B is to be subtracted from A. In *reverse* Polish notation, we put the operational sign after the factors, rather than in front, and write the foregoing as

$$AB-$$

Any operational sign, such as $+$, $-$, $\times$, or $\div$, in RPN always refers to the operation to be performed on the two quantities immediately preceding that sign. Thus, if we write

$$ABC \times - \tag{3.3}$$

this means that B and C are multiplied together, forming a new product P, which replaces BC. Now we are left with

$$AP-$$

so P is subtracted from A.

When expression (3.3) is entered into an RPN calculator, the machine has to know when we have stopped entering A and started on B, and also has to know how to separate B from C. The $\boxed{ENTER\uparrow}$ key is used for this purpose. Thus, we would enter (3.3) into an RPN machine as follows:

$$A \; \boxed{ENTER\uparrow} \; B \; \boxed{ENTER\uparrow} \; C \; \boxed{\times} \; \boxed{-}$$

The reason the foregoing process works is because most RPN machines are equipped with four registers in a sequential array, but only one of them is visible.

Picture the registers as being stacked on one another as in the following diagram:

```
T  ┌─────────────────────┐
   ├─────────────────────┤
Z  ├─────────────────────┤
Y  ├─────────────────────┤
X  └─────────────────────┘
```

These registers are referred to as the "memory stack," or just "stack" for short. They are typically named the X, Y, Z, and T registers, and only the X register is visible. Suppose we key in the number 123. This entry goes into the X register, and we now have the following situation:

```
T  ┌─────────────────────┐
   ├─────────────────────┤
Z  ├─────────────────────┤
Y  ├─────────────────────┤
X  │              123.    │
   └─────────────────────┘
```

When we press the |ENTER↑| key, the value sitting in the X register is "pushed up" into the Y register. In fact, that is why the upward pointing arrow is on the key—to remind you that each time you push on the |ENTER↑| key, everything is pushed upward in the stack.

After pressing the |ENTER↑| key once, our stack has the following in it:

```
T  ┌─────────────────────┐
   ├─────────────────────┤
Z  ├─────────────────────┤
Y  │              123.    │
X  │              123.    │
   └─────────────────────┘
```

Now, if we enter a new number, say, 456, in the keyboard, it "writes over" the value in the X register, giving the following:

```
T  ┌─────────────────────┐
   ├─────────────────────┤
Z  ├─────────────────────┤
Y  │              123.    │
X  │              456.    │
   └─────────────────────┘
```

At this point, if we push any operation key such as +, −, ×, or ÷, the two numbers in the X and Y registers will be combined in accordance with the key we pushed, and the stack will be "pushed down," eliminating the values in the X and Y registers and placing the result in the X register. In our example, if we push the + key, the stack will then have the following in it:

```
T  ┌─────────────────────┐
   ├─────────────────────┤
Z  ├─────────────────────┤
Y  ├─────────────────────┤
X  │              579.    │
   └─────────────────────┘
```

Let us now perform the operation in expression (3.3), assuming $A = 123$, $B = 456$, and $C = 789$.

Keyboard entry: Stack:
   123 ENTER↑

T	
Z	
Y	123.
X	123.

Keyboard entry: Stack:
   456 ENTER↑

T	
Z	123.
Y	456.
X	456.

Keyboard entry: Stack:
   789

T	
Z	123.
Y	456.
X	789.

Keyboard entry: Stack:
   ×

T	
Z	
Y	123.
X	359784.

Keyboard entry: Stack:
   −

T	
Z	
Y	
X	− 359661.

The rule is, each time we press an operation key such as +, −, ×, or ÷, the operation is performed on the two values in the X and Y register, the result is placed in the X register, and the stack is pushed down. If the operation is subtraction, the number in the X register is subtracted from that in the Y register. If we are performing division, then X is divided into Y. For addition and multiplication, the order does not matter because $X + Y = Y + X$, and $XY = YX$.

The foregoing process may seem awkward when compared to algebraic entry, but there are many people who are fiercely loyal to their RPN machines and can point to many practical problems wherein the RPN uses fewer keystrokes than do algebraic machines. The fact is that a person will soon become accustomed to, and facile with, whichever kind of machine is purchased.

Upon reflection, it will be apparent that the RPN process is not as strange as it seems, because this is the way adding machines have always operated. In typical adding machine operation, one enters a number and *then* presses the + key, enters another number, and presses the + key; if a number is to be subtracted, the operator enters the number and *then* presses the − key. This is in fact RPN operation, and for a person who has been trained on an adding machine it is the *algebraic* mode of entry that will seem peculiar.

Exercises

3.5.1 Practice addition and subtraction:

(a) 25.19
  3
  402
  0.357
  ———
  430.547

(b) 0.003 42
  0.102 00
  0.000 59
  0.300 00
  ———
  0.406 01

(c) 0.000 000 346 2
  0.000 000 097 0
  0.000 001 754 0
  ———

Most hand-held calculators will not handle the above numbers without conversion to scientific notation. Convert the above to

3.462×10^{-7}
$9.7 \ \ \times 10^{-8}$
1.754×10^{-6}
———
2.1972×10^{-6}

(d) −42.73
  1.75
  240.
  −19.26
  ———
  179.76

(e) 3.94×10^{-2}
  -2.01×10^{2}
  -43
  1.78×10^{-1}
  ———
  $-2.437\ 826 \times 10^{2}$

3.5.2 In each of the following, make mental estimates of the answers, to two significant figures, then calculate the answer to five significant figures.

(a) $\dfrac{67.8 \times 6.9}{1007 \times 3478} = 1.3357 \times 10^{-4}$

(b) $\dfrac{342 \times 0.073}{32.2 \times 2 \times 62.3} = 6.2226 \times 10^{-3}$

(c) $\dfrac{65\,700 - 42\,500}{25 \times 3470} = 0.267\,44$

(d) $\dfrac{643 - 1072}{35.2 - 43.6} = 51.071$

(e) $\dfrac{3 - 12 \times 7}{36(0.43) - 12} = -23.276$

3.5.3 Calculate the following to four significant figures:
(a) $\sin 23.2° \times \sin 257.3° = -0.3843$
(b) $\cos 35° \times \tan 97.2° = -6.484$
(c) $\cos 35° \times \tan 90° = $ overflow   (Why?)
(d) $\cos^{-1} 0.913 + \sin^{-1} 0.201 = 35.67°$
(e) $\sin \pi/4 \times \sin \pi/6 = 0.3536$

(The following involve radian measure; consult your calculator manual for entering radian measure directly, or convert to degrees by using relation 2π radians $= 360°$.)

(f) $\sin^{-1} 1.013 + \sin^{-1} 0.052 = $ error   (Why?)

(g) $\dfrac{\cos 0.05°}{\sin 0.05°} = 1146$

(h) $\dfrac{\sin \pi/2}{\cos \pi/2} = $ overflow   (Why?)

3.5.4 Calculate, to four significant figures:

(a) $\dfrac{7!(7-2)!}{3!} = 100\,800$

(b) $3^6 = 729.0$

(c) $3.019^{6.28} = 1032$

(d) $(3.019)^{1/6.28} = \sqrt[6.28]{3.019} = 1.192$

(e) $\sqrt[5]{3} = 1.246$

(f) $\ln 273 = 5.609$

(g) $e^{23} = 9.745 \times 10^9$

(h) $e^{-23} = 1.026 \times 10^{-10}$

(i) $528^{-0.313} = 0.1405$

(j) $\dfrac{1}{528^{0.313}} = 0.1405$

(k) $\log_{10} 373 = 2.572$

(l) $(-3)^5 = -243$

(*Note:* some calculators will not accept negative quantities for y when using the y^x or $\sqrt[x]{y}$ key.)

(m) $\sqrt[3]{-27} = -3$

(n) $(0.001)^{25} = 1.0 \times 10^{-75}$

(o) $(1.37 \times 10^{-8})^5 = 4.826 \times 10^{-40}$

3.5.5 Find x:

(a) $x^{3.01} = 8$

Solution

$$x = \sqrt[3.01]{8} = 1.995$$

(b) $5^{0.6x} = 42$

Solution

$$\ln 5^{0.6x} = \ln 42$$

$$(0.6x)\ln 5 = \ln 42$$

$$x = \frac{\ln 42}{(0.6)\ln 5} = 3.871$$

(c) $x^{-0.8} = 0.25$

(d) $25^{2.7x} = 501.9$

(e) $0.90^{-5x} = 20$

3.5.6 Evaluate the following:

(a) $x^2 + 2x + 3$ at $x = 2.7$ (*Ans.:* 15.69)

(b) $6.9x^3 + 3.2x - 7.3$ at $x = 3$ (*Ans.:* 188.6)

(c) $\dfrac{3x^3 + 2x^2 + x + 1}{x + 4}$ at $x = -3$ (*Ans.:* -65)

(d) $\dfrac{(s+2)(s-3)}{s(s+1)}$ at $s = -1$ (*Ans.:* error; why?)

3.5.7 Calculate the following to four significant figures:

(a) $45(7.7 + 3.2) = 490.5$

(b) $45/(7.7 + 3.2) = 4.128$

(c) $45(7.7) + 3.2 = 349.7$

(d) $45/7.7 + 3.2 = 9.044$

(e) $(3.02 \times 10^{-3} + 1.91 \times 10^{-2})/3.93 \times 10^{-3} = 5.628$

(f) $6\left(\dfrac{16.27 - 3.96}{17.25}\right) = 4.282$

(g) $2\pi(92.36 - 17.41) = 470.9$

(h) $\dfrac{-87.34 + 127.15 - 32.32}{17.25 - 9.91} = 1.020$

3.5.8 Calculate the following:
(a) $26\frac{3}{8} \times 3\frac{27}{32} = 101.3789$ **(b)** $2\frac{63}{64} + 3\frac{7}{16} = 6.4219$
(c) 12 ft, 6 in. $\times$ 21 ft, 10 in. $= 272.9167$ ft^2
(d) 60 mi/hr $\times$ 2 hr, 36 sec $= 120.6$ mi

3.5.9 Calculate:
(a) $16°36'59'' + 42°17'40'' = 58.9108°$ or $58°54'39''$
(b) $253°05'15'' + 163°54'45'' = 417.0000°$
$= 57.0000°$
(c) 3 hr, 5 min, 26 sec + 12 hr, 36 min, 48 sec = 15.7039 hr or 15 hr, 42 min, 14 sec

3.5.10 Convert the following quantities to metric units (see the conversion table in Appendix L):
(a) 32 ft, 9 in. to meters (m)
(b) 60 mi per hr to km per hr (km/h)
(c) 8500 pounds-mass (lbm) to kilograms (kg)
(d) 8500 pounds-force (lbf) to newtons (N)
(e) 700 cubic ft (ft^3) to cubic meters (m^3)
(f) 1000 British thermal units (BTU) to joules (J)
(g) 30 000 lb per in.^{2} (psi) to pascals (Pa)
(h) 70°F to degrees Celsius (°C)
(i) 120°C to kelvin (K)
(j) 500 horsepower (hp) to watts (W)
(k) 280 slugs to kilograms (kg)

■ 3.6 THE TRANSITION TO PROFESSIONAL LIFE

To most of the readers of this book, who are assumed to be first-year students, the transition to a professional career may seem to lie in the remote future. But the time will eventually arrive: Graduation will be approaching, and the transition will be imminent. An entire book has been written by this author for students at that stage of their careers.[2] However, a few points are worth emphasizing here.

First, as the transition approaches, most students feel a certain amount of apprehension—mostly centered around the fact that they have spent several years studying engineering but they fear that their knowledge is not equal to the tasks that lie ahead. The apprehension is entirely natural. Although one would not suggest a careless attitude toward such things, the fact is that tens of thousands of new graduates successfully make the transition each year and live to tell the tale. The transition does not seem nearly so bad looking back as when one is approaching it. Employers know very well that they must help new graduates through the transition. In addition, new graduates, by and large, will discover they are more capable than they believed possi-

[2] J. D. Kemper, *Engineers and Their Profession* (Philadelphia, Saunders College Publishing, 4th ed., 1990).

ble. After a year or two on the job, they will generally be amazed at the things they have succeeded in doing and at the responsibilities they are carrying.

As mentioned in a boxed item in this chapter, another fact new graduates learn is that there are two enormously accelerated learning periods in most people's lives. One is college; the other is the first real full-time job. If anything, the second period is more intense than the first. It can be breathtaking and exhilarating. A college education provides the foundation, and one's first job puts that foundation into practice. An engineering education generally does not try to prepare an individual for an exact slot in industry, because the probability of landing in that exact slot is too remote. Instead, college education gives the student the necessary background for a whole range of possible careers. Then the first job completes the task of preparation for an exact slot, building on the foundation provided by a college education. In this connection, here is what some engineers have said regarding the way their educations fit into their lives:

> After graduation, I felt my engineering education had not sufficiently prepared me for work in the real world; that is, I felt it was too theoretical, and with not enough practical application. However, over the years I have seen that a good technical background is important to fall back on when common sense alone cannot provide the solution.
>
> (Project engineer, geotechnical engineering)

> Study the fundamentals. They are what you use the most.
>
> (Senior engineer, aerospace)

Finally, many engineering graduates move into their first jobs almost with a sense of relief. This is not to say that professional engineering jobs are without pressure or do not have sustained periods of overtime. They often have those things. But an engineering education program does seem to have a certain unremitting pressure that continues for four years. Each term consists of a host of unyielding deadlines: classes for which preparation is necessary, homework assignments to be handed in, deadlines for reports and term papers, and exams, exams, exams. True, professional engineering jobs also have many deadlines to be met, and one's performance is constantly being evaluated, or "examined." Yet graduates generally report they feel, following graduation, that they are more in charge of their own time than they were in college.

In spite of this, graduates generally look back on their college experiences as "special." College is not all hard work; there are other things to college besides exams and deadlines. Friendships are often made, for example, which may endure for life. Even the hard work part is usually looked back upon with a sense of gratitude, because along with that hard work comes an acquisition of knowledge and insights that are counted among one's most precious possessions.

■ 3.7 GRADUATE STUDY

Students often worry about whether they should go to graduate school after receiving their B.S. Actually, there is little point in worrying about this until one's junior year. Surveys show that most students do not make this decision until they are seniors.

Figure 3.14
Research and graduate study are tightly coupled in modern universities. Here a geotechnical engineer performs laboratory research relating to the liquefaction behavior of sands and clays under dynamic loading, using a triaxial cyclic testing device.

By this time they will know if graduate school is even possible, because at least a B average is usually required for admission. Also, by this time a student's own individual tastes and abilities will be more evident, and the decision regarding graduate school will be easier.

About one-fifth of engineering graduates go directly on to graduate school after graduation. But another fifth postpone this decision, getting their graduate degrees eventually, but on a part-time basis. In most metropolitan areas, part-time engineering graduate programs are available, sometimes through the delivery of live-TV graduate courses at the place of employment. Almost all employers provide some kind of financial assistance to their engineer employees who seek graduate degrees.

Many young people ask if it would be wise to get a master of business administration (M.B.A.) degree if they are interested in management. Only a very general answer is possible: If one feels destined for top management, an M.B.A. would be extremely useful. But if one is likely to be satisfied with lesser aspirations than the very top, an M.B.A. usually will not offer any advantages other than the obvious one that the more one studies, the more one learns. Many companies seem to believe that a B.S. is just the ticket for management aspirants. In any event, it should be remembered that an individual's characteristics and drive are more important than possession of degrees.

Exercises

3.1 Assuming you are already enrolled in college, examine the catalog for your school to see if its engineering programs resemble those described in this text. If there are important differences, contact your adviser to see if there are reasons for the differences that might be important to you.

3.2 Examine the curriculum that interests you, and see how many courses in the humanities and social sciences are required. Make a list of courses you might select to meet this requirement, and then read the course descriptions to see whether they are likely to fulfill the objective as stated in the text, "... to give [engineers] some feeling for the great ideas and movements that have shaped our culture...."

3.3 Lay out a time budget for your own schedule, showing class time, study time, travel time to and from class, leisure time, time for sleeping, getting dressed, and so on, and being sure to make allowance for "lost" time, which seems to disappear without one's being sure just where it went. What does your week look like? If you are carrying 16 units of work, have you provided 32 hours or more of homework time per week, recognizing that some courses are virtually certain to require more than the traditional 2 hours per unit of homework? If you must work part-time, have you cut down the number of units so there is adequate time for both work and study?

3.4 See if you can find a copy of the curriculum that was used by your school in the field you are interested in, say, 30 years ago. Discuss any changes that have taken place since that time.

3.5 Contact someone who is working in the field you are interested in, and learn what you can concerning the need for graduate-level education in that field. Write a brief report on your findings.

3.6 It is sometimes said that an engineering curriculum provides a good educational basis for a person, even though that person may not actually become an engineer. Discuss this statement, either agreeing or disagreeing.

Photographs courtesy of Grumman Aerospace Corporation (top), Lord Industrial Automation (center), and Kaiser Engineers (bottom).

4
The Engineering Profession

In deciding to study engineering, you have taken the first step toward joining a remarkable profession. There is no other quite like it, because engineering is so broad and diverse that it is almost impossible to give it a neat and concise description. In fact, engineering is not a single kind of activity, but many kinds. Because of this, it is markedly different from the professions of medicine, law, and teaching. The public usually has a reasonably clear—and accurate—perception of what people in those fields do. But the public is usually somewhat confused about what engineers do, and one of the principal reasons for this is the fact that they do so many different kinds of things. One of the purposes of this textbook is to give you some insight into the great variety of things engineers do.

A relatively unknown and somewhat surprising fact about engineers is that there are so many of them. It is estimated that the United States has nearly 2 million engineers, outnumbered only by teachers (see Table 4.1). About one-third of the nation's engineers do not have engineering degrees, demonstrating the fact that engineering historically has been an "open" profession. It traditionally has been accessible to persons with ability, even though they lacked formal education. With the passage of time, however, it has become increasingly difficult to attain full professional status as an engineer without possessing at least a bachelor's degree.

There is a great attractiveness to the title *engineer*. Many groups (and individuals) have sought the right to use the title, and those who have historically possessed the right have striven hard to keep it. For example, locomotive engineers and stationary engineers (power plant

99

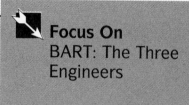

Focus On
BART: The Three Engineers

BART stands for Bay Area Rapid Transit, the organization that has operated an electrified railway transit system in the San Francisco Bay area since 1972. The system has been a controversial one, partly because it escalated in cost from an estimated $700 million to $1.7 billion before it was done and partly because of the publicity accorded the firing of three engineers employed by BART.

BART was set up to oversee the system and to contract with others to do the design and construction. Thus, it would employ a staff of engineers to monitor the design and assume engineering responsibility for the system when it was finished. The engineering firm that was awarded the design responsibility was a consortium of three well-known firms, given the composite name Parsons Brinckerhoff–Tudor–Bechtel, or PBTB for short. PBTB contracted with Westinghouse Electric to design an innovative computer system, known as the Automatic Train Control system (ATC), which would operate the entire railway. It was controversy over this system that led to the firing of the three engineers.

About three years before the system actually began operation, one of the three engineers was assigned to work closely with Westinghouse to learn more about the ATC. After a period of time, he became concerned that BART's organizational system was inadequate to properly monitor the development of the ATC. He wrote a number of memos and reports expressing his concerns and recommending the creation of a Systems and Programming Division. His recommendations were not implemented.

As time progressed, it became apparent that Westinghouse was having trouble with the ATC system, and three of BART's engineers, including the one who had worked closely with Westinghouse, became increasingly concerned that a defective system might be placed in operation. Simultaneously, PBTB was pressing Westinghouse for a testing schedule that would prove out the suitability of the system. The three engineers felt that these efforts were inadequate and made attempts to get the attention of higher management. One of the three called in an outside consultant, who met with the three for a day and who wrote a report agreeing with their concerns. The three, armed with the outside report, then made contact with a member of BART's board of directors.

The board member decided to go public with the information given him, and the resulting conflicts made newspaper headlines. In a showdown meeting of the full board, however, the members voted 10 to 2 to express their confidence in BART's management and to reject the criticisms of the three engineers and their consultant. The names of the three engineers had been kept confidential through this period, but the BART management learned who they were and questioned them about their participation. The three denied any involvement, but they were discharged. The reasons given were insubordination, lying, impairment of staff morale, failure to perform the jobs for which they were employed, and repeated refusal to follow understood procedures.

The three sought assistance from their professional society, the California Society of

Professional Engineers (CSPE). The CSPE unsuccessfully tried to meet with BART management. When their attempts failed, they took the issue to the state legislature, using methods that later created conflict within the ranks of the CSPE. (See boxed item entitled *BART: Conflict over Professional Ethics.*) The legislature set an investigation in motion that resulted in numerous criticisms of BART. Simultaneously, the Public Utilities Commission issued a report critical of BART's lack of attention to safety. Shortly thereafter a "blue ribbon" committee that investigated BART reported likewise, and so did an outside consultant employed by BART.

About a year after they were fired, the three engineers brought suit against BART, claiming they had been wrongfully discharged. The suit was settled out of court and never came to trial. Each engineer received $25,000, of which $10,000 went for attorney's fees. By this time, BART had been in operation for more than two years and had logged 200 million passenger-miles without a passenger death or serious injury.

At the time BART began operation, special precautions were taken, including the stationing of dispatchers on station platforms. Shortly after, the legislature issued its report citing, among other things, the following: (1) The detection system for identifying the presence or absence of trains in track sectors was unreliable; (2) the ATC was unreliable; in one instance a train had speeded up instead of slowing down at the end of the line and ran off the track, injuring five people, though none seriously; (3) the programmed braking circuits were unreliable; (4) many cars had been placed in service without adequate testing. BART agreed to a complete redesign of the ATC, in the process incorporating a system called Sequential Occupancy Release, or SOR. In the SOR, once a train enters a track control section, it is considered by the computer to be in that section until a positive detection is made that it has left the section. This eliminated the unreliability of the detection system, which had given rise to a problem referred to as "ghost trains."

An important question remains. BART has compiled an excellent safety record since its operation began in 1972, although it has not been completely free from accidents. The question is, did the actions of the three engineers, which certainly catalyzed numerous official investigations, have a direct influence on this favorable outcome? Or would the BART management have taken the steps that it did anyway, having recognized the safety problems? No one will ever know, of course, because events actually went the first way and not the second. There are passionate believers both ways: those who insist that the BART management would never have taken the precautions it did without outside stimulus, and those who assert that any prudent management would have done the same without the stimulus and that PBTB would have forced the issue, being reputable engineering consultants with excellent records. A different issue, however, is whether the three engineers were ethically justified in doing what they did, and this aspect is taken up in the case *BART: Conflict over Professional Ethics.*

References

An Assessment of the Bay Area Rapid Transit (BART) Impact Program (Washington, D.C.: National Academy of Sciences, 1980).

R. M. Anderson, L. Trachtman, R. Perrucci, and D. Schendel, *The Three BART Engineers* (Purdue University, 1978).

G. D. Friedlander, "Fixing BART," *IEEE Spectrum,* February 1975, pp. 43–45.

Table 4.1	Comparison of Engineers with Other Professionals	
Occupation		**Number**
Teachers (elementary and secondary schools)		3,773,000
Engineers		1,805,000
Accountants and auditors		1,329,000
Physicians		818,000
Lawyers and judges		757,000
Teachers (college and university)		700,000
Dentists		541,000

Source: *Statistical Abstract of the United States, 1990* (Washington, D.C.: U.S. Bureau of the Census, January 1990), p. 389.

operators) wish to preserve their traditional names, and it must be granted that they probably have as much of a historically established right as anyone to use the title *engineer*.

Today's typical engineer is a college graduate (increasingly likely to have an advanced degree), is at home with science and mathematics, is a resident of suburbia, is a commuter, and is absent from home on business trips fairly often. Most engineers— 80 percent—work for private industry. About 12 percent work for government (see Table 4.2).

■ 4.1 SUPPLY AND DEMAND

Two somewhat contradictory images exist concerning the engineering profession. One is that it is subject to periodic cycles of "boom or bust." The other is that recruiters are always active on campuses, trying to snap up engineering graduates. Both images are somewhat exaggerated.

The boom-or-bust image is the product of certain occasions in the past when there have been major changes in federal funding policy, one of the major ones occur-

Table 4.2	Distribution of Engineers by Type of Employer	
	Type of Employer	**Percent**
	Industry	80
	Federal government	8
	Other government	4
	Education	4
	Other	4
		100

Source: *Science and Technology Data Book—1989* (Washington, D.C.: National Science Foundation, 1989), p. 22.

Figure 4.1
A full-size jet fighter in one of the world's largest anechoic (without echoes) chambers. The chamber permits a maximum power reflection of only 1 percent for frequencies between 1 and 60 GHz. (Courtesy of Grumman Aerospace Corporation)

ring about 1970 when the Apollo space program was cut back. Many engineers have experienced considerable distress as a result of such program reductions, when they had to find new jobs. On the other hand, most of the country's engineers have been unaffected by such cutbacks. Even at the worst of the period following the Apollo cutbacks, the nation's unemployment level for engineers was only 3 percent. Furthermore, it was learned that engineers with college educations were more likely to have jobs than those without.[1]

The image of aggressive campus recruitment has been created during three or four periods in the past few decades when it was indeed true. Stories would circulate of engineering graduates who received as many as 5, 10, or 15 employment offers each. Accounts like these made good newspaper stories, of course. Unfortunately, high school seniors, having read such articles, often found that the image did not hold true in their own cases four years later. Now, what *is* true is that employment prospects for new engineering graduates have remained among the most favorable, when compared to other fields, ever since 1950. In some years, graduates have had to be more aggressive than in others in their job searches and have had to show some flexibility in their preferences. No one can guarantee the future, but capable young people with engineering backgrounds who are willing to work have always been in demand and will probably continue to be.

An important development of the 1970s was the "discovery" of engineering by women. Prior to that time, engineering was considered pretty much "for men only," and women were a rarity. No doubt this circumstance was reinforced by a belief that engineering was heavily involved in construction and manufacturing and that physical strength might be a factor. It is true that engineering often *is* involved with construction and manufacturing, but rarely is physical strength required. Engineering is primarily an activity that calls for creativity, logic, and management abilities, and these are talents as readily expressed by women as by men. The result is that young women have entered engineering schools by the thousands and go into the same kinds of jobs upon graduation as young men do. To aid in helping women enter the engineering profession, the Society of Women Engineers was founded in 1952.

---

[1] *Engineering Employment and Unemployment, 1971* (New York: Engineering Manpower Commission, October 1971).

Figure 4.2
This unusual highway structure is located on the Izu peninsula, Island of Honshu, Japan. Its purpose is to avoid a dangerous stretch of mountainside that is subject to landslides. In the 1970s a large earthquake caused a landslide that buried alive the passengers of a bus traveling along the highway. Rather than rebuild the road in its original location, this bridge was built to spiral the highway down a vertical distance of 220 feet to the valley floor. (Courtesy of Toyota Motor Corporation)

Ethnic minority groups are also "discovering" engineering, although not so fast as women did. Minority enrollment in engineering schools is considerably less than their proportions in the population. To help remedy this situation, minority engineering students have organized themselves into associations such as the National Society of Black Engineers (NSBE) and the Chicano and Latino Engineering and Science Society (CALESS) and offer assistance and encouragement to entering minority students. At the high-school level, organizations such as MESA (Mathematics, Engineering, and Science Achievement), MITE (Minority Introduction to Engineering), and ME[3] (Minority Engineering Education Effort) offer counseling, tutoring, and motivation to encourage young minority students considering engineering careers.

One of the refreshing things about engineering is that so many people in the profession enjoy what they do. Not all, of course. That would be impossible with a group of almost 2 million people. But most seem to. Some of the brief cases described later in Chapter 6 reflect this fact, and surveys have tended to confirm it. For example, in a survey of a group of young electrical engineers, the magazine *IEEE Spectrum* found that most of the respondents enjoyed what they were doing, but wished that the public were more aware of their roles as developers of new ideas and products. One said, "I enjoy going to work in the morning. My wife finds this incredible, because she hates her job and wants to quit.... But, you know, money doesn't have a lot to do with it. I think that even if I were getting paid a little less, I'd still enjoy it and probably still go...."[2]

In another survey, the following comments were made by practicing professional engineers about engineering as a career:

The major thing is the creative aspect. There is a personal satisfaction in seeing a job completed.

---

[2] "We Look at Ourselves: Young EEs on the Way Up," *IEEE Spectrum*, July 1980, pp. 40–43.

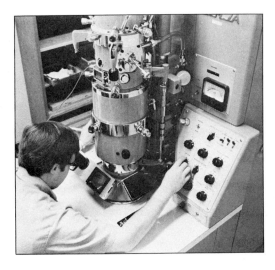

Figure 4.3
Scientists are primarily engaged in the search for new knowledge, a process we call *research*. This scientist uses an electron microscope for basic research into materials behavior. (Courtesy of University of California, Davis)

Mainly, there is the feeling of accomplishment, the challenge. . . .

. . . There is considerable variety in the work.[3]

■ 4.2 THE TECHNOLOGICAL SPECTRUM

Many people must work together to carry out engineering projects. Typically a full spectrum of individuals are involved, ranging through the following occupations—listed in order of their dependence upon knowledge of science and mathematics:

Scientist	Technician
Engineer	Craftsman
Engineering technologist	

Scientist

Scientists characteristically engage in the activity of research. To be accepted as a full-fledged professional scientist, it is almost obligatory to obtain a Ph.D. degree; graduates from science programs with B.S. and M.S. degrees usually are not accorded full professional status. In this respect, engineering is markedly different from the science professions, because it is quite normal for engineering graduates with B.S. degrees to become full-fledged professionals. In fact, it is not unusual for B.S. graduates in physics or chemistry to leave those fields, where they cannot gain full professional status, and enter engineering, where they can.

By and large, scientists produce information in the form of either reports or publications in recognized scientific journals. In order to be published, an article must

---

[3] *Career Satisfactions of Professional Engineers in Industry* (Washington, D.C.: The Professional Engineers Conference Board for Industry), p. 11.

undergo a process called *refereeing*, which means careful review by experts to determine whether the article is worthy of publication. By means of publication, the scientist's work is communicated throughout the world to others who are doing similar kinds of things, perhaps to aid them in their own work or perhaps to be subjected to further critical analysis. Thus, a scientist looks upon a body of research publications as evidence of a productive career, whereas an engineer feels the same way about a set of completed projects that have been brought into physical reality. It has been said that scientists produce *knowledge*, and engineers produce *things*.

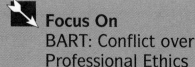

Focus On
BART: Conflict over
Professional Ethics

In the case titled *BART: The Three Engineers*, we have told how three engineers employed by the Bay Area Rapid Transit (BART) were fired for alleged improper actions. The three appealed to the California Society of Professional Engineers (CSPE) to help them. The president of CSPE then asked the local CSPE chapter (Diablo Chapter) to look into the matter. Numerous meetings ensued, and the CSPE made repeated attempts to meet with BART managers. However, those attempts were rebuffed on the grounds that CSPE had no proper role in the matter. CSPE then threatened legal action. The CSPE attorney met with the BART attorney, but nothing came of the meeting.

Two months later, at the annual meeting of the full membership, the CSPE president urged that the society support the three fired engineers and presented an award to the Diablo Chapter for its efforts in the case. In the meantime, members of the Diablo Chapter continued to investigate the case and produced a report highly critical of BART, which they transmitted to the state legislature, apparently acting on their own authority and without formal approval of CSPE. The report concerned itself pri-

marily with allegations of mismanagement and financial waste and was especially critical of the consulting engineers, Parsons Brinckerhoff–Tudor–Bechtel (PBTB). There was less emphasis on safety problems, and most of the safety references were to problems that had already been solved (with costly redesign). However, concern was expressed over the problem of false train detection (the "ghost train" problem) and the fact that proper documentation of the Automatic Train Control (ATC) system was not available. There was also concern over faulty speed signal commands the trains had received during the testing stages and poor braking under wet conditions. Fear was expressed that the BART management would hurry the system into public use without having corrected these conditions.

The Diablo Chapter members found, after they presented their report to the legislature, that most of the legislators were skeptical. Therefore, they decided to "go public" and mounted a massive petition drive through the local newspapers. This campaign was dramatically successful and resulted in an official legislative investigative report that was highly critical of BART. However, one of the allegations in the petition campaign, raising the question as to why the "... *TAXPAYER PAYS* for all the *ERRORS* of the Consulting Engineers," was to cause considerable backlash. The Golden Gate Chapter of the CSPE, based in San Francisco, formally charged the Diablo Chapter with violation of the National Society of Professional Engineers' Code of Ethics. The sections of the

Engineer

The engineer stands next to the scientist in the technological spectrum because of the dependence of engineering upon science and mathematics. The definition of engineering adopted by the Accreditation Board for Engineering and Technology (ABET) is the following:

> Engineering is the profession in which a knowledge of the mathematical and natural sciences gained by study, experience, and practice is applied with judgment to develop ways to utilize, economically, the materials and forces of nature for the benefit of mankind.

ethical code in question were those dealing with (a) avoidance of conduct that would unfavorably reflect on the profession, (b) protecting the engineering profession from misrepresentation and misunderstanding, (c) expressing engineering opinions founded only on adequate knowledge, and (d) avoidance of injuring the reputation of another engineer or indiscriminately criticizing another engineer's work in public. In the case of (d), the Code says that any evidence of unethical or illegal practice should be given to the "proper authority" for action.

The issues split the CSPE into two camps—those who felt the Diablo Chapter should be commended for its actions, and those who felt the chapter had made accusations in public against another engineer before the accused party had had a chance to reply to the charges. The president of CSPE appointed a special committee to conduct hearings on the charges brought by the Golden Gate Chapter.

The special committee, in its report, exonerated the Diablo Chapter on all counts, stating they had not violated the society's Code of Ethics. The committee said the chapter (a) had not reflected unfavorably on the profession, (b) had not caused public misrepresentation or misunderstanding, (c) had conducted an extensive investigation and therefore was expressing opinions based on adequate knowledge, and (d) had not *indiscriminately* criticized anyone—in fact, conversely, the criticisms were based upon specific facts. Further, the special committee decided that the criticisms had indeed been communicated to the "proper authority," i.e., the state legislature. The special committee added that there is another section in the Code of Ethics requiring an engineer to regard "... duty to the public welfare as paramount." Its final finding was that the actions of the Diablo Chapter had been in accord with the requirement to protect the public welfare and that the chapter should be commended for this.

As a final development, when the three engineers brought suit against BART for reinstatement of their jobs (and damages), the Institute of Electrical and Electronic Engineers (IEEE) also entered the case. The IEEE filed an *amicus curiae* (friend of the court) brief. In that brief, the IEEE asked the trial judge to enter a finding that professional codes of ethics for engineers are admissable evidence in trial proceedings, that "... every contract of employment of an engineer contains within it an *implied term* to the effect that such engineer will protect the public safety," and "... that a discharge of an engineer solely or in substantial part because he acted to protect the public safety is a breach of such implied term." However, the suit of the three engineers was settled out of court and never came to trial. Thus, the interesting points raised in the IEEE *amicus curiae* brief were left unresolved.

References
R. M. Anderson, L. Trachtman, R. Perrucci, and D. Schendel, *Professional Societies and Ethical Conflict* (Purdue University, 1978).
G. D. Friedlander, "The Case of the Three Engineers vs. BART," *IEEE Spectrum*, October 1974, pp. 69–76.

The classic role of the engineer is that of design and development, although we will see later that engineers frequently move into research and just as frequently move into manufacturing and construction.

It should be realized that not every aspect of every engineering problem is solved with the use of science and advanced mathematics. Many problems are simply not amenable to scientific solutions, and experienced judgment is used instead. Such would be the case, for example, in problems involving suitability of manufacture, assembly, and maintenance. On the other hand, many engineering projects are impossible to complete without the use of science and advanced mathematics, some examples being jet aircraft, digital computers, suspension bridges, nuclear reactors, and space satellites.

Engineering Technologist

Since about 1970, a new occupation, that of the *engineering technologist*, has entered the scene. Four-year educational programs for engineering technologists, culminating in Bachelor of Science degrees, have been developed in the United States. They do not include as much science and math as the usual engineering curriculum does. The education of technologists emphasizes current design practice rather than the conceptual skills and theoretical knowledge needed for advancement of the state of the art, which is presumed to be the domain of the engineer. Technologists typically work in that part of the engineering spectrum which lies between the engineer and the technician, in product development, manufacturing planning, construction supervision, and technical sales.

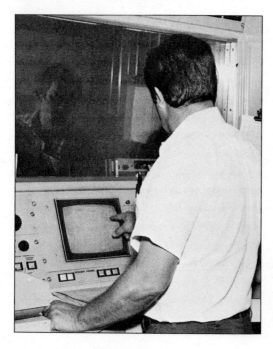

Figure 4.4
Engineers are found in all sorts of roles in addition to their classic ones of research, design, and development. This engineer operates a control console for a test of a new experimental engine. Engine durability, power, heat distribution, fuel economy, and emissions are all being evaluated. (Courtesy of General Motors Corporation)

Figure 4.5

A technician prepares a model of an experimental aircraft for testing in a large wind tunnel. (Courtesy of Grumman Aerospace Corporation)

The advisory committee for a study of engineering technology education stated, "The technologist should be a master of detail; the engineer, of the total system. . . . The development of methods or new applications is the mark of the engineer. Effective use of established methods is the mark of the technologist."[4]

The Accreditation Board for Engineering and Technology accredits technology programs as well as engineering programs, using somewhat different criteria. ABET's definition of engineering technology is the following.

> Engineering technology is that part of the technological field which requires the application of scientific and engineering knowledge and methods combined with technical skills in support of engineering activities; it lies in the occupational spectrum between the craftsman and the engineer at the end of the spectrum closer to the engineer.

Since the field of engineering is so broad and diverse, it is not surprising to find a great deal of overlap in the kinds of positions that engineers and engineering technologists take after graduation. Studies of bachelor-level technology graduates have shown that about half of them have taken jobs that classified them as engineers. On the other hand, many graduates of engineering programs have entered jobs that would be presumed to be normal targets for technology graduates.

Technician

In addition to the four-year B.S. programs in engineering technology just described, there are also many two-year technology programs in the United States, usually resulting in associate degrees rather than baccalaureate degrees. Graduates of such two-year programs are classified as *engineering technicians*. Technicians generally do

---

[4] *Engineering Technology Education Study: Interim Report* (Washington, D.C.: American Society for Engineering Education, June 1971), p. 16.

not possess the knowledge of mathematics and science that engineers or engineering technologists do, but they are highly skilled in certain of the support functions needed by R&D teams. Typical examples are electronics technicians, who assemble experimental "bread-board" circuits; mechanical drafters, who prepare crisp, clear drawings from the engineer's sketches; and computer drafters, who create drawings on a computer faster than can be done on a drawing board. The following are some kinds of activities in which engineering technicians engage:

Drafting	Experimental model assembly
Estimating	Surveying
Field service	Technical writing
Inspection	Time study
Installation	Maintenance

Craftsman

An indispensable member of the technological team is the *craftsman*. Craftsmen are the people who make the first units that physically embody the engineers' designs, commonly called experimental models. Sometimes such individuals are titled *mechanicians* or *experimental machinists*. Generally, they possess the skills of tool and die makers and are often resourceful and creative. They are the discoverers of the engineers' mistakes. Every engineer makes mistakes, of course, and these inevitably show up when someone tries to make a part in accordance with the engineer's sketches and instructions. But that is exactly why experimental models are made—to discover mistakes before the parts go into production. At such a point in a project's history, the engineers work very closely with the experimental model makers—the craftsmen—and these experienced people can often make significant contributions to the design of a project by suggesting simplifications, easier machining methods, avoidance of potential service problems, and the like.

■ 4.3 THE FUNCTIONS OF ENGINEERING

Research

Research is the process of learning about nature and codifying this knowledge into usable theories. (The term *nature* includes the topics of physics and chemistry as well as natural history, the study of living things.) In the common way of looking at things, it is supposed that scientists first do research in order to provide a scientific basis for what engineers will do later, and then engineers apply this knowledge. Many times things have happened just this way, but sometimes they happened the other way around. For example, the early development of the steam engine was accomplished well before any scientific theory was established concerning steam. The same thing happened in the case of electrical generators.

Sometimes the terms *basic research* and *applied research* are encountered. The former means the search for knowledge for its own sake, and the latter implies that there is a potential use for the knowledge being sought. Engineers rarely engage in basic research; if they do, they would more properly be looked upon as scientists. But engineers often engage in applied research. In many instances, during the course of developing new devices or systems, it is found that not enough is known scientifically to make the projects successful. It is then necessary to initiate a special research program to gain the needed information. This is clearly *applied* research, and it occurs frequently in engineering. Engineers engaging in such projects often bear the title *research engineer*.

Design and Development

In many instances, attempts have been made to make a clear distinction between the functions of *design* and *development*. For example, a committee of faculty members from the Massachusetts Institute of Technology defined *design* as ". . . defining a device, a process or a system in sufficient detail to permit its physical realization."[5] The National Science Foundation defines *development* as "the engineering activity required to advance the design of a product or a process to the point where it meets specific functional and economic requirements and can be turned over to manufacturing units."[6] These two definitions are pretty hard to tell apart. In practice, the term *development* is likely to refer to the early stages of a project, where the various methods through which the project could be accomplished are analyzed, compared, and tested. *Design* usually refers more to the later phases of a project, when the basic method of achievement has been decided and it is now necessary to establish the exact shapes and relationships of the various parts. However, even these distinctions may get mixed up, because real-life projects have a way of getting their phases intertwined, and design and development may be mixed together inextricably.

Generally, companies refer to the whole spectrum of their technical activities as *research and development*, usually abbreviated R&D. In most companies, however, very little true research is done, although there are notable exceptions such as Bell Laboratories and many of the R&D divisions of the big aerospace firms. Mostly, the activities carried out under the heading "R&D" are actually design and development.

An activity closely related to design is *drafting*. In fact, much mechanical and civil engineering design actually begins on the drafting board with what is called a *layout*. Layouts continue to play a prominent role in many projects, for it is on these master sheets that the physical relationships between parts are worked out. But layout is not the same thing as drafting, which begins after all the design decisions have been made. Drafting, then, is the final stage of design in which the detailed drawings are made that will govern the actual manufacturing of the parts. In many organizations, this drawing no longer involves putting pencil to paper, but is done on the graphics display terminal of a computer and is referred to as *computer-aided drafting*. This is described in more detail in Chapter 18, along with *computer-aided design*.

---

[5] "Report on Engineering Design," *Journal of Engineering Education*, April 1961, pp. 645–660.
[6] *Instructions for Survey of Industrial Research and Development During 1964* (Washington, D.C.: U.S. Dept. of Commerce, 1965), p. 4.

Figure 4.6

Much research involving harbors and waterways is performed with large-scale models. This photograph and the one on page 113 show some experiments being conducted on the San Francisco Bay model operated by the U.S. Corps of Engineers. The area shown is about 25 miles upstream from San Francisco, near Benicia. Shown here is a preliminary design for new docks to be constructed; note the heavy deposits of black material in the stream channel at the left, which are indicators of the amount of shoaling the docks as designed would probably experience. The photograph on page 113 shows an alternative design, and the near absence of black material in the channel shows that shoaling would probably not be a problem. (Courtesy of University of California, Davis)

Testing

Some organizations have special test departments that are organizationally separate from their design departments. The reason for separation is that the test department can be more objective than designers testing their own creations. Test departments may also conduct tests of new parts or materials and qualification tests of products furnished by others. Extreme examples of the latter are the special centers set up by the armed services for the express purpose of testing the airplanes, tanks, ships, and so forth furnished to them by their suppliers. Engineers and engineering technologists are frequently employed in testing functions.

Figure 4.6
Continued.

Manufacturing

Before a product can reach the public, it must be manufactured (or, in the case of civil engineering, constructed). Many engineers are employed in manufacturing—so many that a special organization called the Society of Manufacturing Engineers (SME) was created. Also, a number of universities have established special educational programs in either manufacturing engineering or manufacturing engineering technology. Such persons are sometimes, but not usually, in direct charge of the production personnel. Some of the chemical companies, oil companies, and semiconductor

Figure 4.7
An industrial robot drilling mounting holes in an F-16 aircraft fuselage panel. The robot first determines by means of an optical character reader which of several different panels it is to drill; this enables it to select the proper routine for drilling the holes in that particular panel. (Courtesy of Cincinnati Milacron)

companies start brand-new graduates as supervisors of production operations, as part of their training programs. Mostly, however, manufacturing engineers have the responsibility for the product, rather than for the people, and are daily engaged in solving the problems that inevitably arise in manufacturing anything. They are also concerned with developing and improving the production processes themselves, including the tools and machines. Some may be in charge of the inspection process, or, as it is often called, *quality control.*

Closely related to the manufacturing engineer is the *plant engineer.* Whereas the former is concerned with the product and the means of manufacture, the latter is concerned with the buildings and utilities that sustain the manufacturing process.

Construction

The civil engineering analog to manufacturing is *construction.* Again, as with manufacturing, construction engineers may be directly in charge of the construction personnel, or may instead have responsibility for the quality of the process. If the former, they will usually be employees of the construction contractor, who may either deliberately hire civil engineering graduates for the purpose or hire graduates of the increasingly available construction technology programs. If, instead of being directly in charge, these individuals have the responsibility for the quality of construction, they will usually be employees of the consulting engineering firm that designed the structure. Under such circumstances, they are typically civil engineering graduates and bear a title such as *resident engineer,* meaning that they spend all their time on the construction site. Needless to say, construction engineers have to go where the construction is; hence, they travel a lot.

Figure 4.8
Not all engineering is carried out inside an office or laboratory. Here a seismic survey crew laboriously obtains geologic data in a Louisiana bayou to determine the potential for petroleum production. (Courtesy of Standard Oil Company of California)

Sales

It is likely that few students, when starting their study of engineering, ever thought it would lead them into sales activities. Yet many employers deliberately recruit engineering or technology graduates for this purpose. Although it is true that many engineers move into straight sales work and thus completely lose contact with engineering, that is not what is generally meant by *sales engineering*. Sales engineering is a province truly intermediate between sales and engineering, occasionally involving engineering design. Such opportunities normally arise in enterprises that sell and produce custom-designed systems. In a typical case of this nature, a fully operating system put together from off-the-shelf components may be offered in a way that is unique to the customer. In some instances, it may be necessary to include a special component that has not yet been designed. The sales engineer works with the customer and essentially makes the sale, but also designs the system to meet the customer's needs and, when necessary, works with the home engineering office to develop hitherto nonexistent components.

Consulting

Consulting engineering is the activity which most closely resembles the mode of operation of doctors and lawyers, but only a very small percentage of engineers are engaged in consulting. It is erroneous to think of a consulting engineer as an individual who typically offers services to the public for a fee, like a doctor. There are some who function like this, of course, especially among those who are just starting out as consultants. But the "consulting engineer" usually is an organization that hires engineers, architects, accountants, drafters, clerks, and people of similar skills. Some consulting organizations are very large indeed and hire hundreds, or even thousands, of engineers of all kinds: chemical, civil, electrical, mechanical, and nuclear, among others.

A person beginning consulting work is much more likely to fail because of a lack of business ability than because of a lack of technical ability. Virtually all consultants warn the prospective newcomer about "that depressing first year." Some even declare that the difficult period is likely to be three years instead of one. Another quality that is necessary is a willingness to work long hours. One successful consulting engineer has warned:

> If you object to working long hours and if you intend to dismiss all the business problems from your mind when you leave the office, don't try to be a consulting engineer, for the problems will be with you 24 hours a day.[7]

Government

Many engineers in the federal government, and some in state governments, are engaged in research, design, and development. They probably do not perceive their

[7] J. B. McGaughy, "So You Want to Open a Consulting Office—By Way of Qualifications," *American Engineer*, October 1955.

Table 4.3 Engineers' Management Responsibility as a
Function of Age

	Age		
	30–35	40–45	50–55
General management	4 percent	12 percent	15 percent
Manager of major division	12	24	27
Project supervisor	25	24	20
Unit supervisor	16	12	9
Indirect supervisor	21	16	16
No supervisory responsibility	22	12	13
	100	100	100

Source: *The Engineer as a Manager* (New York: Engineering Manpower Commission, September 1973), p. 3.

situation as being greatly different from that of engineers in industry. Yet there are areas of engineering activity in government that have few or no counterparts in private industry; these contribute to the formation of public policy and law enforcement, as in environmental regulation, operation of public utilities, and public transportation. Although jobs in such areas require an engineering background, they frequently do not involve the basic engineering function of design. Much of the typical activity of government engineers has to do with the preparation of functional specifications for public works and with supervision of the resulting construction and operations. In the federal government, engineers have participated at high levels in policy matters, since the government has increasingly come to recognize that many decisions of national importance rest primarily upon technical considerations.

Management

Statistics show that, sooner or later, most engineers go into management. In fact, Table 4.3 shows that 40 percent are in management (defined as project supervisor or higher) by their early thirties, and 60 percent by their early forties. Most of these positions, of course, are in *engineering* management, although many engineers do go on to general management positions.

In a series of interviews with 21 successful electrical engineers in industry in their twenties and thirties, *IEEE Spectrum* found that most of the interviewees felt that their move into management was inevitable. As one put it:

> Ninety percent of the guys I interview who are coming directly out of school want to design. They have spent four years getting geared up for design. That's the way colleges present engineering. Now, I've been out of school for nearly 12 years, and I spent only the first 6 in design jobs. I'd hoped to survive in what I think of as pure engineering for 30 years, but after 6 years I found myself on the management payroll.[8]

---

[8] "We Look at Ourselves: Young EE's on the Way Up," *IEEE Spectrum*, July 1980, pp. 40–43.

Anyone planning on a management career should be aware of at least two things: (1) Managers tend to move around a lot, because a promotion usually means moving to a new location; managers must be geographically "flexible"; (2) management, especially general management, entails long working hours and unusual pressures. J. Irwin Miller, chairman of the board of Cummins Engine Company, has graphically described the immense multiple pressures that operate on an executive as follows:

> To illustrate, let us suppose we can see inside the head of the president of a large manufacturing organization. His company employs 20,000 persons and operates half a dozen plants. It distributes its products in every state and in many foreign countries, and—most frightening of all—it has competitors.
>
> Now let us suppose that these competitors are extremely vigorous, and that our president knows that to maintain his share of the market and to make earnings which will please his directors, he must accomplish the following very quickly: design and perfect a brand-new and more advanced line of products; tool up these products in such a way as to permit higher quality and lower costs than his competitors; purchase new machinery; arrange major additional long-term financing. At the same time his corporation's labor contract is up for negotiation, and this must be rewritten in such a way as to obtain good employee response and yet make no more concessions than do his competitors. Sales coverage of all customers has to be intensified, and sales costs reduced. Every one of these objectives must be accomplished simultaneously, and ahead of similar efforts on the part of his competitors—or the future of his company is in great danger. Every head of a corporation lives every day with the awareness that it is quite possible to go broke. At the same time he lives with the awareness that he cannot personally accomplish a single one of these vital objectives. The actual work will have to be accomplished by numerous individuals, some actually unknown to him, most of them many layers removed from his direct influence in the organization.[9]

Teaching

Finally, we should note that some engineers become teachers. The teaching of engineering almost always is at the college level; engineering is not normally taught in high schools. If one wishes to teach at the two-year community college level, a master's degree, plus a teaching credential, is the usual preparation. Professional experience also is highly desired. A master's degree plus professional experience is also considered optimum for many four-year colleges, unless faculty members are required to maintain active research programs in addition to their teaching. Schools of the latter type are called research universities, and a Ph.D. is almost universally required for their faculty members. Generally, the teaching loads in such universities are lighter than in other schools to allow time for research. Research and graduate studies are usually closely coupled in these institutions, because it is believed that the very best training for graduate students is in a research environment. Educational programs, if coupled with research, will be right at the cutting edge of technological developments.

---

[9] J. Irwin Miller, "The Dilemma of the Corporation Man," *Fortune*, August 1959, p. 103.

Figure 4.9
An engineer uses a submicron scanning Auger (pronounced oh-zhay) microprobe for research in materials. The instrument can provide an analysis of the elements present in a selected area less than a micron (10^{-6} m) in diameter.

■ 4.4 PROFESSIONAL REGISTRATION

Most people are surprised to discover that the first state to adopt a registration law (in 1907) was not a big industrial state, but Wyoming. That state's registration law was passed to protect the public from a flood of persons of doubtful qualifications who were representing themselves as engineers during an era of great water resources development. Today, all 50 states, the District of Columbia, and three territories have engineering registration laws, and more than a half million engineers are registered.

The case for registration has been most eloquently stated by a Utah court of law:

> It has been recognized since time immemorial that there are some professions and occupations which require special skill, learning, and experience with respect to which the public ordinarily does not have sufficient knowledge to determine the qualifications of the practitioner. The layman should be able to request such services with some degree of assurance that those holding themselves out to perform them are qualified to do so. For the purpose of protecting the health, safety, and welfare of its citizens, it is within the police power of the State to establish reasonable standards to be complied with as a prerequisite to engaging in such pursuits.[10]

It is commonly taken for granted that state engineering registration laws apply only to those engineers who offer their services directly to the public. In many states this is indeed the case, and the laws in those states contain clauses that specifically exempt engineer employees of manufacturing companies from registration. But some states do not have such exemption clauses in their laws, and engineers in those states who assume they are exempt may find themselves in an equivocal position. The Model Law, prepared by the National Council of Engineering Examiners (NCEE) as a guide for state law-making bodies, does not contain such an exemption.

The requirements for professional registration in most of the states are:

Graduation from an ABET-accredited school, plus four years' engineering experience acceptable to the board, plus passage of a 16-hour written examination

or

Eight years of engineering experience acceptable to the board, plus passage of a 16-hour written examination

---
[10] *Clayton v. Bennett,* 298 P. 2d 531.

The usual pattern is for the 16-hour written examination to be divided into two equal parts. The first is generally known as the Fundamentals Examination (sometimes referred to as the Engineer-in-Training Examination, or EIT) and the second, as the Professional or Principles and Practices Examination. Persons who successfully pass these examinations and become registered are entitled to use the title "Professional Engineer" and to place the initials "P.E." after their names. It is illegal for unregistered persons to use the title.

Nearly all the states have made provision for an EIT status and will allow persons to take the first eight-hour (EIT, or fundamentals) portion of the written examination immediately before or immediately after graduation. EIT status conveys no legal privileges and is offered primarily as a convenience to new graduates so that they may take the examination in fundamentals at a time when the material is fresh in their minds. Almost all of the states use a uniform national EIT examination, administered through the NCEE, and the great majority also use a uniform national examination for the professional portion.

There are compelling reasons why every young engineer should become registered as quickly as possible after graduation, including the following:

1. Registration may become more important in the future.

2. No one can foresee the future course of one's own career. An individual may intend to have a career that lies exclusively in areas not requiring registration; but a change of mind could occur, or unexpected opportunities could arise for which registration is required, or activities currently unaffected by the law might become affected in the future.

3. A court of law generally will not recognize an individual as an engineer unless he or she is registered. This could be frustrating—should the engineer want to testify as an expert witness, for example.

4. If a nonregistered person engages in practice required by law to be performed by registered engineers, then at the very least the courts may not aid in attempts to collect fees.

5. Many companies believe it is desirable for members of their engineering management to be registered; hence, registration could be an aid in promotion.

It was mentioned earlier that a requirement for registration is "graduation from an ABET-accredited school" plus four years of experience and passage of an examination. This raises a natural question: Since ABET accredits *both* engineering and engineering technology programs, do registration boards treat them as equivalent? So far, the answers have varied from state to state. Some states give full equivalency; some do not and require that applicants with technology degrees submit a full eight years of experience in addition to the four years spent in earning the degree. Some states go halfway. The situation is extremely fluid and is subject to change. Individuals will have to check with each state board of registration to see what the current practice is in that state.

■ 4.5 ENGINEERING ETHICS

The complete text of the Code of Ethics of the National Society of Professional Engineers is reproduced in Appendix A. (A summary of this code, consisting of the preamble and fundamental canons, is given in the box below.) The code has been prepared and adopted by leaders in the engineering profession and shows the expectations that engineers have for themselves and their colleagues. Note that matters such as human welfare, integrity, and professional competence dominate the code. Any individual who faithfully uses these three elements—human welfare, integrity, and competence—as daily guidelines will rarely go wrong as a professional.

Sometimes cases arise that seem confusing, particularly when it comes to ethical principles governing the relationships between employees and employers. However, there can be no confusion concerning the ethical obligations of engineers when it comes to out-and-out violations of the law or safety codes or when known safety hazards are concealed. Even if moral obligations were set aside for a moment and the most selfish personal motives invoked, engineers should ask whether their own long-range interests are well served by continued employment with a company that would knowingly conceal a safety hazard or violate the law. The appropriate course

Focus On
Summary of Code of Ethics for Engineers (National Society of Professional Engineers)

Preamble

Engineering is an important and learned profession. The members of the profession recognize that their work has a direct and vital impact on the quality of life for all people. Accordingly, the services provided by engineers require honesty, impartiality, fairness and equity, and must be dedicated to the protection of the public health, safety and welfare. In the practice of their profession, engineers must perform under a standard of professional behavior which requires adherence to the highest principles of ethical conduct on behalf of the public, clients, employers and the profession.

Fundamental Canons

Engineers, in the fulfillment of their professional duties, shall:

1. Hold paramount the safety, health and welfare of the public in the performance of their professional duties.
2. Perform services only in areas of their competence.
3. Issue public statements only in an objective and truthful manner.
4. Act in professional matters for each employer or client as faithful agents or trustees.
5. Avoid improper solicitation of professional employment.

Note: The complete text of this Code of Ethics is provided in Appendix A.

of action for the engineer is best phrased by the Board of Ethical Review of the National Society of Professional Engineers (NSPE):

> The engineer should make every effort within the company to have the corrective action taken. If these efforts are of no avail, and after advising the company of his intentions, he [or she] should notify the client [customer] and responsible authorities of the facts.[11]

Some people take the view that an engineer's career is finished if the route recommended by the NSPE is taken. This is not necessarily the case. Sometimes the violation is the result of overzealousness on the part of lower echelons, and top management may be horrified to learn of the unethical and illegal acts that are being committed by their employees, presumably in the company interest. A good example is that of the Ford Motor Company, which in 1973 reported extensive violations—by its own employees—of the pollution control test procedures of the federal government. Evidently, in order to help test cars meet the emissions standards of the Environmental Protection Agency, Ford employees had given the cars unauthorized maintenance, such as replacement of spark plugs, cleaning of carburetors, and adjustment of timing. When Ford's higher management discovered these violations, they reported the facts to the government and withdrew their application for certification. Even these principled acts did not protect Ford from receiving a $7 million fine, however—brought on by overzealous employees acting contrary to the public interest.[12]

A distinction must be made between *product safety* and *product quality*. The former is the business of the public at large, whereas the latter is between the company and its customers. In clear-cut cases involving safety, the engineer has an obligation to the public, but in matters involving product quality, the engineer has only the obligation to make the necessary facts known to management. As the NSPE's Board of Ethical Review has said: "If the public is misled as to the product's quality, the unfavorable reaction will be directed against the company."[13] Inevitably, cases occur in which issues regarding quality and safety seem to be interrelated. The individual engineer must decide if the safety aspects involve violation of the law or serious concealment of hazard; if this is the case and if the company knowingly proceeds, no ethical alternative exists except to notify the customer and the authorities.

A 1977 study of 1200 readers of *Harvard Business Review* produced some interesting results regarding business ethics and social responsibility. One surprising result was that the respondents—nearly all executives—held their own ethics in high regard, but were inclined to be skeptical regarding those of others. For example, in response to a hypothetical question involving the padding of expense accounts, 89 percent stated that they personally found such a practice to be unacceptable, regardless of circumstances, but only about half thought that other executives would also find the practice to be unacceptable. Another surprise was that, contrary to the conventional wisdom, "pressure for profit" was placed very low on the list of factors that the respondents felt cause a lowering of business ethical standards. Overwhelmingly

---

[11] *Opinions of the Board of Ethical Review* (Washington, D.C.: National Society of Professional Engineers, 1965), pp. 41–42.
[12] *San Francisco Chronicle*, 14 February, 1973.
[13] *Opinions of the Board of Ethical Review*, op. cit., p. 43.

they blamed society, citing the following reasons: "Society's standards are lower; social decay; more permissive society; materialism and hedonism have grown; loss of church and home influence; less quality, more quantity demands." The most surprising result of all—and another blow to conventional wisdom—was that the respondents placed the *customers* far ahead of either stockholders or employees as the group to whom their companies felt the greatest sense of responsibility. As the *Harvard Business Review* stated, "We may be observing a return to the original capitalist doctrine of the customer as the client whom production is intended to serve, and the replacement of the doctrine of 'long-run profit maximization' with the 'long-run customer satisfaction' doctrine."

In the same survey it was apparent that the ethical concerns of the respondents were not the same ones that had been expressed in similar surveys taken in the 1960s. For example, price-fixing was less of an issue than it had been earlier, and concern over honesty in communication, and over bribes and kickbacks, headed the list of worries. For example, a vice-president of engineering in one company stated that he was worried about misrepresentation of products, and another was concerned about deliberate understating of delivery times in order to get contracts. Ninety-eight percent of the respondents expressed agreement with the statement "In the long run, sound ethics is good business."[14]

---

[14] S. N. Brenner and E. A. Molander, "Is the Ethics of Business Changing?" *Harvard Business Review*, January–February 1977, pp. 57–71.

References

G. C. Beakley and H. W. Leach, *Engineering: An Introduction to a Creative Profession* (New York: Macmillan, 4th ed., 1982).

J. D. Kemper, *Engineers and Their Profession* (Philadelphia: Saunders College Publishing, 4th ed., 1990).

R. J. Smith, B. R. Butler, and W. K. LeBold, *Engineering as a Career* (New York: McGraw-Hill, 4th ed., 1983).

Exercises

4.1 See if you can find some examples of engineering projects, in addition to those mentioned in the text, that would be impossible to carry out without the help of science and mathematics. Conversely, see if you can find some examples that probably do not require the application of much scientific knowledge or mathematics.

4.2 Contact your campus placement center and prepare a brief report on the current situation regarding the demand for engineers. Consult periodicals in your library, such as *Professional Engineer* and *IEEE Spectrum*. Also, look at the classified ad sections of large metropolitan newspapers, especially the Sunday editions.

4.3 Examine the course catalog at your college to see if the apparent objectives of the courses can be related to the various functions of engineering, such as research, design and development, testing, manufacturing, construction, sales, consulting, government, management, and teaching. Are there some courses that do not seem to fit in this classification but which appear to be so fundamental that they relate to many of the functions? If so, then classify these as "basic."

4.4 Assess your own talents and personal interests and see where you fit best in the technological spectrum of Section 4.2. (If your interests overlap one or more of the categories, this should not be a cause for concern. Rarely do things fit perfectly into pigeonholes, especially where human beings are concerned.)

4.5 From your state Board of Engineering Registration, obtain a copy of the law relating to the registration of engineers in your state. By examining this law, determine whether the requirements for registration resemble the prevailing pattern for most states as described in this chapter. What are the differences, if any? Is special consideration given to graduates of ABET-approved curricula? How is the practice of engineering defined? Are there gaps in the definition? Is there an exemption clause for certain kinds of engineers? Are there any special provisions for engineers with graduate degrees or for eminent persons?

4.6 Some have argued that *all* engineers should be registered, regardless of whether they offer their services directly to the public or not. Write a brief report on this topic giving the pros and cons.

4.7 Make a list, in priority order, of those things you believe are important to you in professional employment, such as salary, location, potential for advancement, and type of work. Then make a similar list, putting down the things the employer is likely to be seeking from you as a new employee. Compare the two sets of expectations and notice any differences.

4.8 There is a common notion among many students about to enter engineering school that engineers are generally oriented in their job activities to "things" and very little to people. Yet statistics show that more than two-thirds of engineers become managers sooner or later, and management is a very people-oriented activity. How do you account for this inconsistency? Is it still possible that engineering, in some restricted sense, is indeed more thing-oriented than people-oriented? How might this restricted view of engineering be described?

4.9 Most engineers eventually have to decide whether they want to move into management or stay in purely technical work. List for yourself the pros and cons of each alternative. Having done so, consider whether the way you made your list has been influenced by your own inclinations toward one route or the other. If you do have such a pronounced inclination, see if you can put down in written form just what it is that attracts you one way or the other.

4.10 Examine the yellow pages of a metropolitan-area telephone directory, and make note of the different kinds of engineers listed there. Consider the scope of activities that some of the more unfamiliar engineering fields might cover and the kinds of clients it would take to keep these consulting firms in business.

4.11 Contact an engineer who works for a governmental agency—whether federal, state, county, or city—and learn how his or her activities differ from those of engineers in industry. Write a brief report on what you have learned.

4.12 Write a statement on what engineering professionalism means to you.

4.13 Suppose that, upon graduation, you are offered a job with a company in another state and you accept. Subsequently, you receive an offer from a different company that is more attractive in every way than the one you already accepted. Suppose it not only offers a greater salary but is in exactly the kind of activity you have been seeking, offers great opportunity, and will make it possible for you and your spouse to live reasonably close to your family (assuming the latter is a positive and not a negative factor for you). What are your ethical responsibilities? According to your personal ethical code, are you bound to keep your word to the first employer? If so, for how long? One month? One year? Ten years? What obligations might you have to yourself or your spouse to select a career that will provide a happier life for your family, assuming that the second offer gives you such a prospect? If you chose one route over the other, would you somehow have to rationalize to yourself a modification in your ethical code?

4.14 Suppose, in your employment as an engineer, you discover that some of your colleagues have been involved in kickback schemes with suppliers, which are in violation of state law. Do you have an ethical obligation to do something, or should you keep silent since you are not directly involved? Suppose you decide something should be done and go to your boss about it, but you are told to keep quiet? What do you do then, since you know a violation of law has occurred?

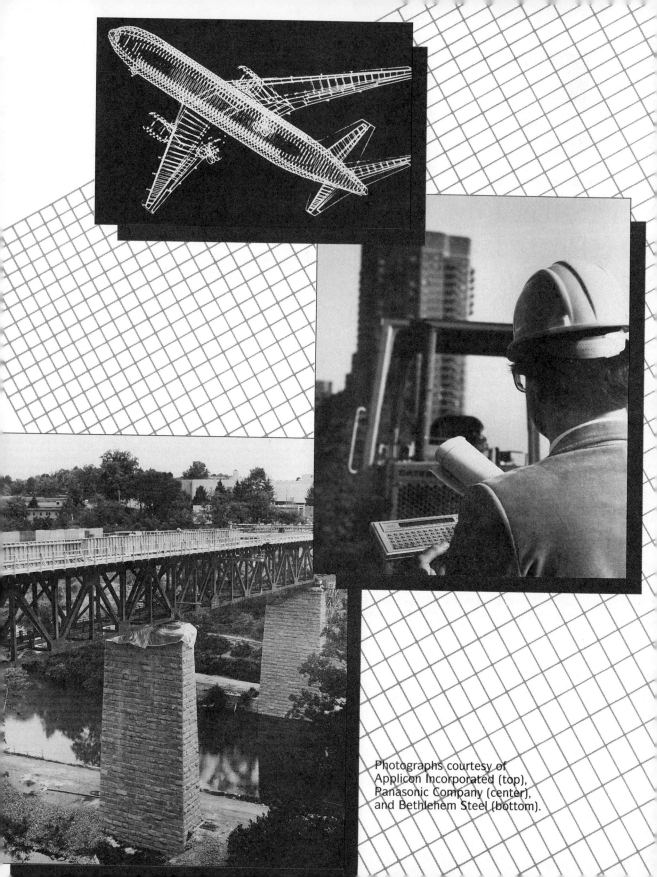

Photographs courtesy of
Applicon Incorporated (top),
Panasonic Company (center),
and Bethlehem Steel (bottom).

5

Safety and Professional Responsibility

■ 5.1 SAFETY

Since the mid-1960s concern has increased over the legal responsibility of manufacturers for product safety. The landmark event that is generally accepted as the beginning of this trend is the case of *Greenman v. Yuba Power Products, Inc.*, which established the principle of "strict product liability." (See the next section.) Under this principle, a manufacturer is held liable for a product that proves to have a defect resulting in injury, even though there may be intervening sellers. Also since the 1960s, a debate has mounted over the moral obligations of engineers, not only with regard to safety but also relating to other matters such as pollution, nuclear weapons, and war.

It frequently is not clear where the line is to be drawn between "safe" and "unsafe." Is a kitchen knife safe? Nearly 200,000 Americans are injured by kitchen knives every year. Is taking a bath safe? Over 100,000 are injured yearly in tubs and showers. Should one ride a bicycle? It is estimated that each year 370,000 people are injured seriously enough by using bicycles to require emergency-room hospital treatment.

Such hazards are commonly accepted by the public with little question. Each instance obviously carries great benefit, and the degree of risk is considered acceptable. The safely built into a product must be matched with the level of public acceptance. In cases that are sufficiently controversial, the acceptance level finally must be arbitrated by governmental action.

The individual engineer cannot be expected to assume a hard-line moral position

125

that moves very far ahead of public opinion. If engineers as a group presumed to take on themselves the authority to act as moral judges for the rest of society and to provide certain items to, or withhold them from, the public, they would be assuming totalitarian powers over others. The justification for applying the word *totalitarian* in this context lies in the fact that a small group—the engineers—would be making decisions for society that were not subject to review by elected officials or by the functioning of the market system. This arrogation of power would act to cancel the authority of the body politic, which is supposed to decide questions of public policy. Furthermore, engineers possess no special qualities that make them superior moral authorities and justify such a presumption of power.

Many safety issues involve differences of opinion, and some individuals may become exceedingly militant on seemingly irrelevant grounds. Even here, however, we must be careful, for something that seems fanciful today may not seem so tomorrow. An example is the issue of double insulation for electric power tools. Not many years ago it would have seemed a foolish precaution to provide power tools with two layers of insulation, yet today the industry seems to be moving toward this requirement as standard. In the past, prevention of electrical shocks in the United States relied mostly on a third wire for grounding, but a study by Underwriters' Lab-

Disasters
The St. Francis Dam Disaster

Compared to the Johnstown Flood in 1889, the death toll resulting from the failure of the St. Francis Dam was small. The Johnstown Flood killed nearly 3000 people; the St. Francis Dam killed 385. However, the dam whose failure wiped out Johnstown was an old, neglected earth-fill dam that had recently been repaired and heightened without adequate engineering supervision. On the other hand, the St. Francis Dam, built by the Los Angeles Water Department in 1926, was a modern concrete arch—and was brand-new.

Almost as soon as it was completed and had begun to fill, leakage developed between the dam and the foundation. During its second year of service, several large cracks formed in the dam. These were simply caulked, as was the usual practice. In March 1928, the water below the dam turned muddy, and William Mulholland, chief engineer of the Los Angeles Water Department, inspected the situation. He was relieved to find that the muddy water was caused by nearby road construction and decided the dam was not in immediate danger.

At about midnight the same night, the dam collapsed. A wall of water 100 feet high surged down the narrow canyon, carrying huge blocks of concrete as far as a half mile. Once out of the canyon, the wall of water receded to 25 feet and rolled at a speed of 18 miles per hour down the Santa Clara River valley, toward the towns of Piru, Fillmore, and Santa Paula. Piru had no warning, but Fillmore and Santa Paula were

oratories found that grounding was effective for only about 13 percent of the power tools in use. Many electrocutions have resulted from power tools in the United States, whereas in some European countries where double insulation has been broadly adopted, *no* electrical shocks were reported for double-insulated tools.

The foregoing discussion has related mostly to the responsibilities of manufacturers and thus is relevant particularly to chemical, electrical, and mechanical engineers. However, civil engineers obviously have an enormous responsibility for public safety, too, because they design the structures upon which we all depend. They have an enviable record of safety overall, but disasters have occurred.

One of the most sensational of these occurred in 1928 when the St. Francis Dam in California burst without warning in the middle of the night and killed nearly 400 people. (See boxed item.) A more recent incident of almost exactly the same type occurred in France in 1959; the Malpasset Dam collapsed, killing 421 people. In 1961, a government engineer was charged with involuntary homicide through negligence, and in 1964 his case came to trial. He was acquitted. The prosecution stated the engineer had been responsible for seeing that tests of the foundations were adequately carried out, and charged that he had not carried them out correctly. However, one of the witnesses testified that the blame really should fall on the designer of the dam, who had been absolved of responsibility for the collapse.

---

alerted by telephone from Los Angeles. As a result, most of their residents escaped, but many had no warning other than the terrifying roar as the flood approached. Finally, 5 hours after the break, the flood reached the ocean near Ventura and subsided.

Official inquiries followed. Los Angeles paid all damage claims submitted without contesting them—a total of $15 million—but this did not return the dead to life. Mulholland, 73 years of age, publicly took responsibility for the disaster and soon resigned, a broken man. Investigations showed that the dam site was intersected by at least one earthquake fault, but, as one observer has noted, "you can hardly turn over a shovelful of dirt in California without finding an earthquake fault." The real cause of the failure was found to be the nature of the "bedrock" material on which the dam was built. When this material was tested by immersion in water, it was observed to change into a mushy mass. In the urgency of Los Angeles's need for water, Mulholland had overlooked ordinary engineering precautions. An investigating committee concluded that ". . . the dam was constructed without a sufficiently thorough examination and understanding of the foundation materials upon which the dam was constructed." A coroner's jury found Mulholland responsible for an error in engineering judgment, but no charge of criminal negligence was made. The jury added that construction of a great dam "should never be left to the sole judgment of one man, no matter how eminent. . . ."

References
"Essential Facts Concerning the Failure of the St. Francis Dam," *Proc. Am. Soc. Civil Engrs.*, October 1929, pp. 2147–2163.
R. A. Nadeau, *The Water Seekers* (Bishop, Calif.: Chalfant Press, 1974).

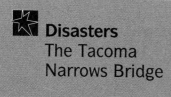

Disasters
The Tacoma
Narrows Bridge

The Tacoma Narrows Bridge was completed on July 1, 1940. Four months later, it tore itself apart during a windstorm and crashed into the waters of Puget Sound.

The bridge had been opened to traffic amid a great fanfare of publicity. It was the third-longest suspension bridge in existence, with a central span of 2800 feet. The piers for its suspension towers reached 247 feet below the water's surface, and their construction had been very difficult because of the strong tidal currents. The bridge was graceful, beautiful, and slender. In fact, it was the slenderest bridge ever built, and therein lay the source of the trouble.

The width-to-span ratio for the Tacoma Narrows Bridge was 1 to 72. The ratios for some other large bridges are: Verrazano Narrows, 1 to 41; Golden Gate, 1 to 47; Mackinac, 1 to 56. Why was the bridge so slender? The answer given by one of the investigators, after the collapse, was: "The builders of this bridge, being limited in funds, were anxious to build as inexpensive a bridge as possible in order to build any bridge at all." Ten years later, when a replacement bridge was finished, its width-to-span ratio was 1 to 47.

Besides being slender and flexible, the bridge's design exposed a flat side to the winds sweeping through the narrows. Motorists crossing the bridge had early noticed that the bridge tended to "gallop" in a cross-wind, and some motorists drove across the bridge just for the thrill. On the day of the collapse, the bridge's undulations became so violent that it was closed to traffic. Thus, when it broke, no one was killed.

Not until wind-tunnel tests had been carried out using scale models did it become clear what had happened. The tests showed that with a moderately strong wind blowing against a flat surface, so-called Karman vortices would be formed, first at the upper edge and then at the lower, in an alternating pattern. At a particular wind velocity, the alternations of the vortices coincided with one of the natural frequencies of the bridge, causing it to oscillate. When the wind velocity held steady at the right velocity long enough the oscillations were able to build up to destructive levels.

When the bridge was rebuilt in 1950, not only was it less slender than before, it was 58 times stiffer. It also no longer presented a flat surface to the wind, but had an open structure instead. In addition, it had a system of open slots in the deck, to let the wind pass through. The bridge no longer galloped in the wind like the earlier one, which had been nicknamed "Galloping Gertie." In fact, it was so solid, it soon was given the name "Sturdy Gertie."

References
J. P. Den Hartog, *Mechanical Vibrations*, 4th ed. (New York: McGraw-Hill, 1956), p. 308.
Engineering News-Record, November 14, 1940, p. 10, and *Engineering News-Record*, August 23, 1963.
"Failure of the Tacoma Narrows Bridge," *Proc. Am. Soc. Civil Engrs.*, December 1943, p. 1568.
"Tacoma Narrows Bridge Number II," *Pacific Builder and Engineer*, October 1950, pp. 54–57.

An interesting perspective on risk is provided by the list in Table 5.1. Whether or not one agrees with the ranking of risks, it is hard to avoid the inference that the greatest risks in our lives are not necessarily those on which we focus most of our attention.

Figure 5.1
The Tacoma Narrows Bridge in the process of breaking up. The first three photographs dramatically show how violent the oscillations were. The last photograph shows the catastrophic collapse. (United Press International Photos)

■ 5.2 PRODUCT LIABILITY

Not too many years ago, the doctrine of *caveat emptor* prevailed in seller-buyer relationships. *Caveat emptor* means "Let the buyer beware." In other words, once you bought something, you were stuck, whether the object was defective or not. Today matters have shifted so far the other way that some have said the doctrine has now become "Let the *seller* beware."

The issue that concerns us in this chapter is not whether the product does or does not function as intended. We have to come to expect proper function as a matter of course: A washing machine should be capable of getting clothes clean; a refrigerator should be capable of keeping food cold. (Such matters come under the doctrine of *implied warranty*, to which we will refer only briefly.) Our major concern is with the doctrine of *strict liability*, which covers product defects and consumer safety.

Engineers obviously have a great deal to do with safety: As described earlier, codes of ethics have been deeply occupied with such matters. What is not so evident

Table 5.1 Loss of Life Expectancy (Days) Due to Various Causes

	Days Lost
Being an unmarried male	3500
Being a coal miner	1100
Being 20 percent overweight	900
Dangerous job—accidents	300
Motor vehicle accidents	207
Average job—accidents	74
Job with radiation exposure	40
Accidents to pedestrians	37
Natural radiation	11
Medical x-rays	8
All catastrophes combined	3.5
Reactor accidents (UCS)	2
Reactor accidents (Rasmussen)	0.02

Source: B. L. Cohen and I-Sing Lee, *A Catalog of Risks* (unpublished manuscript, University of Pittsburgh).
The last two items assume that all U.S. power is nuclear. *UCS* refers to Union of Concerned Scientists. *Rasmussen* refers to the Reactor Safety Study (WASH-1400, Nuclear Regulatory Commission, October 1975), commonly referred to as the "Rasmussen Report" after the director of the study. Note that the risk associated with the first line, "Being an unmarried male," probably is partially a reflection of the facts that highway accidents are the leading cause of death for people ages 1 to 44 (and are more likely for males than for females) and that, if one dies young, one obviously has lost a large number of days in terms of life expectancy. [Among people of all ages, the leading causes of death are, in order, (1) heart disease, (2) cancer, (3) stroke, and (4) auto accidents. The first three involve mostly older people; if one dies at an advanced age, very few days of life expectancy are lost.] It is probably also worthwhile to point out that there has been a marked improvement in highway safety over the years. The fatality rate per billion vehicle miles shows the following trend between 1970 and 1982: 47.4 (1970); 33.5 (1975); 33.8 (1980); 28.0 (1982). (See K. L. Campbell, "Risk Assessment: Highway Travel," *Mechanical Engineering*, November 1984, pp. 46–51.)

is that the engineer's responsibility to the public for safety coincides with a responsibility to protect the engineer's employer. Up to the present time, legal suits involving product liability have been brought not against product *designers* (unless they were negligent, of course), but against manufacturers and sellers. Some of these suits have been enormously costly, so an action by an engineer to protect the public is at the same time an action to protect the engineer's employer from great financial loss.

Since engineers bear such heavy responsibilities, it is appropriate here to look into the kinds of things courts have said in the past about product safety, to provide at least a small measure of guidance to engineers in making their design decisions.

Strict Liability

Most people have a general idea of what is meant by the term *negligence*: A certain standard of conduct is implied, the accused party did not conform to this standard, and the failure to conform caused injury to someone. But in the case of *strict liability*, no questions of negligence arise. The American Law Institute has published the following rules regarding *strict tort liability* (a "tort" is a wrongful civil act committed by one person against another, other than a breach of contract):

(1) One who sells any product in a defective condition unreasonably dangerous to the user or consumer or to his property is subject to liability for physical harm thereby caused to the ultimate user or consumer, or to his property, if
   (a) the seller is engaged in the business of selling such a product, and
   (b) it is expected to and does reach the user or consumer without substantial change in the condition in which it is sold.

(2) The rule stated in Subsection (1) applies although
   (a) the seller has exercised all possible care in the preparation and sale of his product, and
   (b) the user or consumer has not bought the product from or entered into any contractual relation with the seller.

A number of things should be noted about the foregoing rules. For one thing, they apply to *sellers*. But since manufacturers are necessarily sellers, they apply to manufacturers too. Note the use of the words *defective* and *unreasonably dangerous*. These are difficult words to pin down in court. What seems to one person to be a defect may not appear so to another, and people also rarely agree on whether something is "reasonable" or "unreasonable." For example, knives are dangerous, but are they unreasonably so? A car moving at a legal speed on a freeway is certainly dangerous if it hits something, but is it unreasonably dangerous?

Unreasonable Danger

Any attempt to define the word *unreasonable* merely leads to the need for more definitions. For example, the American Law Institute attempts to define "unreasonably dangerous" as

> ... dangerous to an extent beyond that which would be contemplated by the ordinary consumer ... with the ordinary knowledge common to the community. ...

Now we are left with the job of finding out what *ordinary* means, instead of *unreasonable*. A couple of cases may help to show where some courts have drawn the line between reasonable and unreasonable. In a Georgia case, a child was injured while riding after dark on a bicycle that was not equipped with a headlight or reflector. The child's father attempted to collect damages from the seller of the bicycle but lost. The court ruled that the father was aware of the danger and that the seller had no duty to protect against obvious, common dangers. On the other hand, in a South Carolina case, a child was injured by lawn mower blades, and the plaintiff did collect. The court agreed that the danger was obvious, but said that the seller of the lawn mower nevertheless had a duty to improve its safety. In this case, the court was swayed by the gravity of the danger and by the relative ease with which the danger could have been reduced, either by better design or through warnings. For engineers, it is especially important to note that the viewpoint of the courts may be undergoing a significant change—away from judging cases on the basis of what an "ordinarily prudent person" might do and toward judging on the basis of what an "occasionally careless person" might do.

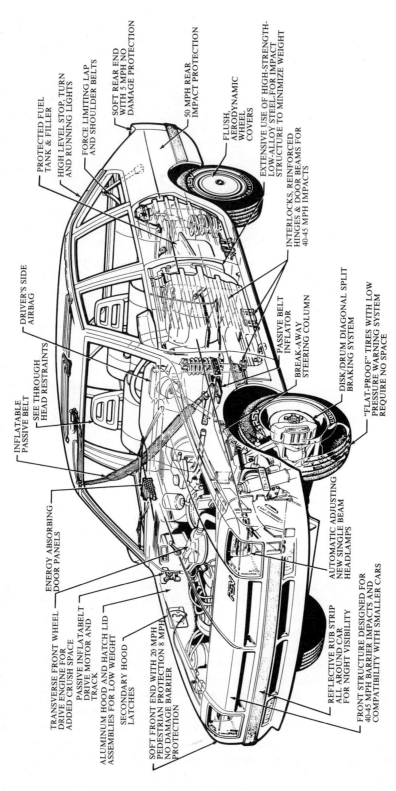

PROTECTED FUEL TANK & FILLER

HIGH LEVEL STOP, TURN AND RUNNING LIGHTS

FORCE LIMITING LAP AND SHOULDER BELTS

SOFT REAR END WITH 5 MPH NO DAMAGE PROTECTION

50 MPH REAR IMPACT PROTECTION

FLUSH, AERODYNAMIC WHEEL COVERS

EXTENSIVE USE OF HIGH-STRENGTH-LOW-ALLOY STEEL FOR IMPACT STRUCTURE TO MINIMIZE WEIGHT

INTERLOCKS, REINFORCED HINGES & DOOR BEAMS FOR 40-45 MPH IMPACTS

DRIVER'S SIDE AIRBAG

SEE THROUGH HEAD RESTRAINTS

INFLATABLE PASSIVE BELT

PASSIVE BELT INFLATOR

BREAK-AWAY STEERING COLUMN

DISK/DRUM DIAGONAL SPLIT BRAKING SYSTEM

"FLAT-PROOF" TIRES WITH LOW PRESSURE WARNING SYSTEM REQUIRE NO SPACE

ENERGY ABSORBING DOOR PANELS

TRANSVERSE FRONT WHEEL DRIVE ENGINE FOR ADDED CRUSH SPACE

PASSIVE INFLATABELT DRIVE MOTOR AND TRACK

ALUMINUM HOOD AND HATCH LID ASSEMBLIES FOR LOW WEIGHT

SECONDARY HOOD LATCHES

SOFT FRONT END WITH 20 MPH PEDESTRIAN PROTECTION 8 MPH NO DAMAGE BARRIER

REFLECTIVE RUB STRIP ALL AROUND CAR FOR NIGHT VISIBILITY

FRONT STRUCTURE DESIGNED FOR 40-45 MPH BARRIER IMPACTS AND COMPATIBILITY WITH SMALLER CARS

AUTOMATIC ADJUSTING NEW SINGLE BEAM HEADLAMPS

Figure 5.2

Developed jointly by Chrysler Corporation and Calspan Corporation, this research safety vehicle features a reinforced structure; soft front end, rear, and interior; run-flat tires; driver air bag; and passenger "inflatabelt." The vehicle weighs less than 3000 pounds. (Courtesy of Chrysler Corporation)

Concealed Dangers

Courts are likely to be especially critical when it comes to concealed dangers. In a Florida case, the manufacturer of a folding aluminum lounge chair was held liable in a case in which a person lost a finger. If the chair was not fully unfolded, and if the user sat in it and simultaneously had a finger in the danger spot, the further unfolding of the chair under pressure would neatly amputate the finger. In another case, a worker became entangled with a manufacturing robot, had difficulty reaching the on–off switch, and then died of a heart attack. In a somewhat more complicated case, one person died and another was severely injured by accidental overdoses of radiation from a machine used to treat cancer. The problem here was in the software. The operator initially set the machine for too high an energy level. She recognized her error and entered the instructions to reset for a lower level, but she did this so quickly that the operating system for the machine was still engaged in setting itself up for the higher level. The result was that the machine ignored a part of her input, and the patient received 2.5 times as much radiation as intended. There was no reason why the operator should have been aware of such a possible outcome; this was an error of the software designer.

Another case involved a tractor operator who slipped on a step while dismounting, injuring his back. It was shown that the operator could not see the step in dismounting, that mud had been thrown up by the tractor treads and had collected on the step, and that the step had no antiskid material on it. The tractor manufacturer lost. The case contains an especially powerful message for design engineers. During the trial it was revealed that the tractor manufacturer had violated its own design guidelines, which required, among other things, that steps and ladders be designed to minimize the accumulation of mud and debris and that steps be provided with antiskid material.

Obvious Dangers

It has already been pointed out that, for the manufacturer of a knife, there is no duty to protect against the sharp edge because the danger is obvious. However, just because a danger is obvious does not mean that the designer is relieved of responsibility. In a California case involving an earth-moving machine, a person standing behind the machine was run over and killed. The machine had a blind spot in the rear caused by the presence of a large engine box, and no rear-vision mirrors gave a view of the blind spot. The manufacturer was held liable, the court saying that, even though the danger was obvious, it was unreasonable.

The manner in which the doctrine of product liability has evolved over the years becomes apparent as one reflects on what has happened with automobiles during recent decades. As recently as the 1960s, plaintiffs often lost cases involving injury during an auto accident. The doctrine in such cases essentially said that if a user was injured in an accident, the injury occurred because of an unintended use: The intended purpose of an automobile does not include its collision with other objects. Today, however, the doctrine has moved far beyond that. Any use of an automobile creates a certain risk of accidents, and manufacturers are expected to design them in such a way as to protect occupants from aggravated harm.

Figure 5.3

This huge tunneling machine, developed for use in German coal mining, bores a hole 6 meters in diameter. A cylindrical roller thrust bearing 1.66 meters in diameter takes up the cutting force of 6400 kilonewtons. (Courtesy of Mannesmann-Demag and FAG Ball and Roller Bearing Engineering)

Foreseeability

It sometimes seems as if the designer of a product is expected to foresee everything and to detect every manner in which a product might be used or misused. As unfair as it may seem, this is approximately the case. One student of product liability has said:

> Many engineers and designers assume their products are safe if they meet all regulations and standards and if moving parts are protected with a guard when the product leaves the factory. In fact, a review of litigation resulting from product failure shows that accidents are caused less often by mechanical failure than by the designer's failure to consider how the product would be used.

A case that illustrates this point involved a "printer-slotter" machine. It was frequently necessary to separate the machine into two freestanding halves with a 30-inch passage between them, to gain access to the inside of the machine for changing dies. When this occurred, one half of the machine was "dead," whereas the other half continued to be supplied with power so that an automatic roller-washing operation could

Figure 5.4
A space shuttle on the launching
pad. Space projects are usually
thought of as utilizing mostly mechanical and electrical engineers,
but engineers of nearly every kind
are involved. Some of the launching
facilities, for example, have required the development of unusually
challenging civil engineering structures. (Courtesy of NASA)

continue while the dies were being changed. The object, of course, was to minimize
changeover time; otherwise, the machine would have to be shut down completely,
lengthening the time required for changeover. During such a changeover, a workman
walked between the two halves of the machine. He was carrying a rag, and the rag
was caught by the moving rollers on one half of the machine. His arm was drawn
into the rollers and subsequently had to be amputated. He won his court case on
the grounds that the machine should have been equipped with a safety switch (an
"interlock" switch) that would cut off all power to *both* halves of the machines whenever it was opened.

There are conflicting views regarding the state of affairs with product liability.
Some feel that the whole matter has gone too far, but the prevailing view is that the
current system compensates victims who have suffered grievous harm and discour-
ages bad design. From the standpoint of the individual engineer, the following might
represent a reasonable course of action to guard against product liability:

1. Be aware of, *and adhere to*, your company's own design standards. In court,
   the most damaging testimony could be the admission that your company did
   not follow its own standards.
2. Be aware of, *and adhere to*, any industry design standards. Among these might
   be such things as Underwriters' Laboratories procedures and ASME (American Society of Mechanical Engineers) pressure vessel codes. Even if the standards are unwritten and informal, it is important to know about and follow
   them.

3. Make design choices that lean toward safety. Often these come at little or no cost penalty and merely require foresight.

4. If a severe cost penalty is associated with safety, make sure the issue is thoroughly understood and resolved at the highest levels in your organization. It is tempting to adhere to the view that an unsafe product should never be placed on the market, but one should remember that even the word *unsafe* requires amplification and definition. (Imagine, for example, that you work for an automobile manufacturer, and you are aware that autos kill 40,000 people a year. This means that autos are probably the most dangerous objects with which we commonly come in contact. As an engineer bearing a great responsibility for public safety, must you demand of your employer that your autos be built like tanks in order to be safe? Then, recognizing that your tanklike autos will probably mangle quite a few pedestrians, does it now follow that you must demand that your employer cease manufacturing autos altogether?)

The obligation to protect the public from needless harm exists. So does the obligation to provide the public with useful goods and services. Some products are indeed

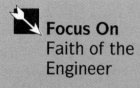

Focus On Faith of the Engineer

I AM AN ENGINEER. In my profession I take deep pride, but without vain-glory; to it I owe solemn obligations that I am eager to fulfill.

As an Engineer, I will participate in none but honest enterprise. To him that has engaged my services, as employer or client, I will give the utmost of performance and fidelity.

When needed, my skill and knowledge shall be given without reservation for the public good. From special capacity springs the obligation to use it well in the service of humanity; and I accept the challenge that this implies.

Jealous of the high repute of my calling, I will strive to protect the interests and the good name of any engineer that I know to be deserving; but I will not shrink, should duty dictate, from disclosing the truth regarding anyone that, by unscrupulous act, has shown himself unworthy of the profession.

Since the Age of Stone, human progress has been conditioned by the genius of my professional forebears. By them have been rendered usable to mankind Nature's vast resources of material and energy. By them have been vitalized and turned to practical account the principles of science and the revelations of technology. Except for this heritage of accumulated experience, my efforts would be feeble. I dedicate myself to the dissemination of engineering knowledge, and especially to the instruction of younger members of my profession in all its arts and traditions.

To my fellows I pledge, in the same full measure I ask of them, integrity and fair dealing, tolerance and respect, and devotion to the standards and the dignity of our profession; with the consciousness, always, that our special expertness carries with it the obligation to serve humanity with complete sincerity.

(Prepared by the Ethics Committee, Engineers' Council for Professional Development)

so dangerous that they should be banned altogether, although judgment must always play a part in reaching such decisions. If the issues become large enough, law-making bodies will make the final judgments, as they have with certain pesticides and chemicals. In the meantime, this discussion describes the environment in which engineers must honorably pursue their profession.

■ 5.3 WHISTLE BLOWING

"Whistle blowing" has received a great deal of attention in recent years. The whistle blower is the courageous person who discovers negligence, fraud, or waste in his organization and goes to battle against the establishment, risking job and reputation on behalf of the greater good.

Perhaps the classic whistle-blowing case is that of Ernest Fitzgerald, who exposed the huge cost overruns on the C-5A cargo plane. Fitzgerald's forthright testimony before a congressional committee earned him the wrath of his superiors in the U.S. Air Force. He was stripped of his duties, assigned to trivial projects, and, four months later, laid off. Fitzgerald fought back, but it took him four years of legal battles before a federal court concluded that he had been wrongfully fired and ordered the Air Force to rehire him. He continued to press through the courts for full reinstatement to his former position and, 13 years after the testimony that got him into trouble, he finally succeeded.

The B.F. Goodrich Case

A celebrated case of whistle blowing involved alleged falsification of data by B.F. Goodrich Co. in the design of a wheel brake for the A7D light attack aircraft. The case is especially noteworthy because it appeared as the lead chapter in a book entitled *In the Name of Profit*, published by Doubleday & Co. The book was a selection of the Book-of-the-Month Club and was read by hundreds of thousands—perhaps millions—of people, who were thereby influenced in their views of engineers and their corporate employers. The book presented a number of cases in which corporations purportedly placed their allegiance to profit ahead of their responsibility to society. The chapter involving the B.F. Goodrich brake was entitled "Why Should My Conscience Bother Me?" and was written by one of the participants in the incident, Kermit Vandivier. Subsequently, the case was included in several texts on engineering ethics, which gave it special prominence.

We will start with Vandivier's account of what occurred.

In 1967 B.F. Goodrich received a contract to supply wheels and brakes for the A7D. The prime contractor was Ling-Temco-Vought (LTV). Goodrich is a major supplier of aircraft wheel brakes and was very anxious that the brake be successful. A part of Goodrich's proposal that made the brake attractive was that it had only four disks instead of the usual five. This made it very compact and light—premium advantages for a military aircraft brake. It also meant that the development program would be very demanding and fraught with risk.

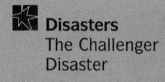

Disasters
The Challenger
Disaster

On January 28, 1986, the space shuttle *Challenger* blew up, killing seven people and putting the country's space program into deep freeze. After exhaustive investigations, it was finally decided that the accident had been caused by hot combustion gases blowing past one of the seals in the rocket motors. It was concluded that the seal had been unable to do its job properly because of the unusually low temperature in Florida on the day of the launch—about 18°F. Claims were made that the management of Morton Thiokol, the manufacturer of the motors, had possessed engineering information that cast doubt on the seals but had decided to launch anyway.

This much is well known because the accident had heavy treatment in the news media. There is a great deal more to the story. Nothing about it is secret; everything is revealed in the report of the presidential commission on the accident.

In an earlier flight, a postflight examination of the solid rocket motors (these are recovered from the ocean after each flight) showed that hot gases had blown past the primary seal in one of the joints, although the gases had been stopped by a secondary seal. (Each joint was provided with two seals because the motor sections could distort under pressure; if the primary seal failed, then the secondary seal would maintain joint integrity—or so it was thought.) Attempts had been made to improve the situation, but without much result.

What seems to not be generally known is that this distortion of the joint was revealed by testing as far back as 1978, eight years before *Challenger* blew up. Furthermore, the joint was labeled as unacceptable by NASA engineers, and a redesign was demanded. Morton Thiokol argued back, apparently on the grounds that the joint design was derived from that used in many successful flights on the *Titan* rocket and had proven to be very reliable. Furthermore, Thiokol argued, the new design was better than *Titan*'s because it used two O-rings for sealing instead of one. This was perceived as conservative, with safety as the rationale.

Throughout the whole sorry tale that follows, it is worth bearing in mind that the proven reliability of the *Titan* system dominated the thoughts of almost everyone and finally proved to be the source of the self-mesmerization that was the core of the problem. One of the witnesses before the presidential commission put it

A young engineer $1\frac{1}{2}$ years out of college, Searle Lawson, was assigned to have test models of the brake constructed and to run the tests that would be necessary to qualify the brake for service. Vandivier, a technical writer, was assigned to prepare the qualification report.

During the tests, the brake temperatures often reached 2000°F, causing the linings to fail. When Lawson reported this, he was told that success would simply be a matter of finding the right lining material. He was directed to continue testing.

After nearly a year, the brake was still a failure, in spite of the fact that irregular procedures had been used in an effort to get the brake to qualify, such as dismantling the brake between simulated stops to repair warpage, using fans to cool the brakes, and making deliberate miscalculations to indicate lower braking pressures than were

this way: "... All of the people in the program I think felt that this solid rocket motor ... was probably one of the least worrisome things in the program."

But there were a couple of important differences from the *Titan* design. First, the *Challenger* joint was more flexible than the one in the *Titan,* so more distortion took place. Second, when the sections of the rocket were put together, there was a gap leading from the O-rings right up into the combustion area. Zinc chromate putty was placed into this gap to protect the O-rings from the hot gases. After a number of tests in which the joints apparently were successful, the viewpoint gradually became "After all, it's worked for years in the *Titan,* so isn't it the safest thing to do?"

Everything might still have come out all right if it had not been for another step taken to ensure safety: The joints had to be tested under pressure to make sure the O-rings were sealed in properly. But the people conducting the tests wanted to be sure they were really testing the O-rings and not just the putty, so they put on enough pressure to blow a hole through the putty. There you have it: The putty was essential to protect the O-rings from the flames, but to test the rings a hole was put in the putty that would lead directly from the flames to the O-rings. Even after these facts of the test became known, there was this belief: If the first

ring does not hold, the second one will. Besides, there was the comforting image of *Titan*'s reliability.

Some observers have declared that it was not just the low temperature before *Challenger*'s flight that caused the accident; it was a fundamentally bad joint design. Furthermore, complaints about the joint had been lodged eight years before. Why was nothing done?

In reflecting on this issue, engineering students can learn the following lessons: (1) in an enormously large organization, it is easy for decision-making to get lost in the cracks; in fact, this was the principal conclusion drawn by the members of the presidential commission, and they directed their harshest criticisms at the NASA administrative structure; (2) it is very easy for engineers to fall into the comforting belief that they are following a conservative course (the Thiokol engineers apparently thought so), when in fact they are deviating into dangerous territory. Disasters are easy to create; safety comes hard.

References
R. M. Boisjoly, "Ethical Decisions—Morton Thiokol and the Space Shuttle Challenger Disaster," *Winter Annual Meeting of American Society of Mechanical Engineers,* Boston, December 13–18, 1987.
Report to the President by the Presidential Commission on the Space Shuttle Challenger Accident (Washington, D.C.: 1986).

actually applied. Both Vandivier and Lawson complained to their superiors about these irregularities but were told to continue with the tests. Nothing about all this was communicated to LTV, according to Vandivier.

After 13 failed qualification tests, the word came down from above that, on the fourteenth attempt, "it's going to qualify." Following the fourteenth test, a formal qualification report was issued by Goodrich that certified that the brake had qualified. Vandivier asserted that this report contained nearly 200 pages of graphics and other displays, nearly all of which he had falsified. His motive, he said, was fear for his job and the welfare of his family if he refused to cooperate.

During the flight tests that followed, difficulties developed with the brake. On one occasion the brake plates welded themselves together, causing the plane to skid.

Upon receiving this news, Vandivier went to the FBI and told them his story, to be followed by Lawson a few days later. Subsequently, both Vandivier and Lawson resigned. (Vandivier, incidentally, became a newspaper reporter, and Lawson went to work for LTV.)

At about the same time that Vandivier and Lawson resigned, Goodrich withdrew the four-disk brake and substituted a five-disk brake at its own expense.

This accounting of events would be enough to shock almost anyone, and Senator William Proxmire was shocked enough when he heard of the case that he requested the General Accounting Office (GAO) to investigate. Senator Proxmire also subsequently conducted hearings before the Subcommittee on Economy in Government, of which he was chairman. Vandivier and Lawson both testified at this hearing, as did representatives of Goodrich.

The GAO sent a team to the Goodrich plant; they spent a week going over the raw test data and examining the procedures that had been employed. The GAO report, in brief, said the following:

1. In some instances, Goodrich's test procedures did not appear to comply with specification requirements.
2. Some discrepancies in the data had occurred that "might be considered significant."
3. Opinions differed as to the degree of danger that might have been caused to the pilot; no significant aircraft damage had been reported because of the brake.
4. Goodrich had replaced the four-disk brake with a five-disk brake at no additional cost to the government. No schedule delays had taken place.
5. The procedures used by both the prime contractor (LTV) and the Department of Defense had been inadequate to protect the interests of the government.
6. The Air Force had protected the interests of the government by withholding approval of the qualification report.

During the hearings before Senator Proxmire's committee, representatives from Goodrich made the following statements in reply to Vandivier's allegations:

1. Goodrich was the manufacturer of the brakes for the Boeing 707, 720, 727, Lockheed L-1011, C-5A, General Dynamics F-111, North American XB-70, and many other civilian and military aircraft. It was a leading manufacturer of brakes and could have no possible incentive to falsify information or produce a defective brake.
2. A total of 267 test flights had taken place. In two of the flights a problem developed with the four-disk brake, in which the brake plates "fused slightly" (Goodrich's words) at low speeds. Following the test flights Goodrich conducted further tests at its factory, employing the full brake system (including the hydraulic system and antiskid mechanism, which were not made by Goodrich) instead of just the brake alone. From these tests, Goodrich concluded that the four-disk brake was not going to work and substituted the five-disk brake.
3. There were indeed some deviations from the specifications. For example, some of the stop times were longer than allowed by the specifications. These devia-

tions had been discussed with and approved by LTV, and LTV had notified the Air Force, contrary to Vandivier's assertion that LTV had been kept in the dark. Vandivier simply had not known what was going on.

4. It was true that "rolling stops" had been employed, which was contrary to the specifications. The Air Force itself had established such a precedent when testing brakes at Wright Field. This variation had been discussed with LTV in advance and approved by them.

5. Test data had not been changed or falsified, but in evaluating the data, the project design engineer had arrived at judgments regarding the validity of the data.

6. Some of the discrepancies noted by GAO came about because certain data points were recorded by two different methods: one set of data was visually recorded, whereas the other was recorded by computer. If the data points differed, the computer-recorded points were used. This had caused some revisions in the report.

7. In one case, the test brake had been disassembled and a spacer plate replaced by a pressure plate. This was done so that adjusters could be installed and the correct pressures could be applied.

8. Some of the pressure values had been changed because the use of adjusters during some of the stops had caused incorrect values to be recorded. These values had to be revised to compensate for the adjuster pressures and thus show the correct values.

9. In another case, some of the temperature figures had to be revised because the thermocouple leads had become interchanged on the slipring.

So there we have it. A situation that, when presented from only one point of view, looked like a shocking case of corruption instead turned into a confusing set of complexities. If Goodrich did something wrong, it probably was to deviate from the specifications, but here it apparently had the willing cooperation of LTV and the Air Force. Of course, some are ready to believe that anything said by a whistle blower is true and anything said by a corporation is false, but most people take a more

Figure 5.5
The wing of the Boeing 757 in this load test fixture was purposely deflected upward 11.6 feet above its normal position before it finally broke. (See the scale, calibrated in feet, in the center of the picture.) The final breaking load was 68 percent higher than the design limit load. (Courtesy of Boeing Commercial Airplane Company)

balanced view. Clearly, Goodrich, in its proposal for a four-disk brake, had moved right to the edge of technical feasibility and was trying very hard to produce a technological breakthrough. In this it failed and had to fall back on the proven technology of the five-disk brake.

Perhaps the most we can learn from the case is that there are always two sides to every story. In fact, it may sometimes be difficult to figure out who the "good guys" are. Vandivier and Lawson certainly felt that they had done the right thing in taking their information to the FBI, and indeed they had done exactly what they were supposed to do, according to the duty spelled out in codes of ethics: "notify the proper authorities and withdraw from further service on the project."

But if we can believe the testimony of the Goodrich official, Vandivier and Lawson had only a part of the story. It is clear that the lodging of a charge of wrongdoing against other persons carries with it an enormous responsibility, because those other persons have rights too.

The DC-10 Case

A prime example of a case in which a danger was clearly perceived but the whistle was *not* blown is that of the DC-10. On March 3, 1974, 9 minutes after a DC-10 took off from Paris, a cargo door blew open. The resulting decompression of the cargo compartment caused the cabin floor to collapse. In the DC-10 the control systems are routed through the floor (instead of through the ceiling, as in the 747), so when the floor collapsed, control of the ailerons and rudder was lost. The plane crashed, killing all aboard.

The case revolves around the design of the cargo door latches. In 1968, McDonnell Douglas, the manufacturer of the DC-10, gave a subcontract to Convair Division of General Dynamics to perform the detail design of the fuselage of the DC-10, including its cargo doors. (Such arrangements are common in the aircraft industry, whereby one manufacturer subcontracts portions of the design to others.) Initially, the specifications required the use of hydraulic actuators to drive the cargo door latches, but later Douglas told Convair that electric actuators were to be used instead, since they were lighter.

The distinction between hydraulic actuators and electric actuators is important to the case. If, for some reason, the latches failed to seat properly, a fairly moderate degree of internal fuselage pressure would force a hydraulic latch to open. Such a sequence of events, if it occurred, would take place at fairly low altitude and pressure differential, and the resulting decompression would not be disastrous. The aircraft could land safely. But with electric latches, if they did not fully seat, the cargo doors would probably be forced open at a time when a much higher pressure differential had developed, so that the resulting decompression would be catastrophic. In the year prior to the certification of the DC-10 for service, there had been five instances in which cargo doors on DC-8's and DC-9's had blown open during flight. Because the DC-8's and DC-9's were equipped with hydraulic actuators, the accidental openings occurred under moderate pressure differentials, and the planes landed safely.

During the design of the fuselage, Douglas asked Convair to prepare a Failure Mode and Effects Analysis (FMEA) for the cargo door system. In the FMEA, Convair pointed out that little reliance could be placed on the use of warning lights to indicate the presence of an improperly latched cargo door, because the warning lights themselves were subject to malfunction. Even less reliance could be placed upon the alertness of ground crews to check such things as fully closed latches, because any such procedure was subject to human error. Beyond this, the FMEA described a number of scenarios of events that could produce a hazard to life. One of the scenarios involved the failure of a latch to seat properly, producing an explosive decompression of the cargo hold and leading to collapse of the cabin floor and loss of control of the aircraft.

In 1970, during the ground tests of the first DC-10, the aircraft was being subjected to pressurization tests when a cargo door suddenly blew open. The accident was blamed on the failure of a mechanic to close the door properly. But even before the accident Douglas had decided that further safeguards were needed to prevent a cargo door from blowing open. A hand-driven locking handle was provided, together with a small door near the locking handle. The door was supposed to stand open until the locking handle was operated. A member of the ground crew would look through the door to be sure that the locking pins had fully seated; then the door would be closed. However, there was no necessary connection between the two. If the locking handle failed to seat properly, the door could still be closed. Thus, the DC-10 system depended upon a member of the ground crew to understand and follow the correct steps.

The DC-10 was certified for use in 1971. In 1972, a cargo door blew out on a DC-10 over Windsor, Ontario, and part of the cabin floor collapsed. Miraculously, the pilot was able to land the aircraft safely.

But now the Federal Aviation Administration (FAA) was officially in the picture. FAA personnel prepared a draft of an "airworthiness directive," which would have ordered Douglas to take certain actions before the DC-10 could resume operation. But the airworthiness directive was not issued. Instead, an informal agreement (referred to later as the "gentlemen's agreement") was worked out between Douglas and the head of the FAA, detailing certain steps that Douglas would take to correct the situation. (These steps apparently included the provision of small inspection windows, deeper latch engagement, and stiffer latch linkages.) The aircraft that crashed near Paris did not receive these modifications, even though the inspectors responsible for the aircraft had certified that the modifications had been done. Also relevant is the fact that the instructions printed on the aircraft were in English. The ground crew member who was responsible for cargo door closure knew several languages, but English was not one of them.

Shortly after the Windsor blowout, Convair's director of product engineering wrote a memo to his superiors declaring that the safety of the cargo door latching system had been progressively degraded since the inception of the program. After the first blowout (the one that occurred during ground testing), Convair had discussed possible corrective action with Douglas, including the possibility of providing "blowout panels" in the cabin floor: If the cargo area suddenly lost pressure, these panels would be blown out, but the cabin floor would not collapse. Instead of that alternative, however, Douglas had opted for the small doors that would provide visual observation of the manual latching system. The memo also suggested strengthening the cabin floor so that it could resist a sudden decompression, but this would add 3000 pounds in weight—a serious matter for an aircraft. The memo contained this paragraph:

> My only criticism of Douglas in this regard is that once this inherent weakness was demonstrated by the July 1970 test failure [the ground test], they did not take immediate steps to correct it. It seems to me inevitable that, in the twenty years ahead of us, DC-10 cargo doors will come open and I would expect this to usually result in the loss of the airplane.

Convair management responded to the memo essentially as follows:

1. Exception to Douglas' design philosophy had not been registered by Convair at the beginning of the program. By not taking exception, Convair in effect had agreed that a proper design philosophy was the design of a safe cargo door latching system, in lieu of designing a stronger floor or providing "blowout panels."
2. A design philosophy involving a safe latching system would satisfy FAA safety requirements.
3. Douglas had unilaterally redesigned the cargo door latch system and had previously rejected the proposal of "blowout panels."
4. Convair management had been informally advised that Douglas was making corrections to the latching system and was reconsidering the provision of "blowout panels."

As a result of the above, Convair management made no formal communication to Douglas. Their justification was that the arguments advanced in the memo were already well known to Douglas, and it was not likely that any additional actions would be produced beyond those that they understood were already taking place. In addition, an adversarial relationship between Douglas and Convair had developed, in which Douglas was seeking to shift the costs of redesign to Convair. It was feared that if Convair now questioned a design philosophy with which it had originally concurred, Douglas would use this as a further pretext for shifting all the costs of redesign to Convair.

The case makes sorry reading. There certainly were opportunities to provide a safe locking system. For example, when the inspection door was provided to go with the manual latching system, it could have also been provided with an absolute interlock, which would have prevented the closure of the door unless the locking pins were fully seated. Probably the saddest feature of the story is that the FAA failed so badly in its role of safety watchdog.

From the viewpoint of engineering students, the case has value not only because it underscores the responsibility for safety borne by engineers, but because it introduces the idea of "design philosophies" and what those philosophies might mean with respect to safety. Also, the case brings home a point of view that many people overlook: A safety system that depends upon human beings to execute a series of actions, as in making visual checks to see if latches are fully seated, is almost guaranteed to fail sometimes. Safety needs to be designed into the system itself.

Figure 5.6
The Interstate 880 viaduct in Oakland, California, after the 1989 Loma Prieta earthquake. The photograph dramatically shows how the steel reinforcements failed at the points where the columns holding the upper deck were attached to the lower deck, allowing the two decks to "pancake" together. (Courtesy of CALTRANS)

Exercises

5.1 Examine newspapers and magazines—especially magazines such as *IEEE Spectrum* and *Mechanical Engineering*—for articles describing product failures. If these cases have not already been decided in the courts, try to predict their likely outcomes, based upon what you now know about product liability.

5.2 Using the information gained from Exercise 5.1, work out design solutions that are better than those

actually employed and that might have prevented the subsequent troubles.

5.3 In his book *Unsafe at Any Speed*, Ralph Nader said engineers "subordinate whatever initiatives might flow from professional dictates in favor of preserving their passive roles as engineer-employees." Write a short essay on this statement, saying whether you think it is justified. Before preparing your essay, it might be wise to obtain a copy of Nader's book from your library so that you can see the full scope of his arguments.

5.4 In the discussion of negligence in this chapter, it was pointed out that conforming to a standard of conduct is a critical issue. In establishing in court just what a particular standard might consist of, evidence regarding customary practices is often introduced. See if you can think of some cases in which customary practices might not be acceptable to a court. (One example: On almost every street and highway in America, it is customary for a large fraction—maybe even a majority—of the motorists to exceed the speed limit. If you had an accident and were speeding at the time, do you believe a court would accept your argument that your speeding should not be a factor because it is "customary"?)

5.5 In one of the landmark cases establishing strict product liability, a combination power tool was being used as a lathe. A piece of wood being held in the lathe by setscrews came loose, flew up, and hit the operator in the head, inflicting serious injuries. Other product-liability cases have also involved setscrews that did not do their jobs. Examine the issue of employing setscrews as hold-down devices. (Some engineers believe that setscrews should never be used under any circumstances.) See if you can come up with some ideas for hold-down devices that would be safer and more reliable than setscrews but would not be excessively costly.

5.6 A case once occurred in which the plaintiff's shoes slipped on a wet laundromat floor, causing injury. The plaintiff brought suit against the seller of the shoes, alleging that they were dangerously inclined to slip. The plaintiff did not win this case, because it was shown that the shoes were no more slippery than any others and thus were not defective. Suppose you are a judge, and a case comes before you in which the shoes *are* more slippery than most. As the judge, what would you do? Since you are a good judge, you worry about how your ruling might stand up under further appeal, and you also worry about how any "slipperiness standard" you establish might be used for future guidance of shoe manufacturers. Since you know that some kinds of shoes, such as dancing shoes, are *supposed* to be slippery, what would you do about all this?

5.7 Examine magazines in your library, such as *IEEE Spectrum, Engineering News-Record,* and *Mechanical Engineering,* to see if you can find some examples in which the liability of engineers is involved in matters of product safety, structural design, and so on.

5.8 Every year between 40,000 and 50,000 people are killed in auto accidents in the United States alone, yet people throughout the world, and not just in the United States, show every sign that they consider the benefits of the auto to be worth the risk. In the case of nuclear power, only three people have been killed in the United States to date (by a test reactor in Idaho), yet many people are frightened of nuclear power, knowing that a major reactor accident could kill thousands. Many people die every year in the process of producing food (farming is dangerous), and thousands of coal miners have been killed so far in the twentieth century. In each case cited there are benefits, but there are also terrible risks involving the deaths of many. Discuss these issues. Are we entitled, as a matter of moral behavior, to accept any benefits whatsoever, if the benefits entail risks to others? How can we continue to live practical lives in such an imperfect world and satisfy ourselves that we are also living in accordance with an acceptable code of ethical behavior?

5.9 An incident that allegedly occurred in the aircraft industry goes as follows: An aeronautical research engineer from Company A conducted tests of a certain aircraft tail assembly configuration in his company's wind tunnel and knew that devastating vibrations could occur with that configuration under certain circumstances, leading to destruction of the aircraft. Later, at a professional meeting, Company A's engineer hears an engineer from Company B, a competitor, describe a tail assembly configuration for one of Company B's new aircraft that runs the risk of producing just the kind of destructive vibrations Company A's engineer had discovered in his tests. What are the ethical obligations of the engineer from Company A? Presumably there is an obligation, as a matter both of morals and of law, to maintain company confidentiality regarding Company A's proprietary knowledge. On the other hand, engineers are supposed to bear a responsibility for public safety and welfare. If the engineer from Company A remains silent, Company B may not discover the possibility of destructive vibrations until a dreadful crash occurs, killing many people. What would you do if you were the engineer from Company A?

5.10 Suppose you are employed as a junior engineer by a company that is designing a brand-new rail transit system, involving new technology, such as computer control of trains. You hear from an older engineer that the company is rushing ahead too fast with the program, acting under the pressure of completion deadlines, and is not conducting a sufficient number of tests. As a result, according to the older engineer, latent flaws in the system may not be uncovered until the system is in public operation, perhaps causing death or injury to innocent passengers. Deeply concerned, you seek the counsel of your supervisor, who tells you that the other engineer, although competent, is an alarmist and that if his counsel were heeded, so much testing would occur that the project could be delayed to the point where it would be cancelled. Cancellation, the supervisor points out, not only would result in the loss of jobs for a lot of people but would also deprive the public of the benefits of the transit system. He also adds that all technological progress entails uncertainty and risk and that insistence upon absolute safety might have the effect of stopping further technological progress. However, you (having read this book) know that there *have* been examples in which the engineers pressed too hard on the margin of safety, with resultant disasters, such as the St. Francis Dam, the Malpasset Dam, the Tacoma Narrows Bridge, *Challenger*, and the DC-10. What is your opinion of this kind of controversy, which appears frequently in real life? Admittedly, as a junior engineer you could not do much about the circumstances in this particular example, but someday you will be in a position of authority, perhaps a chief engineer. How would you, as chief engineer, resolve such a controversy into a wise course of action for your company? Or, better,

how might you be able to influence the course of events at the beginning of a project so that such issues do not arise at all?

5.11 Write a statement of your philosophy as to the goal of human activity. Is it to maximize economic gain? Is it to survive? Is it to maximize individual pleasure? If so, what kind of pleasure? Is it to ensure the survival of the species? Is it to ensure a balanced natural ecosystem on Earth? Something other than any of these?

5.12 There is much concern in the world about unemployment and the role labor-saving devices such as computers, industrial robots, and farm machinery may have played in contributing to such unemployment. As a consequence, some groups have argued that the creation of labor-intensive jobs should be a priority for our society and that programs to improve productivity will merely result in increasing unemployment. Make a list of jobs that, if made more labor-intensive, would improve the quality of life for those job-holders; also make a list in which a loss in quality of life would result. As an alternative to the possibility that automation will always increase unemployment, consider the realistic prospects of retraining displaced workers. For example, if farm automation produces a surplus of migrant farm workers, but there is simultaneously a shortage of factory workers in semiconductor plants, what are the obstacles to be overcome in retraining farm workers to be semiconductor workers? Include in your analysis such factors as relocation from farm to city, education levels, training costs, and the like. Consider also whether you believe the retrained, displaced worker would agree that an improvement in quality of life had indeed occurred. What other, perhaps more beneficial, alternatives to the foregoing scenario can you think of?

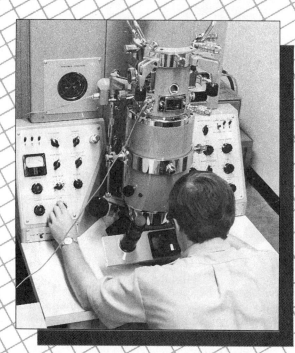

Photographs (top right) courtesy
of University of California, Davis;
Bethlehem Steel Company (center); University of California,
Davis (bottom right).

6

The Branches of Engineering

It has been said that there are really only four basic branches of engineering—chemical, civil, electrical, and mechanical—and that all the other fields are mixtures of the "Big Four." The truth in this statement is stretched a bit. For example, persons in the fields of computer engineering, industrial engineering, materials science, mining engineering, and nuclear engineering may feel strongly that there are important subject areas in their fields that do not derive from the Big Four.

In this chapter, brief descriptions are provided to give a feel for what the fields, or branches, of engineering and engineering technology are like. Also, some information is presented regarding the principal engineering societies that are related to those fields.

No book can examine everything all kinds of engineers do, because the variety of things they do is almost without limit. Also, the field of engineering studied by a student in college may not turn out to be the field he or she winds up in afterwards. To give some inkling of the variety of engineering, and also of the surprise of some people concerning where their careers took them, several boxed items in this chapter describe the true experiences of some engineering graduates. (The names of the individuals and their employers have been changed or purposefully left imprecise so as to protect their privacy.)

■ 6.1 THE FIELDS AND BRANCHES OF ENGINEERING

Aeronautical and Aerospace Engineering

Aeronautical engineering is the application of scientific principles to flight or other movement in the atmosphere. Aerospace engineering is concerned with flight outside the atmosphere. The specific focus of the aeronautical engineer is upon the design, development, and manufacture of airplanes, VTOL (vertical take-off and landing) aircraft, helicopters, hovercraft, and high-speed ground transportation systems. Special fields within aeronautical engineering are aerodynamics, which includes the lift, drag, and loading effects of air acting on aircraft surfaces; propulsion, including propellers, jets, and rockets; controls; and aircraft structures. The last-named field—structures— uses the same basic principles as structural engineering, but since weight is so important in a structure that must fly, there is a tendency to use much more sophisticated analysis techniques for an aircraft than for an earthbound structure. The field of *wind energy* often is considered to be a part of aeronautical engineering because it depends upon the aerodynamics of large propeller-like structures.

The term *aerospace engineering* is sometimes used more to designate all engineering activities in the broad industrial sector known as *aerospace* than to denote a specifically defined field of engineering. Vehicles that must fly into space are designed through an intimate and inseparable blending of virtually every known engineering field, with the possible exception of such areas as water storage and crop production. In fact, there are probably more electronics and computer engineers involved in the design of aerospace vehicles than any other kind of engineer, because of the importance of those two activities in space flight. But engineers engaged in the design of

Figure 6.1
Engineers of almost every kind—aeronautical, mechanical, structural, electrical, and materials—must combine forces to design a new airplane. (Courtesy of Boeing Commercial Airplane Company)

rockets, ultralight structures, exotic materials, robotics, control instruments, and re-entry aerodynamics are all equally vital. In all of the activities listed, the development of aerospace systems almost always lies at the cutting edge of technical knowledge and therefore is closely associated with scientific research.

Agricultural Engineering

Agricultural engineers apply engineering principles to problems of food and fiber production, storage, and processing; animal and plant environments; agricultural waste management; irrigation and drainage; and other phases of agriculture and related industries. The field is unique in that it requires a general understanding and appreciation of the biological, soil-management, and environmental aspects of agriculture, plus a knowledge of engineering fundamentals. Some of the subfields are *food engineering and agricultural processing*, which is concerned with the development of systems for food manufacturing and operations such as sorting, cleaning, size reduction, storage, refrigeration, and drying; *irrigation and drainage*, emphasizing water quality, on-farm irrigation and drainage system design, water law, hydrology, and hydraulics; *packaging and handling*, which involves preservation of product quality during handling, shipment, and storage and requires knowledge of the physical properties of foods, as well as of packaging materials; *power and machinery*, which

Figure 6.2
Modern harvesting machines are marvels of engineering. Not only does the machine perform several tasks simultaneously, but the machine operator can run it from an air-conditioned cab. (Courtesy of Deutz-Allis Corporation)

has to do with the development of field machines and their effects upon soils, crops, and people; and *structures and environment*, in which structures are analyzed for their effectiveness in providing optimum environments for animal production, product storage, and greenhouse production and which includes agricultural waste management, environment modification, and micrometerology.

Forest engineering is a relatively new field which is generally treated as an option within agricultural engineering. Besides completion of a program in agricultural engineering, the student is expected to do a substantial amount of course work in forestry. The purpose of a forest engineering program is to prepare the graduate to work in forest management, with emphasis upon physical installations and forest materials management. Included are tree nursery practices, reforestation, brush control, waste management, recreational development, soil and water conservation, and aesthetics. The purpose throughout is to develop systems to maintain forest lands in permanent production, help to preserve or produce a desirable environment, and ensure a supply of wood products to society. Forest engineers are quick to point out

**Focus On
Forest Engineer
(Forest Products
Company)**

Theresa graduated with a B.S. in forest engineering, which was an option within an agricultural engineering program. Her required course work included several courses in forestry, with summer forestry camp, as well as all of the regular courses in agricultural engineering. While she was an undergraduate student, she worked part-time for her university on research projects involving forest tree seed handling and forest soil compaction studies. She even became a coauthor of a published research paper on these topics—a rare accomplishment for an undergraduate.

Her first job following graduation was with a large forest products company in the Northwest. Some of her early assignments were on projects involving forest regeneration. She worked on equipment to control and conserve pollen and ran productivity and survival tests on prototype seedling containers. Subsequent assignments included a major analysis of her company's ability to become self-sufficient in energy use by using forest biomass residues and advanced technology.

Theresa's work drew favorable attention from management, and about four years after graduation she was selected for the company's management training program. A part of this program involves participation in direct first-line supervision. Hence, she was transferred to a southern state and became a foreman in one of the company's sawmills, supervising 29 people on the "green drops and stackers."

Theresa has continued to work part-time on a master's degree in wood science, although her ability to complete the degree obviously will depend upon the locations to which her company sends her. In the meantime, her supervisor has indicated that her performance has been outstanding, and it is clear that Theresa, like many engineers before her, is on her way to a management career.

Figure 6.3
A prototype tree-climbing machine utilized by forest engineers in their research. A tree with superior genetic characteristics is selected in the forest. The tree-climbing machine is then sent aloft to cut off limbs in order to access the seed-bearing cones. The machine can climb a pine or fir tree 2 feet in diameter, cutting off limbs up to $2\frac{1}{2}$ inches in diameter as it goes. (Courtesy of University of California, Davis)

that forests represent one of our few *renewable* natural resources and thus deserve special attention.

Aquacultural engineering is the application of engineering principles to fisheries and the culture of aquatic organisms. It deals with the environment that must be created in order to grow aquatic animals for food, including elements such as water temperature, dissolved chemicals, nutrients, and the like. It also deals with the structures and systems required to produce those environments, such as tanks, filters, pumps, and aerators. Some of the species of interest are salmon, trout, catfish, carp, shrimp, oysters, mussels, clams, lobsters, crayfish, and seaweed. At this writing, there are no educational programs leading to degrees in this field, which is usually included within agricultural engineering.

Automotive Engineering

Automotive engineering is concerned with the design, construction, and utilization of self-propelled mechanisms, prime movers, and related equipment. It is an interdisciplinary field, encompassing the traditional fields of mechanical, electrical, and chemical engineering as they apply to land, sea, air, and space vehicles. In addition to passenger cars, trucks, and buses, automotive engineers are active in aircraft, aerospace systems, farm machinery, off-highway vehicles, fuels and lubricants, transportation safety, vehicle emissions, marine propulsion, automotive electronics, air cargo, and energy conversion.[1] Educational programs in automotive engineering usually are treated as options in mechanical engineering.

---

[1] Description courtesy of the Society of Automotive Engineers.

Biomedical Engineering

Biomedical engineering is a relatively new field, although its origins extend back into the 1940s. It consists essentially of three segments: (1) the application of engineering concepts to biological phenomena—a basic research activity sometimes referred to simply as "bioengineering"; (2) the application of engineering to the development of instrumentation, materials, artificial organs, prosthetic devices, and the like, frequently called "medical engineering"; and (3) the application of engineering to the improvement of health delivery systems, sometimes called "clinical engineering." Of these, medical engineering is the oldest. Very few biomedical engineering programs at the *undergraduate* level exist in the United States. Mostly, preparation for these fields requires graduate work, at least to the master's level. Hybrid programs at the Ph.D. level, with equal emphasis in engineering and in the life sciences, have been developed for the research-type biomedical engineer.

Clinical engineers may be called on to offer the following kinds of solutions to unexpected problems: development of gas scavenging systems to remove residual anesthetic gases from operating rooms, decontamination of compressed air lines in a hospital to get rid of infectious organisms, and provision of adequate electric power for upgrading of x-ray laboratories. More often, however, clinical engineers are concerned with the totality of instrumentation (frequently meaning computers) needed for the operation of a modern hospital.

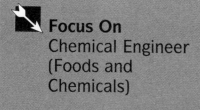

Focus On
Chemical Engineer (Foods and Chemicals)

Immediately after graduation, Anne went to work for one of the country's largest firms manufacturing foods, soap products, and chemicals. Her first job was as a project engineer, but at the end of two years she was moved into a supervisory position.

As a project engineer, Anne quickly became responsible for projects with budgets totaling a half million dollars, among them the installation of a programmable logic controller for a detergent-making operation, coordination of changes in an acid-handling process, and improvement of the company's sewer system. Her responsibilities included cost estimating, "interfacing" with vendors and construction workers, budgeting, and scheduling.

Currently, Anne is a first-line supervisor of a large industrial chemical process plant. She directly supervises the 23 operators who run the plant, which operates 24 hours a day, 7 days a week. Anne is responsible for personnel matters, such as scheduling and disciplinary problems, as well as for tracking and optimizing the production rate and yield of the process plant. On a typical day, Anne may be found troubleshooting a distillation column, making safety audits, consulting the home office on a technical matter, writing a report on some aspect of the system, deciding what to do with a batch of off-quality product, or doing just about anything else that needs to be done to keep her plant running.

Anne says: "In general, I find that manufacturing management is quite challenging. The days pass quickly, and it seems that every day I learn something either about the plant or about management."

Chemical Engineering

Chemical engineers apply chemical, physical, and engineering principles to solve problems and supply materials for our technology-based civilization. Their work ranges from pharmaceuticals to fuels to industrial chemicals. It includes energy conservation and pollution control. The emphasis on chemistry and the chemical nature of everything we use are what make chemical engineers different from other kinds of engineers. A rapidly growing field for chemical engineers is the semiconductor industry. This field is usually thought of as the domain of electronics and computer engineers, but the actual production problems in the semiconductor industry are overwhelmingly chemical in nature.

Commonly, chemical engineers first define a problem or a product; next, develop a process to do what is needed; and then design the plant to carry out the process. After the plant is started up, chemical engineers typically manage operations, oversee equipment maintenance, and supervise control of product quality. They troubleshoot the "bugs" that hinder smooth operations, and they plan for future expansions or improvements. Their training and knowledge qualify them to market the products from a plant, the processing equipment for it, or even the complete plant itself. Chemical engineers are largely responsible for the production of the fuels we burn and the food we eat, the purification of water and air, and the recovery and use of the raw materials found in the oceans—and perhaps, in the future, in space. The chemical engineer develops industrial processes worth millions of dollars and economically works with tons of material. Often, the commercial success or failure of a product depends on the chemical engineer's designs for a pilot plant and the subsequent full-scale plant.

Some current and future areas for employment of chemical engineers include conversion of fossil fuels, petroleum processing, nuclear processing, solar and geothermal energy, electrochemical energy sources, water and air pollution abatement, metals refining and processing, water demineralization, food processing, fertilizer and

Figure 6.4

An engineer in a large oil refinery uses portable equipment to verify noise readings from permanent monitors mounted overhead. Corporations are obliged to conduct noise reduction and monitoring programs in their installations. In this refinery, noise levels were reduced below those associated with ordinary freeways. (Courtesy of Standard Oil Company of California)

pesticide production, pharmaceutical products, paints and inks, corrosion prevention, plastics, electronic components, vacuum technology, propulsion systems, instrument development, clinical engineering, artificial limbs and organs, and oceanographic research.[2]

Biochemical engineering is a discipline within chemical engineering that is concerned with microbial and enzyme processes. Thus, it is a link between biology and chemical engineering. Such things as industrial fermentation, wastewater treatment, and genetic manipulation (so-called genetic engineering) are included. Industrial examples are food processing, the manufacture of alcoholic beverages, and the production of vitamins, hormones, and antibiotics.

Civil Engineering

Civil engineers are responsible primarily for planning the design and construction of the nation's constructed facilities. They plan, produce, and help operate the nation's transportation system. They must develop, yet conserve, our water resources. They have a large role in designing the country's environmental protection relating to water, air, and solid wastes. They are involved in housing and urban development. They study the Earth's soils and oceans to better serve the human race.[3]

Highly visible results of the work of civil engineers are apparent in major structures such as highways, bridges, dams, and tall buildings. These are the fruits of *structural engineering*, which includes activities such as earthquake analysis, and *geotechnical engineering*, the analysis of the load-bearing qualities of foundation soils. *Architectural engineering* is closely associated with structural engineering. One of the main things architectural engineers do is combine the architectural design of building (form, appearance, function) with the structural needs, particularly in cases where the function and structure may be in conflict.

Another highly important field of civil engineering is the design and operation of wastewater treatment plants. Plants are required not only to handle domestic sewage—the domain of *sanitary engineering*—but also to treat the effluents of industrial plants, which often contain heavy metals and other toxic materials. A more modern version of sanitary engineering is *environmental engineering*, which is concerned not only with water quality, but also with air quality and land use. Environmental engineers typically deal with air pollution, water pollution, solid waste management, radiological hazard control, pesticide hazard control, and water supply. Their responsibilities include the management of the quality of rivers, lakes, estuaries, and ocean shore waters.

Transportation engineering involves the design of not only highways, railroads, and rapid transit systems, but also airports, harbors, waterways, and pipelines. "Transportation" includes the structural portions of the systems in question, as well as system analysis to determine the most effective ways of meeting the needs of the public, conserving resources (especially energy), and identifying the impacts

---

[2] Description courtesy of the American Institute of Chemical Engineers.
[3] Description courtesy of the American Society of Civil Engineers.

Figure 6.5
Floating a span into place on a new bridge across the Hudson River in New York. Final vertical positioning is accomplished by pumping water into the barges that carry the span, sinking them enough to lower the span onto the piers. The bridge is made of USS COR-TEN weathering steel. This material turns an earth brown color as it forms its own protective oxide coating, which minimizes maintenance. (Courtesy of United States Steel)

upon living patterns, land use, and the environment. In this context, transportation engineering overlaps the fields of urban and regional planning.

Water resources engineering obviously includes the design of dams and water delivery systems such as aqueducts and canals, but also is concerned with the total water cycle, including precipitation and runoff, groundwater resources, and drainage.

Construction engineering is sometimes considered a subfield of civil engineering and sometimes a separate field. Civil engineers often go into construction engineering, usually after they have had a few years of experience in an office that does structural design. Construction engineers are managers of construction and usually do not handle the design; nevertheless, their knowledge of civil engineering principles and of design practices is extremely useful to them in their management roles.

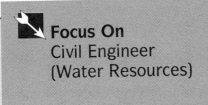

Focus On
Civil Engineer
(Water Resources)

Edward obtained a B.S. degree in civil engineering and has spent all of his ten years since graduation in the water resources field. Immediately after graduation, he went to work for a small consulting firm consisting of three engineers and three support personnel. During the nine years he spent with the firm, he was given increasingly responsible assignments, which included design, cost estimation, and construction supervision of earth-fill dams, canals, and pipelines. He was directly involved with his firm's clients and their

contractors and had to coordinate his projects with various governmental agencies, such as the Water Resources Control Board and the Division of Dam Safety.

In his tenth year, Edward moved to his present job, which is with the Water Resources Control Board in his state. His new job requires him to investigate complaints, unauthorized water diversions, cases of water pollution, and violations of permits. He makes field investigations of the alleged violations, prepares reports with recommendations, and gives testimony at hearings of the board.

Like many other engineers, Edward decided that a master's degree in business administration would be useful to him, and he earned his M.B.A. degree by attending classes part-time, in the evenings.

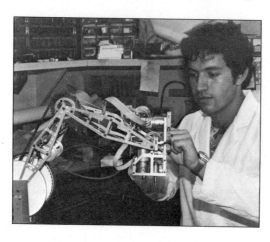

Figure 6.6
Many engineers are engaged in the design and development of new products. This electrical engineer is checking out the operation of a prototype robotics device.

Computer Engineering and Computer Science

The closely related fields of computer engineering and computer science overlap a great deal, and it is probably impossible to make a clear separation between them. Both are concerned with hardware and software (computer programs), although computer engineering has a greater emphasis upon the organization and design of computers themselves, whereas computer science has a greater emphasis upon computer use and applications. In the broad sense, computer science is the study of algorithms (procedures for solving problems), computability, and systems of computation. The mathematical field of *numerical analysis* overlaps computer science. In numerical analysis, approximate methods of analysis are substituted for the exact methods of calculus and differential equations. Often, an exact solution to a set of equations is not possible, but the step-by-step approximations of numerical analysis may give an approximate solution. Such procedures were impossibly lengthy and cumbersome before the advent of digital computers.

Cybernetics is a coined word and refers to the organization of mechanical and electrical systems for stability and purposeful action. Thus, it includes the field we commonly think of as *control systems* and overlaps the fields of *robotics* and *artificial intelligence*. Cybernetics should actually not be considered a subfield of computer science; it is included here because of its overlap with artificial intelligence, which *is* usually considered to be a subfield of computer science.

Electrical and Electronics Engineering

These two fields are difficult to separate, although the term *electrical engineering* is often used to embrace electronics as well as all other kinds of electrical engineering, whereas the field of *electronics engineering* clearly excludes such typical electrical engineering activities as power generation, power transmission, electric motors, and electrical systems for buildings and civic use. Electrical engineering is extremely broad, and the Institute of Electrical and Electronics Engineers (IEEE), to which most electrical engineers belong, is our largest engineering society. Perhaps the best way to indicate the enormous scope of the activities of electrical engineers is to list the many subfields in which the IEEE publishes regular periodicals:

Acoustics, speech, and signal processing
Aerospace and electronic systems
Antennas and propagation
Broadcast, cable, and consumer electronics
    Broadcasting
    Consumer electronics
    Cable television
Circuits and systems
Communications
Components, hybrids, and manufacturing technology
Computers
    Pattern analysis and machine intelligence
    Software engineering
    Computer graphics and applications
    Microprocessor technology
Control systems
Education
Electrical insulation
Electromagnetic compatibility
Electron devices
Engineering management
Engineering in medicine and biology
Geoscience and remote sensing
Industrial electronics and control instrumentation
Industry applications
Information theory
Instrumentation and measurement
Magnetics
Microwave theory and techniques
Nuclear science
Oceanic engineering
Plasma science
Power apparatus and systems
Power engineering
Professional communication
Quantum electronics and applications
Reliability
Solid-state circuits
Sonics and ultrasonics
Systems, man, and cybernetics
Technology and society
Vehicular technology

(Courtesy of Institute of Electrical and Electronics Engineers)

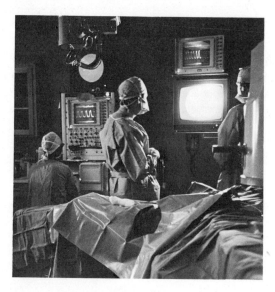

Figure 6.7

An operating-room scene showing electronic medical equipment. Electronics and computer engineers play an important role in the segment of biomedical engineering known as *clinical engineering*. (Courtesy of University of California, Davis)

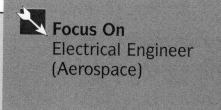

Focus On
Electrical Engineer
(Aerospace)

Richard has spent the four years since graduation working for one of the largest aerospace firms in the United States. His degree was in electrical engineering, and he coupled this with a second major in materials science. While he was an undergraduate, Richard had summer co-op jobs in the aerospace industry, working in stress analysis, avionics (aviation electronics), and reliability/maintainability.

His first job as a full-time engineer after graduation gave him responsibility for a set of specialized computer peripheral equipment used by the engineering department of his employer. Richard's job was to keep the equipment functional, correct any design flaws that cropped up, and make modifications from time to time. He points out that the underlying purpose of the assignment was to familiarize him with the entire system so that he could participate in future design work. Part of his responsibility was to "keep the paperwork moving," such as drawings, operations manuals, specifications, and procedures. His time was split about 50-50 between the hardware and the paperwork.

Richard finds his work personally rewarding and says that the knowledge he acquired in college was essential to carrying out his duties. However, there are certain things he wishes he could have learned more about while he was in college:

1. Digital noise (including both causes and remedies)
2. Planning and scheduling projects
3. Company organization and politics

Since going to work full-time, Richard has taken advantage of some of the courses and seminars offered by his employer. He has joined the IEEE and the AIAA (see Section 6.3 for information on engineering societies) and has become a registered electrical engineer in his state.

Engineering Mechanics

The topic of engineering mechanics is studied by all engineers because it is one of the branches of physics. Engineering mechanics is usually subdivided into *fluid mechanics* and *solid mechanics*. The former is basic to the study of energy production, pumps, compressors, turbines, jet engines, and aerodynamics. If the fluid is water and is in open channels, such as canals, rivers, and harbors, it is of special interest to civil engineers. The understanding of solid mechanics, on the other hand, underlies our knowledge of structures and machines. Solid mechanics is further divided into *statics* (forces at rest, as in a building or a bridge) and *dynamics* (forces coupled with motion, as in an automobile or a rocket ship). It can be seen that engineering mechanics is vital to both mechanical and civil engineers.

Educational programs in engineering mechanics frequently tend to be more theoretical than programs in mechanical or civil engineering. Mostly, they are at the graduate level. However, some universities also offer programs in engineering mechanics at the undergraduate level. Many of the graduates of these programs continue on to graduate school, but many also go directly to work in industry, usually as mechanical engineers.

Environmental Engineering

Environmental engineering may crop up as a subtopic in agricultural engineering, chemical engineering, civil engineering, or mechanical engineering. The agricultural engineer's concerns may include irrigation water quality, pesticide control, and disposal of animal wastes. Chemical and civil engineers are both interested in problems of water quality: The chemical engineer is more likely to be concerned about controlling pollution *inside* a plant and preventing its escape, and the civil engineer about handling of wastewater in municipal or industrial treatment plants. Civil engineers, in addition, often are called upon to solve problems of solid waste disposal, as in sanitary landfills. Mechanical engineers and chemical engineers share interests in air pollution problems. Here the focus is primarily upon preventing the escape of pollutants into the environment, with mechanical engineers tending to focus on engines and energy conversion processes while chemical engineers deal with those processes as well as refineries and chemical plants. Most environmental engineering programs are at the graduate level.

Industrial Engineering

Industrial engineering is an interdisciplinary field involving applications in industrial, service, commercial, and government activities. The American Institute of Industrial Engineers, Inc., views its members as engineers "concerned with the design, improvement, and installation of integrated systems of people, materials, and energy." Some of the major fields in which industrial engineers work are *plant design and engineering*, including planning of machines, equipment, labor supply availability, transportation of raw materials and finished goods, and provision of utilities and services; *systems engineering*, including analysis or design of production facilities, health care delivery

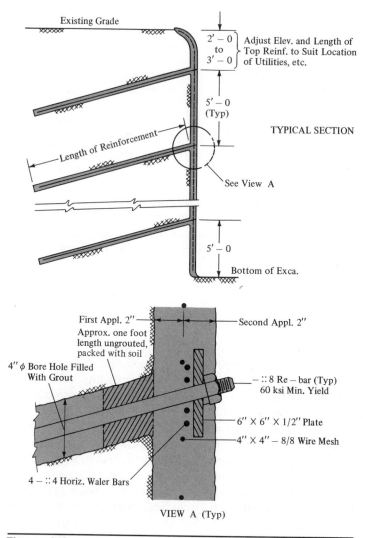

Existing Grade

2' – 0
to
3' – 0

Adjust Elev. and Length of
Top Reinf. to Suit Location
of Utilities, etc.

5' – 0
(Typ)

TYPICAL SECTION

Length of Reinforcement

See View A

5' – 0

Bottom of Exca.

First Appl. 2"

Approx. one foot
length ungrouted,
packed with soil

4" φ Bore Hole Filled
With Grout

Second Appl. 2"

– :: 8 Re – bar (Typ)
60 ksi Min. Yield

6" × 6" × 1/2" Plate

4" × 4" – 8/8 Wire Mesh

4 – :: 4 Horiz. Waler Bars

VIEW A (Typ)

Figure 6.8

Reinforced earth has recently been introduced in both Europe
and the United States as a way to maintain soil embankments at
steeper slopes than was previously possible. In this technique,
long steel reinforcing rods are placed in the soil embankment,
each surrounded by a 4-inch protective concrete casing. (Courtesy
of University of California, Davis)

systems, and other kinds of organizational operations; *production and quality*, with
special emphasis upon costs, productivity improvement, and quality control; *performance and operational standards*, with particular attention to the setting of goals and
the measurement of progress in achieving those goals; and *operations research*, which
is the technique of reducing facts to numerical values that can then be manipulated

Figure 6.9
An electronics engineer checks a highly magnified image of a new integrated circuit design.

mathematically as a help in making management decisions. Clearly, industrial engineering is a field that is closely allied with management.[4]

Manufacturing Engineering

The manufacturing engineer is the person who plans, develops, and optimizes the processes of production, including methods of manufacture, and designs tools and equipment for manufacturing. Typical areas of competence are tool design, manufacturing planning, numerical and computer control, computer-aided manufacturing, factory automation, robotics, and development of new production techniques. Examples of the latter are laser and electron beam welding, new structural adhesives, hot isostatic pressing (a method for consolidating metal and metal-ceramic powders to nearly final shape for high-performance parts), new cutting tools that may greatly increase cutting speeds, use of new composite materials for unusual strength-to-weight ratios, and increased use of high-performance abrasives that will improve quality and decrease costs.[5]

Marine Engineering

It has been said that marine engineering combines the fields of transport, warfare, exploration, and natural resource retrieval, which have one thing in common: operation in or on the water. *Naval architecture*—that is, the design, construction, and

---

[4] Description courtesy of the American Institute of Industrial Engineers, Inc.
[5] Description courtesy of the Society of Manufacturing Engineers.

operation of naval vessels—is a field closely related to marine engineering. Marine engineers have concerns such as propulsion of ships, steering, vibration, cargo handling, electrical power distribution, and air conditioning. Naval architects, on the other hand, are more interested in the hydrodynamic characteristics of ships, hull forms, control, stability, and the general usefulness of vessels for their intended purposes. The field of *ocean engineering* is considered, by some, to be a subfield of marine engineering, but it is described separately in this text.

Materials Science and Engineering

Materials *science* is concerned with the properties and behavior of materials. It seeks to understand, on a fundamental basis, why some materials are strong and others are not, why some are good conductors, why some corrode the way they do, and so on. Materials *engineering*, on the other hand, is directed toward the development of new or improved materials and to using the knowledge gained by materials scientists to achieve certain ends. Materials engineers and scientists are involved with the investigation of fractures, fatigue, corrosion damage, and radiation damage, as well as methods of production of new materials. Problems of this sort are spread throughout industry; materials are used everywhere, from automobiles to semiconductors, from surgical devices to nuclear power plants. *Corrosion engineering* is a subfield of materials science.

    Metallurgical engineering is, of course, concerned with metallic materials in the contexts described above. *Extractive metallurgy* is the process of producing metallic materials from their natural ores—for instance, making steel from iron ore, aluminum from clay, and magnesium from seawater. Among other things, metallurgists are concerned with material strength and workability, and especially with methods that can be used to improve these characteristics, such as alloying, heat treatment, and work hardening.

 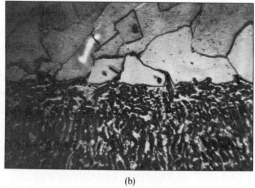

(a)   (b)

Figure 6.10

Photomicrographs of a fusion zone in electron-beam welding. A clean interface at the surface of the fusion zone tells the materials engineer working on this project that the weld is likely to produce little disturbance of the parent material. Magnification of (a) is 49×; of (b), 520×. (Courtesy of University of California, Davis)

Ceramic engineering is concerned with the development, production, and application of engineering materials such as glass, refractories, electronic coatings, and abrasives. Also included in this category are homely materials such as concrete, brick, tile, and porcelain. Ceramics can be described in a general way as materials that are nonmetallic and inorganic and that require high temperatures in their processing or use. (Materials scientists, however, usually prefer to classify materials as metals, ceramics, or plastics by considering their bonding structures.) Unusual applications for ceramics include glass fibers used in furniture, fabrics, boat hulls, and insulation; mammoth glass mirrors; glass- and ceramic-coated metal bearings and pistons; linings for furnaces; refractory materials for rocket nozzles; voltage insulators; record-player pickups; substrates for microcircuitry; porcelain surfaces for medical equipment; fuel elements for nuclear reactors; and artificial bones and prosthetic devices.[6]

Mechanical Engineering

Mechanical engineers are present in virtually every industry. Most fundamentally, they apply the principles of two of the fields of physics—mechanics and heat—to the design of machines. Heat is the principal form in which we use energy, so mechanical engineering is fundamental to all processes in which energy is created and used. The field of mechanics is divided into the subfields of *solid mechanics* and *fluid mechanics*, and both of these subfields are basic to mechanical engineering. A few of the myriad activities in which mechanical engineers engage are engine design, whether for autos, jets, diesel locomotives, or lawn mowers; rocket propulsion; combustion research; rapid transit systems, such as the new subways under construction in some U.S. cities; earth-moving machinery; air-conditioning systems; wind energy and solar energy devices; aerospace vehicles; turbines for electric power generation; automatic control for rolling mills; farm machinery; typewriters; computer input–output devices; prosthetic devices and artificial limbs; artificial hearts; precision measuring equipment; printing presses; food processing systems; and pumps, whether to circulate the water in a swimming pool or to drive the coolant through a nuclear power plant. The list is almost endless, but wherever a machine is needed to create a motion, to move a load, to create energy, or to convert it, there you will find mechanical engineers at work.[7]

One type of mechanical engineer is the *air-conditioning and refrigeration engineer*, who designs systems to produce controlled indoor environments. Most often, the designers of these mechanical systems are members of consulting firms and must work closely with architects and structural engineers in developing building plans. Closely related are the activities of the *sales engineer*, who works with designers to supply the best equipment for their needs, and the *building systems engineer*, who is responsible for the continued effective performance of the systems after they are installed.[8]

---

[6] Description courtesy of the American Ceramic Society.
[7] Description courtesy of the American Society of Mechanical Engineers.
[8] Description courtesy of the American Society of Heating, Refrigeration and Air-Conditioning Engineers.

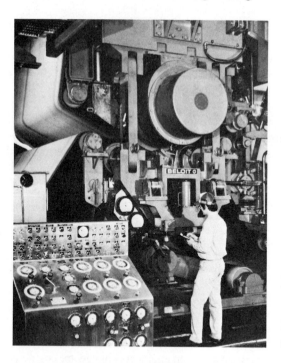

Figure 6.11
Energy conservation is often thought to pertain primarily to home insulation, lowered thermostat settings, and smaller cars. However, enormous amounts of energy are saved by industrial firms that regularly introduce energy-saving ideas. For example, mechanical engineers designed this huge piece of papermaking machinery specifically to remove extra water without slowing down the machinery. By increasing water removal by 5 to 7 percent, at least 20 percent less energy is required in the following dryer stage. (Courtesy of Weyerhaueser Company)

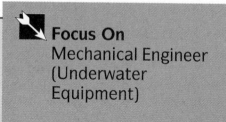

Focus On
Mechanical Engineer (Underwater Equipment)

Fifteen years after receiving his B.S. degree in mechanical engineering, Robert had progressively advanced through positions as mechanical design engineer, project engineer, program manager, and director of engineering, all with the same company, and had then resigned in order to start his own firm.

The projects Robert worked on were sophisticated underwater viewing systems and remotely controlled underwater vehicles. These devices are used to inspect underwater structures, including offshore oil platforms and pipelines; recover Navy torpedoes; and the like. One of his most rewarding experiences, he says, was the development of an advanced underwater vehicle, for which he served as project engineer and program manager. Subsequently, after he became director of engineering and thus was responsible for the entire engineering activity of his company, he expressed regret that he could no longer be as directly involved in technical activities as he would like.

When asked whether his education had equipped him well for his career, he replied that it had, and he expressed special gratitude that his studies had provided him with breadth and had avoided too much narrow specialization. He did have one regret, however. In his own words: "I have often wished that my education had included greater emphasis on electronics, especially with regard to the many traditionally mechanical problems which are now being solved with electronics."

Mining Engineering and Geological Engineering

The *mining engineer* works with mineral deposits of all kinds from the time of their discovery through their evaluation and production. Although the chief job of mining engineers is usually to get the most ore out of the ground for the least cost, they often work in many other areas, such as research, safety, design, environment, pollution control, and management. The work of the *geological engineer* is closely related to that of the mining engineer but may be more concerned with the exploration and mapping of ore bodies than with the structures and systems for ore recovery. Once an ore body is located, it must be precisely mapped by a series of test drillings to approximate its size, which way it is situated, and what it contains, and to establish the many other factors that determine which method of mining should be used.

In addition to the foregoing, mining engineers must plan for such things as the construction of roads, airstrips, mine buildings, and homes for the mine's employees and their families; ventilation systems for the mine; and safety features. As a final responsibility, after the mining ceases, provision must be made to restore the land for reuse for other purposes.[9]

Nuclear Engineering

Nuclear engineering is concerned with the release, control, and use of all aspects of energy from nuclear sources. It draws on all of the conventional branches of engineering, with the connecting link of nuclear radiation running among them. Most often, the term *nuclear engineering* applies to the design or operation of nuclear power plants; in the staff of each nuclear plant are nuclear engineers whose primary job it is to maintain technical surveillance over the operation of the nuclear reactor core and the handling of the nuclear fuel. However, the term is often expanded to include engineers concerned with the production or reprocessing of nuclear fuel, development of radiation standards, or the use of radiation materials for biological or industrial purposes. Some other areas in which nuclear engineers are active are development and testing of fast breeder reactors, control-rod interaction effects in liquid sodium coolants, instrumentation systems for nuclear plant safety, radioactive waste disposal, nuclear propulsion of naval vessels, and nuclear fuel management. *Controlled fusion* is an especially important new field in nuclear engineering because of the potential future importance of fusion energy sources to the world.

Ocean Engineering

Ocean engineering is a relatively new field and has strong connections to marine engineering. Broadly, the field is defined as the application of engineering principles to anything connected with the ocean. More specifically, ocean engineers are concerned with ocean exploration (development and use of submersible vehicles); ocean structures, such as deep-water drilling platforms; wave action on beaches, docks, buoys, moorings, and harbors; underwater pipelines; pollution control in oceans and estuaries; and resource recovery. Ocean engineers must learn about oceanography,

---

[9] Description courtesy of the Society of Mining Engineers of AIME.

of course, but also need to be knowledgeable about hydrodynamics, soil mechanics, corrosion, materials science, and structures. An ocean engineer who wants to emphasize underwater structures needs to know most of the things that civil engineers know. One who is primarily interested in undersea exploration needs to possess much of the knowledge of a mechanical engineer or, if the emphasis is upon instrumentation, an electrical engineer.

Petroleum Engineering

Petroleum engineering is concerned with the planning, development, and production of oil and gas fields, including economic factors and optimal recovery. Normally, the field does not encompass related functions such as refining and distribution, which also employ many engineers, but it does include the development of storage systems at the oil fields and of pipeline networks to deliver the product to refineries. The development of an oil and gas field requires gaining knowledge of the geological reservoir through exploratory drilling and working this information into an overall plan, which also takes into consideration economic factors such as capital availability, transportation requirements, and the like. Petroleum engineers require knowledge of the geology of reservoirs, characteristics of crude oil, flow of fluids through porous media, drilling techniques, estimation of reserves, probable recovery of total resources, and engineering economics.

Figure 6.12
Offshore oil-drilling platforms are complex structures that require the efforts of almost every kind of engineer—civil, mechanical, electrical, petroleum, and ocean. This platform is located in the Santa Barbara Channel, off California. (Courtesy of Standard Oil Company of California)

Some of the problems of recent concern to petroleum engineers have been secondary and tertiary oil recovery using chemical, water, and/or steam injection; retorting of oil shales; hydraulic fracturing of oil-bearing strata to improve flow; in situ recovery of oil from oil shales and coal fields; computer simulation of petroleum reservoirs; offshore structures for drilling and production in deep waters; branch drilling of directional holes from a single well; radio-frequency heating of oil shales for in situ recovery; measurement techniques for estimating the amount of oil remaining in developed fields; and design of new and more effective pumping systems.

Since most of the world's energy comes from the combustion of petroleum, the field of petroleum engineering is a vital one. Many employers hire mechanical and chemical engineers to work as petroleum engineers and then provide them with the additional education they need in order to function effectively in the field.

Systems Engineering

The field of systems engineering exists because of the fact that many outwardly different systems actually have common relationships which can be dealt with in similar ways. A complex system may contain a large mixture of elements, such as instruments, computers, control units, motors, pumps, and reactors. Yet somehow a system is usually expected to meet pre-established performance criteria, such as cost,

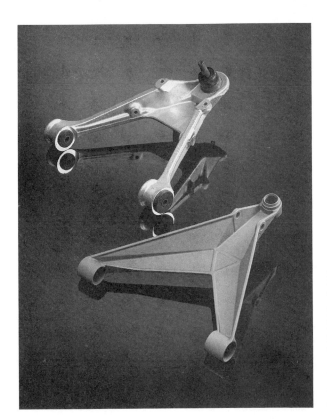

Figure 6.13
The steel control arm shown in the lower part of the photograph was produced by means of advanced vacuum-casting technology. It offers a 39 percent saving in weight over the forged aluminum arm shown in the upper part of the photograph (used in the Chevrolet Corvette) and meets the same performance requirements. (Courtesy of CWC Castings Division of Textron Inc.)

reliability, and maintainability. Analyzing a system and comparing it against the performance criteria *before the system is built* is the domain of the systems engineer. The methodologies typically employed are statistics and probability, network analysis, linear and nonlinear system modeling, control theory, computer programming, economic analysis, and formalized program management techniques.

■ 6.2 THE FIELDS AND BRANCHES OF ENGINEERING TECHNOLOGY

Generally speaking, the fields of engineering technology tend to parallel those of engineering. However, in recognition of the fact that engineering technologists often are more highly specialized in certain topics than are engineers and that they have more of an applied orientation, some American universities have adopted accredited technology curricula that do not have direct parallels in engineering. The following is a condensed list of accredited technology programs in the United States leading to bachelor's degrees.

Aeronautical engineering technology
Air-conditioning and refrigeration technology
Aircraft engineering technology
Apparel engineering technology
Architectural and building construction engineering technology
Biomedical engineering technology
Civil engineering technology
Computer engineering technology
Construction engineering technology
Design and drafting engineering technology
Electrical engineering technology
Electromechanical engineering technology
Electronics engineering technology
Energy engineering technology
Engineering technology
Environmental engineering technology
Fire protection and safety technology
Industrial engineering technology
Manufacturing engineering technology
Marine engineering technology
Mechanical engineering technology
Metallurgical engineering technology
Mining engineering technology
Nuclear engineering technology
Petroleum technology
Production management technology
Welding engineering technology

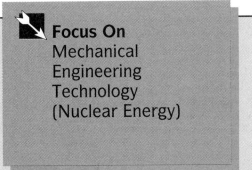

Focus On Mechanical Engineering Technology (Nuclear Energy)

Karl graduated eight years ago with a B.S. degree in mechanical engineering technology. After graduation, he went to work for a large piping fabrication firm. He has found—to his surprise—that his entire career so far has been spent in doing civil and structural work, which was not what he studied in college. He says he found it not too difficult to move from one area to the other, but he obviously had to learn a lot of new things fast.

Karl's first job was the structural design of seismic supports for piping systems in a nuclear power plant. Within a year and a half, he became a field engineer and supervised the installation of these systems in plants under construction. Two years after graduation, he shifted to a new employer and took on gradually increasing responsibilities relating to structural problems on nuclear piping systems, working on new plants in several states.

Karl is presently an engineering section manager for the firm he joined two years after graduation. He lives in the Southeast, and his responsibilities have included marketing, proposal writing, budgeting, scheduling, and development of a CAD (computer-aided design) system for his regional office.

Karl strongly recommends that a new student pick a field on the basis of personal aptitude and interest and not solely on the basis of future job opportunities. He also recommends that students gain some exposure to manufacturing processes, and he looks for this quality in the people he hires. He says, "It is a tremendous asset to any engineer to be aware of how a design matures from a concept to a drawing, and finally to the finished product. It can make the difference between an adequate design and a truly good one."

■ 6.3 ENGINEERING SOCIETIES

There are more than 400 engineering societies or related groups in the United States, including state and local organizations, and another 21 engineering societies in Canada. Probably no other profession is organized into such a great number of societies.

The remarkable proliferation of engineering societies often leads engineers to ask whether some degree of unity is desirable and possible. Actually, several major bodies work toward attaining unified action in the engineering profession. Prominent among these is the American Association of Engineering Societies, which carries out numerous programs for the benefit of the 772,000 members of its affiliated societies. Moreover, the National Academy of Engineering, the Accreditation Board for Engineering and Technology, and the National Society of Professional Engineers are all concerned with the engineering profession as a whole.

Figure 6.14
Electronics engineers often use computers to design their circuits. Here, an engineer is laying out an integrated circuit (IC) on a computer screen, using a stylus to control the shapes superimposed on the layers of the IC, which will later be exposed to the successive steps of diffusion, etching, and metallization.

Numerous personal advantages are derived from membership in a professional society, especially membership in one of the five major societies referred to as the *Founder Societies*:

> American Society of Civil Engineers
> American Institute of Mining, Metallurgical, and Petroleum Engineers
> American Society of Mechanical Engineers
> The Institute of Electrical and Electronics Engineers
> American Institute of Chemical Engineers

The Founder Societies have this name because in 1904 they founded United Engineering Trustees, Inc., in New York, which owns the United Engineering Center. They all have high standards for membership, and they account for a combined membership of 427,000 engineers (including some duplicate memberships).

Some personal advantages of membership relate to the regular receipt of journals and periodicals and to attendance at society meetings at a reduced cost. More important, it has been observed that members of professional societies have higher average earnings than nonmembers, although this particular statistic can probably be attributed to the fact that the more natural ability an engineer has, the more likely he or she is to join a society. The most important reason for joining one or more professional societies, however, is simply this: If the societies had never existed, the progress of the engineering profession (and, therefore, of all civilization) would have been much slower. The societies cannot exist or function unless engineers belong to them and support them. It follows, then, that the fundamental reason for belonging to a society is to participate in the activities of the profession and in producing the social benefits that flow from those activities.

American Association of Engineering Societies (AAES)

Twenty-eight American engineering societies are members of the AAES. The AAES was organized in 1980 as the successor to the Engineers Joint Council (EJC), which had similar purposes but did not have as broad a membership as its successor. This body acts as the representative of the engineering profession as a whole, in cases where such representation is appropriate, and helps develop public policy regarding the profession.

One of the important components of the AAES is the Engineering Manpower Commission (EMC). This group conducts studies on efficient engineer utilization; the future demand for engineers, scientists, and technicians; and the placement of engineering graduates. The EMC regularly conducts the most useful of all engineering salary surveys.

Accreditation Board for Engineering and Technology (ABET)

ABET is made up of 20 participating organizations as follows. (Prior to 1980, ABET was known as the Engineers' Council for Professional Development, or ECPD.)

American Academy of Environmental Engineers (AAEE)
American Congress on Surveying and Mapping (ACSM)
American Institute of Aeronautics and Astronautics, Inc. (AIAA)
American Institute of Chemical Engineers (AIChE)
American Institute of Industrial Engineers, Inc. (AIIE)
American Institute of Mining, Metallurgical and Petroleum Engineers (AIME)
American Nuclear Society (ANS)
American Society of Agricultural Engineers (ASAE)
American Society of Civil Engineers (ASCE)
American Society for Engineering Education (ASEE)
American Society of Heating, Refrigeration, and Air-Conditioning Engineers (ASHRAE)
The American Society of Mechanical Engineers (ASME)
American Society for Metals (ASM)
The Institute of Electrical and Electronics Engineers, Inc. (IEEE)
National Council of Engineering Examiners (NCEE)
National Institute of Ceramic Engineers (NICE)
National Society of Professional Engineers (NSPE)
Society of Automotive Engineers (SAE)
Society of Manufacturing Engineers (SME)
Society of Naval Architects and Marine Engineers (SNAME)

The purpose of ABET is to examine and accredit engineering curricula and technology curricula. In order to become accredited, a school must undergo a careful inspection of its program by an accreditation review team that is composed of educators and engineers from industry. Accreditation is granted by individual *curriculum*, and not for a school or a college as a whole. There are 250 institutions in the United States that have one or more engineering curricula approved by ABET. In Canada, accreditation is carried out by the Canadian Accreditation Board (CAB), which uses criteria similar to those of ABET. Canada has 28 universities with accredited engineering programs.

Curricula in engineering technology also are accredited by ABET. About 200 American institutions have such curricula. Most of those programs are at the two-year level, but about 90 institutions have accredited engineering technology curricula at the four-year level.

National Society of Professional Engineers (NSPE)

The NSPE has, as its primary objective, "the promotion of the profession of engineering as a social and an economic influence vital to the affairs of men and of the United States." As one means to this end, it has its headquarters in the national capital and takes a direct interest in legislation that affects the engineering profession. The NSPE is America's fourth-largest engineering society, with approximately 77,000 members. One of the activities of the NSPE is coordinating the annual Engineers' Week.

Fifty-four different state and territory societies (such as the Illinois Society of Professional Engineers and the California Society of Professional Engineers) are member societies of NSPE. Membership in one of these automatically makes one a member of the NSPE.

National Academy of Engineering (NAE)

It could almost be said that the engineering profession came of age on December 11, 1964, with the announcement that the National Academy of Engineering had been formed. As *Saturday Review* commented, "The half million or more engineers in the country had achieved a voice in public affairs on a prestigious level equal to that of the country's quarter million scientists."

The National Academy of Engineering (NAE) was set up by the joint efforts of the EJC, the Engineering Foundation, and the National Academy of Sciences (NAS) under the original congressional charter granted to the National Academy of Sciences in 1863. The two academies are autonomous and parallel, but representatives from both groups have declared their intention to operate them on a closely coordinated and cooperative basis.

Exercises

6.1 Without referring back to the text, see how many of the following professional engineering organizations you can identify from their initials: AAES, ABET, AIAA, AIChE, ASCE, ASME, IEEE, NAE, NSPE, ECPD, EJC. Which are Founder Societies? Which are concerned with the engineering profession as a whole? Which are obsolete names, and what are the new names that have taken their places?

6.2 Write a report on the advantages and disadvantages of having only a single professional society to which all engineers would belong.

6.3 Do you believe some new fields of engineering are currently taking shape that have not been described in this chapter? Select a possible new field, and write a brief description of it.

6.4 Contact an engineer who is working in a field of engineering that interests you. Write a brief report outlining his or her job activities, and compare it to the section in this book describing the same field.

6.5 For your own field of engineering, make a list of organizations that employ engineers in that field and their locations.

6.6 Contact some employers who hire engineers in the field you are interested in. From this information, prepare a report on job opportunities and trends in starting salaries for your field.

6.7 Make an investigation by consulting published references in your library, talking to members of your faculty, and/or engineers in industry, and write a brief report on the differences and degree of overlap among the fields of electrical engineering, computer engineering, and computer science.

6.8 This book has said that there are probably more electronics and computer engineers employed by the aerospace industry than any other kind of engineer. Investigate, and write a brief report explaining why this should be so, not only for the design of space vehicles, but also for military and civilian aircraft.

6.9 Write a report on the developing field of food engineering and its job opportunities.

6.10 Using your college catalog of courses as a guide, see if you can classify the subject areas of engineering science as being fundamental to one or more of the Big Four fields of chemical, civil, electrical, and mechanical engineering. Then, using such fields as computer engineering, industrial engineering, materials science, mining engineering, and nuclear engineering, see if you can list engineering and science topics for these that are not generally included within the Big Four.

6.11 Pick an engineering project (from magazines, books, or personal knowledge) and write a report showing how it is usually impossible to carry out a project of any size using *only* knowledge from a single field of engineering. Discuss how the knowledge of many fields of engineering will be required for the successful completion of virtually any project.

6.12 Prepare a report for your classmates describing the role of the engineering societies and the reasons why an engineer should become a member of a professional society.

6.13 Locate a large unabridged dictionary and look up the definition of *engineering*. Also, look on page 107 of this book, where a definition is given. Compare these definitions with what you already know about engineering and see if you can think of ways in which the dictionary definition falls short of the reality.

6.14 Examine the brief descriptions of real-life engineers given in this chapter and note how they either fit or fail to fit the dictionary descriptions of engineering.

6.15 Examine the brief descriptions of engineering activities given in this chapter to see if the elements in them fit your own conceptions of what engineers do. Are there any elements in these descriptions that you would find attractive and potentially rewarding? If not, then write a brief description of what would please you in an engineering career.

6.16 Engineers are expected to be concerned about the economics of their activities, as well as safety and environmental effects. Briefly discuss your views of these topics.

6.17 Discuss with a practicing engineer, or with a member of the faculty at your school, the pros and cons of the branch of engineering you are considering. Find out the reasons for the career choices he or she made.

6.18 Make an investigation, and write a discussion of the differences between engineering and other professions such as law, medicine, dentistry, architecture, accounting, and teaching.

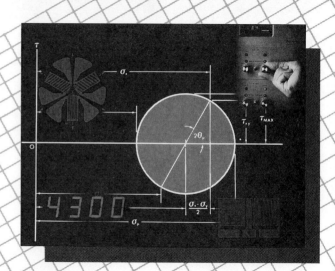

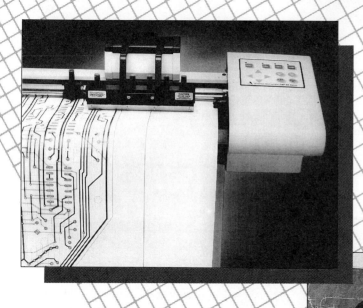

Photographs courtesy of Measurements Group Inc., Raleigh, North Carolina (top right); Houston Instrument Division of AMETEK, Inc. (center); and University of California, Davis (bottom right).

7

Presentation of Engineering Results

■ 7.1 UNITS

Engineers frequently use so-called *scientific notation*, involving powers of ten. Thus, we have

$$1\,000\,000 = 1 \times 10^6 \qquad 0.1 = 1 \times 10^{-1}$$
$$100\,000 = 1 \times 10^5 \qquad 0.01 = 1 \times 10^{-2}$$
$$10\,000 = 1 \times 10^4 \qquad 0.001 = 1 \times 10^{-3}$$
$$1000 = 1 \times 10^3 \qquad 0.0001 = 1 \times 10^{-4}$$
$$100 = 1 \times 10^2 \qquad 0.00\,001 = 1 \times 10^{-5}$$
$$10 = 1 \times 10^1 \qquad 0.000\,001 = 1 \times 10^{-6}$$
$$1 = 1 \times 10^0$$

and so on. First of all, we should note that the designations are mathematically correct. That is, 10^2 means 10×10, which is 100, and 10^6 means 10 multiplied by itself 6 times, which turns out to be 1 000 000. Likewise, 10^{-1} means $\frac{1}{10}$, or 0.1; 10^{-3} means to divide 1 by 10 three times, which certainly is $\frac{1}{1000}$, or 0.001. A simple way to handle this matter is to consider our starting point to be the numeral 1, with the decimal point just to the right of it:

$$1.0$$

Then, a positive exponent of ten represents the number of places the decimal point must be moved to the *right*, and a negative exponent of ten represents the number of places the decimal point must be moved to the *left*.

Using scientific notation, numbers may be expressed, for example, as follows:

$$6\,520\,000 = 6.52 \times 10^6$$
$$628 = 6.28 \times 10^2$$
$$0.00\,012 = 1.2 \times 10^{-4}$$

177

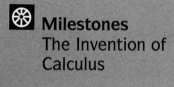

Milestones
The Invention of Calculus

Calculus—that bane of the undergraduate student's life—is the most powerful tool known for understanding how nature behaves. Two thousand years ago, Archimedes was working with methods that resemble the use in modern calculus of "infinitesimals."

As one example, it was known before Archimedes' time that the area of a circle is proportional to the square of the radius. This relationship would be written $A = Cr^2$, where we would recognize C as $\pi = 3.1416...$. But the value of C was then unknown. Archimedes arrived at an approximation for C by drawing two regular polygons, one circumscribing and one inscribed in a circle, as shown in the figure. He knew that the area of the circle was larger than that of the inscribed polygon and smaller than that

of the circumscribing polygon. He could calculate the areas of the polygons because they are made up of triangles, and he knew how to calculate the areas of those.

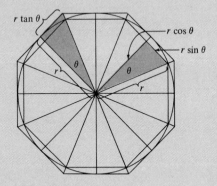

For example, shown in the figure are eight-sided polygons, each containing 16 right triangles. Each triangle of the circumscribing polygon has an area of $\frac{1}{2} \cdot r \cdot r \tan \theta$; each triangle of the inscribed polygon has area $\frac{1}{2} \cdot r \cdot r \sin \theta \cos \theta$. The area of the circle, A, where n is the number

In complicated formulas, scientific notation can be used to simplify the calculations. Suppose, for example, we wish to reduce the following to a single number:

$$\frac{50(7\,280\,000)(0.0012)}{2(0.707)(5280)}$$

We can convert all values to scientific notation:

$$\frac{5 \times 10^1(7.28 \times 10^6)(1.2 \times 10^{-3})}{2(7.07 \times 10^{-1})(5.28 \times 10^3)}$$

Then, collecting the exponents, we have

$$\frac{5(7.28)(1.2)}{2(7.07)(5.28)} \times \frac{10^{1+6-3}}{10^{-1+3}} = \frac{5(7.28)(1.2)}{2(7.07)(5.28)} \times 10^2$$

We can estimate the order of magnitude of the answer by making a mental calculation, observing that 7.28 and 7.07 just about cancel out in the numerator and denominator, and so do 5 and 5.28. This leaves 1.2 divided by 2, which is about $\frac{1}{2}$, so the

of triangles, lies between the areas of the polygons:

$$n \cdot \tfrac{1}{2}r^2 \sin \theta \cos \theta < A < n \cdot \tfrac{1}{2}r^2 \tan \theta$$

Reasoning thus (although he used geometry, rather than trigonometry) and by using polygons with many sides so that the triangles became infinitesimally small, Archimedes decided the constant of proportionality lay between $3\tfrac{10}{71}$ and $3\tfrac{10}{70}$.

Because of his use of infinitesimals on this and other problems, some have declared that Archimedes should be declared the father of calculus. However, that honor is usually accorded to Isaac Newton (1642–1727) and Gottfried Leibniz (1646–1716). Newton, working in England, and Leibniz, in Germany, independently developed calculus into a generalized method and showed that differentiation and integration are inversely related.

It was Leibniz who formulated these ideas in clearest form and who gave us the symbols we use today: dx and dy for differentiation and ∫—an elongated letter "s"—for integration. Leibniz examined sets of "infinite series" (an-other baneful topic for undergraduates) and found that the differences between successive terms in a series bore special relationships to the sums of those terms. The "differences" became "differentials," and the "sums" became "summation," or "integration," which is the same thing. Leibniz applied his ideas to continuous functions as well as to series and found that he had discovered a new analytical tool of enormous power.

Actually, Newton had made the same discovery about 10 years before Leibniz but neglected to publish his findings until after Leibniz did. This led to a disgraceful squabble over priority, with Newton's followers claiming Leibniz was a plagiarist. History has shown Leibniz could not have plagiarized Newton, but Newton's followers became so alienated from mathematicians on the Continent that they broke off contact with them for much of the eighteenth century.

References

Carl C. Boyer, *A History of Mathematics* (New York: Wiley, 1968).

C. H. Edwards, Jr., *The Historical Development of the Calculus* (New York: Springer-Verlag, 1979).

answer is around 0.5×10^2, or about 50. If we work the answer out on a calculator, we get 58.5. Of course, with the availability of modern hand-held calculators, we can put all the numbers into the calculator in their original forms and let the machine take care of the decimal point. But sometimes a calculator may not be at hand, and an engineer should know how to estimate an answer by the method just given.

It is useful to note that units can be manipulated algebraically, and cancellations made, just as if they were numbers. If, for example, we wish to convert a velocity of 60 miles per hour ($v = 60$ mi/hr) to its equivalent in feet per second (ft/sec), we can proceed as follows:

$$v = 60\,\frac{\text{mi}}{\text{hr}} \left(5280\,\frac{\text{ft}}{\text{mi}}\right) \left(\frac{1}{3600}\,\frac{\text{hr}}{\text{sec}}\right)$$

$$= 60\,\frac{\cancel{\text{mi}}}{\cancel{\text{hr}}} \left(5280\,\frac{\text{ft}}{\cancel{\text{mi}}}\right) \left(\frac{1}{3600}\,\frac{\cancel{\text{hr}}}{\text{sec}}\right)$$

We have "miles" in both numerator and denominator, which cancel, and "hours," which do the same. We are left with the units ft/sec, which is what we want. We can then compute the velocity as

$$v = \frac{60(5280)}{3600} \frac{ft}{sec} = 88 \frac{ft}{sec}$$

The foregoing is based upon the fact that, when we have a conversion formula such as

$$1 \text{ mi} = 5280 \text{ ft}$$

we can transpose the "mi" to the right-hand side of the equation, and write

$$1 = 5280 \frac{ft}{mi}$$

Thus, when we multiply a quantity by such an expression, we are essentially multiplying it by unity.

In the metric system (called, more properly, the *Système International*, or SI), a parallel example would be to convert 60 mi/hr, which is 96.54 kilometers per hour (km/h), to its equivalent in meters per second (m/s). The calculation would be

$$v = 96.54 \frac{km}{h} \left(1000 \frac{m}{km}\right)\left(\frac{1}{3600} \frac{h}{s}\right)$$

$$= \frac{96.54(10^3)}{3.6 \times 10^3} \frac{m}{s} = 26.82 \frac{m}{s}$$

You may have noticed some differences in the foregoing regarding the abbreviations employed for different units. In the so-called American customary system of units, also called the English system, "hr" is usually the abbreviation for hour, and "sec" is used for second. But in SI (Chapter 9), "h" is the abbreviation for hour, and "s" is used for second. The "mile" and the "foot" don't appear in SI, of course, but in the English system the abbreviations "mi" and "ft" are commonly used. Note that periods are omitted after these abbreviations, but it is customary to put a period after the abbreviation for "inch" (in.) to distinguish it from the preposition "in."

Exercises

7.1.1 Express the following numbers in scientific notation:
(a) 29 000 000   (b) 4370
(c) 0.139   (d) 1.376
(e) 0.000 132   (f) 0.001
(g) 1 000 000   (h) 0.000 000 693

7.1.2 Make the following unit conversions, showing the factors, with units attached, that must be used (conversion factors are in Appendix L):
(a) 60 mi/hr to ft/hr

Solution

$$60 \frac{mi}{hr} \times 5280 \frac{ft}{mi} = \mathbf{3.168 \times 10^5} \frac{\mathbf{ft}}{\mathbf{hr}} \qquad \text{Answer}$$

(b) 1010 m/s to mi/hr
(c) 96 ft² to m²

Solution

$$1 \text{ m} = 3.281 \text{ ft}$$

$$1 \text{ m}^2 = 10.765 \text{ ft}^2$$

$$96 \text{ ft}^2 \times \frac{1}{10.765} \frac{\text{m}^2}{\text{ft}^2} = \mathbf{8.92 \text{ m}^2}$$ Answer

(d) 25 in.² to mm²

(e) $65 \dfrac{\text{km}}{\text{h}}$ to $\dfrac{\text{mi}}{\text{hr}}$

(f) 0.13 in./sec to m/d (meters per day)

◼ 7.2 SIGNIFICANT FIGURES

We are often confronted by numbers containing many decimal places and wonder just how many of them are useful to us—in other words, how many of the figures in the number are *significant*. For example, if we use a hand-held calculator to compute the product 3π, the answer will be given to us as 9.424778. Do all these decimal places mean something? The answer is, probably not, unless the original value "3" in our product is really 3.000000 ± 0.000001, which it probably is not. So the specter of *significant figures* raises its head, and this leads us to consider the matter of measurement.

The engineering world is a physical world, and when we use numbers we are implying that these numbers either came from measuring something or will be used in the future to measure something. For example, when we measure a short distance carefully with a precision steel scale, we become accustomed to the fact that the eye probably cannot discriminate closer than about ± 0.01″, so we would mark down such a measurement typically as 5.28″ ± 0.01″. It would make no sense whatever to write down a measurement as 5.287″ ± 0.01″ or as 5.2875″ ± 0.01″ because we just can't measure to an accuracy of three or four decimal places with a steel scale. It would even be wrong to write such a measurement as 5.280″, because the presence of the third decimal place would imply an accuracy at that level, or ± 0.001″, which would be misleading. So, the number of significant figures we use in reporting a number should be no greater than the accuracy that is possible in making the reading.

If we add or subtract numbers with different numbers of significant figures, the answer will be no more accurate than the "weakest link." Therefore, in addition or subtraction, each number that goes into the answer should be examined to see which has the greatest uncertainty, and the answer should retain the number of significant figures that will convey the same degree of uncertainty. Thus, if we add the following,

```
  25.1
133.267
   .8755
─────────
159.2425
```

the answer should be reported as 159.2. This is because the number 25.1 conveys an impression that it is accurate to $\pm$ 0.1. Therefore, there is no way in which the answer can be more accurate than $\pm$ 0.1, so all decimals after the first should be omitted.

In multiplication or division, the answer should contain the same number of significant digits as was present in the smaller of the two factors (meaning the one with the lesser number of significant digits). Thus, if we have

$$351.2 \times 0.707106781 = 248.3359016$$

we express the answer as 248.3. This is especially important to note, because the above product could easily be the consequence of the following multiplication:

$$351.2 \cos 45° =$$

and if we do this on a hand-held calculator, the machine will dutifully display an answer of 248.3359016, but only the first four digits are significant.

We also must concern ourselves with *rounding off* as follows. We examine the digit immediately to the right of the last one we plan to keep. If this digit is 5 or greater, we increase the last digit we plan to keep by 1. If the first digit we plan to drop is less than 5, we just drop it and do nothing else. Thus, if we have a number such as 17.67766953 in our calculator and plan to keep only four digits, we round the number *up* and write 17.68, because the first digit to the right of the one we plan to keep is greater than 5. If the number were 17.67500000, our rule would still require us to round this number *up,* and we would write 17.68. But if the number were 17.67499999, we would round *down* and write 17.67.

Sometimes it is not clear how many significant figures a number is supposed to have. In the following examples, there is no ambiguity:

351.2   Four significant figures

0.1   One significant figure

0.00758   Three significant figures

But what about this one?

2500   Two, three, or four significant figures?

To make numbers like the foregoing unambiguous, we should use scientific notation. For example, if the number actually possesses two significant figures, we write it 2.5×10^3. If it has three, we write it 2.50×10^3, and if it has four, we write it 2.500×10^3.

Exercise

7.2.1 Carry out the indicated operations, and express your answers to an appropriate number of significant figures.

(a)   10.75
        0.1342
        1073.
        ‾‾‾‾‾‾

(b) 92.73
     15.60
     1.375
     ‾‾‾‾‾

(c) 0.375
0.0063
0.015
_____

(d) 2.500 × 10³
1.73 × 10²
6.45 × 10³
_____

(e) (10.75)(0.1342) =

(f) (0.015)(0.375) =

(g) 1.73 ÷ 0.012 65 =

(h) 0.012 56 ÷ 675 =

■ 7.3 GRAPHS

Graph paper comes in many types, with different sorts of spacing to suit different tasks. One of the commonest types is that with rectangular grid spacing, shown in Figure 7.1, in this case with ten divisions to the inch in each direction. This kind of

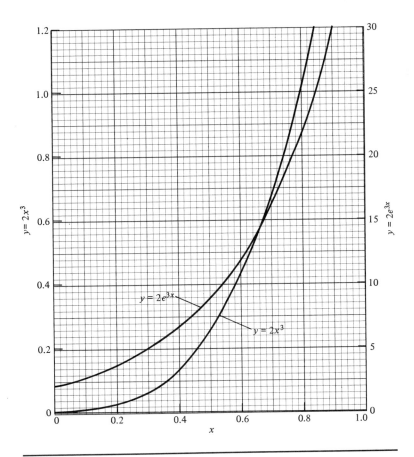

Figure 7.1
The most familiar kind of graph paper has rectangular grid spacing, commonly with ten spaces to the inch. The function $y = 2x^3$ plots as a straight line on *log-log* graph paper (see Figure 7.2). The function $y = 2e^{3x}$ plots as a straight line on *semilog* graph paper (see Figure 7.3).

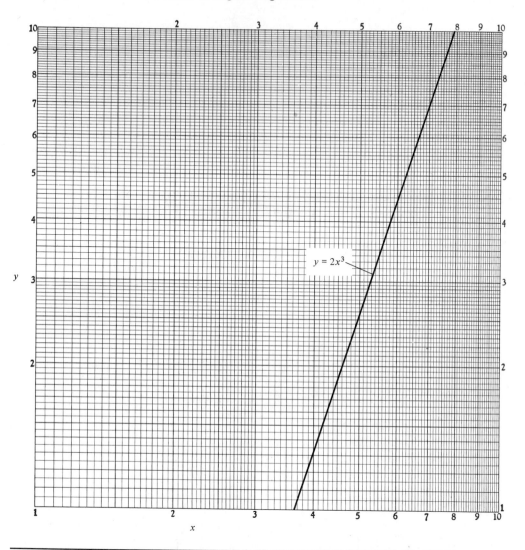

Figure 7.2
An example of log-log graph paper. It is frequently employed when functions of the type $y = bx^m$ are used, because they plot as straight lines on log-log paper. The function $y = 2x^3$ is shown.

paper provides good spacing for most purposes. Paper is available with finer spacing, such as millimeter spacing, with the 5-mm lines accented and the 10-mm lines heavy.

Paper with logarithmic scales is also frequently employed, as in Figure 7.2, which has logarithmic scales in both directions and is called log-log paper. The example shown has only one logarithmic cycle in each direction; other paper is available with multiple logarithmic cycles. The example shown in Figure 7.3 has a logarithmic scale in one direction and a linear grid scale in the other; it is called semilog paper.

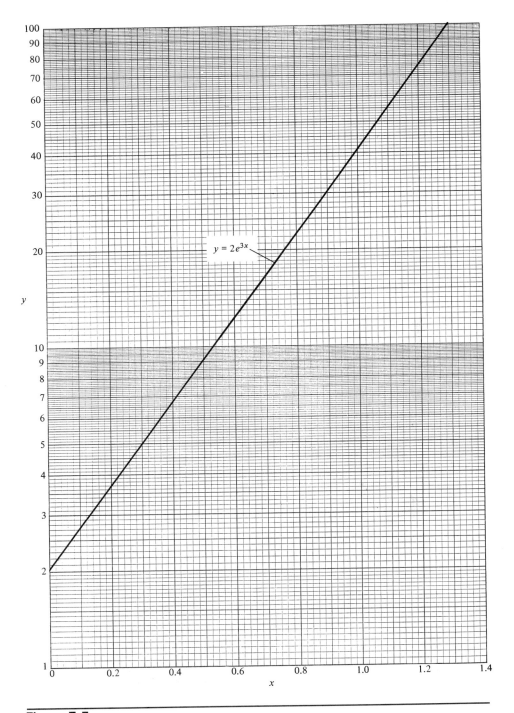

Figure 7.3
An example of semilog paper. It is frequently employed when functions of the type $y = be^{mx}$ are used, because they plot as straight lines on such paper. The function $y = 2e^{3x}$ is shown.

185

Some explanation of logarithmic scales may be in order for the uninitiated. Log-log and semilog graph paper are prepared on the basis of logarithms to the base 10. We can construct a table of numbers and their logarithms as follows:

x	$\log_{10} x$ (4 significant decimal figures)
1	0
2	0.3010
3	0.4771
4	0.6021
5	0.6990
6	0.7782
7	0.8451
8	0.9031
9	0.9452
10	1.0000
20	1.3010
30	1.4771
40	1.6021
50	1.6990
60	1.7782
70	1.8451
80	1.9031
90	1.9452
100	2.0000
200	2.3010
300	2.4771
⋮	
1000	3.0000
2000	3.3010
3000	3.4771
⋮	

Let us set up two scales, side by side, as shown in the following figure. The scale on the left is for values of x, but let us pretend we have not yet marked any numbers on it. The scale on the right is for values of $\log_{10} x$. We mark off equal intervals on it, ranging from 0 to 1.0, which is the range of $\log_{10} x$ for x from 1 to 10. (In the remainder of this discussion we will write simply $\log x$ instead of $\log_{10} x$, and the use of the base 10 will be understood.)

On the $\log x$ scale, we note that $\log 1 = 0$, so we mark the numeral "1" on the x scale, opposite the numeral "0" on the $\log x$ scale. Next we note that $\log 2 = 0.3010$, so we mark the numeral "2" on the x scale, opposite 0.3010 on the $\log x$ scale. Then we note that $\log 3 = 0.4771$, so we mark the numeral "3" on the x scale, opposite 0.4771 on the $\log x$ scale. We proceed up the x scale, marking the values of x in their appropriate positions, until we come to $x = 10$, which is opposite 1.0000 on the $\log x$ scale. Now, if we examine the scale on the left for values of x, we see that it has a peculiar spacing, which we refer to as *logarithmic spacing*. This kind of spacing was used for the logarithmic scales of Figures 7.2 and 7.3.

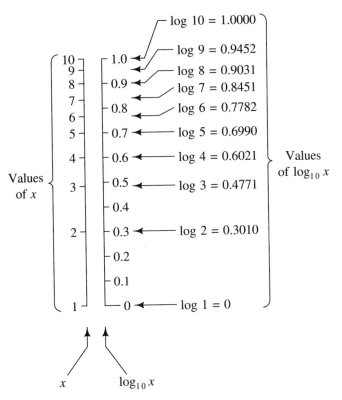

If we look once more at the preceding table giving values of x and log x, we can see that the logarithms from 10 to 100 simply repeat the values of the logarithms from 1 to 10, with the exception that in each case there is a "1" in front of the decimal instead of a "0." We can also see that the numbers 100, 200, and 300 begin to repeat the same pattern, except now there is a "2" in front of the decimal point. Finally, we can see the same pattern beginning for the numbers 1000, 2000, and 3000, except now the numeral "3" appears in front of the decimal point. This pattern repeats itself endlessly, so it means that the logarithmic scale we constructed above for x ranging from 1 to 10 can just as well be used for values of x ranging from 10 to 100, from 100 to 1000, and from 1000 to 10 000 (and onward forever). A portion of such a pattern can be seen in Figure 7.3, where two logarithmic cycles are shown, the first one ranging from 1 to 10 and the second one from 10 to 100.

Keep in mind that "0" will not appear on a logarithmic scale, because log 0 is undefined.

The advantage of logarithmic paper is that certain mathematical functions that plot as curved lines on regular graph paper will plot as straight lines on log paper. For example, curves for $y = 2x^3$ and $y = 2e^{3x}$ have been plotted in Figure 7.1. However, $y = 2x^3$ plots as a straight line on log-log paper, as is shown in Figure 7.2, and $y = 2e^{3x}$ plots as a straight line on semilog paper, as is shown in Figure 7.3.

The reasons behind the foregoing results are simple. For example, we know that the equation for a straight line is

$$y = mx + b \tag{7.1}$$

The variable x is the independent variable, and y is the dependent variable. The constant m is the *slope* of the line, and b is the value of the *intercept* on the y axis.

The general equation for the type of function shown in Figure 7.2 is

$$y = bx^m \tag{7.2}$$

If we take logs of both sides, we get

$$\log y = m \log x + \log b \tag{7.3}$$

The value of $\log b$, of course, is just a constant. If we plot Equation (7.3) on paper scaled to $\log y$ and $\log x$, then it has the form of Equation (7.1) and thus represents a straight line on $\log x/\log y$ paper, with a slope m and an intercept on the $\log y$ axis with value $\log b$.

The general equation for the type of function shown in Figure 7.3 is

$$y = be^{mx} \tag{7.4}$$

If we take logs of both sides of this one, we get

$$\log y = mx \log e + \log b \tag{7.5}$$

The value of $\log b$ is again a constant, and so is the value of $m \log e$. Thus, we have a function consisting of values of $\log y$ and values of x. Semilog paper is set up just this way, so Equation (7.5) takes on the form of Equation (7.1) when plotted on such paper, and again is a straight line with a slope of $m \log e$.

Some good practices should be followed in drawing graphs. One of the most basic is that plenty of space should be left in the margin, if possible. In the case of Figure 7.1, the axes have been set well in from the margins, in order to leave plenty of room for labeling them. A common error is to place the axes at the outer edges of the printed grid and attempt to crowd their labels into the margins. Not only does the appearance suffer, but if it becomes necessary to run off copies of the graph on a photocopier, some of the labeling will probably get cut off by the machine. Also, a generous margin must be left on the left-hand side for binding purposes. A piece of graph paper may lie nice and flat in your loose-leaf binder, but reports typically are stapled or otherwise bound on the left-hand side. If the reader of your report has to pry the binding apart in order to see what you put along that left-hand edge, some hard thoughts will be directed your way.

Close attention should be paid to locating the axes correctly for the independent and dependent variables. In a typical experiment you will deliberately vary one quantity, called the *independent* variable, and measure the result that variation produces in another quantity, called the *dependent* variable. A simple example would be to measure the speeds that result from different accelerator pedal positions in an automobile. The independent variable is the pedal position; the dependent variable is the automobile speed. It is customary to locate the independent variable along the horizontal axis of a graph (called the *abscissa*) and the dependent variable along the vertical axis (called the *ordinate*). Inexperienced people sometimes switch the axes, and the result is usually confusing.

After the axis locations have been selected, the spacing must be chosen. We neither want our graph to crowd the borders, nor do we want it to be too tiny. Also, we want the spacing to bear some convenient relationship to the number ten. If a given space on a graph represents 1, 2, 5, or 10 units of the quantity to be plotted, matters will work out nicely. But if you decide that four spaces will represent ten units, so that one space represents 2.5 units, you will almost surely regret it. Even worse is to let three spaces represent ten units, so that one space represents 3.33 units. In either of these latter two cases, you will find it almost impossible to keep from making errors as you plot your graph.

Sometimes it is convenient to place two or more plots on the same graph, using a different scale for each. This has been done, for example, in Figure 7.1. The scale on the left is to be used for the function $y = 2x^3$, and the one on the right is to be used for the function $y = 2e^{3x}$.

If a graph is that of a mathematical function, then usually the only labeling the axes need will be designators such as x and y, as has been done in Figures 7.1, 7.2, and 7.3. But most engineering graphs are of real physical quantities. The axes should be clearly labeled to describe those quantities and the units used. Figure 7.4 shows an example from a published journal that clearly labels the axes.

If a mathematically derived curve is being plotted, it should be shown as a smooth line, without any of the plotted points being shown discretely. However, if the graph is derived from experimental results, then the experimental points should be plotted. In order to separate different experimental runs from each other, different symbols can be used, such as small circles, triangles, squares, and the like, as in Figure 7.5. A curve that connects the experimental points fairly smoothly can be drawn in, either by visual estimation or by more sophisticated techniques, to be described shortly. In drawing such a curve, the line should not be drawn through the experimental points

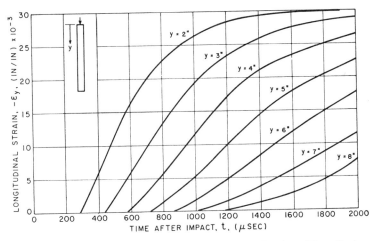

Longitudinal strains at different locations in strut as a function of time after impact

Figure 7.4

This graph, taken from a paper published in a professional journal, tells exactly what is being represented. The variables on the two axes are given clear-cut names, and the units are given. Also, the description below the graph tells just what is going on. (From I. M. Daniel, "Experimental Methods for Dynamic Stress Analysis in Viscoelastic Materials," Paper 65-APM-18, American Society of Mechanical Engineers)

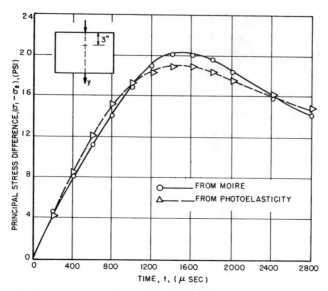

Comparison of principal stress differences at point 3 in. below impacted point of plate obtained independently from moire and photoelastic data

---

Figure 7.5

In this graph, two curves are plotted, derived through the use of different experimental methods. The curves are distinguished from each other by the use of different plotting symbols: a circle and a triangle. (From I. M. Daniel, "Experimental Methods for Dynamic Stress Analysis in Viscoelastic Materials," Paper 65-APM-18, American Society of Mechanical Engineers)

like this:

but should stop at the edges of the points like this:

All work should be neat and clear. Your plotted lines should be drawn with authority and should be heavier and thicker than would seem appropriate at first glance. This does not mean they should cover such a wide swath as to produce ambiguity, but it does mean that they should not be thin and tentative. Figures 7.1, 7.2, and 7.3 provide reasonable examples.

Finally, graphs, tables, and other figures should always be arranged to be readable without the need to rotate the page, if at all possible. If you must label your material in such a way as to require the page to be rotated, be sure all the material is labeled so it is necessary to rotate only 90 degrees clockwise. It is guaranteed to upset your reader if you arrange some of your material to be read by rotating 90 degrees clockwise, and some to be read by rotating the other way.

Exercises

7.3.1 Plot the function $y = 3x^{2.5}$ on rectangular graph paper and on log-log paper.

7.3.2 Plot the function $y = 3e^{2x}$ on rectangular graph paper and on semilog paper.

7.3.3 For the function $y = 2x^3$, plotted on log-log paper in Figure 7.2, what is the slope m, if the function is considered to be of the general type $y = bx^m$? What is the intercept on the log y axis?

Solution

In $y = 2x^3$, if we consider the function to be of the form $y = bx^m$, then obviously

$$m = 3 \qquad b = 2$$

From this, we get immediately that the slope m is equal to 3. If we take logs of both sides of $y = bx^m$, we get

$$\log y = m \log x + \log b$$

where $\log b = \log 2$ is the intercept on the log y axis.

$$\log b = \log 2 = \mathbf{0.30} \qquad\qquad \text{Answer}$$

We can visualize this better if we take the equation

$$\log y = m \log x + \log b$$

and make the substitutions

$$p = \log y$$
$$q = \log x$$
$$\log b = 0.30$$
$$m = 3$$

and convert the equation to

$$p = 3q + 0.30$$

This is the equation of a straight line, and we can plot it on p–q coordinates by making the calculations shown in the following table:

x	$y = 2x^3$	$q = \log x$	$p = \log y$
0	0	Undefined	Undefined
0.2	0.016	−0.70	−1.80
0.4	0.128	−0.40	−0.89
0.6	0.432	−0.22	−0.36
0.8	1.024	−0.10	0.01
1.0	2.000	0	0.30
1.2	3.456	0.08	0.54
1.4	5.488	0.15	0.74

We can plot the p and q coordinates on rectangular grid paper, as has been done in the figure. We can see that the p-axis intercept is indeed 0.30, where $p = \log y$.

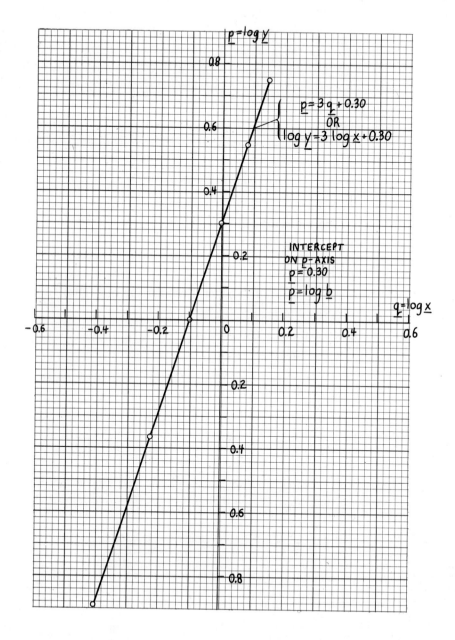

7.3.4 For the function $y = 3x^{2.5}$ of Exercise 7.3.1, show that the slope is 2.5 on log-log paper by dividing the change in log y by the change in log x.

Solution

x	$y = 3x^{2.5}$	log x	log y
0	0	Undefined	Undefined
0.2	0.054	−0.699	−1.268
0.4	0.304	−0.398	−0.517
0.6	0.837	−0.222	−0.077
0.8	1.717	−0.097	0.235
1.0	3.000	0	0.477

For interval $\Delta x = 1.0 - 0.2$, $\Delta y = 3.000 - 0.054$

$$\Delta(\log x) = 0 - (-0.699) = 0.699$$

$$\Delta(\log y) = 0.477 - (-1.268) = 1.745$$

$$\frac{\Delta(\log y)}{\Delta(\log x)} = \frac{1.745}{0.699} = 2.496$$

$$\cong 2.5 \qquad\qquad\qquad \text{Answer}$$

7.3.5 For the function $y = 2x^3$, show that the slope is 3.0 on log-log paper, using the method of Exercise 7.3.4.

7.3.6 For the function $y = 2e^{3x}$, plotted on semilog paper in Figure 7.3, what is the slope of the line if the function is considered to be of the general type $y = be^{mx}$?

7.3.7 Calculate the slope of the line in Figure 7.3, using the method of Exercise 7.3.4.

7.3.8 Plot the following and draw a smooth curve through the points, as closely as you can. (A plastic drafting curve, obtainable at most college bookstores, is a great help.)

Input	Output
1.0	1.35
2.0	1.70
3.0	2.30
4.0	3.25
5.0	4.40
6.0	6.20
7.0	9.00
8.0	12.85

■ 7.4 CURVE FITTING

One of the simplest of ways to fit a curve to a set of experimental points is by visual approximation, or, as it is called, "by eye." As an example, if we are given the experimental points in Table 7.1, we can plot them as has been done in Figure 7.6. If the

Table 7.1

x	y
0	0
1.0	2.7
2.0	3.7
3.0	4.6
4.0	5.2
5.0	6.1
6.0	6.8
7.0	7.8
8.0	8.7
9.0	9.6
10.0	10.4

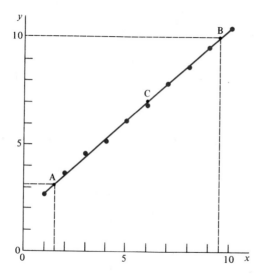

Figure 7.6
A straight-line graph drawn "by eye" through a set of experimental points.

points seem to lie in a straight line, then a line is drawn that leaves as many points to one side of the line as to the other. It is a rare experiment that produces all points exactly in a straight line, even though the relationship is one that *ought* to produce a straight line. There are many reasons for this phenomenon: There will always be "reading errors" on the part of the observer; rarely does anything in nature behave in a straight-line fashion (what we call *linear* behavior); errors are often introduced by the measuring equipment itself—not only may the equipment not be linear, but the use of the equipment may actually alter the behavior of the thing being measured.

At any rate, after we plot the points, we can draw our best estimate of the line that represents these points. But we may want more: We may want a mathematical representation of the line we have drawn. To do this, we remember that the equation of a straight line is

$$y = mx + b \tag{7.6}$$

All we need do is to select the coordinates of two points from the line we have drawn, substitute them into Equation (7.6), and solve the two equations we get for m and b. Two such points are shown in Figure 7.6 and have been marked A and B. They are spaced well apart in order to produce the best accuracy possible. We read the coordinates of point A as ($x = 1.50$, $y = 3.20$) and of point B as ($x = 9.50$, $y = 10.05$). Substituting the values of x and y belonging to point A into Equation (7.6), we get

$$3.20 = 1.50m + b \tag{7.7}$$

Doing the same for point B, we get

$$10.05 = 9.50m + b \tag{7.8}$$

If we solve Equations (7.7) and (7.8) for m and b, we get $m = 0.86$ and $b = 1.91$. Therefore, our equation is

$$y = 0.86x + 1.91 \tag{7.9}$$

To check Equation (7.9), we can select any point on the line other than points A and B and substitute the coordinates of that point into Equation (7.9) to see if it is satisfied. If we select, say, point C, with coordinates $x = 6.00$, $y = 7.00$, we get

$$7.00 \approx 0.86(6.00) + 1.91$$
$$\approx 7.07$$

which is reasonably close, given the accuracy that is possible in reading coordinates from the graph.

There are more accurate ways to find a "best-fit" straight line for a set of experimental points than the judgmental one just described. One of these is the *method of least squares*, described in Chapter 8.

If we have experimental results that do not plot as a straight line, but that we suspect may have the form $y = bx^m$—a so-called power curve—we may proceed as follows. Suppose we have the values in Table 7.2, and we plot them in rectangular

Table 7.2

x	y
1	2
2	11
3	31
4	64
5	112
6	176
7	259
8	362

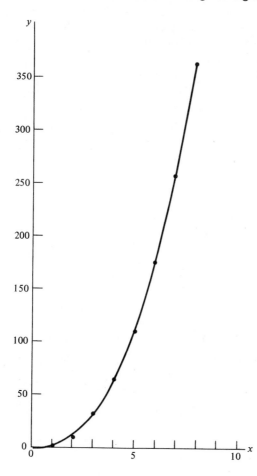

Figure 7.7

A set of experimental points that appear to fit a function of the form $y = bx^m$, with an approximate curve drawn through the points "by eye."

coordinates as has been done in Figure 7.7. From the plot, we feel we are probably dealing with a curve of the form $y = bx^m$. We could, of course, immediately substitute the values of two pairs of coordinates into the equation $y = bx^m$, getting two equations in the two unknowns b and m. But we may not know whether the two pairs of coordinates will give us the best approximation, nor are we convinced at this point that we really are dealing with a power curve. So we plot the values from Table 7.2 onto log-log paper, as has been done in Figure 7.8. The points line up in a good approximation to a straight line, so we conclude that a function of the form $y = bx^m$ will do the job. Note, in Figure 7.8, that we are unable to plot all the points from $(x = 1, y = 2)$ to $(x = 8, y = 362)$ because this would require three cycles along the y axis, and our paper only provides for two cycles in the y direction. Therefore, we omit the first point and start with $(x = 2, y = 11)$. From Table 7.2 we can select two points, such as $(x = 1, y = 2)$ and $(x = 8, y = 362)$. If we substitute the first pair into $y = bx^m$, we get

$$2 = b(1)^m$$

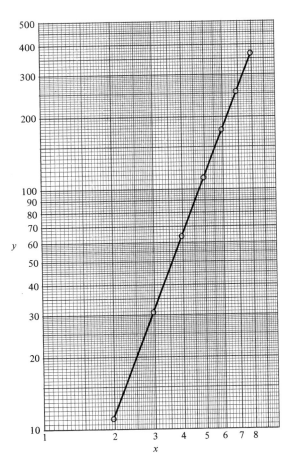

Figure 7.8
The experimental points from Figure 7.7 are redrawn here on log-log paper. Since a straight-line function seems to fit the points quite well, we conclude that the points do indeed satisfy a function of the form $y = bx^m$.

This immediately gives us the result $b = 2$, because $(1)^m = 1$. Then, using the second pair of values, we get

$$362 = 2(8)^m$$

or

$$181 = 8^m$$

$$\log 181 = m \log 8$$

$$m = \frac{\log 181}{\log 8}$$

$$= \frac{2.258}{0.903} = 2.5$$

Therefore, our equation is

$$y = 2x^{2.5}$$

A similar procedure can be used if we suspect that our experimental data can be approximated by an exponential curve of the form $y = be^{mx}$. In a case of this sort, we plot the curve on semilog paper to see if it produces a straight line, and then select two pairs of points for substitution into the equation in a manner analogous to that already described. (See Exercise 7.4.3.)

Exercises

7.4.1 Plot the points in the following table, and if it appears that a straight line can be plotted through them, compute the values of m and b for the equation $y = mx + b$, using the method of this section.

x	y
3.00	2.00
5.00	7.15
7.00	11.20
9.00	16.70
11.00	21.10
13.00	26.75

7.4.2 Plot the following points. Then decide whether the curve can be approximated by an equation of the form $y = bx^m$ by replotting the points on log-log paper. If so, then determine the equation that fits the points.

x	y
1.0	2.2
2.0	15.3
3.0	47.7
4.0	107
5.0	199
6.0	332
7.0	511
8.0	743

7.4.3 Plot the following points. Then decide whether the curve can be approximated by an equation of the form $y = be^{mx}$ by replotting the points on semilog paper. If so, then determine the equation that fits the points.

x	y
0	1.8
0.2	3.6
0.4	7.3
0.6	14.7
0.8	29.6
1.0	59.6

■ 7.5 CIRCLE CHARTS AND BAR GRAPHS

Some kinds of data lend themselves naturally to presentation in the form of circle charts or bar graphs rather than in the form we have described in the foregoing sections. As an example, the information in Table 7.3 for the distribution of engineers by field is not in the form of the familiar x-y functional data we have used so far. The data for the "y axis" could consist of the percentage of engineers in each field, but the "x axis" would consist merely of a listing of the fields of engineering. The result would be a bar graph, as shown in Figure 7.9.

Table 7.3 Distribution of Engineers by Field

Field	Percent
Electrical/electronics	24
Mechanical	20
Civil	14
Chemical	6
Aeronautical/astronautical	5
Other	31
	100

Source: *Science and Technology Data Book, 1988* (Washington, D.C.: National Science Foundation, NSF 87–317).

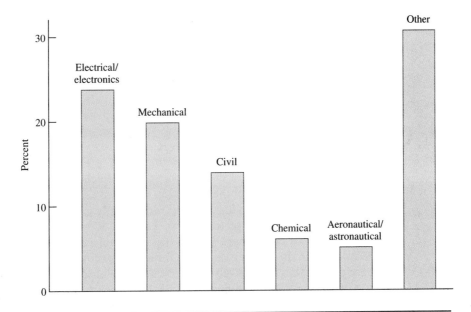

Figure 7.9
A bar graph displaying the same information as that shown by the circle chart, Figure 7.10.

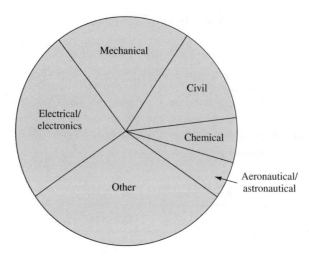

Figure 7.10
A circle chart displaying the same information as that shown by Figure 7.9.

A big advantage of a bar graph, as compared to a simple table of values, is that its visual imagery gives a "feel" for the relative magnitudes of the various sectors. For example, if one looks only at the numbers in Table 7.3, it is easy to overlook the fact that the largest sector of all is the one marked "Other." But in Figure 7.9, the huge relative size of the "Other" sector is immediately apparent, especially when it stands directly alongside the small sector labeled "Aeronautical/astronautical." This graphic method of displaying the data has an immediate impact, which is why it is so heavily favored by the news media and by advertising people. (Incidentally, the reason the "Other" category is so large is that there are many engineering fields, each with a relatively small number of engineers.)

Generally, in constructing a bar graph, you should begin at the left with the largest group and proceed to the right with progressively smaller groups until you have nothing left but "Other." The "Other" category always goes at the far right (assuming you have such a category). This procedure obviously is not a hard-and-fast rule, but it generally produces good results.

If the data are expressed as percentages of a whole, a circle chart is possible. Figure 7.10 shows the same information as that given in Figure 7.9, but in the form of a circle chart. The number of degrees occupied by each sector of the circle chart is of course the same fraction of the whole (360 degrees) as is given in Table 7.3. For electrical/electronics engineers, for example, we multiply 360 degrees by .24 and get 86 degrees for the size of that sector; and so on.

In constructing a circle chart, we use the same general scheme as with the bar graph: We start with the largest category at the left of the circle, and proceed clockwise around the circle with progressively smaller categories until we have nothing left but "Other." Circle charts also are popular with the news media and with advertisers because they are so easily grasped. They are sometimes called *pie charts*.

One special kind of bar graph shows a frequency distribution, or *histogram*. Figure 7.11 is an example of a histogram taken from Chapter 8.

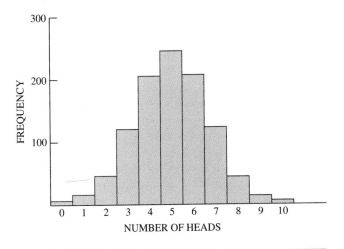

Figure 7.11

Reproduction of a frequency distribution chart from Chapter 8, showing how many combinations containing a given number of heads could occur in ten successive tosses of a coin. There are 1024 possible combinations of ten objects; among these 1024 combinations there are 10 containing only one head, 45 containing two heads, and so on.

Exercises

7.5.1 The materials used in a particular make of automobile are as follows:

Material	Pounds
Steel	1500
Cast iron	330
Aluminum	170
Plastics	250
Other	350
	2600

Show these data as a circle chart and as a bar graph.

7.5.2 The Consumer Price Index (CPI) from 1977 to 1989 changed as follows (1967 = 100): 1977 = 181.5; 1978 = 195.3; 1979 = 217.7; 1980 = 247.0; 1981 = 272.3; 1982 = 288.6; 1983 = 299.3; 1984 = 303.5; 1985 = 322.8; 1986 = 327.4; 1987 = 340.8; 1988 = 354.9; 1989 = 372.6. Show this information in the form of a bar graph.

7.5.3 A series of tests on the yield strength of 450 steel samples gives the following results. Plot this information in the form of a histogram.

Range of Yield Strengths	Frequency*
114,000–117,900	10
118,000–121,900	34
122,000–125,900	73
126,000–129,900	106
130,000–133,900	112
134,000–137,900	78
138,000–141,900	26
142,000–145,900	11
	450

Data adapted from G. E. Dieter, *Engineering Design, A Materials and Processing Approach* (New York: McGraw-Hill, 1983).
* Number of samples failing in the range of yield strengths in the first column.

7.5.4 The information given in the following table shows the six-year summary of the operations of a major U.S. bank. Plot two bar graphs, one for the income and the other for the profit. Note that profits will extend above the line that represents zero, but losses (shown in parentheses) should extend below the line.

Year	Income (millions of dollars)	Profit (Loss) (millions of dollars)
1984	4,418	273
1985	4,525	304
1986	4,364	322
1987	4,239	(604)
1988	4,961	129
1989	5,376	(125)

7.5.5 The registrar at your university reports that in previous terms the distribution of grades for engineering and mathematics were as follows. (In certain courses in which letter grades are not given, students earn either a "pass" or a "not pass.")

	Percent of Grades Given					
	A	B	C	D	F or Not Pass	Pass
Engineering	19	25	27	4	4	21
Mathematics	17	28	33	10	6	6

Show this information in the form of two circle charts, one for engineering and the other for mathematics.

7.5.6 The numbers of scientists and engineers engaged in research and development in the following countries in 1985 were as follows: France = 102,300; West Germany = 143,600; Japan = 381,000; United Kingdom = 98,000; U.S. = 772,500; U.S.S.R. = 1,485,300. Show this information in the form of a bar graph. (Source: National Science Foundation)

7.5.7 The scores on an examination are distributed as follows. Show the information in the form of a histogram.

100–3	91–3	82–0	73–2	64–1	55–3	46–0
99–3	90–4	81–1	72–3	63–4	54–1	45–0
98–3	89–1	80–1	71–5	62–2	53–1	44–0
97–2	88–4	79–1	70–5	61–0	52–2	43–1
96–3	87–1	78–0	69–3	60–2	51–2	42–0
95–1	86–0	77–0	68–2	59–3	50–3	41–0
94–2	85–3	76–5	67–4	58–3	49–0	40–0
93–3	84–0	75–5	66–2	57–4	48–0	39–0
92–2	83–2	74–3	65–6	56–4	47–2	38–1

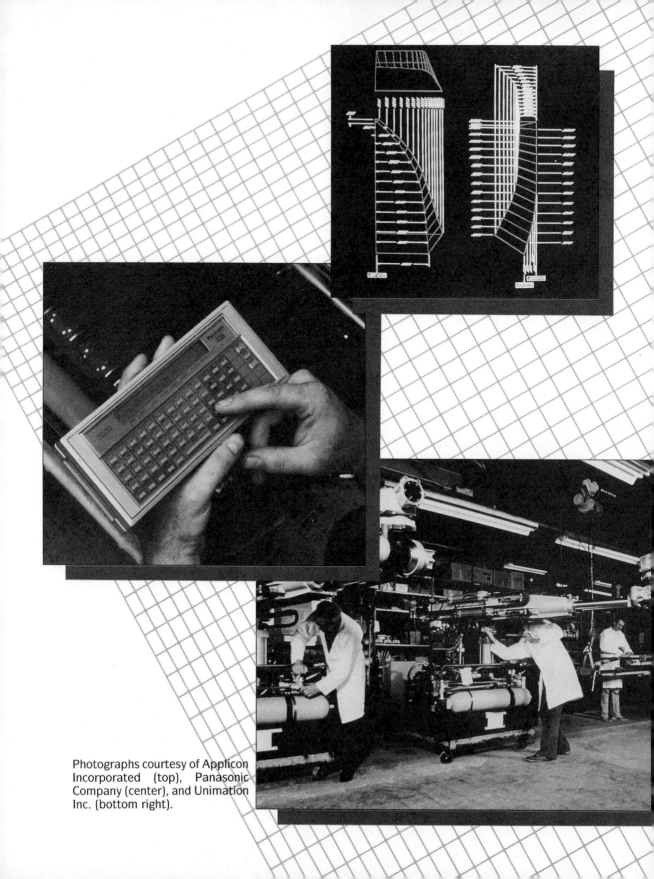

Photographs courtesy of Applicon Incorporated (top), Panasonic Company (center), and Unimation Inc. (bottom right).

8

Statistics

Most engineers deal with statistics in one way or another, because *statistics* is the science of collecting and interpreting numerical information. In some cases the word describes a relatively straightforward process—for example a population census. In other cases, the process is anything but straightforward, as when we attempt to predict the likelihood of getting cancer from consuming a particular food additive. Engineers frequently use the word in both senses: collecting data and making judgments based upon sampling techniques. Some examples of the latter are taking occasional samples of parts being produced by a machine and, by measuring the samples, judging whether the entire production run will have dimensions falling within the usable range; taking samples from several points in a body of water—say, an estuary—and judging whether the entire body of water falls within acceptable quality limits; making a set of measurements involving two or more variables and, by a process called *regression*, determining if there is a mathematical relation between them.

Statistics is treated very briefly here—only enough to acquaint the reader with the basic ideas and the meanings of some of the terms. The major benefit of studying this chapter will be the resultant ability to recognize when a problem requiring statistical analysis has arisen and how to begin to solve it. It should be emphasized, however, that statistical analysis can become quite complex, and some individuals spend entire careers specializing in the topic.

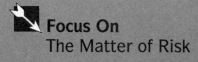

Focus On
The Matter of Risk

Risk exists in the lives of all of us in a myriad of forms. Each day we are exposed to the possibility of injury or death from natural causes: earthquakes, hurricanes, tornadoes, floods, lightning, and disease. We are also exposed to risks from man-made causes: motor vehicles, industrial machinery, air pollution, pesticides, explosions, nuclear power, and sports. The perceptions of the public regarding the seriousness of these risks may not always be in accordance with the facts. For example, researchers have found that most people believe as many deaths are caused by accidents as by diseases, whereas diseases actually cause 15 times as many deaths. It has also been found that many people believe homicides are more frequent than suicides, whereas the reverse is true.

Knowledge of statistical data does not always control an individual's perception of risk. For example, in one research program, the perceptions of risk provided by two different groups were as follows:

Perceptions of Risk

College Students	Risk Assessors
1. Nuclear power	1. Motor vehicles
2. Hand guns	2. Smoking
3. Smoking	3. Alcoholic beverages
4. Pesticides	4. Hand guns
5. Motor vehicles	5. Surgery

The "risk assessors" placed nuclear power at number 20 on their list, whereas college students placed it at number 1. The risk assessors were well informed about the actual data regarding the *statistical* riskiness of various items and ranked them accordingly. But the research showed that the college students as a group were also well aware that, statistically, nuclear power was far down the list, and they ranked it number 1 anyway, for other reasons: (1) dread of the technology, (2) how well understood the risks were perceived to be, and (3) how many people would be exposed. These factors were more important to them than the statistical data.

The foregoing demonstrates a very important issue for engineers. As a group, engineers tend to be influenced by factual data. For them, the economic benefits of a technology such as nuclear power may be sufficient to balance the risks. But economic benefits do not play a part in the judgments of the opponents of nuclear power—they simply are not important. The two sides of such an issue turn out to be operating by different value systems. As a result, their attempts at communication may produce no results except exasperation. In a very real sense, they live in different worlds.

To a large degree, the arguments of engineers are based upon probabilistic and statistical analysis. For example, it has been noted that, for natural disasters, the public seems to accept the risk of the possibility of one death per million per year, or 10^{-6}. The people living behind Holland's dikes have a risk of death of 10^{-7}, but they seem to be opposed to having chemical plants near them with a similar level of risk. Proponents of such plants ask why they should be required to spend large sums of money to make the risk of industrial hazards less than that from the dikes. However, such arguments seem to have little impact on opposing groups, because they introduce the concept that there might be such a thing as an "acceptable risk." To the opponents, the notion of an "acceptable risk," if it is applied to man-made hazards, is itself unacceptable.

References
R. F. Griffiths (ed.), *Dealing with Risk* (New York: Wiley, 1981).
J. Stein, "Assessing the Risk," *Mosaic*, September–October 1982, pp. 17–23.

■ 8.1 MEASURES OF CENTRAL TENDENCY

Whenever we deal with a group of numbers, we almost automatically seek to characterize the group by its *average* as a measure of the *central tendency* of the group. Thus, we speak of the average age of the population, the average income of a group, or the average score on a test. When we do so, we are usually referring to the *arithmetic mean*. There are other measures of central tendency, as we will see.

The *arithmetic mean* is defined as[1]

$$\bar{X} = \frac{X_1 + X_2 + \cdots + X_N}{N}$$

$$= \frac{\sum_{i=1}^{N} X_i}{N} = \frac{\sum X_i}{N} \tag{8.1}$$

where the X's are the values to be averaged and N is their total number. This is the expression we usually mean by the term *average*: We sum up all the values (called *variates*) and divide by the number of variates in the group.

If we wish some variates to be more heavily counted than others, we can associate weighting factors $w_1, w_2, w_3, \ldots$ with the variates $X_1, X_2, X_3, \ldots$ and produce a *weighted arithmetic mean*:

$$\bar{X} = \frac{w_1 X_1 + w_2 X_2 + \cdots + w_N X_N}{w_1 + w_2 + \cdots + w_N}$$

$$= \frac{\sum w_i X_i}{\sum w_i} \tag{8.2}$$

If we wish to determine the *median* of a set of numbers, we first arrange them in order of magnitude. If the number of variates is odd, we pick out the one in the middle as the median; thus, the median is the value for which there are as many variates above as there are below. If the number of variates is even, we pick out the two in the middle and take their arithmetic mean as the median.

Sometimes the mean does a better job of characterizing the central tendency of a group of numbers, and sometimes the median is better. In general, if a group of numbers contains a few that show a striking departure from the rest, the median is to be preferred. To see this, consider the following annual incomes from seven different people. We ask the question, "What is the average income for this group?"

Incomes: $16,000
17,000
17,000
19,000
22,000
24,000
60,000

---

[1] A brief treatment of the summation process, signified by the $\sum$ symbol, is given in Appendix E.

The "average" income for this group, as characterized by the arithmetic mean, is readily calculated to be $25,000. But this number is larger than the incomes of all but one in the group, and the other six might be startled to discover that the "average" income of the group is larger than any of theirs. They might even characterize the seventh member of the group as "that lousy average-raiser." On the other hand, the *median* of the above salaries is $19,000, which does a much better job of representing the central tendency of this group than the mean does. For most sets of data, the median and the mean will be fairly close to the same value, and if the variates are distributed in accordance with the so-called *normal distribution*, the mean and median will be identical. (The normal distribution will be dealt with later.) But any time the data are skewed, as in the above example, a judgment will have to be made regarding the most appropriate measure of central tendency.

The *mode* is a measure that comes up infrequently in engineering. The mode of a group of variates is the one that occurs with the most frequency. In the example of seven annual salaries given above, the mode is $17,000. In a group of numbers such as 2, 3, 5, 6, 9, 10, there is no mode because each number occurs only once. In the group 2, 2, 2, 3, 4, 4, 4, 5, there are two modes: 2 and 4. Such a group is said to be *bimodal*.

As indicated above, if the distribution of the data is skewed, the mean, median, and mode will have different values. In the case of Figure 8.1(*a*), the data are skewed to the right. This is like the "average" income example we discussed earlier. In that example, the mode was $17,000, the median $19,000, and the mean $25,000. In a right-

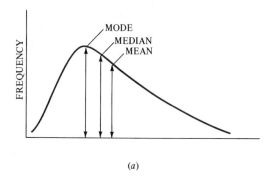

(*a*)

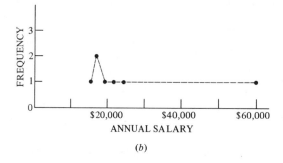

(*b*)

Figure 8.1
(*a*) The mean, median, and mode for a right-skewed distribution curve. (*b*) A plot of salary data to determine the "average" salary for the example on page 207.

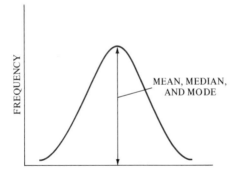

FREQUENCY

MEAN, MEDIAN,
AND MODE

Figure 8.2
Location of the mean, median, and mode for a symmetrical distribution curve, as in the case of the normal distribution curve.

skewed distribution curve the mean will always be larger than the median because there is a disproportionate number of high-valued variates. Also, if the curve is as smooth and regular as the one in Figure 8.1(*a*), then the median and mean will both be larger than the mode. The salary data from our example are plotted in Figure 8.1(*b*), with salary along the horizontal axis and frequency of occurrence along the vertical axis. This is not a very regular-appearing curve; nevertheless, it certainly exhibits right-skewedness, and the relative positions of the mode, median, and mean are the same as those shown in Figure 8.1(*a*). If the distribution curve is symmetrical, as is the "normal" distribution curve in Figure 8.2, then the mode, median, and mean will all be identical.

Another measure of central tendency that may be encountered occasionally is the geometric mean, defined as

$$\text{geometric mean} = \sqrt[N]{X_1 X_2 X_3 \cdots X_N} \qquad (8.3)$$

As a comparison, for the set of numbers 3, 4, 7, 8, the arithmetic and geometric means are

$$\bar{X} = \frac{3 + 4 + 7 + 8}{4} = 5.5$$

$$\text{geometric mean} = \sqrt[4]{3 \cdot 4 \cdot 7 \cdot 8} = \sqrt[4]{672} = 5.09$$

One more indication of central tendency, called the *root mean square* (r.m.s.) or the *quadratic mean*, is sometimes of interest to engineers. It means "the square root of the mean of the squares" and is defined as

$$\text{quadratic mean} = \sqrt{\frac{\sum_i X_i^2}{N}} \qquad (8.4)$$

Exercises

8.1.1 The following numbers are the scores made by a group of students on a test. Find the arithmetic mean, the median, the mode, and the quadratic mean: 12, 42, 23, 25, 41, 100, 30, 16, 45, 60, 0, 32, 35, 18, 25, 27, 41, 35, 25, 21.

Solution

First, the numbers are arranged in numerical sequence, as has been done in the following table.

X_i	X_i^2
0	0
12	144
16	256
18	324
21	441
23	529
25	625
25	625
25	625
27	729
30	900
32	1,024
35	1,225
35	1,225
41	1,681
41	1,681
42	1,764
45	2,025
60	3,600
100	10,000
$\sum X_i = 653$	$\sum X_i^2 = 29{,}423$

$N = 20$, so the arithmetic mean is

$$\bar{X} = \frac{\sum X_i}{N} = \frac{653}{20} = \textbf{32.65} \qquad \text{Answer}$$

The number of variates is even, so the median is the average of the central pair:

$$\text{median} = \frac{27 + 30}{2} = \textbf{28.5} \qquad \text{Answer}$$

The number that appears most frequently is **25** (three times), so that is the *mode*. The quadratic mean is

$$\text{quadratic mean} = \sqrt{\frac{\sum X_i^2}{N}} = \sqrt{\frac{29\ 423}{20}} = \textbf{38.36} \qquad \text{Answer}$$

8.1.2 The following numbers are the sizes of a group of shoes. Find the arithmetic mean, the median, and the mode. Which number best represents the group of shoes (bearing in mind that shoes don't come in sizes such as 9.86)? 8, $8\frac{1}{2}$, 9, 9, 9, 9, $9\frac{1}{2}$, 10, $10\frac{1}{2}$, $10\frac{1}{2}$, 11, 11, $11\frac{1}{2}$, $11\frac{1}{2}$.

8.1.3 The following numbers are test scores: 15, 20, 25, 25, 25, 30, 35, 40, 40, 40, 40, 40, 50, 55, 60, 60, 60, 75, 75, 75. Compute the arithmetic mean and the median.

Next, recognizing that 25 occurs three times, 40 occurs five times, and so on, treat these as weighting factors, w_1, w_2, w_3, ... according to the following table, and recompute the *weighted* arithmetic mean, using Equation (8.2). Compare the two answers you got for the arithmetic mean.

Weighting Factor w_i	Variate X_i
1	15
1	20
3	25
1	30
1	35
5	40
1	50
1	55
3	60
3	75

8.1.4 Compute the arithmetic mean and the median for the following numbers, which might represent the ages of a group of young people: 18, 18, 17, 17, 17, 10, 19, 20, 17, 20, 20, 19, 18, 17, 17. Do you notice anything different about the relationship between the mean and the median, compared to the examples in the text? Would you characterize this group of numbers as right-skewed or left-skewed?

8.1.5 The annual salaries of a group of people are as follows: $21,600, $21,600, $21,600, $22,500, $22,500, $23,400, $24,300, $24,300, $25,200, $25,200, $27,000, $50,400. Compute the mean and the median. Which gives a better representation of the data?

8.1.6 A precision timer is used to record the time it takes a pendulum to swing from one extreme position through a complete swing and back to the original position. Many readings are taken, with the following results:

Elapsed Time (sec)	Number of Times the Elapsed Time Is Recorded
1.996	2
1.997	5
1.998	9
1.999	12
2.000	16
2.001	15
2.002	12
2.003	7
2.004	3
2.005	1

Make a histogram for this information. (See Chapter 7 for the definition of a histogram.) Compute the weighted arithmetic mean.

■ 8.2 MEASURES OF DISPERSION

Just knowing a measure of central tendency is not enough to characterize a group of variates. We also need to know something about how the variates are scattered. For example, the mean of the group 5, 5, 5, 5, 5, 5, is 5, which obviously characterizes the group very nicely. But the mean of the group 1, 1, 2, 7, 9, 10 is also 5, which does not do a very good job of characterizing the group because the group is so scattered.

The *range* of a group helps to give an indication of its scatter. The range is defined as the difference between the largest and smallest variates in the group. Thus, the range of the first group in the previous paragraph is 0, and that of the second is 9.

Measuring all the individual *deviations* $X_i - \bar{X}$ of a group would also give us a feeling for the dispersion of the group. But if we tried to take an average of these deviations in order to get a single number that would characterize the dispersion, we would discover that the average is zero. This is so because the data points are all scattered evenly, plus or minus, about the mean. All the plus deviations cancel out the minus deviations. So we calculate an average of the deviations by taking the mean of their absolute values. Thus, the *mean deviation* of a group of N numbers is defined as

$$\text{mean deviation} = \frac{\sum |X_i - \bar{X}|}{N} \tag{8.5}$$

The mean deviation gives a nice measure of the dispersion of a group but is not as widely used as the standard deviation. The reason for the preference is that the standard deviation fits in neatly with the use of normal distribution curves, as will be demonstrated later. The *standard deviation* σ of a group of N numbers is defined as

$$\sigma = \sqrt{\frac{\sum (X_i - \bar{X})^2}{N}} \tag{8.6}$$

Going back to the groups 5, 5, 5, 5, 5, 5, and 1, 1, 2, 7, 9, 10, the mean deviation and standard deviation for the first group are obviously both zero. For the second group of numbers, the mean and standard deviations are calculated as

$$\text{mean deviation} = \frac{|1-5| + |1-5| + |2-5| + |7-5| + |9-5| + |10-5|}{6}$$

$$= \frac{4 + 4 + 3 + 2 + 4 + 5}{6} = 3.67$$

$$\sigma = \sqrt{\frac{4^2 + 4^2 + 3^2 + 2^2 + 4^2 + 5^2}{6}} = 3.79$$

Exercises

8.2.1 Compute the range, mean deviation, and standard deviation for the variates given in Exercise 8.1.1.

Solution

We have already computed $\bar{X} = 32.65$, so we construct a new table:

X_i	$|X_i - \bar{X}|$	$(X_i - \bar{X})^2$
0	32.65	1066.02
12	20.65	426.42
16	16.65	277.22
18	14.65	214.62
21	11.65	135.72
23	9.65	93.12
25	7.65	58.52
25	7.65	58.52
25	7.65	58.52
27	5.65	31.92
30	2.65	7.02
32	0.65	0.42
35	2.35	5.52
35	2.35	5.52
41	8.35	69.72
41	8.35	69.72
42	9.35	87.42
45	12.35	152.52
60	27.35	748.02
100	67.35	4536.02
	$\sum|X_i - \bar{X}| = 275.60$	$\sum(X_i - \bar{X})^2 = 8102.50$

$$\text{range} = 100 - 0 = \mathbf{100} \qquad \text{Answer}$$

$$\text{mean deviation} = \frac{\sum|X_i - \bar{X}|}{N} = \frac{275.60}{20} = \mathbf{13.78} \qquad \text{Answer}$$

$$\sigma = \sqrt{\frac{\sum(X_i - \bar{X})^2}{N}} = \sqrt{\frac{8102.50}{20}} = \mathbf{20.13} \qquad \text{Answer}$$

These represent rather large deviations, which might be expected for a set of numbers that are so widely scattered.

8.2.2 Compute the range, mean deviation, and standard deviation for the variates given in Exercise 8.1.2.

8.2.3 Compute the range, mean deviation, and standard deviation for the variates given in Exercise 8.1.3.

8.2.4 Compute the range, mean deviation, and standard deviation for the variates given in Exercise 8.1.4.

8.2.5 Compute the arithmetic mean, median, and standard deviation for each of the two following groups of variates. Compare the results and explain any differences between the two groups.

Group A: 5, 8, 9, 10, 11, 17, 19

Group B: 2, 3, 9, 10, 11, 12, 13

8.2.6 In a manufacturing process, the following set of measurements, in inches, is recorded for the diameters of a group of shafts that have just been completed. Compute the range, weighted arithmetic mean, and standard deviation.

2.4990	2.5000	2.4980	2.5015	2.4985	2.4990	2.4975
2.4985	2.4990	2.5010	2.4985	2.5010	2.5005	2.4980
2.4990	2.4990	2.4995	2.4975	2.4980	2.5000	2.4995
2.4985	2.5005	2.4985	2.4995	2.4980	2.4975	2.4985

8.2.7 The records of a small shop show that it used the following amount of electricity, in kilowatt-hours, in each of the months in a particular year. (The marked rise in the summer months is because of air conditioning.) Compute the arithmetic mean and the standard deviation.

J	F	M	A	M	J	J	A	S	O	N	D
750	650	705	780	840	1020	1500	1200	1750	950	820	705

■ 8.3 PROBABILITY

The *probability* of an event is its frequency of occurrence divided by the number of trials, when the total number of trials is very large. Statistical theory and probability theory are closely linked; statistical sampling is often used to predict the probability of an event occurring in the total population.

If there are n possible ways in which an event can occur and s of these are the looked-for outcomes, then we say that the probability of the looked-for event is

$$p = \frac{s}{n} \tag{8.7}$$

We define the probability of all the other events (the ones not looked for) as

$$q = \frac{n - s}{n} \tag{8.8}$$

It follows, then, that

$$p + q = \frac{s}{n} + \frac{n - s}{n} = 1$$

If $p = 0$, we say the looked-for event cannot occur; if $p = 1$, the event is certain to occur.

Two simple probability laws are worth remembering. One says that if p_1, p_2, p_3, ..., p_n are the probabilities of n mutually exclusive events, then the probability

Figure 8.3
Modern technology received a major boost when ways were found to mass-produce ball bearings to extremely close tolerances. This is possible only through application of the principles of statistical quality control. (Courtesy of FAG Kugelfischer)

of one of these events occurring is $p_1 + p_2 + p_3 + \cdots + p_n$. The other says that if $p_1, p_2, p_3, \ldots, p_n$ are probabilities of n independent events, then the probability of all of them occurring at once is $p_1 p_2 p_3 \cdots p_n$.

The probability that an event will occur X times in N trials is

$$_N C_X = \frac{N!}{X!(N - X)!} \tag{8.9}$$

which is the same as the expression for the number of possible combinations of N things taken X at a time. $N!$ and $X!$ are to be read as "N factorial" and "X factorial." $N!$ is equal to $1 \cdot 2 \cdot 3 \cdot 4 \cdots N$, and $X! = 1 \cdot 2 \cdot 3 \cdot 4 \cdots X$.

Example: Frequency Distribution

A simple example of the application of Equation (8.9) will lead us into the important topic of *frequency distribution* and eventually into the special case of the *binomial frequency distribution*. Let us use the example of tossing a coin. Each time we toss it (assuming it is evenly balanced, i.e., is a "fair" coin), we have equal probabilities of getting a head or a tail. Therefore, $p = \frac{1}{2}$ and $q = \frac{1}{2}$. Suppose we toss the coin twice. There are four possible results: (1) two tails in a row, (2) first a tail and then a head, (3) first a head and then a tail, and (4) two heads in a row. We can write all the possible combinations as

T T
T H
H T
H H

Suppose we consider all the possible combinations resulting from three coin tosses. There are eight, or 2^3:

T T T
T T H
T H T
T H H
H T T
H T H
H H T
H H H

And if we consider the combinations resulting from four tosses, we get 16, or 2^4:

T T T T
T T T H
T T H T
T T H H
T H T T
T H T H
T H H T
T H H H
H T T T
H T T H
H T H T
H T H H
H H T T
H H T H
H H H T
H H H H

If we count the number of heads in each of the combinations that can result from four consecutive tosses, we find the following results:

Number of Heads	Frequency of Occurrence
0	1
1	4
2	6
3	4
4	1

The numbers in the right-hand row are exactly the coefficients in the binomial expansion to the fourth power:

$$(q + p)^4 = q^4 + 4q^3p + 6q^2p^2 + 4qp^3 + p^4$$

But this should not be surprising, because the coefficients are the result of calculating the probability that an event will occur X times in N trials, as given by Equation (8.9), and the general expression for the binomial expansion is

$$(q + p)^N = q^N + Nq^{N-1}p + \cdots + \frac{N!}{X!(N - X)!} q^{N-X}p^X + \cdots + p^N \qquad (8.10)$$

where the coefficient of the general term is also seen to be identical to Equation (8.9).

To underscore the foregoing, let us calculate the probabilities for $N = 4$, using Equation (8.9). For $X = 0$ (0 heads in 4 trials), we get

$$_4C_0 = \frac{4!}{0! \, 4!} = 1$$

Note that 0! is defined as equal to unity. For $X = 1$, and so on,

$$_4C_1 = \frac{4!}{1! \, 3!} = \frac{4 \cdot 3 \cdot 2 \cdot 1}{1 \cdot 3 \cdot 2 \cdot 1} = 4$$

$$_4C_2 = \frac{4!}{2! \, 2!} = \frac{4 \cdot 3 \cdot 2 \cdot 1}{2 \cdot 1 \cdot 2 \cdot 1} = 6$$

$$_4C_3 = \frac{4!}{3! \, 1!} = \frac{4 \cdot 3 \cdot 2 \cdot 1}{3 \cdot 2 \cdot 1 \cdot 1} = 4$$

$$_4C_4 = \frac{4!}{4! \, 0!} = 1$$

We have computed exactly the same results as we achieved by the previous laborious tabulations. If we had gone on, by the laborious method, to $N = 5, 6, 7,$ and so on, we would soon have exhausted our patience. For example, for as few as ten tosses, we would have to examine $2^{10} = 1024$ combinations. It is much easier to use Equation (8.9). In the next section, we will see how this leads into the subject of the binomial distribution.

■ 8.4 BINOMIAL AND NORMAL DISTRIBUTIONS

Let us suppose we toss a coin ten times, knowing that we will get 1024 possible combinations. The binomial expansion for $N = 10$ is

$$(q + p)^{10} = q^{10} + 10q^9p + 45q^8p^2 + 120q^7p^3 + 210q^6p^4 + 252q^5p^5$$
$$+ 210q^4p^6 + 120q^3p^7 + 45q^2p^8 + 10qp^9 + p^{10} \qquad (8.11)$$

The coefficients are the frequencies with which the successive numbers of heads will occur in the combinations. That is, in the first term, for $X = 0$, the coefficient is unity; this is the one combination out of 1024 that consists of all tails, or zero heads. Its probability of occurrence is obviously one out of 1024, or 0.000 977. In the second term, for $X = 1$ (that is, the term for which the exponent of p is unity), the coefficient is 10. In other words, ten combinations out of 1024 will have exactly one head; so the probability of getting exactly one head in ten tosses of a coin is $10/1024 = 0.009\ 766$. Going to the next term, where the exponent of p is 2, i.e., $X = 2$, we see that the frequency with which combinations contain exactly two heads in ten tosses of a coin is $45/1024 = 0.043\ 945$. The highest frequency in the group is the 252 combinations that each contain exactly five heads; and so on.

We can now construct a *frequency distribution diagram* for the relative frequencies of combinations containing specified numbers of heads, as has been done in Figure 8.4. Each of the coefficients in the 11 terms of Equation (8.11) has been plotted as a vertical bar, as an indicator of how many combinations, out of the total of 1024 possible combinations, contain the specified number of heads. Since the frequency distribution is derived from the coefficients of a binomial expansion, it is referred to as a *binomial distribution*. The distribution shown here is symmetrical about the center, but this only occurs when $p = q = \frac{1}{2}$, as is the case with a coin toss. The kind of frequency distribution chart in Figure 8.4, consisting of vertical bars, is called a *histogram*, as explained in Chapter 7.

It will be recalled that the value of N that was used in the binomial distribution of Figure 8.4 was 10. If N increases without limit, the histogram of Figure 8.4 will approach a smooth curve known as the *normal distribution curve*, or sometimes called a bell-shaped curve. One is shown in Figure 8.5, drawn with proportions to match those of Figure 8.4. The equation of the normal distribution curve is

$$Y = \frac{1}{\sigma\sqrt{2\pi}}\ e^{-0.5(X-\mu)^2/\sigma^2} \tag{8.12}$$

where Y is the frequency, X is the event, σ is the standard deviation, and μ is the mean.

If we change the variables in Equation (8.12), we can produce a so-called *standard normal distribution curve*, as is often done in engineering and science. The advantage of a standardized curve is that the units along the horizontal axis become even mul-

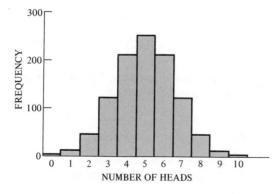

Figure 8.4

Frequency distribution showing how many combinations containing a given number of heads could occur in ten successive tosses of a coin. There are 1024 possible combinations of ten objects; among these 1024 combinations there are 10 containing only one head, 45 containing two heads, and so on.

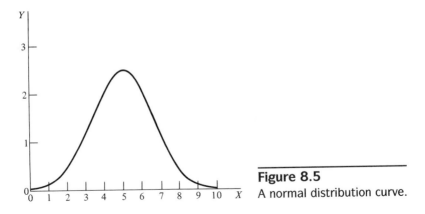

Figure 8.5
A normal distribution curve.

tiples of σ, the standard deviation, and the area under the curve is purposely caused to become 1. In the case at hand, the value of the procedure is that the same curve is then usable for all problems involving normal distribution curves.

The changes in variables that we will use are

$$y = Y\sigma, \qquad z = \frac{X - \mu}{\sigma} \qquad (8.13)$$

By simple substitution of the variables defined by Equations (8.13), we change Equation (8.12) into

$$y = \frac{1}{\sqrt{2\pi}} e^{-0.5z^2} \qquad (8.14)$$

Equation (8.14) is plotted in Figure 8.6. Note that $z = 0$ occurs at the center line, so the mean of this curve is zero. Also, the horizontal axis is divided into units that are directly interpretable as multiples of σ. For example, we can write, using the

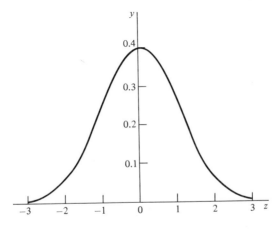

Figure 8.6
A standardized normal distribution curve.

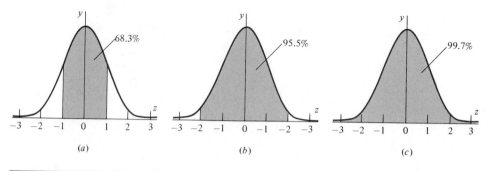

Figure 8.7
(a) 68.3 percent of the data points lie within the range $z = -1$ to $z = +1$, that is, within $\pm\sigma$ of the center point, or mean. (b) 95.5 percent of the data lie within $\pm 2\sigma$ of the mean. (c) 99.7 percent of the data lie within $\pm 3\sigma$ of the mean.

second of Equations (8.13):

z	X
-2	$\mu - 2\sigma$
-1	$\mu - \sigma$
0	μ
+1	$\mu + \sigma$
+2	$\mu + 2\sigma$

Thus, the values of z on the standardized curve can be interpreted as integral multiples of σ, measured from the center point. In Figure 8.7, we note that approximately 68.3 percent of the area under the standardized normal distribution curve lies between $z = -1$ and $z = 1$; in other words, 68.3 percent of the data points of a normally distributed population will lie within $\pm\sigma$ of the center point. In like fashion, 95.5 percent of the data points will lie within $\pm 2\sigma$ of the center, and 99.7 percent will lie within $\pm 3\sigma$ of the center. These values, plus a couple of others we will be interested in, are provided in Table 8.1.

Table 8.1   Area Under Normal Distribution Curve for Selected Values of z

z	Percent of Area Under Normal Distribution Curve Lying Between $-z$ and $+z$
1	68.3
1.65	90.0
1.96	95.0
2	95.5
2.58	99.0
3	99.7

Exercises

8.4.1 Show that the area under the normal distribution curve of Figure 8.6 between plus-or-minus one standard distribution ($\pm\sigma$) of the center point is approximately 68 percent of the total area under the curve. (Assume the total area under the curve equals 1.)

Solution

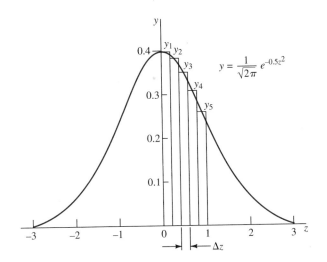

$$y = \frac{1}{\sqrt{2\pi}} e^{-0.5z^2}$$

We will use an approximation method to compute the area under the curve by dividing it up into narrow strips of rectangular cross section. The curve in the figure has five narrow strips drawn between the origin and $z = 1$. (Remember that $z = 1$ is one standard deviation from the origin. See Figure 8.6.) Of course, if we were to use more strips than five, we would get a closer approximation to the area under the curve, but at the expense of extra labor; our decision is to use five. The width of each strip is Δz, and the heights $y_1 = 0.395$, $y_2 = 0.378$, $y_3 = 0.350$, $y_4 = 0.310$, and $y_5 = 0.266$ are scaled off the figure. Since we are dividing the distance from 0 to 1 into five parts, $\Delta z = 0.2$. Summing the areas,

$$y_1 \cdot \Delta z = 0.395(0.2) = 0.079$$
$$y_2 \cdot \Delta z = 0.378(0.2) = 0.076$$
$$y_3 \cdot \Delta z = 0.350(0.2) = 0.070$$
$$y_4 \cdot \Delta z = 0.310(0.2) = 0.062$$
$$y_5 \cdot \Delta z = 0.266(0.2) = \underline{0.053}$$
$$\text{area} = \overline{0.340}$$

$$2 \times \text{area} = \textbf{0.680} \qquad\qquad \text{Answer}$$

8.4.2 Continue the preceding example. Plot the curve, and show that the area under the curve between $\pm2\sigma$ is approximately equal to 95.5 percent of the total area under the curve.

8.4.3 Carry out the details of substituting Equations (8.13) into Equation (8.12), getting Equation (8.14).

■ 8.5 APPLICATION: SAMPLING

One application of the ideas presented in the previous section is *sampling*. Suppose you are an employee of a tire manufacturer. Your company manufactures large numbers of tires and wishes to include in its advertising a statement about the average mileage that can be expected of its product. You could get an average by testing *all* the tires your company makes, but that would use up all the tires in your inventory. So you test a sample—100 tires, say—and get an arithmetic mean of 21,000 miles, with a standard deviation of 1050 miles, for the sample. The question to be answered is whether you can include this average of 21,000 miles in your advertising and be confident that it is reasonably representative of your total production.

The total number of tires produced is referred to as the *population*, with symbol N. The 100 tires are referred to as the *sample*, represented by n. The symbol μ will be used for the *population* mean, and the symbol $\bar{x}$ for the *sample* mean. The symbol σ will be used for the *population* standard deviation, and the symbol s for the *sample* standard deviation. We know $\bar{x}$ and s; these came from our tests of $n = 100$ tires. We don't know μ or σ.

First we must address the properties of $\bar{x}$, the sample mean. Hypothetically, if we tested *all* the tires in the population, we would expect to get a normal distribution curve for the results that resembled that in Figure 8.6. We would know the exact value of μ and would not have to make inferences about it from our sample of 100 tires. But we also would be left without any tires, so we will use the sampling process instead.

If we divide the entire population into sample groups of $n = 100$ and test them (another hypothetical case), we will get a sample mean for each group, which we can call $\bar{x}_1, \bar{x}_2, \bar{x}_3, \ldots$. If we plot these as a frequency distribution, we will find that the sample means $\bar{x}_1, \bar{x}_2, \bar{x}_3 \ldots$ will also arrange themselves into a normal distribution curve, provided the sample size is reasonably large, say, more than $n = 30$. The dis-

Figure 8.8
Engineering research programs frequently produce huge masses of data, which must then be statistically analyzed. (Courtesy of University of California, Davis)

tribution curve for $\bar{x}_1, \bar{x}_2, \bar{x}_3 \ldots$ will of course have its own mean, called the mean of the sample means and designated $E(\bar{x})$, and its own standard deviation, called $\sigma_{\bar{x}}$. If n is less than 10 percent of N, then $E(\bar{x})$, the mean of the sample means, will be approximately equal to μ, the mean of the total population. Furthermore, $\sigma_{\bar{x}}$, the standard deviation of the sample means (called the *standard error of the mean*), will be approximately equal to $\sigma/\sqrt{n}$, where σ is the standard deviation of the total population and n is the sample size. We can summarize this by writing

$$E(\bar{x}) \cong \mu \tag{8.15}$$

and

$$\sigma_{\bar{x}} \cong \sigma/\sqrt{n} \tag{8.16}$$

provided $n < 0.10N$ and $n > 30$.

We assume the production is very large, so the condition $n < 0.10N$ is easily met. Also, the condition $n > 30$ is met, because $n = 100$. The known values from our sample tests are

$\bar{x} = 21{,}000$ miles

$s = 1050$ miles

$n = 100$

Let us assume that Figure 8.9 represents the normal distribution of all the values of $\bar{x}$ for all our hypothetical sample groups of 100 each. Also, because of Equation (8.15), we can replace $E(\bar{x})$, the mean of the sample means, by μ, the mean of the total population, although we do not actually know the numerical value of either one. We do know from Table 8.1 that 95 percent of the values of $\bar{x}_1, \bar{x}_2, \bar{x}_3 \ldots$ will lie between $z = \pm 1.96$, that is, will lie in the interval described by $\mu \pm 1.96\sigma_{\bar{x}}$. For each value of $\bar{x}$, we could also establish an interval of $\bar{x} \pm 1.96\sigma_{\bar{x}}$. Since the values of $\bar{x}$ are distributed in the form of a normal distribution curve, then the intervals $\bar{x} \pm 1.96\sigma_{\bar{x}}$ will also be distributed in the form of a normal distribution curve. And because 95 percent of these intervals will have values of $\bar{x}$ that lie within the shaded region of Figure 8.9, 95 percent of the intervals will also overlap the value of μ, whatever it is. Therefore,

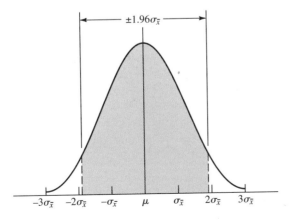

Figure 8.9

A distribution curve for the means $\bar{x}_1, \bar{x}_2, \bar{x}_3 \ldots$ of a set of samples.

we can say with a 95 percent degree of confidence that the interval $\bar{x} \pm 1.96\sigma_{\bar{x}}$ contains μ. This interval is called the *.95 confidence interval estimate of* μ.

We know from Equation (8.16) that $\sigma_{\bar{x}} = \sigma/\sqrt{n}$. We do not know the value of σ, but we do know the value of $s = 1050$ miles. For a sample size $n > 30$, we can use the approximation $\sigma \cong s$. Thus, we can say that the approximate .95 confidence interval estimate of μ for our case is

$$\bar{x} \pm 1.96\sigma_{\bar{x}}$$
$$\cong \bar{x} \pm 1.96\sigma/\sqrt{n}$$
$$\cong \bar{x} \pm 1.96s/\sqrt{n}$$
$$= 21{,}000 \text{ miles} \pm 1.96 \ (1050 \text{ miles}/\sqrt{100})$$
$$= 21{,}000 \text{ miles} \pm 206 \text{ miles}$$

If, instead of 95 percent, we want a 99 percent confidence interval, we can use Table 8.1 to show that the approximate .99 confidence interval estimate of μ is

$$\bar{x} \pm 2.58s/\sqrt{n}$$
$$= 21{,}000 \text{ miles} \pm 2.58 \ (1050 \text{ miles}/\sqrt{100})$$
$$= 21{,}000 \text{ miles} \pm 271 \text{ miles}$$

So our average mileage of the total tire population might vary between 20,729 and 21,271 miles, with a 99 percent degree of confidence. In our advertising we might be wise to state that the average mileage for our tires is 20,700 miles rather than the 21,000 we obtained from our test. Even so, a small fraction of our tires might fall below our published claim. (One-half percent will fall below 20,729 miles and one-half percent above 21,721 miles, with 99 percent in between.)

Exercise

8.5.1 You are a manufacturer of electrical resistors, and you wish to establish a confidence interval for the average resistance of one of your products. Its nominal rating is 200,000 ohms. You test 200 of the units and get the values $\bar{x} = 200{,}900$ ohms and $s = 1750$ ohms. What is the .95 confidence interval estimate of μ? The .99 confidence interval estimate? (Assume $n < 0.10N$.)

■ 8.6 APPLICATION: MANUFACTURING CONTROL LIMITS

Another use of statistics is *control charts*, which are widely used in quality control by manufacturing firms. Let us say we want to set up a control chart for the parts being produced in a manufacturing process. Figure 8.10 shows how such a chart is laid out. A center line (CL) is assumed to be equal to μ, the population mean. An upper control limit (UCL) and a lower control limit (LCL) are then selected. A common practice for selecting the locations of the UCL and LCL is to place them a distance of $\pm 3\sigma$ from CL. From what we already know about normal distribution curves, we can see that 99.7 percent of the parts produced will lie between these control limits (see Table 8.1).

There are various kinds of control charts, but the ones we will describe deal with $\bar{x}$, the mean of the samples we will test, and R, the range of the samples. (These are

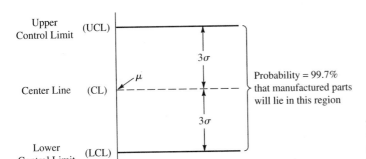

Figure 8.10
A control chart, with upper and lower control limits set at $\mu \pm 3\sigma$.

typically called "x-bar and R charts.") To set up the charts we would really like to know the value of μ, but the only way we could find it would be to test every one of the parts. Even if the parts could be tested without destroying them, testing the entire population would be too expensive. So, as in the previous section, we will use the mean of the sample means $E(\bar{x})$ as an approximation for μ. In general, this is acceptable if the number of sample groups is reasonably large, say, 20 or more.

We compute an average range $\bar{R}$ for our group of samples, using the following relation:

$$\bar{R} = \frac{R_1 + R_2 + R_3 + \cdots + R_m}{m}$$

Once $E(\bar{x})$ and $\bar{R}$ are known, Table 8.2 gives values that can be used directly to calculate UCL, CL, and LCL, as follows. For the $\bar{x}$ chart, we can use the equations

$$\text{UCL} = E(\bar{x}) + A_2\bar{R}$$

$$\text{CL} = E(\bar{x})$$

$$\text{LCL} = E(\bar{x}) - A_2\bar{R}$$

(8.17)

Table 8.2   Selected Values of Factors to Be Used in Setting Up Control Charts

n	A_2	D_3	D_4
2	1.880	0	3.267
3	1.023	0	2.575
4	0.729	0	2.282
5	0.577	0	2.115
6	0.483	0	2.004
7	0.419	0.076	1.924
8	0.373	0.136	1.864
9	0.337	0.184	1.816
10	0.308	0.223	1.777

Adapted from *ASTM Manual on Presentation of Data and Control Chart Analysis* (Philadelphia, Pa.: American Society for Testing and Materials, 1976), p. 83.

And for the R chart, we can use the equations

$$UCL = \bar{R}D_4$$

$$CL = \bar{R}$$

(8.18)

$$LCL = \bar{R}D_3$$

where A_2, D_3, and D_4 are taken from the table. (Establishing the validity of the values of A_2, D_3, and D_4 would require material far beyond the scope of this text. Our purpose is simply to give an example of how the subject of statistics is used.)

Now we will put some actual numbers into our example. Let us assume we are manufacturing electrical resistors, each with a nominal rating of 20 ohms. We take a sample of $n = 5$ each day for 20 days, and record the resistance measurements as shown in Table 8.3. (In the table we will only use the measurements for the first ten days, to keep our example simple.) The sample means $\bar{x}_i$ and the ranges R_i have been computed for each group and entered in the last two columns of the table. We can now compute $E(\bar{x})$ and $\bar{R}$ as follows:

$$E(\bar{x}) = \frac{19.9 + 20.0 + 20.2 + 20.0 + 19.7 + 20.1 + 19.7 + 19.5 + 20.1 + 19.9}{10}$$

$$= 19.9$$

$$\bar{R} = \frac{1.4 + 2.8 + 3.1 + 2.8 + 2.0 + 3.4 + 3.4 + 2.0 + 3.0 + 2.2}{10}$$

$$= 2.6$$

We can now compute UCL, CL, and LCL for the $\bar{x}$ chart by using Equations (8.17) and taking A_2 from Table 8.2 for $n = 5$:

$$UCL = E(\bar{x}) + A_2\bar{R} = 19.9 + 0.577(2.6) = 21.4$$

$$CL = E(\bar{x}) = 19.9$$

$$LCL = E(\bar{x}) - A_2\bar{R} = 19.9 - 0.577(2.6) = 18.4$$

Table 8.3   Data for Samples of Resistors

Sample Group	Individual Readings (Ω)					Mean of Sample Group ($\bar{x}_i$)	Range of Sample Group (R_i)
	1	2	3	4	5		
1	19.5	20.2	20.8	19.5	19.4	19.9	1.4
2	19.2	18.8	19.9	21.6	20.5	20.0	2.8
3	18.8	20.8	21.9	19.5	19.9	20.2	3.1
4	20.8	18.6	20.2	18.9	21.4	20.0	2.8
5	19.1	19.2	20.2	20.9	18.9	19.7	2.0
6	22.0	18.6	19.2	20.1	20.5	20.1	3.4
7	18.5	21.9	18.8	20.2	18.9	19.7	3.4
8	18.9	19.5	20.2	20.5	18.5	19.5	2.0
9	20.5	19.6	21.9	18.9	19.5	20.1	3.0
10	19.3	19.8	19.7	19.3	21.5	19.9	2.2

We can compute UCL, CL, and LCL for the range (R) chart by using Equations (8.18), where the values of D_3 and D_4 are also taken from Table 8.2 for $n = 5$.

$$\text{UCL} = \bar{R}D_4 = 2.6(2.115) = 5.50$$

$$\text{CL} = \bar{R} = 2.6$$

$$\text{LCL} = \bar{R}D_3 = 2.6(0) = 0$$

The $\bar{x}$ and R charts are shown in Figure 8.11, and the ten sets of values of $\bar{x}$ and R are plotted on those charts. The values all lie within the control limits we have established, which assures us that no out-of-control conditions exist in our process. Because of this assurance, we plan to use the charts in our production process from now on, by taking samples at regular intervals and plotting the values of $\bar{x}$ and R from those samples onto our control charts. As long as the plotted values remain within the control limits, we will permit production to continue. If the plotted values

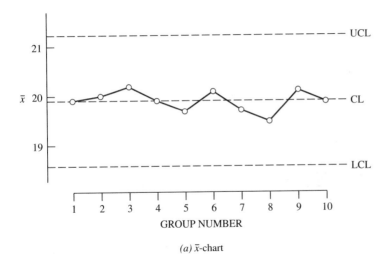

(a) $\bar{x}$-chart

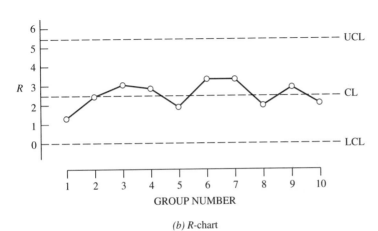

(b) R-chart

Figure 8.11
Charts for the resistor example.
(a) $\bar{x}$ chart. (b) R chart.

begin to drift outside the control values, we will order production to stop and investigate the cause.

Exercise

8.6.1 In Table 8.3, assume you took samples of $n = 4$ instead of $n = 5$. (In other words, ignore the values in column 5 under "Individual Readings.") Compute new values for x_i and R_i, and prepare x and R charts corresponding to these new values.

■ 8.7 REGRESSION

In an earlier chapter we discussed fitting curves to a set of experimentally obtained points by using judgment. Now we will be more scientific about fitting a straight line to a set of scattered points.

Suppose we have a set of data points involving two variables x and y, as in Table 8.4. The points are plotted in Figure 8.12 in what is called a *scatter diagram*. The job now is to find the equation of the straight line that best fits the data points. This process is termed *linear regression*. (It is possible to fit curves other than a straight line to data points—a process called *polynomial regression*. Here we will deal only with linear regression.)

From algebra, we know that the typical equation for a straight line is in the form $y = a + bx$, where we call x the independent variable and y the dependent variable. The numerical value of b is the slope of the line, and a is the y-axis intercept.[2] From Figure 8.13 we can see that $b = \tan \alpha = \Delta y/\Delta x$. In our analysis we will reserve the symbol y for the actual measured data points. (Each value of y is associated with a particular value of x.) We will use the symbol y_e to represent the *estimated* values of y that we will compute from the equation for the straight line we are seeking. (Again, each value of y_e will be associated with a particular value of x.) The equation we are seeking, then, has the form

$$y_e = a + bx \tag{8.19}$$

In the regression method we are going to use, called the *method of least squares*, we will find the line that produces the smallest sum of the values $(y - y_e)^2$. Each

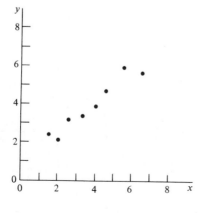

Figure 8.12
Points from Table 8.4, plotted in a scatter diagram.

Table 8.4   Paired Values of x and y*

x	y	x	y
1.50	2.50	4.00	4.00
2.00	2.25	4.50	4.75
2.50	3.25	5.50	6.00
3.25	3.50	6.50	5.75

* Determined from a hypothetical experiment and plotted in Figure 8.12.

value $y - y_e$ is the difference, for any given value of x, between the actual measured value y and the value y_e estimated by the yet-to-be-found Equation (8.19).

In mathematical terms, what we wish to do is find the values of a and b that will cause $\Sigma (y - y_e)^2$ to be a minimum. This is a standard problem in differential calculus. We will not give the details here, but only the result, which is in the form of two simultaneous equations called *normal equations*:

$$an + b \sum x = \sum y, \qquad a \sum x + b \sum x^2 = \sum xy \qquad (8.20)$$

where n is the number of pairs of values (x, y). The values of a and b can be found by solving Equations (8.20) simultaneously, which initially gives us

$$b = \frac{n \sum xy - \sum x \sum y}{n \sum x^2 - (\sum x)^2} \qquad (8.21)$$

Having found b from Equation (8.21), a can be found from the first of Equations (8.20) as

$$a = \frac{\sum y - b \sum x}{n} \qquad (8.22)$$

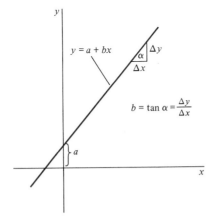

Figure 8.13
Graph of a line with the equation $y = a + bx$.

---

[2] In Chapter 7, the equation of a straight line was given as $y = mx + b$, where m was the slope and b the y-axis intercept. However, it should be apparent that $y = a + bx$ represents the same thing as does $y = mx + b$ and that the choice of symbols is arbitrary.

Table 8.5	Calculation of Quantities for Finding the Equation $y_e = a + bx$ for Data in Table 8.4		
x	y	x^2	xy
1.50	2.50	2.25	3.75
2.00	2.25	4.00	4.50
2.50	3.25	6.25	8.13
3.25	3.50	10.56	11.38
4.00	4.00	16.00	16.00
4.50	4.75	20.25	21.38
5.50	6.00	30.25	33.00
6.50	5.75	42.25	37.38
29.75	32.00	131.81	135.52

$$\sum x = 29.75$$
$$\sum y = 32.00$$
$$\sum x^2 = 131.81$$
$$\sum xy = 135.52$$

In Table 8.5, we have repeated the x and y values from Table 8.4 and have added the columns necessary to complete the calculation of a and b. From Equations (8.21) and (8.22) we get

$$b = \frac{8(135.52) - (29.75)(32.00)}{8(131.81) - (29.75)^2}$$

$$= 0.78$$

$$a = \frac{32.00 - 0.78(29.75)}{8}$$

$$= 1.10$$

Thus, the equation we are seeking is

$$y_e = 1.10 + 0.78x$$

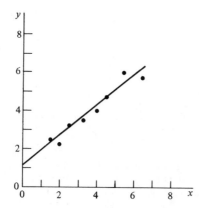

Figure 8.14

A regression line with the equation $y_e = 1.10 + 0.78x$, which best fits the scatter diagram of Figure 8.12, based on the least-squares method.

which has been plotted in Figure 8.14 together with the scatter diagram of Figure 8.12.

Exercises

8.7.1 In an experiment, the following values of voltage and current were recorded. Determine the equation of the best straight line that will represent the points, using the method of least squares. Plot the line and the points on graph paper. *Hint:* Represent the voltage (V) by x and the current (mA) by y.

V	1	2	3	4	5	6	7	8	9	10
mA	6	12	20	24	32	35	38	45	50	53

8.7.2 A group of 30 college freshmen are identified, all of whom are taking engineering at the same university and all of whom came from the same high school. A table is made as follows, comparing each student's overall high-school grade point average (GPA) with the GPA earned by that student during the freshman year in college. Prepare a scatter diagram for these 30 students with high-school GPA along the horizontal axis and freshman-year college GPA along the vertical axis. Using the method of least squares, determine the equation of the best straight line that will represent the points.

High-School GPA	Freshman-Year College GPA	High-School GPA	Freshman-Year College GPA
3.00	2.45	3.65	3.20
3.05	2.55	3.65	2.80
3.05	2.70	3.70	3.80
3.15	3.45	3.75	3.75
3.15	2.80	3.75	2.95
3.25	2.70	3.80	3.65
3.25	2.55	3.80	2.70
3.25	2.40	3.85	3.30
3.30	3.10	3.90	4.00
3.40	2.80	3.90	3.45
3.50	3.50	3.95	3.60
3.50	2.65	4.00	3.80
3.55	3.05	4.00	3.45
3.65	3.20	4.00	3.25
3.65	3.15	4.00	2.20

■ 8.8 CORRELATION

In statistics, a value called the *coefficient of correlation* appears repeatedly. It is a numerical measure of how well a regression line fits a set of scattered data. The coefficient of correlation, usually designated as r, is defined by the following expression:

$$\text{coefficient of correlation} = r = \sqrt{\frac{\sum (y_e - \bar{y})^2}{\sum (y - \bar{y})^2}} \qquad (8.23)$$

From the section on measures of dispersion, we find that the deviation of a data point with respect to the mean of a group of data points is given as $X_i - \bar{X}$ or, in the case at hand, $y - \bar{y}$. We also saw, in Equation (8.6), that the standard deviation is defined (we have substituted y's for the X's and n for N) as

$$\sigma = \sqrt{\frac{\sum (y - \bar{y})^2}{n}} \tag{8.24}$$

The square of the standard deviation is called the *variance*:

$$\text{variance} = \sigma^2 = \frac{\sum (y - \bar{y})^2}{n} \tag{8.25}$$

We will give the name *total variation* to the quantity

$$\text{total variation} = \sum (y - \bar{y})^2$$

and the name *explained variation* to the quantity

$$\text{explained variation} = \sum (y_e - \bar{y})^2$$

Thus, we can see that the coefficient of correlation is

$$r = \sqrt{\frac{\text{explained variation}}{\text{total variation}}}$$

If the explained variation $\sum (y_e - \bar{y})^2$ is exactly the same value as the total variation $\sum (y - \bar{y})^2$, then r will be equal to unity. This means that all the estimated values y_e have exactly the same relationship to the mean value $\bar{y}$ as do the actual data points y. In other words, all the data points lie exactly on the line $y_e = a + bx$, and the representation of the data points by the line is perfect. If the values y_e estimated by our regression line show less dispersion than do the actual data points, then the "explained variation" will be less than the total (actual) variation, and our correlation coefficient will be less than unity. As the actual dispersion of the data points becomes worse and worse, the total variation in the denominator of Equation (8.23) gets larger and larger, causing the coefficient of correlation to get smaller and smaller. Obviously, our regression line becomes less and less adequate for representing the scattered data under such circumstances, and this inadequacy is measured by the coefficient of correlation.

Returning to our previous set of scattered data points given in Tables 8.4 and 8.5, we can compute the coefficient of correlation for the regression line in Figure 8.14 by computing the necessary quantities as in Table 8.6. The coefficient of correlation, using Equation (8.23), is

$$r = \sqrt{\frac{12.883}{13.752}}$$

$$= 0.968$$

This is a very high correlation coefficient, and it gives a numerical value to our subjective evaluation of Figure 8.14, which is that the line fits the data points quite well. A coefficient $r = 1$ indicates perfect correlation; that is, all the points lie exactly

Table 8.6   Calculation of Quantities for Finding
Coefficient of Correlation, Using Values
from Table 8.4 and Regression Line
$y_e = 1.10 + 0.78x$

x	y	y_e	$y - \bar{y}$	$y_e - \bar{y}$	$(y - \bar{y})^2$	$(y_e - \bar{y})^2$
1.50	2.50	2.27	−1.50	−1.73	2.250	2.993
2.00	2.25	2.66	−1.75	−1.34	3.063	1.796
2.50	3.25	3.05	−0.75	−0.95	0.563	0.903
3.25	3.50	3.64	−0.50	−0.36	0.250	0.130
4.00	4.00	4.22	0	0.22	0	0.048
4.50	4.75	4.61	0.75	0.61	0.563	0.372
5.50	6.00	5.39	2.00	1.39	4.000	1.932
6.50	5.75	6.17	1.75	2.17	3.063	4.709
	32.00				13.752	12.883

$$\sum y = 32.00$$

$$\bar{y} = \frac{\sum y}{n} = \frac{32.00}{8} = 4.00$$

$$\sum (y - \bar{y})^2 = 13.752$$
$$\sum (y_e - \bar{y})^2 = 12.883$$

on the regression line. A coefficient $r = 0$ indicates uniform dispersion; that is, no line can represent the data. A value greater than $r = 0.70$ would show that the regression line represents the data to some degree, and a value greater than 0.90 would be considered quite good.

Exercises

8.8.1 Compute the coefficient of correlation for the data given in Exercise 8.7.1. Do you think these data show a good degree of correlation?

8.8.2 Compute the coefficient of correlation for the data given in Exercise 8.7.2 for high-school and college GPAs. Do you think there is a good correlation between these GPAs?

References

W. S. Messina, *Statistical Quality Control for Manufacturing Engineers* (New York: Wiley, 1987).

D. C. Montgomery, *Introduction to Statistical Quality Control* (New York: Wiley, 1985).

F. H. Zuwaylif, *General Applied Statistics* (Reading, Mass.: Addison-Wesley, 3d ed., 1979).

Photographs courtesy of Grumman Aerospace Corporation (top) and University of California, Davis (center and bottom left).

9

SI and Other Unit Systems

There has been a great deal of discussion in the United States since about 1960 regarding conversion to the metric system. Actually, the United States has been partially on the metric system for years. Film, lenses, and ball bearings have all been standardized in millimeter sizes for a long time, and the familiar units of volts, amperes, and watts are all from the metric system. Much of the recent dialogue has emphasized units of measure for length, volume, and mass. We now see road signs with kilometers on them, we occasionally buy gasoline in liters, and we purchase some bulk commodities in kilograms instead of in pounds. We are also regularly assailed by arguments that the familiar English system of units is irrational—that we should not be using a system involving conversion factors of 12 inches to the foot, or 5280 feet to the mile, when we could be using a rational system in which all the conversions are done with factors of ten.

Actually, these arguments miss the point. Emphasis upon converting to metric units of length, volume, and mass tells only a small portion of what conversion is really all about. Similarly, the "factors of ten" argument misses the point, because this is basically an argument concerning *decimalization*, not metrication per se. Measurements in virtually all U.S. factories have been based on the decimalized inch for years. All measurements are in inches and decimal fractions of inches, even long measurements such as the wingspan of an airplane. Decimalization indeed is an advantage, but the advantage can be achieved by dealing with the decimalized inch as well as by the decimalized meter. Also, we should realize that not all factors in the metric system are decimalized. Any

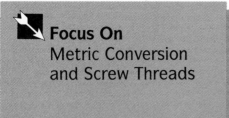

Focus On
Metric Conversion and Screw Threads

Much criticism has been directed at the United States for clinging to its "archaic" system of pounds and inches. But the criticism is largely misguided and confuses the advantages of the metric system with those of decimalization, as is described in this chapter. Actually, the most basic issue in conversion comes down to the matter of screw threads.

Learning to think in terms of kilometers, meters, and kilograms can be troublesome if one has been raised under a different system, but it is not impossible. However, if one has a machine tool in the shop that is constructed with gears to cut screw threads with an integral number of threads per inch, it may literally be impossible to use that machine to cut threaded parts to metric standards. To do so might require a gear with a fractional number of teeth—something that does not exist. Similarly, machine tools come equipped with measuring dials that produce some convenient measure, such as 0.100 inch per complete revolution. Just to change the dial to metric measure would mean causing one rotation of the dial to be worth 2.54 mm. If we marked off 254 divisions on the dial, with each division worth 0.01 mm, we would find the dial very inconvenient to use if we wanted to make a measurement requiring multiple rotations. To make it satisfactory, we would have to change not only the dial, but also the measuring screw to which it is attached. And so it goes. All mea-

suring devices would have to be scrapped and replaced. Most of them use precision screws, which are expensive. Estimates of the cost to the United States of conversion have ranged from $20 billion to $30 billion. Nevertheless, as the old equipment wears out, the United States is gradually equipping itself with new tools that can go either way.

Metric bolts and inch bolts are not interchangeable. With the world on two different screw-thread standards, duplicate sets of screws, bolts, and wrenches have to be kept on hand, certainly a costly process. A further problem occurs with certain metric and inch bolts that are almost—but not quite—interchangeable.

For example, a $\frac{3}{8}''$-16 (0.375-inch diameter, 16 threads per inch) U.S. bolt will almost fit a tapped hole intended for an M10 (10-mm-diameter) metric bolt. The metric thread is 0.393 inches in diameter and has 16.9 threads per inch. Thus, the $\frac{3}{8}''$ bolt will enter the hole, and the first few threads will engage before interference starts to develop. After that, as the bolt is tightened, it simply strips out the threads. The mechanic may discover the error too late, perhaps after an expensive part—an engine block, say—has been ruined.

Metric conversion is occurring very gradually. Metric standards cannot be ignored in a world that becomes constantly more dependent upon international trade. Reluctance to convert does not rest so much upon clinging to an old familiar system as it does upon a simple thing like screw threads.

References
"Inches, Meters, and the Facts," *Mechanical Engineering*, March 1965, pp. 31–37.
"The Metric System—Should We Convert?" *Mechanical Engineering*, July 1962, pp. 41–47.

time we have a calculation involving mass and the force we call "weight," the value of the acceleration of gravity comes into the picture, and this acceleration (at the Earth's surface) is equal to 9.807 meters per second squared.

■ 9.1 STANDARDIZATION AND COHERENT UNITS

It is important to realize that the two basic forces behind conversion to metrication (or rather, conversion to the SI, or *Système International*[1]) are (1) standardization and (2) coherence. The first force is commercial, whereas the second is scientific. But we will come to that in a moment. First, let us look at a bit of history.

The United States actually legalized the metric system in this country in 1866, but did not mandate its use. It still has not done so, but in recent years it has passed legislation intended to provide for smooth conversion. By 1893, the United States had gone so far as to drop any independent reference standards for the foot and the pound, instead defining them legally in terms of their relationships to the meter and the kilogram.

In the late nineteenth century, the international metric system was based on three fundamentally defined quantities: the centimeter (for length), the gram (for mass), and the second (for time). However, this system, called the *cgs system*, was cumbersome because some of the basic units were too small for convenient use. Thus, a new system came into effect, using the meter, kilogram, and second as the fundamental units, called the *mks system*. This is the system that has been recognized as "the metric system" by most generations of Americans, but it is not exactly the same system as the modern SI.

During the decades when the industrial power of the United States dominated international trade, there was little incentive in this country to convert. United States products made to the inch standard were widespread throughout the world, and users in other countries were forced to stock two sets of parts—one set made to inch standards, the other to metric. It might not be obvious why two sets of parts would be needed, until one looks into the nature of threaded fasteners, such as bolts and machine screws. In developing thread standards, there is a natural tendency to deal in convenient quantities. Under the inch system, for example, we might establish a standard for a particular bolt of, say, 20 threads to the inch, which works out to a pitch distance (the distance from the top of one thread to the next, measured parallel to the bolt axis) of 0.05 in., or 1.27 mm. But, if one were working out a set of standards under the metric system, the number 1.27 probably would not come up. Instead, a pitch distance of 1.25 mm might be a convenient number. Obviously, a bolt from one system will not fit a hole that has been threaded under the other system. Also, wrenches for one system do not fit properly on the heads of bolts made under the other system, as anyone can testify who has had to buy two sets of wrenches, one for American standard bolts and the other for metric bolts.

---

[1] The full name is *Le Système International d'Unitès*, or The International System of Units.

Figure 9.1
Many of the presumed advantages of the metric system are in fact advantages of decimalization. American industry has used the decimalized inch for decades. Note the steel scale in background. (Courtesy of Applicon Incorporated)

It was apparent to U.S. manufacturers during the 1950s and 1960s that conversion to the metric system should be avoided if at all possible. It was estimated that it would cost the country billions of dollars to convert because of duplicate inventories of fasteners and tools and the replacement of measuring instruments and machines with gearing systems that could not readily be converted to metric equivalents. Duplicate inventories would be required for as long as parts had to be stocked for equipment still in use, although the duplicates could be dropped once the whole world was on one compatible standard. Obviously, the United States hoped that the standard that won out would be the one it was currently using, but matters did not turn out that way. As the industrial might of other countries grew (and almost all of them were on the metric system) and the relative position of the United States declined, it finally became a matter of industrial survival for the United States to convert. Otherwise, its products would experience sales resistance abroad, and international sales have become vital for all manufacturers, no matter where they are located. So it is *standardization*, and for commercial reasons, that is the most compelling force behind the drive to convert.

The other compelling reason for conversion is *coherence*, and the reasons are scientific. Over the years, many different ways of defining fundamental units have been developed, leading to slightly different results for units that were supposed to be the same. For example, from the beginning the *volt* has been defined in terms of the metric system, but a new international definition reached in 1948 changed its value by about 300 parts per million. Another agreement, in 1969, changed it again by 8.4 parts per million. Such small changes do not usually affect engineers, but in precision scientific measurement it is essential that everyone use the same standard, or interpretation will suffer. One of the principal advantages that the adoption of SI gives us is a coherent and internally consistent system.

In the SI there are only nine independent, fundamentally defined units, and they are based on the most precisely reproducible reference standards we can devise. All units are derivable from these nine by simple formulas involving conversion factors of exactly unity. Other reference standards that produced conversion factors of almost—but not quite—unity were dropped. The coherence thus produced, where all units refer back to nine universally accepted fundamental quantities, is of great scientific importance. Although much of the international activity leading to the adoption of the SI was motivated by this consideration, it would not, by itself, have been sufficient to force the United States to convert. After all, scientists have always preferred to use the metric system and could easily have used the SI whether it came into commercial use in the United States or not. But when the standardization factor is added, then conversion becomes inevitable.

The United States has been on the so-called English system for centuries.[2] As a result, enormous quantities of records—blueprints, surveys, engineering reports, journals, and books—have English units. Now, with conversion to the SI, American engineers will have no choice but to learn both systems. In this book, some examples will be worked out in SI units and others in English units. You will have to become thoroughly familiar with both in order to function successfully later as an engineer.

Extensive conversion tables have been provided in Appendix L. They provide not only for conversion from English to SI units, but also for numerous other kinds of conversions, such as SI to English, and certain conversions within systems, both English and SI.

■ 9.2 ABSOLUTE AND GRAVITATIONAL UNIT SYSTEMS

A complication in using SI and the familiar English unit system is that they are defined in fundamentally different ways. The SI is an *absolute* system, meaning that *mass* is a fundamentally defined quantity. The unit of force is not a fundamental one, but is a so-called *derived* unit. The usual English system is a *gravitational* system, meaning that the unit of *force* is the fundamental unit, and the unit of mass is a derived one. (We are careful to say here that the *usual* English system is a gravitational system, for there is also an English absolute system. There is also a metric gravitational system. Both will be described later.)

In what follows, for the SI we will use the symbols m (meter), kg (kilogram), and s (second). For the English system we will use ft (foot), lb (pounds-mass), lbf (pounds-force), and sec (second). The reason for distinguishing between lb (for mass) and lbf (for force) will become apparent later.

The equation that establishes the relationship between mass and force is called *Newton's second law of motion* and is written

$$F = ma \tag{9.1}$$

where F is the force acting on an object possessing mass m, and a is the acceleration of the object resulting from the action of the force.

In textbooks on physics, it is customary to present Equation (9.1) in a fundamental way as

$$F = kma$$

---

[2] The English system is also referred to as the U.S. Customary system or the British system.

The equation is written this way because, from a physical point of view, we are required to say that force is proportional to mass and acceleration, and the proportionality constant k expresses this point. However, in the SI, the units have been defined in such a way that $k = 1.0$, so we omit it whenever we use the SI.

Absolute System

Those who favor an *absolute system of units* reason as follows: the mass of an object is a fundamental property of that object and is invariant. We can compare the mass of an object with that of a standard mass by the use of a balance scale and state that it has such-and-such a value. Also, if a force is applied to an object, causing it to accelerate, we can use the fundamental definitions of length and time to make direct measurements of displacement, which can then be used to compute the acceleration. Thus, m and a in Equation (9.1) can be determined by direct experimental measurement. But what about force? We cannot measure force directly, but only indirectly. If we measure a force by observing how far it deflects a spring scale, we are really measuring a deflection (a length) and inferring something about the force through our knowledge of the spring material and shape. And, if we try to calibrate the spring by seeing how far it deflects under the weight of a known mass, then we have allowed a further complication to creep in, because the weight of an object depends upon the force of gravity, which is not constant but varies slightly at different points on the Earth's surface and is far different on the moon than on Earth.

Scientists universally avoid the use of force as a fundamental quantity because of the foregoing difficulties. They say, "Mass is invariant, force is not; therefore, let us use mass as a fundamental quantity." By virtue of this choice, force becomes a derived quantity and is expressed in terms of the basic quantities—mass, length, and time.

In the SI, which is an absolute system, the force F is measured in newtons (symbol N), mass m is measured in kilograms (kg), and acceleration a is measured in meters per second squared (m/s^2, or ms^{-2}). If we substitute these units into Equation (9.1), as if they were numerical values, we get

$$F = ma$$

$$\text{newtons} = \text{kilograms} \cdot \frac{\text{meters}}{(\text{seconds})^2}$$

or

$$N = \frac{\text{kg} \cdot \text{m}}{\text{s}^2} \tag{9.2}$$

which tells us immediately that 1 newton is 1 kg·m/s^2, where kg, m, and s are all fundamentally defined quantities.

You may have noticed in the foregoing that the letter "m" was used to indicate both mass and meters. This is a situation you will simply have to get used to. However, it should also be noted that, in typeset material, italic letters are used for algebraic variables in equations, whereas roman (nonitalic) letters are used for abbreviations of units. Thus, *m* in Equation (9.1) is the algebraic variable signifying mass, and m in Equation (9.2) is the abbreviation for meter. These are the accepted conventions.

While on the subject of confusion of units, perhaps we should note the ambiguity of such English terms as *ounce, gallon,* and *ton.* Note these conversions:

ounce (avoirdupois) $= 2.835 \times 10^{-2}$ kg

ounce (troy) $= 3.110 \times 10^{-2}$ kg

ton (long, 2240 lb) $= 1.016 \times 10^3$ kg

ton (short, 2000 lb) $= 0.907 \times 10^3$ kg

gallon (U.K. liquid) $= 4.546 \times 10^{-3}$ m^3

gallon (U.S. dry) $= 4.405 \times 10^{-3}$ m^3

gallon (U.S. liquid) $= 3.785 \times 10^{-3}$ m^3

One of the attractions of the SI is that there is an exact, unique definition and symbol for each unit. Well, almost. The term *ton* does crop up in SI, where it means the *metric ton,* or 1000 kg. In some countries, the term *tonne* is used to designate the metric ton. However, the usage of either *ton* or *tonne* in the SI is now being discouraged because of possible confusion, and the recommendation is that masses be designated in either kilograms (kg) or megagrams (Mg). Another possible source of confusion is the use of a lowercase letter "l" for liter, because it could be mistaken for the numeral 1. The liter is not an official SI unit, but its use is so widespread that it is considered preferable to the "official" SI equivalent, the cubic decimeter. In the United States, the symbol recommended for the liter is L, to avoid the confusion with the numeral 1.

Gravitational System

Those who favor a *gravitational system of units* reason as follows: If we look at Equation (9.1) and use SI units, then when we compute the weight of an object with 1-kg mass, we get a value in *newtons* as the weight. This seems peculiar to those who favor gravitational systems. It comes about as follows. By the term *weight* we mean the force, caused by the Earth's gravity, acting upon an object. From experimental observations we know that an object at sea level that is dropped in a vacuum accelerates with $a \cong 9.807$ m/s^2 (about 32.2 ft/sec^2 in the English system), commonly designated as *g*, the acceleration of gravity. If we substitute $m = 1$ kg and $a = g = 9.807$ m/s^2 into Equation (9.1), we get the force (weight is after all, by definition, the force acting on a mass because of gravity) as follows:

$F = ma$

$\quad = 1 \text{ kg}(9.807 \text{ m/s}^2)$

$\quad = 9.807 \text{ N}$

This, say the proponents of a gravitational system of units, is not a sensible result. One kilogram of a substance should weigh 1 kilogram, they assert, not 9.807 newtons.

It should be observed, however, that when a person goes into a store to buy a kilogram of potatoes, there will be no confusion. In such a request, it is implied that a kilogram *mass* of potatoes is what is wanted, and that is what will be delivered. However, if a civil engineer wishes to design a beam to support an object whose mass is known in kilograms, it will be necessary to convert the mass into newtons in order to arrive at the force that is exerted on the beam by the mass. Therefore, civil engineers, and most others who analyze objects attached to the Earth's surface, have historically preferred a gravitational system of units.

In the English gravitational system (which is the system that was used almost exclusively by engineers in the United States until the 1960s), the force F in Equation (9.1) is measured in pounds-force (symbol lbf), mass m is measured in slugs (lbf-sec^2/ft), and acceleration a is measured in feet per second squared (ft/sec^2). If we substitute these units into Equation (9.1), as if they again were numerical values, we get

$$\text{lbf} = \frac{\text{lbf-sec}^2}{\text{ft}} \cdot \frac{\text{ft}}{\text{sec}^2}$$

If we cancel units, again treating them as if they were numbers, we get

$$\text{lbf} = \frac{\text{lbf-\cancel{sec}}^{\cancel{2}}}{\cancel{\text{ft}}} \cdot \frac{\cancel{\text{ft}}}{\cancel{\text{sec}}^{\cancel{2}}}$$

or

$$\text{lbf} = \text{lbf}$$

Thus, we see that the result is dimensionally consistent, because the dimensions on both sides are in pounds-force (lbf). The addition of the "f" on the symbol "lb" is simply to remind us that the quantity we are speaking of is a force, because "lb" by itself has sometimes been used to mean either mass *or* force, and if we tried to use it both ways at once in Equation (9.1), we would get nonsense, as follows:

$$\text{lbf} = \text{lb} \cdot \frac{\text{ft}}{\text{sec}^2}$$

This result cannot be true, because the units are not consistent on both sides of the equals sign. Unfortunately, this is just the kind of error that can occur in the English system, because we have never been precise concerning whether we mean mass or force when we use the unit "pound." Sometimes we have used it to mean one, and sometimes the other.

The student can avoid these problems by always thinking in terms of $F = ma$ when uncertain about mass and force. In the English gravitational system, where we are accustomed to saying that such-and-such an object *weighs*, say, 500 lb when we are really referring to its mass, we can think of Equation (9.1) as if it were written

$$F = \frac{W}{g} \cdot a \tag{9.3}$$

Then we can regard our object as having a weight $W = 500$ lbf (by which we mean that a force of 500 lbf is exerted by gravity on the object at the Earth's surface) and convert it to mass by dividing by g:

$$m = \frac{W}{g} = \frac{500 \text{ lbf}}{32.2 \text{ ft/sec}^2}$$

$$= 15.5 \frac{\text{lbf-sec}^2}{\text{ft}}$$

$$= 15.5 \text{ slugs}$$

If we always make a simple check like this for units, many mistakes will be avoided. One of the most common errors made by students (and sometimes by experienced engineers, too) is to omit g when it is needed and thus produce an answer with inconsistent units. The foregoing problems do not occur in the SI, where mass is always in kilograms and force is always in newtons. This is one more reason why the SI is favored universally by scientists and by many engineers, too. However, not *all* problems disappear with SI, because in that system, whenever a force results from the weight of a mass and we are working at the Earth's surface, we have to remember that the applied force is equal to mg, as the following example shows.

Example: Stress in a Beam

In Figure 9.2(*a*) we show a 10 000-lb (4536-kg) object in the center of a simple beam 20 ft (6.1 m) long. We wish to calculate the maximum stress in the beam. (Stress is defined as force per unit area and is the quantity that usually interests us the most when we try to find out how strong a structural member is. See Chapter 15 for further details.) We will make the calculations first in the English gravitational system and then in SI.

The formula for the maximum stress in a simple beam with a concentrated load in the center is

$$\sigma = \frac{PL}{4S} \tag{9.4}$$

where σ is the stress, P is the load, L is the length of the beam, and S is a quantity known as the *section modulus*, which is a property of the cross-sectional geometry of the beam. (This formula is given here without derivation; the development of beam

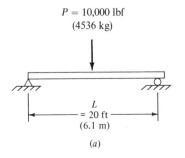

$P = 10,000$ lbf
(4536 kg)

L
= 20 ft
(6.1 m)

(*a*)

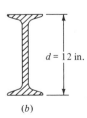

$d = 12$ in.

(*b*)

Figure 9.2

(*a*) A simple beam supporting 10 000 lbf (4536 kg). (*b*) A typical cross section of a standard I-beam.

formulas is a topic in courses on mechanics of materials, which typically come in the second or third year of an engineering education.)

We have assumed that the object can be treated as a concentrated load, i.e., that its dimensions are small relative to the size of the beam. Also, to simplify the problem we have deliberately ignored the weight of the beam itself. We tentatively choose a standard steel I-beam of depth $d = 12$ in., as shown in cross section in Figure 9.2(b). From a table in a handbook, we select a beam weighing 40.8 lbf/ft, which has a section modulus $S = 44.8$ in.^{3} We can now compute the stress:

$$\sigma = \frac{10\ 000\ \text{lbf} \cdot 20\ \text{ft} \cdot 12\ \text{in./ft}}{4 \cdot 44.8\ \text{in.}^3}$$

First we check the units:

$$\frac{\text{lbf} \cdot \cancel{\text{ft}} \cdot \cancel{\text{in.}}/\cancel{\text{ft}}}{\cancel{\text{in.}} \cdot \text{in.} \cdot \text{in.}}$$

which gives us lbf/in.^{2} (pounds-force per square inch, or psi), which is correct, for stress units—force per unit area. The answer, then, is

$$\sigma \cong 13\ 400\ \text{lbf/in.}^2$$

which, incidentally, is within reasonable limits for a steel I-beam.

To make the calculations in SI, the mass of 4536 kg must first be converted into newtons:

$$P = mg$$
$$= 4536\ \text{kg} \cdot 9.807\ \text{m/s}^2$$
$$= 44\ 480\ \text{N}$$
$$= 4.448 \times 10^4\ \text{N}$$

The section modulus must also be converted to SI. We know that 1 in. = 0.0254 m, so we write

$$(1\ \text{in.})^3 = (0.0254\ \text{m})^3$$
$$1\ \text{in.}^3 = 0.000\ 0164\ \text{m}^3$$
$$= 1.64 \times 10^{-5}\ \text{m}^3$$

Therefore,

$$S = 44.8\ \cancel{\text{in.}^3} \cdot 1.64 \times 10^{-5}\ \frac{\text{m}^3}{\cancel{\text{in.}^3}}$$
$$= 7.35 \times 10^{-4}\ \text{m}^3$$

Now we can compute the stress, from Equation (9.4):

$$\sigma = \frac{4.448 \times 10^4\ \text{N} \cdot 6.1\ \text{m}}{4 \cdot 7.35 \times 10^{-4}\ \text{m}^3}$$

Figure 9.3
Many older American machine tools are calibrated in decimal parts of the inch. Converting them to the metric system is expensive, and the only practical alternative may be complete replacement.

We see that the units cancel out to N/m², which is correct for stress in SI, and the answer is

$$\sigma \cong 92\ 290\ 000 \text{ N/m}^2$$
$$= 92.29 \times 10^6 \text{ N/m}^2$$

We can check this result by looking up the conversion from lbf/in.² to N/m² (see Appendix L). We note, also, that the units N/m² are designated as the pascal (symbol Pa) and find

$$1 \text{ lbf/in.}^2 = 6895 \text{ Pa}$$

Then, using our previous answer of $\sigma = 13\ 400$ lbf/in.², we get

$$\sigma = (13\ 400 \text{ lbf/in.}^2)\left(6895 \, \frac{\text{Pa}}{\text{lbf/in.}^2}\right)$$

$$\cong 92.39 \times 10^6 \text{ Pa}$$
$$= 92.39 \times 10^6 \text{ N/m}^2$$

which checks, with reasonable accuracy.

Other Systems

We said earlier that there is an *absolute* English system; it has received a certain amount of use in the United States. In this system, the unit of mass is the pound (in this case symbolized by "lb," since we are talking about mass), the unit of acceleration is ft/sec², and the unit of force is the *poundal*. The poundal is analogous to the newton in SI. Its units are found from application of Equation (9.1),

$$F = ma$$

$$\text{poundals} = \text{pounds} \cdot \frac{\text{feet}}{(\text{seconds})^2}$$

or

$$poundal = \frac{lb \cdot ft}{sec^2} \qquad (9.5)$$

which can be seen to be similar to Equation (9.2) in unit structure, with the English units taking the place of the metric ones.

Interestingly, there is also a metric *gravitational* system, which has been widely used by engineers in some European countries. In this system, the numerical value of the *mass* in kg is also used as the *weight* in kg. The acceleration is in m/s², and the resulting force unit is called the kilogram-force (kgf). In Germany, the kgf has also been called the kilopond, with the symbol "kp." This last symbol is most unfortunate for American engineers, for it resembles the unit "kip," meaning a force of 1000 lb.

Comparison of Systems

So far we have defined four different systems of units, which dramatize the confusing mess we have gotten ourselves into in the past regarding units. At some time in the future these four will give way to just one, the SI, but Americans will have to continue to deal with the English gravitational (U.S. Customary) system for an indefinite period. The remaining two—the English absolute system and the metric gravitational system—are of less importance, although engineers must know of their existence. Table 9.1 may help to place these unit systems in perspective.

Table 9.1   Comparison of Unit Systems

	SI	English Gravitational (U.S. Customary)	English Absolute	Metric Gravitational
Type of system	absolute	gravitational	absolute	gravitational
Length	m	ft	ft	m
Time	s	sec	sec	s
Mass	kg	$slug = \frac{lbf\text{-}sec^2}{ft}$	lb	kg
Force (name)	newton (N)	pounds-force	poundal	kilograms-force
Force (units)	$\frac{kg \cdot m}{s^2}$	lbf	$\frac{lb\text{-}ft}{sec^2}$	kgf
Acceleration	m/s²	ft/sec²	ft/sec²	m/s²
Form of equation	$F = ma$	$F = \frac{W}{g}a$	$F = ma$	$F = \frac{1}{g_c}ma$

Note: $g = 9.807 \text{ m/s}^2 = 32.2 \text{ ft/sec}^2$

$$g_c = 9.807 \frac{(m/s^2)kg}{kgf} = 32.2 \frac{(ft/sec^2)lb}{lbf}$$

Examples of calculations in each of the four systems follow. Note that all the calculations are for exactly the same physical system, in which the mass "weighs" 10 pounds (i.e., has a force of gravity of 10 pounds exerted on it at the Earth's surface) and has a force applied to it that results in an acceleration of 10 feet per second per second. (In metric terms, 10 lb converts to 4.536 kg, and 10 ft/sec² converts to 3.048 m/s².)

SI

Form of equation: $F = ma$

$m = 4.536$ kg

$a = 3.048$ m/s²

$F = ma = 4.536$ kg$(3.048$ m/s²$) = 13.83$ kg·m/s²

$$= 13.83 \text{ N}$$

English Gravitational

Form of equation: $F = \dfrac{W}{g}a$

$W = 10$ lbf

$a = 10$ ft/sec²

$m = \dfrac{W}{g} = \dfrac{10 \text{ lbf}}{32.2 \text{ ft/sec}^2} = 0.3106$ lbf-sec²/ft

$$= 0.3106 \text{ slugs}$$

$F = \dfrac{W}{g}a = ma = 0.3106 \dfrac{\text{lbf-sec}^2}{\text{ft}}\left(10 \dfrac{\text{ft}}{\text{sec}^2}\right) = 3.106$ lbf

We can check this answer by using the unit conversion table in Appendix L:

$$3.106 \text{ lbf}\left(4.448 \dfrac{\text{N}}{\text{lbf}}\right) = 13.83 \text{ N}$$

which gives the same result we got using SI directly, as it should.

English Absolute

Form of equation: $F = ma$

$m = 10$ lb

$a = 10$ ft/sec²

$F = ma = 10$ lb$(10$ ft/sec²$) = 100$ lb-ft/sec²

$$= 100 \text{ poundals}$$

Again, we can check our answer by using the unit conversion table in Appendix L:

100 poundals (0.1383 N/poundal) = 13.83 N

Metric Gravitational

Form of equation: $F = \dfrac{1}{g_c} ma$

Note that this is an awkward equation to use, because in the metric gravitational system the mass of an object and the force required to lift it against gravity are numerically equal and are both given in kilograms. This leads to great confusion for the student who is struggling to understand what units are all about, and points up the advantage of the SI, which unambiguously uses the kilogram for mass and the newton for force. Nevertheless, we will carry out the example of the metric gravitational system, noting that it is necessary to use the value

$$g_c = 9.807 \, \frac{(\text{m/s}^2) \, \text{kg}}{\text{kgf}}$$

to make the equation come out with the correct units.

$m = 4.536 \text{ kg}$

$a = 3.048 \text{ m/s}^2$

$$F = \frac{1}{g_c} ma = \frac{1}{9.807 \, \dfrac{(\text{m/s}^2) \, \text{kg}}{\text{kgf}}} (4.536 \text{ kg})(3.048 \text{ m/s}^2)$$

$$= 1.409 \text{ kgf}$$

We can check this by using the unit conversion table:

1.409 kgf(9.807 N/kgf) = 13.83 N

■ 9.3 SI UNITS AND SYMBOLS[3]

There are nine fundamental units in SI, of which seven are defined physically and are called *base units*; the other two are called *supplementary units* and are defined geometrically. The nine units are listed in Table 9.2. Their definitions follow:

---

[3] The sources of the information in this chapter are *Standard Practice for Use of the International System of Units (SI)* (E380-89a) (Philadelphia, Pa.: American Society for Testing Materials, 1989), *Metric Guide for Educational Materials* (Washington, D.C.: American National Metric Council, 1977), and *The Metric System of Measurement* (Washington, D.C.: *Federal Register*, vol. 41, no. 239, December 10, 1976, pp. 54018–54019).

Table 9.2   Fundamental Units in SI

Quantity	Unit	SI Symbol
Base Units		
Length	meter	m
Mass	kilogram	kg
Time	second	s
Electric current	ampere	A
Thermodynamic temperature	kelvin	K
Amount of substance	mole	mol
Luminous intensity	candela	cd
Supplementary Units		
Plane angle	radian	rad
Solid angle	steradian	sr

Meter   Originally, the size of the meter was calculated as 1/10 000 000 the distance from the equator to the North Pole, and it was defined as the distance between two finely scribed lines on a platinum alloy bar kept in France. Now it is defined as the length of the path traveled by light in a vacuum during a time interval of 1/299 792 458 of a second.

Kilogram   The kilogram is equal to the mass of the international prototype of the kilogram, which is maintained at the International Bureau of Weights and Measures in France.

Second   The second was originally defined as a fraction of the mean solar day, essentially using the Earth's rotation as a clock. This is referred to as the *ephemeris* definition of the second. (An ephemeris is a table showing the positions of astronomical bodies as a function of time.) It is now defined as the duration of 9 192 631 770 periods of the radiation corresponding to the transition between the two hyperfine levels of the ground state of the cesium-133 atom.[4]

Ampere   At one time, the ampere was defined as a rate of flow of 1 coulomb per second, where 1 coulomb was defined as equal to 3×10^9 electrostatic units of charge. Subsequently, an attempt was made to establish the so-called *international ampere* as the current that, when passed through a solution of silver nitrate, would deposit silver at the rate of 0.001 118 g/s. This latter definition differs from the one in current use by a factor of 0.999 843. According to the present-day SI definition, the ampere is the constant current that, if maintained in two straight parallel conductors of infinite length and of negligible cross section, and placed 1 meter apart in vacuum, would produce between these two conductors a force equal to 2×10^{-7} newtons per meter of length.

---
[4] The meanings of the technical terms given in the definition of the second are developed in the study of spectral analysis, a topic in chemistry.

Kelvin   The kelvin is 1/273.16 of the thermodynamic temperature of the triple point of water.

Mole   The mole is the amount of substance of a system that contains as many elemental entities (atoms, molecules, ions, electrons, or other particles) as there are atoms in 0.012 kg of carbon-12. It is necessary, in using the mole, to specify the elemental entities under discussion.

Candela   The candela is the luminous intensity, in a given direction, of a source that emits monochromatic radiation of frequency 540×10^{12} hertz and that has a radiant intensity in that direction of 1/683 watt per steradian.

Radian   The radian is a unit of measure of a plane angle, equal to the measure of the angle with its vertex at the center of a circle and subtended by an arc equal in length to the radius.

Steradian   The unit of measure of a solid angle, equal to the measure of the angle with its vertex at the center of a sphere and enclosing an area of the spherical surface equal to that of a square with sides equal in length to the radius.

There are also numerous *derived units*, which are listed in Table 9.3. Detailed definitions of these derived units will not be given here, although it should be noted that they are all based upon the fundamental units just defined. Many of the derivations come about through the application of simple physical laws. For example, the *newton* is defined by $F = ma$, as has already been described; the unit of electrical resistance, the *ohm* (symbol Ω), is defined by Ohm's law, $v = iR$ (see Chapter 17); the unit of capacitance, the *farad* (F), is defined by the equation F = C/V; and so on. As can be seen in Table 9.3, some derived units have been given special names, such as the newton (kg·m/s^2), the pascal (N/m^2), the joule (N·m), and the watt (J/s). Others have not been given special names, but retain their unit forms, such as the expressions for acceleration (m/s^2), volume (m^3), and density (kg/m^3). Note that the joule (J) is used throughout for energy and work, replacing units such as Btu, ft-lb, calorie, horsepower-hour, kilowatt-hour, and so forth.

In addition to the units that are officially part of the SI, others have been accepted for use with SI units, some only within limited fields. These units are listed in Appendices J.1, J.2, and J.3. Note that the *liter* (L), which is equal to 10^{-3} m^3, may be used for volumetric capacity.

Prefixes for use with SI are given in Table 9.4. They are used to indicate orders of magnitude. Thus, 1000 g becomes 1 kg, and 1 000 000 g becomes 1 Mg. (Note that 1 mg is 10^{-3} g, but 1 Mg is 10^6 g.) Also, 0.001 m = 10^{-3} m becomes 1 mm, and 0.000 000 001 m = 10^{-9} m becomes 1 nm, or one nanometer. A list of units that were used formerly, but that are now either approved only for a limited time or considered undesirable, are given in Appendix J. SI conversion values are given in Appendix L.

The Kelvin scale is the basic one for the measurement of temperature, although the Fahrenheit and Rankine scales have been widely used in the English system. The Celsius (formerly called centigrade) scale uses the same definition of a degree as does

Table 9.3 Derived Units in SI

Quantity	Unit	SI Symbol	Formula
Absorbed dose	gray	Gy	J/kg
Acceleration	meter per second squared	—	m/s²
Activity (of a radionuclide)	becquerel	Bq	1/s
Angular acceleration	radian per second squared	—	rad/s²
Angular velocity	radian per second	—	rad/s
Area	square meter	—	m²
Density	kilogram per cubic meter	—	kg/m³
Dose equivalent	sievert	Sv	J/kg
Electric capacitance	farad	F	C/V
Electric conductance	siemens	S	A/V
Electric field strength	volt per meter	—	V/m
Electric inductance	henry	H	Wb/A
Electric potential difference	volt	V	W/A
Electric resistance	ohm	Ω	V/A
Energy	joule	J	N·m
Entropy	joule per kelvin	—	J/K
Force	newton	N	kg·m/s²
Frequency	hertz	Hz	1/s
Illuminance	lux	lx	lm/m²
Luminance	candela per square meter	—	cd/m²
Luminous flux	lumen	lm	cd·sr
Magnetic field strength	ampere per meter	—	A/m
Magnetic flux	weber	Wb	V·s
Magnetic flux density	tesla	T	Wb/m²
Moment of force	newton-meter	—	N·m
Power	watt	W	J/s
Pressure	pascal	Pa	N/m²
Quantity of electricity	coulomb	C	A·s
Quantity of heat	joule	J	N·m
Radiant intensity	watt per steradian	—	W/sr
Specific heat	joule per kilogram-kelvin	—	J/kg·K
Stress	pascal	Pa	N/m²
Thermal conductivity	watt per meter-kelvin	—	W/(m·K)
Velocity	meter per second	—	m/s
Viscosity, dynamic	pascal-second	—	Pa·s
Viscosity, kinematic	square meter per second	—	m²/s
Volume	cubic meter	—	m³
Work	joule	J	N·m

the Kelvin scale, although the Celsius scale is not an official part of the SI. The definitions of all follow, given in terms of T, the "absolute" temperature in Kelvin:

Celsius temperature (°C): $t_C = T - 273.15 \text{ K}$ (9.6)

Rankine temperature (°R): $T_R = \frac{9}{5}T$ (9.7)

Fahrenheit temperature (°F): $t_F = T_R - 459.67°R$ (9.8)

Table 9.4   Prefixes for Use with SI

Multiplication Factor	Prefix	SI Symbol
1 000 000 000 000 000 000 = 10^{18}	exa	E
1 000 000 000 000 000 = 10^{15}	peta	P
1 000 000 000 000 = 10^{12}	tera	T
1 000 000 000 = 10^{9}	giga	G
1 000 000 = 10^{6}	mega	M
1 000 = 10^{3}	kilo	k
100 = 10^{2}	hecto*	h
10 = 10^{1}	deka*	da
0.1 = 10^{-1}	deci*	d
0.01 = 10^{-2}	centi*	c
0.001 = 10^{-3}	milli	m
0.000 001 = 10^{-6}	micro	μ
0.000 000 001 = 10^{-9}	nano	n
0.000 000 000 001 = 10^{-12}	pico	p
0.000 000 000 000 001 = 10^{-15}	femto	f
0.000 000 000 000 000 001 = 10^{-18}	atto	a

* To be avoided wherever practical.

In most thermodynamic calculations, *absolute* temperatures must be used. A temperature of absolute zero is the point where all molecular motion ceases. Both the Kelvin and Rankine scales are absolute scales. A temperature of absolute zero on one is equal to absolute zero on the other, but from that point the Kelvin and Rankine scales diverge in accordance with Equation (9.7). Failure to use absolute temperatures when they are required is a frequent source of error. If a temperature is given in °F, it can be converted to °R by use of Equation (9.8), and then to K by use of Equation (9.7). If °C is also needed, it can be computed from Equation (9.6).

For example, starting with 68°F and using Equations (9.8), (9.7), and (9.6) in succession, we get

$$T_R = 68°F + 459.67°R$$
$$= 527.67°R$$
$$T = \tfrac{5}{9}(527.67°R)$$
$$= 293.15 \text{ K}$$
$$t_C = 293.15 \text{ K} - 273.15 \text{ K}$$
$$= 20°C$$

This in itself is a rather interesting result. It may be helpful to remember that 68°F (the so-called ideal human comfort temperature) is precisely 20°C. Incidentally, in case you wondered about the propriety of adding °F and °R in the first computation of this example, this is possible because both units describe the same interval, even though zero degrees on one is not the same on the other.

Some remarks on pronunciation may be helpful. The American Society for Testing and Materials (ASTM) recommends the following:

giga: JIG-ah
kilo: KIL-oh
nano: NAN-oh (a as in "ant")
pico: PEEK-oh
candela: kan-DEL-ah
joule: jool (rhymes with "pool")
kilometer: KIL-oh-mee-tur
pascal: PASS-kal (rhymes with "rascal")
siemens: SEE-mens (pronounced like "seamen's")

■ 9.4 RULES FOR USING SI

Much of the value of SI lies in its consistency and the unique meaning of each symbol. However, ambiguity does result if precise rules are not followed. For example, what does "Nm" stand for, and is it different from "mN"? The answer is that the latter symbol stands for "millinewtons," and the former one doesn't mean anything at all in SI. Perhaps the person who wrote "Nm" meant "newton meters," in which case the symbol should be written "N·m" or, alternatively, with a space instead of a centered dot, as in N m.

There simply is no way to deal with the topic of rules except to list the important ones, as briefly as possible:

1. Unit symbols are printed in upright (roman) letters. Sloping (italic) letters are used for quantity symbols. Thus, m means meter, but *m* means mass.

Figure 9.4
For many years to come, engineers will have to work with both SI metrics and with the English, or U.S. Customary, system. Engineering drawings, reports, and technical articles using either system may be encountered.

2. A period is *not* used after a symbol, except at the end of a sentence.

3. Symbols for units are the same in singular and plural, as 1 m, 100 m.

4. The millimeter (mm) is recommended as the basic unit to be used for all dimensions on architectural and mechanical engineering drawings. The kilopascal (kPa) is recommended for tire pressure, atmospheric pressure, and the like. Material tensile properties would commonly be expressed in megapascals (MPa), and moduli of elasticity in gigapascals (GPa).

5. Capital and lowercase letters must be used exactly as prescribed. Otherwise we might confuse things like G (giga) and g (gram), K (kelvin) and k (kilo), N (newton) and n (nano), and M (mega) and m (milli).

6. The correct symbols should be used. Abbreviations such as amp for ampere or m.a. for milliamperes should not be employed. The former would be shown correctly as A, the latter as mA.

7. When prefixes are used with unit symbols, no space is left between. Thus, for megawatts, we write MW, not M W.

8. Do not use double prefixes. We would write, for example, GN, not MkN.

9. A single space should be left between a numeral and the unit symbol. Thus, 22 mm is correct, but 22mm is not. An exception is made for degrees Celsius, which is written, for example, 20°C. However, in kelvin, this same temperature would be written with a space, 293.15 K.

10. A dot on the line should be used as a decimal marker in the United States, as 1.000. It should be noted that some countries use a comma as a decimal marker (1,000 for 1.000) and some use a raised dot (1·000 for 1.000).

11. Commas should not be used to separate groups of digits, because some countries do use the comma as a decimal marker. Large numbers should be grouped in threes without commas. Thus, we would write a number *not* as 1,019,000, but as 1 019 000. For decimal numbers less than 1, we would write not 0.32463, but 0.324 63. Four-digit numbers can (optionally) have the space left out, as 7354 or 0.1225.

12. A zero should always be written before the decimal point for numbers less than 1: for example, 0.01, not .01.

13. In writing symbols for compound units, the slash should not be used more than once, or ambiguity will result. For example, does m/s/s mean meters divided by seconds per second? Symbolically, that would be

$$\frac{m}{s/s}$$

which, of course, turns out to be just meters, once the units of seconds in the denominator cancel out. Apparently, what m/s/s means is

$$\frac{m/s}{s} = \frac{m}{s^2}$$

which is the familiar unit for acceleration. Thus, m/s^2, or $(m/s)/s$ would be all right, but not m/s/s. Also, a notation like $m \cdot s^{-2}$ would be acceptable.

14. For units that are the result of a product, like the newton meter, we would use a multiplication dot, as in $N \cdot m$ (or, alternatively, a space). Of course, we

might immediately note that 1 N·m is the definition of the joule and write J instead of N·m.

15. As an exception to Rule 14, the dot in W·h may be omitted, and we can write Wh.

16. Note that 10^6 is 1 million everywhere, but that, although 10^9 is called a billion in the United States, it is a milliard in other countries, or sometimes a thousand million. Americans call 10^{12} a trillion and 10^{18} a quintillion, but others call them a billion and a trillion, respectively.

17. It should be noted that a quantity designated as mm³ means 10^{-9} m³, not 10^{-3} m³. To see this, note that

$$1 \text{ mm} = 10^{-3} \text{ m}$$

If we raise both sides of the preceding equation to successively higher powers, we get

$$1 \text{ mm}^2 = 10^{-6} \text{ m}^2$$

$$1 \text{ mm}^3 = 10^{-9} \text{ m}^3$$

and so on.

■ 9.5 CONVERTING UNITS

A couple of examples will serve to show how one kind of unit can be converted to another.

Suppose we wish to convert 62.4 lb/ft³ to its equivalent value in kg/m³. From Appendix L we get

$$1 \text{ lb} = 0.4536 \text{ kg} \tag{9.9}$$

$$1 \text{ ft}^3 = 0.028\ 32 \text{ m}^3 \tag{9.10}$$

The first thing we will do is convert the left-hand sides of both Equations (9.9) and (9.10) to unity, by transposing unit symbols as if they were algebraic variables:

$$1 = \frac{0.4536 \text{ kg}}{\text{lb}} \tag{9.11}$$

$$1 = \frac{0.028\ 32 \text{ m}^3}{\text{ft}^3} \tag{9.12}$$

Now, if we multiply any quantity by the *left-hand* side of, say, Equation (9.11), we are multiplying it by unity, and therefore are not changing it. But the left-hand and right-hand sides of Equation (9.11) are equal, because of the "=" sign, so this means we can also multiply any quantity by the *right-hand* side of Equation (9.11) without changing that quantity, except for a change in the units in which it is expressed. Let us do that.

$$62.4 \frac{\text{lb}}{\text{ft}^3} \times \frac{0.4536 \text{ kg}}{\text{lb}}$$

The units "lb" in numerator and denominator cancel, giving us

$$28.30 \frac{kg}{ft^3}$$

We do not want ft^3 in the denominator, of course; instead, we want m^3. We try to eliminate ft^3 by using the right-hand side of Equation (9.12). If we set this up as follows, we immediately see there is a problem:

$$28.30 \frac{kg}{ft^3} \times \frac{0.028\ 32\ m^3}{ft^3}$$

The units "ft^3" appear twice in the denominator, and we cannot cancel them. To remedy this, we go back to Equation (9.10); and this time, instead of transposing the units "ft^3" into the denominator on the right, we do just the reverse, transposing the expression $0.028\ 32\ m^3$ into the denominator on the left. (Remember, we are treating the units as if they were algebraic quantities and transposing them along with numerical values, according to the usual rules of algebra.) We get

$$\frac{1\ ft^3}{0.028\ 32\ m^3} = 1 \tag{9.13}$$

The right-hand side of Equation (9.13) is equal to unity, so we can multiply any quantity by it without changing that quantity. And, because of the equals sign, we can do the same with the left-hand side. Thus, we can get

$$28.30 \frac{kg}{ft^3} \times \frac{1\ ft^3}{0.028\ 32\ m^3}$$

Now the units "ft^3" in numerator and denominator cancel, and we are left with the units we want: kg/m^3. Carrying everything out, we get

$$999.3\ kg/m^3$$

So we can finally write

$$62.4\ lb/ft^3 = 999.3\ kg/m^3$$

Note that $62.4\ lb/ft^3$ is the approximate density of water. If we round off $999.3\ kg/m^3$ to $1000\ kg/m^3$, we see that the latter value is also the approximate density of water, expressed in SI.

As another example, let us convert 10^6 J to kWh. Of course, we could go immediately to Appendix L and look up the conversion factor, but let us do it the hard way. (One reason for doing it the hard way is to emphasize the fact that 1 hour contains 3600 seconds; it is a common error to act in haste and put down the number 60, which of course is the number of *minutes* in an hour, not the number of seconds.) From the definitions of the watt and the hour, we write

$$1\ W = 1\ \frac{J}{s} \tag{9.14}$$

$$1\ h = 3600\ s \tag{9.15}$$

We convert Equations (9.14) and (9.15) to

$$\frac{1 \text{ W} \cdot \text{s}}{\text{J}} = 1 \tag{9.16}$$

$$\frac{1 \text{ h}}{3600 \text{ s}} = 1 \tag{9.17}$$

Now, multiplying the quantity 10^6 J by the left-hand sides of Equations (9.16) and (9.17), and making appropriate cancellations, we get

$$10^6 \text{ J}\left(\frac{1 \text{ W} \cdot \text{s}}{\text{J}}\right)\left(\frac{1 \text{ h}}{3600 \text{ s}}\right) = 277.8 \text{ Wh}$$

$$= 0.2778 \times 10^3 \text{ Wh}$$

$$= 0.2778 \text{ kWh}$$

Exercises

9.1 Develop a conversion factor for each of the following. (Refer to the conversion table in Appendix L and the unit symbols in Appendix I.)

(a) lb/in.³ to kg/m³

Solution

$$1 \text{ lb} = 0.4536 \text{ kg}$$

$$1 \text{ in.} = 2.54 \times 10^{-2} \text{ m}$$

$$1 \text{ in.}^3 = 16.39 \times 10^{-6} \text{ m}^3$$

$$\frac{\cancel{\text{lb}}}{\text{in.}^3} \times 0.4536 \frac{\text{kg}}{\cancel{\text{lb}}} \times \frac{1}{16.39 \times 10^{-6}} \frac{\cancel{\text{in.}^3}}{\text{m}^3}$$

$$1 \frac{\text{lb}}{\text{in.}^3} = 2.768 \times 10^4 \frac{\text{kg}}{\text{m}^3} \qquad \textbf{Answer}$$

(b) gal (U.S. liquid) to L
(c) acres to ha
(d) mm of mercury to atm
(e) Btu/min to ft-lbf/sec
(f) ft/sec² to m/s²
(g) lbf-sec²/in. to kg
(h) tons/cubic mile to kg/m³ (Note: "ton" = short ton)
(i) tonnes to tons
(j) MWh to J
(k) rad/s to rpm (revolutions per minute)
(l) kW to hp
(m) psi to GPa

9.2 Convert the following:
(a) 100°F to °C

Solution

See "Degrees Fahrenheit" in Appendix L.

$$t_c = (100°\text{F} - 32)/1.8 = \textbf{37.8°C} \qquad \textbf{Answer}$$

(b) 100°F to K
(c) 7°R to K
(d) 7°R to °C
(e) 3000°F to K

9.3 What is the mass, in slugs of an object which weighs 1000 lbf? What is the mass in kg?

Solution

$$m = \frac{W}{g} = \frac{1000 \text{ lbf}}{32.2 \text{ ft/sec}^2} = 31.1 \frac{\text{lbf-sec}^2}{\text{ft}}$$

$$= \textbf{31.1 slugs} \qquad \textbf{Answer}$$

$$= 31.1 \text{ slugs} \times 14.59 \frac{\text{kg}}{\text{slug}}$$

$$= \textbf{453.7 kg} \qquad \textbf{Answer}$$

9.4 For each of the following cases, convert the given quantity first into slugs, and then into kg.
(a) 1 lb
(b) 100 tons (1 ton = 2000 lb)
(c) 25 oz (avoirdupois ounces)
(d) 1 gal (U.S. liquid) of water (Densities are in Appendix K.)

9.5 Convert the following quantities, expressed in obsolete units, into the appropriate SI units. (See Appendices I, J.1, J.2, J.3, and L.) Compute your answers to four significant digits.
(a) 1000 gausses
Solution
See "gauss" in Appendix J.3.

$$1000 \text{ gausses} \times \frac{10^{-4} \text{ T}}{\text{gauss}} = \textbf{0.1 T} \qquad \textbf{Answer}$$

(b) 4132 angstroms
(c) 53 torr
(d) 53 mmHg
(e) 1.75 Oe
(f) 186.3 Mx
(g) 0.173 mho
(h) 5736 dyn
(i) 5736 erg

9.6 In the older metric system, the *calorie* (cal) was the amount of heat it took to heat 1 cc (cubic centimeter) of water 1 degree centigrade. (Note that the abbreviation "cc" is not provided for in the modern SI. The proper designation would be cm³.) Unfortunately, there was another unit, the *Calorie* (Cal) (note the capital letter), which was worth 1000 cal. Convert the following quantities into proper SI units:
(a) 4156 cal
Solution
See "calorie" in Appendix J.3.

$$4156 \text{ cal} \times \frac{4.1868 \text{ J}}{\text{cal}} = 17\,400 \text{ J} = \textbf{17.4 kJ} \qquad \textbf{Answer}$$

(b) 4.156 Cal

9.7 A *capacitor* is a widely used device that stores electricity. The unit used for a quantity of electricity is the *coulomb* (C). The size of a capacitor (called *capacitance*) is expressed in farads, and a *farad* (F) is defined by the equation F = C/V, where C is the number of coulombs stored in the capacitor, and V is the number of volts measured across the electrical terminals of the capacitor. If a current of 28 A (average) is applied for 10 s to a capacitor of 0.01 F, what voltage will we measure across the terminals of the capacitor?
Solution
(First of all, we said the current had an *average* value of 28 A because, when the current is first supplied to the capacitor, it has a very large value, and then it

starts to diminish as the capacitor becomes charged.) Using Table 9.3, we see that C = A·s, and that F = C/V, or capacitance = coulombs/voltage.

$$\text{coulombs} = 28 \text{ A}(10 \text{ s}) = 280 \text{ C}$$

$$\text{voltage} = \frac{\text{coulombs}}{\text{capacitance}} = \frac{280 \text{ C}}{0.01 \text{ F}}$$

$$= \textbf{28\,000 V} \qquad \textbf{Answer}$$

9.8 If a voltage of 200 V is applied for 1 s to a capacitor of 1 mF, how many coulombs will have been stored in the capacitor? What would the average current be, in amperes?

9.9 A dose of 1000 rad of radioactivity is enough to cause death. What is the magnitude of this dose in grays? In J/kg? (See Table 9.3 and Appendix J.2.)

9.10 Convert the following masses into weights (in newtons), assuming conditions at the Earth's surface.
(a) 100 kg
(b) 100 lb
(c) 1 metric ton

9.11 Convert the following:
(a) 3760 TW to nW
(b) 92 Gm to km
(c) 8377 μN to kN
(d) 0.376 kJ to MJ
(e) 0.000 076 4 GW to kW
(f) 4400 μm to m
(g) 33 mm² to m²
(h) 1010 cm³ to L
(i) 530 μA to mA

9.12 The *period* of a simple pendulum is given by the approximate formula:

$$\tau(\text{period}) = 2\pi \sqrt{\frac{L}{g}}$$

where L is the length of the pendulum and g is the acceleration of gravity. (The period is the time it takes for the pendulum to swing away from a given point and back again.) If a pendulum has a length of 1 m, what is its period? Repeat the problem using English units.

9.13 The formula for kinetic energy of a mass is K.E. = 0.5 mv^2, where m is mass and v is its velocity. Substitute the appropriate units for m and v, and reduce the result to joules.

Solution

$$\text{K.E.} \propto \text{kg}\left(\frac{m}{s}\right)^2 = \frac{\text{kg} \cdot m \cdot m}{s^2}$$

$$1\,N = 1\,\frac{\text{kg} \cdot m}{s^2}$$

$$\therefore \text{K.E.} \propto N \cdot m$$

$$1\,J = 1\,N \cdot m$$

$$\therefore \textbf{K.E.} \propto \textbf{J} \qquad\qquad \textbf{Answer}$$

9.14 What is the kinetic energy (see Exercise 9.13) of a 3-kg mass, moving with a velocity of 3 m/s? Express the answer in base units (see Table 9.2), in N·m, and in J.

9.15 Redo Exercise 9.14 using English units. Convert the mass and velocity from SI to English and carry out the entire problem using English units. Then, when you have your answer, convert it back to SI, and check the result against the answer you got in Exercise 9.14.

9.16 What is the kinetic energy (see Exercise 9.13) of an object that weighs 10 lbf, moving with a velocity of 5 ft/s? Express your answer in ft-lbf. (Hint: Check your units carefully. How should g enter?)

9.17 Redo Exercise 9.16 using SI units, and then, by appropriate conversion, compare your answer to the one you got in 9.16 to see if they check.

9.18 A handbook gives you a formula for the quantity of heat Q added to a given mass M of a substance, when the temperature increases from t_1 to t_2:

$$Q = Mkc(t_2 - t_1)$$

The coefficient c is the *specific heat*, and k is a constant that depends upon the units of measurement. Let us assume we are dealing with 10 kg of steel, and the temperature differential $t_2 - t_1$ is 10°C. The value for c for steel is given in calories per gram per degree C, as 0.118. A table in the handbook tells us that $k = 1.16 \times 10^{-3}$ when M is given in kg and $t_2 - t_1$ is in degrees C. Under these conditions, Q will be in kWh. Find Q, and convert the answer to J.

9.19 With respect to the conditions in Exercise 9.18, the handbook tells us that, if the mass M is in pounds and $t_2 - t_1$ is in degrees F, then $k = 1$. The value for c is given as 0.118 Btu per pound per degree F. Under

these conditions, the answer for Q will be in Btu. Find Q in Btu, then convert to J, to check your answer in 9.18.

9.20 The formula for the horsepower (hp) being transmitted by a shaft that is rotating n rpm (revolutions per minute), carrying a torque of T inch-pounds, is

$$\text{hp} = \frac{2\pi n T}{33\,000 \times 12}$$

Suppose a shaft is carrying $T = 100$ foot-pounds (force) at a speed of 1800 rpm; what is the horsepower? Convert your answer to J/s and then to W.

9.21 A beam 30 ft long has an object in the center that weighs 5000 lbf. (See Figure 9.2.) The beam has a section modulus, $S = 40$ in³. Using Equation (9.4), compute the stress in the beam. Ignore the weight of the beam.

9.22 In Exercise 9.21, what is the length of the beam, in meters? What is the mass of the object, in kilograms? What is the weight of the object, in newtons? Working entirely in SI units, compute the stress in the beam, in MPa. Convert your answer to English units to see if it matches the answer you got in Exercise 9.21.

9.23 Tell what is wrong with the way in which each of the following quantities is written, with respect to the "Rules for Using SI."
(a) 10,130 k · W
Solution
The comma should be omitted, and a space used instead. Also, there should be no dot between "k" and "W" (there also should not be a space). The quantity should be written 10 130 kW.
(b) 45.3 Nm
(c) 393.15°K
(d) 20 C
Solution
This quantity is ambiguous. It could possibly be correct as is, assuming the writer intended it to stand for 20 coulombs. (See Table 9.3.) But it is also possible the writer intended it to represent a temperature of 20°C. In an actual case, the context would probably tell which of these two interpretations is correct.
(e) 5.03 KN
(f) 5.03 mN
(g) 14 kMPa

Photographs courtesy of University of California, Davis (top and center), and Measurements Group, Inc. (bottom).

10

Communication: Written, Oral, and Graphical

For reasons that have always remained mysterious, many engineering students acquire the false notion that the ability to write and speak well is not important for engineers. Here is what some professional engineers have to say on this topic:

> Many engineers are well prepared technically, but cannot do a good job of expressing their ideas in written and oral presentations. Good ideas have to be sold to others, particularly via group presentations, to have any chance for success. (From a principal design engineer employed by a major automobile manufacturer)

> I find as a supervisor that clear, concise technical writing is absolutely essential to every engineer, to convey ideas. New graduates are weak in this area. (From a group leader employed by a major oil company)

> I've come to appreciate the value of technical report writing. My job is 60% report writing and 40% design work. (From a civil engineer doing hydroelectric power design for a large consulting firm)

Written and oral skills are especially critical for those who hope to move up the management ladder. It has been noted that the lack of such skills inhibits advancement more often than does lack of technical knowledge.

During a massive nationwide study of engineering education called *The Goals Study*, opinions regarding subjects taken in college were obtained from approximately 4000 engineers in 129 companies. The ten highest-rated subjects are shown in Table 10.1, ranked according to the degrees to which the engineers (a) recommended that the subjects be emphasized for future engineers, (b) had used them in the past

Focus On
The Abstraction Ladder

In communicating with others, one needs to be concerned with whether intended meanings are the same ones that are perceived. Words do not have the same meanings for all persons, and many words come loaded with unsuspected emotional connotations. The author recalls one time when he told his boss that a certain problem would be "taken in stride." By this, he meant that it would be overwhelmed by an irresistible application of force. But to the boss, the phrase meant that the problem would be treated casually and perhaps even ignored. Needless to say, some explanations were required by this misunderstanding. But sometimes we do not get the opportunity to make explanations, and all we get instead is bad communication.

The study of the meanings of words is called *semantics*. One of America's best-known semanticists is S. I. Hayakawa, who at one time was a professor at San Francisco State University. Later, he served a term as a U.S. Senator, but he is remembered by many people for his best-selling book, *Language in Thought and Action*. One of Hayakawa's central concepts is that of the "abstraction ladder." Almost all human speech involves abstractions. For example, when we use the word *chair*, we bring together in that word all the features chairs have in common, ignoring the differences that exist between folding chairs, swivel chairs, easy chairs, and infant highchairs. Hayakawa uses the example of Bessie, the cow, to illustrate his meaning. Bessie, as an individual cow, is unique. Yet, we can never be precisely sure of what we mean by referring to Bessie, because Bessie consists of an ever-changing assemblage of electrons, atoms, and physiological processes. Bessie today is not the

same as Bessie tomorrow. Yet we are not troubled by this, and if Bessie should even lose something highly visible, such as her tail, she is still Bessie. We have selected, from the swirl of electrons, a useful abstraction called "Bessie."

Moving up the abstraction ladder, Bessie belongs to another abstraction called "cow." Cows come in many sizes and colors, yet we have abstracted from all of these the characteristics that "cows" have in common and ignored the differences. Moving farther up the abstraction ladder, we can define a group called "livestock" and, yet farther, a group called "farm assets." At this point, we would have to deal with the question of what tractors, barns, wheat, and cows have in common.

Hayakawa points out that many of our troubles in communication lie in the tendency to conduct discussions at high abstraction levels. Worse, in trying to establish communication, we may move *up* the abstraction ladder when we should be moving down. He gives a sample conversation that moves up the abstraction ladder, thus:

"What is meant by the word red?"
"It's a color."
"What's a color?"
"Why, it's a quality things have."
"What's a quality?"
"Say, what are you trying to do, anyway?"

Some groups—politicians, for example—may deliberately seek to obscure their meanings by talking at high abstraction levels. Most of us, however, would prefer to clarify our meanings, rather than obscure them. The best way to do this, when we find ourselves moving up the abstraction ladder, is to stop, slide down the abstraction ladder instead, and provide a concrete example of our meaning.

Reference
S. I. Hayakawa, *Language in Thought and Action* (New York: Harcourt Brace Jovanovich, 4th ed., 1978).

Table 10.1 Results of Survey of 4000 Engineers with Respect to Selected College Subjects

Subject	Recommended for Future		Used During Past Month		Used During Career		Taken in College	
	%	Rank	%	Rank	%	Rank	%	Rank
Algebra	99	01	62	02	98	01	100	01
Physics (general)	99	02	41	06	95	04	99	03
English composition	99	03	78	01	98	02	97	07
Trigonometry	99	04	51	03	96	03	100	02
Calculus	97	05	23	18	81	12	99	04
Speech	96	06	51	04	92	05	77	22
Mechanics of solids	95	07	28	13	83	09	88	13
Chemistry (general)	95	08	26	14	78	19	98	05
Analytical geometry	95	09	19	25	82	11	98	06
Electric circuits	94	10	33	09	82	10	91	10

Source: *Goals of Engineering Education: Final Report of the Goals Committee* (Washington, D.C.: American Society for Engineering Education, January 1968).

month, (c) had used them during their careers, and (d) had taken them in college. English composition and speech ranked in the "top ten" of the engineers' lists.

■ 10.1 WRITTEN REPORTS

These are some of the faults commonly found in the written work of engineering students:

1. They have not thought carefully, in advance, about what they want to say. In other words, they do not know what their purpose is. Admittedly, for some students the purpose is just to get the writing assignment out of the way. As a result, they frequently sit down and start recording whatever words first come to mind. The usual outcome is poor writing.
2. They have not organized their ideas in advance. The result, unless something is done, is likely to be muddy thinking, and the writing will show it. A brief outline usually helps to get the writer's basic ideas in the proper, logical order. Clarity improves as a result.
3. They do not adopt the reader's point of view. The purpose of writing is to convey information to someone else. The writer should periodically step back from the written work and think about the prospective reader who is supposed to be influenced by the writing. Such reflection should include an evaluation of the probable gaps in the reader's knowledge and some estimates of the reader's interests.
4. Their writing tends toward lengthy sentences and cumbersome constructions. Unfortunately, the way some engineering instructors grade lab reports, students get the idea that length and bulk are virtues. But in professional engineering, the truth is just the opposite: Brevity is a virtue. But then, so is

completeness. A new engineer might understandably be somewhat perplexed if the boss says, "I want *everything* to be in the report, but I want it to be short!" But the same new engineer might discover, upon carefully re-editing a just-finished report, that a surprising number of the words are not doing anything useful. Also, it may be discovered that many long sentences can profitably be broken into two, yielding improved readability. Editing can materially shorten a report and improve its clarity at the same time.

5. They often write ungrammatically, with the result that clear communication is hampered. Sometimes students do not seem to know what constitutes a complete sentence or are vague in their references. Here are a few examples, with suggested improvements:

> I met him at the New York airport, coming home from London.

We do not know which person is coming home from London. It is better to recast the sentence: *When I was coming home from London, I met him at the airport.*

> After the records are tabulated, the manager knows the day's results. Such as completed assemblies, partially completed assemblies, and spare parts.

The first statement is a complete sentence, but the second is only a sentence fragment. What must be done here is to replace the period after *results* with a comma. The error in this example may appear so obvious as to not merit comment, but professors, with sorrow, know that errors of this type appear with distressing frequency in students' writing. They suggest carelessness on the part of the writer, and a professional engineer should be quite concerned about the possibility of being labeled careless.

> Occasionally a person will come along, like Edison or Howe, who have made important inventions.

The problem here is with the word *have*. The writer of this sentence apparently was looking at the words *Edison or Howe*, and thus chose the plural form. But the sentence should be: *Occasionally a person will come along, like Edison or Howe, who has made important inventions*, because the verb applies to the words *a person* rather than to the words *Edison or Howe*. This kind of error does not interfere with the meaning of the sentence, but it is likely to make the reader feel that the writer does not know what he or she is doing.

> While sleeping in the garden, a snake bit me.

This famous example has the obvious fault that the reader may not know who was doing the sleeping—me or the snake. The idea of being bitten by a sleeping snake is mildly humorous, but the thought that a snake might creep up and bite me while I am sleeping is, to choose a word, creepy. The example may appear obvious, but students make this kind of error regularly by not being careful about which of two possible subjects a verb is supposed to refer to.

6. Misspelling is very common in the writing of engineering students. In fact, some students seem to take the position that correct spelling is not of great

importance because secretaries will take care of the errors, or spelling checkers in word processing programs will fix them. Sometimes a misspelling will be picked up by secretaries or word processors, and sometimes it will not. Following are two paired lists of words that are frequently misused because of lack of attention to spelling. All the spellings in both columns are correct, but the meanings of each pair of words are quite different. Word processors generally let the misused words pass right on by; many secretaries will also fail to notice the errors. If errors appear in your final draft, readers may be irritated by the misuse and may begin to harbor doubts about you. If you are uncertain about the distinctions between some of these pairs of words, now is the time to get out your dictionary.

principal	principle
effect	affect
advise	advice
accept	except
cite	site
its	it's
their	there
their	they're
to	too
who's	whose
passed	past
breech	breach
credible	creditable
desert	dessert
planned	planed
hopping	hoping
dinning	dining
latter	later
pinned	pined
gripping	griping
sitting	siting
access	excess
allusion	illusion
angel	angle
allude	elude
course	coarse
council	counsel
later	latter
loose	lose
your	you're

The choice of the wrong word from the above lists sometimes produces ludicrous results, as in: *I am going to sit in the dinning room.* The reader does not know whether you customarily go into that room with the intention to dine, or to create some din.

The foregoing may look as though it is just going to lead to an awful lot of hard work. That is correct. Writing *is* hard work, and practice is necessary in order for some of the good habits to become semiautomatic. The rules are as follows:

1. Have a clear idea of what you want to say.
2. Make an outline.
3. Think about your reader.
4. Seek brevity, while not omitting important material.
5. Pay attention to grammar and spelling. If in doubt about the spelling of a word, look in a dictionary.

■ 10.2 ORGANIZATION OF A TYPICAL REPORT

In the case of technical reports, there is a more or less standard organizational outline that should be followed. The outline presented here is not necessarily perfect for every situation, but an engineer who uses it will not go far wrong. The outline is as follows:

Letter of transmittal
Title page
Abstract
Introduction
Procedure
Results
Discussion
Conclusions
Recommendations
List of references
Appendix

The report should be broken into sections, each of them bearing a heading from the above list. Upon occasion, some of these sections may be combined, especially in a short report. For example, a section headed "Results and Discussion" or a section headed "Conclusions and Recommendations" might be appropriate for some reports.

Some persons like to see an additional section titled "Summary" at the end of the report, following the recommendations section. However, if the abstract is well done, it may serve as a summary, rendering a separate summary unnecessary.

A table of contents might be needed for a long report, but it is usually omitted for short and medium-length reports. Indexes are rarely used in technical reports.

Letter of Transmittal

Students are sometimes surprised by the fact that a letter of transmittal is needed, because it would seem that the arrival of a report on someone's desk is its own evidence of transmittal. However, in an organization there may be many people who will see the report, and they will not all necessarily know why the report exists, where it came from, or what it is for. A letter of transmittal answers these questions.

(Writer's Address)

(Date)

Dean John A. Smith
School of Engineering
(Name of University)
(City, State, Zip Code)

Dear Dean Smith:

I am transmitting to you herewith a report entitled,
"Faculty Advising in the School of Engineering." The
report gives the results of a project undertaken during
the last academic year by the local chapter of Tau Beta
Pi, in accordance with discussions which the officers
of our group had with you in your office.

The report gives the results of an opinion poll conducted
by our chapter, involving all of the students of the school.
We sought information on the students' views of the present
status of faculty advising, and recommendations concerning
the process.

After you read the report, we would be pleased to meet
with you to see if our group can be of further service
to the School in connection with our report.

                          Sincerely yours,

                          Stephen Q. Doe

                          Stephen Q. Doe
                          President
                          Tau Beta Pi

Figure 10.1
A sample letter of transmittal.

The letter of transmittal should be short, should deal with the question of why the report exists, and should give some insight into its subject matter. A typical example is shown in Figure 10.1. Note that an address is given where the writer of the letter can be reached. This is important; otherwise, the recipient of the letter must go to a lot of effort to find out where a response, if any, is to be sent. Also, note that the letter closes with a short paragraph that appears to be there only as a formality. In fact, it is just that—a formality—but the absence of such a courteous closing sentence or short paragraph strikes most people as rude. Almost universally, writers of business letters (and a letter of transmittal is a business letter) are advised to close with such a statement—offering to provide further information if wanted, to answer questions, to meet with the recipient, or to be of help in other ways. Circumstances dictate the kind of closure that is appropriate.

Title Page

The title page contains not only the title and the authors' names, but also the date and the name and address of the organization from which a copy of the report can be obtained. In the case of a student report, the organization no doubt is the student's academic department. But in professional life, the organization most likely will be the engineer's employer or a professional society. The date is important. Your report may wind up in a library, and someone may come across it years later. It may be vital for that user to know where your report fits in, whether it was produced last month or 20 years ago.

Abstract

The abstract comes right in the front of the report, but is usually the last thing to be written. Actually, the writing of a report begins in the middle and works both ways. The first things a writer usually prepares are tables and graphs—in other words, the results. The discussion, conclusions, and recommendations probably come next. No doubt the introduction and procedure sections will be next, then the references and appendix, and last of all the abstract.

Writing a good abstract is more difficult than it would seem. The abstract must be short, certainly no more than 200 words, and it should give the reader a pretty good idea of the report's contents. It should contain the following essentials:

1. Description of the problem
2. Description of your method
3. Statement of your principal result(s)

Obviously, not much detail can be given on these, or your abstract will turn into the report itself. The major problem with most abstracts is that they dwell too much on point 2, the method used. Your reader certainly does want to know something about the method, but if some information about your results is missing, you are likely to have a disappointed reader on your hands.

Following is an example of an abstract that contains all three of the essential elements. It is an actual abstract taken from the professional literature, and it contains about 125 words.

Bear in mind that many readers of your report will read only the abstract and perhaps the conclusions and recommendations sections.

Optimum Damping and Stiffness in Nonlinear
Single-Degree-of-Freedom Systems.   II. Velocity Shock

ABSTRACT

Cases of landing impact (velocity shock), such as the dropping of a packaged item or the landing of an aircraft or spacecraft, are investigated. The packaging material, or the shock-absorbing landing gear, is assumed to be nonlinear (exponentially "hardening" or exponentially "softening") in both stiffness and damping. A large range of the system parameters is investigated, using a digital computer. It is found that, for the range of parameters investigated, significant advantages are possible by using nonlinear systems, as compared to what can be achieved with purely linear systems. If an allowable displacement is specified, it appears that the peak acceleration can be reduced as much as 35%. With a given allowable acceleration, the peak displacement can be reduced as much as 25% to 30%.

(From *Journal of the Acoustical Society of America*, vol. 47, no. 3 [part 2], March 1970, pp. 852–856.)

One thing to remember about abstracts is that they are frequently reproduced separately in other publications, as in periodicals that make a regular business of publishing collections of abstracts. Persons who come across your abstract in such a publication and who have an interest in your topic can then decide whether they want to get a copy of your complete report.

Introduction

The introduction should be brief. It brings your reader into the picture, so to speak. It tells what your purpose is, describes the problem to be solved, and gives any other relevant background, such as results found on similar problems by earlier investigators.

Procedure

A section on procedure will probably make dull reading, but is vital nonetheless. The validity of your conclusions is likely to rest on the kind of procedure you used. If your procedure consisted of a mathematical analysis, that is where your assumptions and calculations should be focused. The same is true if your report is based upon a computer analysis, although your actual computer program should be placed in the appendix. If your report describes a physical experiment, then the procedure section must contain a description of your apparatus and experimental method, although such things as equipment lists and wiring diagrams should go in the appendix. If your report concerns an opinion survey, then you tell in the procedure section exactly how you did the survey and the assumptions you made.

Results

This is the centerpiece of the report. It tells what you learned.

Discussion

In this section you give your opinions about your results. If you have any doubts about the validity of the results, you express them here and tell why you have them. Obviously, the opinions you express should be backed up by logical argument. If your doubts stem from the fact that your investigation is not complete or was carelessly conducted at some point, then a serious question exists whether you are ready to write the report. Perhaps more investigation is needed. But your doubts may rest upon other things. For example, if your report is based upon a computer analysis,

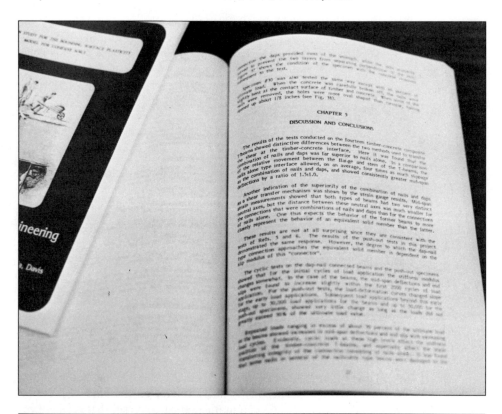

Figure 10.2
Engineering reports are clear, precise documents, presented in attractive formats.
(Courtesy of University of California, Davis)

you know that many computer procedures produce certain unavoidable errors; you discuss them in this section. You also discuss the magnitude and significance of measurement errors, which always exist. If your investigation is based upon statistical methods, then this is where you would discuss sampling errors.

Conclusions

The purpose of conducting an investigation is to reach some conclusions about the topic. The difference between *results* and *conclusions* is that the former are more detailed and mostly tell what happened, whereas the conclusions bring the results together into an overall judgment and give them meaning and value. For example, your report may have to do with the safety of a new kind of electrical insulation. The results section might give data regarding the voltage breakdown levels you observed in your tests of the new insulation, resistance to abrasion, or behavior under various environmental conditions such as heat and humidity. The conclusions section would place these results into context, to show whether the data imply greater (or lesser) safety, economy, and durability when compared to other products.

Recommendations

A recommendations section is not always appropriate; sometimes all that is wanted at the end of a report is a set of results and conclusions. More often, though, recommendations are in order. They should follow logically from the report's results and should not be irrelevantly inserted by the writer. If a recommendation seems contrary to a report's results, a serious reflection upon the writer occurs. Every recommendation should be justified by what went before it in the report. Usually it is desirable to number the recommendations so they can be conveniently referred to in subsequent correspondence.

List of References

If you have relied, even in part, on the work of others, you should make reference to that fact in your report. A typical way to do this is as follows. In the body of the text, perhaps in the introduction, a sentence such as the following might appear:

The case of a flexible disk rotating next to a surface has been investigated by Pearson (1).

Then, in your list of references, you place the full reference as follows:

1. R. T. Pearson, "The Development of the Flexible Disk Magnetic Recorder," *Proceedings of the IEEE*, Vol. 49, 1961, pp. 164–174.

Throughout the report, the references are numbered sequentially, each one matching an identically numbered item in the list of references.

Incidentally, the specific methods for making references to the literature are highly varied. The method shown is commonly used by engineers, but many other methods exist. When you get ready to write a report, you will probably be given a style manual that must be followed. In the absence of such a manual, the preceding style is a good choice. The example already given was for a journal article. If the referenced item is a book, the listing is as follows:

2. F. H. Zuwaylif, *General Applied Statistics*, 3rd ed. (Reading, Mass.: Addison-Wesley, 1979).

Appendix

Material that is necessary to the report, but so detailed that the reader might be bogged down by it, should be placed in the appendix. Many times a nicely balanced act of judgment is required to decide what to put in the main text and what to put in the appendix. No hard rules can be given, except to keep the reader in mind. If you put yourself in the place of the reader and conclude that he or she may lose patience with your material or lose track of it, then the material in question probably belongs in the appendix.

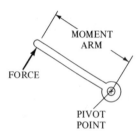

Figure 10.3

Sketch showing definition of the term "moment arm."

Exaggeration

In almost everyone's writing, words like "very," "extremely," and "greatly" seem to creep in by themselves. It is wise to take nearly all of them out unless you feel strongly that they are justified. There is also a temptation to underline words and phrases, to emphasize them. But underlining should be used sparingly, or the effect is dulled and the desired response in the reader does not take place. Sometimes there is a temptation to emphasize a statement by using an exclamation point, but an exclamation point is almost always out of place in a business memorandum, a technical report, or, for that matter, a textbook!

Jargon

Jargon is technical language that is unique to a particular group and understood only by them. (Another definition of *jargon*, of course, is confused talk, or gibberish.) A simple example of a term that would be understood by all engineers is *moment arm*. A moment arm is related to a force that tends to rotate an object about a pivot point and is defined as the perpendicular distance between the force and the point (see Figure 10.3). It is similar to the term *lever arm*. A lay audience might understand the term *lever arm*, but probably not the term *moment arm*. To them, the latter is jargon. Similarly, the terms *stress* and *strain* have precise (and distinctly different) meanings for engineers (see Chapter 15), but lay audiences are likely to think stress and strain mean approximately the same thing. Once more, to non-engineers these terms are jargon.

In preparing reports, whether written or oral, one should be careful to consider the probable audience. If it consists of engineers, then normal engineering jargon is probably all right, although one should be aware that language that is clear to those in one branch of engineering may not be clear to those in another. Language that is suitable for the proposed audience should be chosen.

■ 10.3 PROGRESS REPORTS

A special kind of report, known as a *progress report*, does not use the formal organizational structure defined in the foregoing section. The purpose of a progress report is to let your boss or your sponsoring organization know how things are going. It usually takes the form of a memorandum, and it should be brief. Such formal items as a letter of transmittal, an abstract, a detailed description of procedure, and a list of references are all unnecessary.

What *is* necessary in a progress report is some identification of the project, so the reader will know what you are talking about, and a mention of whether this is the first progress report, the second, or whatever. Then a brief description of the results to date should be given, along with a statement of whether unforeseen difficulties have emerged that might interfere with your planned completion date. If you *are* experiencing difficulties and require some sort of action in order to proceed on course, then some recommendations are appropriate; otherwise, recommendations are probably not needed. A summary is almost always desirable, and this should be placed right at the beginning of your progress report. Then the boss can simply scan the summary and will not need to read the entire report unless the summary mentions that some trouble has occurred.

■ 10.4 RÉSUMÉS

A *résumé* (pronounced REZ-oo-may, and sometimes spelled *resume*, without the accents) is a brief summary of biographical information with an emphasis upon education and experience. Each time you apply for something, such as for a scholarship or a job, you will probably be expected to provide a résumé. There is no "official" format for a résumé, but the arrangement shown in Figure 10.4 is widely used.

A common error in résumés is verbosity. People sometimes write résumés with 10 or even 20 pages. Usually, the effect of such length is negative. A résumé should be complete, yet brief; it is the same old problem of finding the right balance. Good judgment is important, but as a general guide, most people can say everything that needs to be said in three or four pages. For a college student, a résumé longer than one or two pages is probably too long.

■ 10.5 ORAL REPORTS

The need to present an oral report arises frequently in an engineer's professional life. The most common occasions are design and manufacturing conferences, where the engineer needs to make clear, useful, and dependable statements. But occasions also often arise wherein a more organized presentation is needed. For example, some companies expect their engineers to give regular oral progress reports on their projects. A typical audience may be a group of colleagues in the engineering department, although it also may be necessary to address a group of company executives or even the board of directors.

It is only normal for a person to experience some nervousness when preparing to make an oral report. But nervousness should not stand in the way. Usually it will disappear after the first few minutes, and it also tends to become less troublesome as a person develops experience. If the nervousness is so great that it causes trembling of the hands or a shaky voice, then the thing to do is to acquire experience in a less

(Date)

John A. Doe

Address: 1234 Third Street
            Anyplace, CA 95616
            (916) 752-0553

Objective: Full-time employment in system design and development
            in computer industry.

Education: Anywhere University
            B.S. in Computer Engineering expected, June 19XX
            G.P.A.: 3.4/4.0

Honors and Awards: Dean's List: 3 terms
                    Tau Beta Pi, National Engineering Honor Society

Experience: Anywhere University, Department of Electrical and
             Computer Engineering: January 19XX--present; reader,
             correcting homework and lab reports.

             Hydrologic Engineering Inc.: Summer, 19XX; computer
             programmer, for data storage and retrieval systems.

             Aerospace Systems, Inc.: Summer, 19XX; computer trainee.

Activities: Recording Secretary--Tau Beta Pi, 19XX-XX.
             Member, Institute of Electrical and Electronics Engineers
             Intramural sports: baseball, basketball
             Hobbies: skiing, hiking

Personal: Citizenship: United States of America

References: Professor Richard A. Roe
             Department of Electrical and Computer Engineering
             Anywhere University
             Anyplace, CA 95616

             Professor John W. Smith
             Department of Mechanical Engineering
             Anywhere University
             Anyplace, CA 95616

             Mr. Robert A. Jones
             Director, Computer Center
             Aerospace Systems, Inc.
             P.O. Box 5000
             Los Angeles, CA 95698

Figure 10.4
A typical résumé.

threatening environment than on the job—for example, in clubs, church groups, or evening classes. For most people, though, the problem of nervousness will not be that serious. For them, the best defense against nervousness is to be well prepared. If you know your topic inside and out, your nervousness will probably disappear soon after you start speaking.

Formal Speeches

Engineers are not often called upon to make formal speeches before large audiences. However, if this should happen to you, the best thing to do is to write out your speech and read it to the audience. This way, you will know exactly what you are going to say and will be certain you have chosen the words you want. This method is much superior to memorization. On the other hand, if one is in the regular business of making speeches, memorization may be appropriate. Presented by a talented and experienced person, a memorized speech can even sound spontaneous. (After all, professional actors speak memorized lines.) But for most of us, an attempt at memorization invites the catastrophe of forgetting what comes next.

In presenting and reading a written speech, it is important to remember that spoken words and written words have different styles. As you write your speech, speak it to yourself. Hear it in your "inner ear," and imagine the pauses and kinds of emphasis that you will use. Then practice it—not once, but several times. Go in a room by yourself and read it out loud. Put in your pauses and your changes of emphasis just as if you were speaking to a real audience. If you can get a friend to listen to it, by all means do so. The object is to make the speech second nature. When it comes to the actual presentation, you will find that whole sentences will come to you as if memorized, and it will only occasionally be necessary to look down at your text.

Informal Presentations

For every formal speech you make, you will probably make a hundred informal ones. Presuming you know in advance that you must make an informal presentation to a group, the first thing you will want to know is something about your audience and their background in your subject. You certainly do not want to insult their intelligence by presenting your topic at too elementary a level, nor do you want to "snow" them by speaking over their heads. This point is vital and will require a little advance investigation.

In addition to knowing your audience, you should be the master of your subject, because you almost surely will have to answer questions following your talk. Next you should prepare a brief outline of what you want to say, and then you should think through your outline, going over each sentence in your mind as if you were actually giving the talk. You should be prepared to speak from notes alone. Reading a speech in a case like this would definitely be the wrong choice.

An oral presentation, by nature, is less detailed and complex than a written report. Typically you will have only 15 to 20 minutes, and you can make only a few points in such a limited time. Those who want more detail can always be referred to your written reports.

A typical procedure in organizing a talk is to divide it into three parts: introduction, main body, and summary—or, in the oft-quoted form,

First: Tell them what you're going to tell them.
Second: Tell them.
Third: Tell them what you just told them.

These three points should not be taken too literally, of course, but they do get across the point that an introduction and a summary can be important to your presentation.

As a final matter, there is the question of whether you should try to insert jokes into your presentations. The conventional wisdom seems to be that you should open a speech with a joke, but much treacherous ground lies in that direction. If you are not a good joke teller, then the best advice is—don't try it. Furthermore, jokes are not appropriate in many circumstances. Generally, if you are making a progress report to a group of executives, jokes are inappropriate. On the other hand, if you are making an after-dinner speech, jokes are acceptable, although they should never be in bad taste. Remember, you will have many people in your audience, and their ideas of good taste will range from one end of the spectrum to the other. Your own ideas of the limits of good taste may leave as much as a third of your audience on the wrong side of the limit. You don't need to have that many people angry at you. Stay on safe ground. Also, even though your joke may be in good taste, it still runs the risk of falling flat, and then you would have a handicap to overcome. If you have any doubt concerning your ability to tell jokes with skill, pass them up.

Visual Aids

Almost any oral presentation is improved by visual aids. If the group is small, a chart on a large piece of poster board is often appropriate. For larger groups, the charts should be converted to $8\frac{1}{2} \times 11$ transparencies or 35-mm slides.

A simple rule to follow with respect to your visual aids is to put yourself literally in the place of your audience, to see if you can read them. For example, put your chart up in the room you expect to use (or one like it) and then go to the back of the room to see if it is readable from there. Remember that other persons' heads will be in the way of those in the back rows. Think about your audience.

If you are using transparencies or 35-mm slides, test them on equipment of the type you will use. (The author once had a shattering experience when he had a number of charts drawn up on the kind of transparent plastic material that is often used by drafters to make drawings. The charts looked beautiful, but the author neglected to test them on an overhead projector before setting out to make a presentation at a national conference. When the charts were placed on an overhead projector for the first time—during the actual presentation—nothing appeared on the screen but a white blur. The problem was that the drafting material had a "toothed" surface to receive and retain pencil or ink. In the hand, the material looked transparent. But on the projector, the toothed surface scattered the light in all directions, producing a blur. It takes only one such experience to make a true believer out of a person.)

Keep your charts simple. Audiences tend to grit their teeth because of the frequency with which they hear speakers say, "Well, this chart is pretty complicated, and you won't be able to read it from your seats, but...." Do not let that happen to you. Your audience will go to sleep. Remember, too, that a graphical chart or slide is much superior to one with a table containing lots of numbers.

A vital rule is to arrive at your presentation room well before your audience and check the equipment. Put your slides into the projector and cycle them through to be sure none of them is reversed or upside down. Also, you may discover the projector has to be repositioned. Sometimes you may discover the projector has not even been plugged in, and you will need to locate an outlet. You might even find that an extension cord is needed, and someone must be sent in haste to get one. You certainly do not want to discover any of these problems for the first time *after* you have started your talk. It is amazing how often a speaker does not even know in advance how to turn on the projector, or where the room's light switch is, and must discover these things after the talk has begun.

One great advantage to having a set of slides for your talk is that they structure your speech for you, reminding you of what you should say next. In this way they substitute for a set of notes. However, avoid the temptation to have your outline notes essentially duplicated on slides. Some speakers do this, and the result is often deadly. To be effective, your visual aids should portray something the spoken word cannot, as is the case with graphs and photographs, for example.

Some Bad Habits

Some speakers do things that interfere with the communication of their topics and are completely unaware that they are doing them. In fact, most of us do this at one time or another. If you can arrange to have a presentation recorded on videotape and watch it later, a number of bad habits may become apparent of which you had no knowledge. Alternatively, a friend may be willing to listen to your oral presentation and point out bad habits. One should not take personal offense when such traits are found; we all have them.

A few typical bad habits are listed here. See if you can discover whether you have any.

Not speaking loudly enough, or not pronouncing words clearly.

Putting "and-uh" after almost every sentence, or putting "you know" into nearly every phrase.

Speaking so fast that the listener is unable to follow.

Jingling, fidgeting, paper shuffling, foot shuffling, or anything else that distracts the listener. It is amazing how common it is for a male speaker to shove his hand into his pocket, almost as soon as he starts to speak, and jingle the coins in his pocket throughout the entire speech. You can be sure he is totally oblivious to the existence of this habit, or he would not do it. Women speakers usually do not have pockets with coins in them, so this particular offensive habit does not plague them.

Standing in front of whatever the speaker wants the audience to see or, if writing on a blackboard, writing so low on the board that most of the audience cannot see the material. (Interestingly, students are generally very critical of teachers who have these faults, and then usually commit the same errors themselves when it is their turn to speak.)

Failing to maintain eye contact. You should look at the people in your audience. Look them in the eye. Look successively at different ones, in different parts of the audience. This way, you are speaking directly to individuals rather than to a blob. You are much more likely to retain your audience's interest this way, and also, you are more likely to know if your listeners are receiving your message.

■ 10.6 ENGINEERING GRAPHICS

Written and oral communication are both of importance to engineers, but there is another mode of communication that has special importance, because engineers can scarcely convey their ideas to other persons without it. This is graphical communication through the use of *engineering graphics*.

When we talk about engineering graphics we mean something more than just the preparation of graphs. Graphs are important, as has already been pointed out in Chapter 7. However, engineering graphics includes not only graphs, but also such things as sketches, diagrams, schematics, manufacturing drawings, and construction drawings.

The making of sketches, diagrams, and schematics comes naturally to engineers, because engineering college courses and textbooks are stuffed full of these graphics. Following the completion of a four-year engineering curriculum, a graduate is usually fairly proficient at them. In contrast, the preparation of manufacturing and construction drawings is a special art, usually acquired through formal course work. Some individuals acquire the necessary knowledge through on-the-job training, but, one way or another, special skills must be learned if one is to deal properly with the kinds of drawings that are employed in manufacturing or construction.

The major "trick" in using manufacturing or construction drawings is to be able to visualize three-dimensional objects through the use of two-dimensional media. Real objects in real space are three-dimensional, but any time an object is rendered on paper, the rendition necessarily is two-dimensional. Even if we employ pictorial means, such as artistic perspective drawings, the drawings are still on paper, which is two-dimensional. The use of computer-aided drafting systems does not change this situation, because the computer screen is also two-dimensional. We can get around this problem either by constructing three-dimensional models of everything we wish to discuss, which is obviously cumbersome and expensive, or by learning the principles of *orthographic projection*. The almost universal choice is for the latter.

The theory of representing three-dimensional objects in orthographic projection is called *descriptive geometry*. But it is also vital for engineers to learn the standardized conventions that are understood by both the makers and the users of drawings. These conventions include the proper use of lines (visible lines, hidden lines, centerlines, section lines, and extension lines), dimensioning, tolerancing, cross-sectioning, designation of materials and surface finishes, and representation of special features such as welds, rivets, and screw threads. These matters are so important that they are prescribed in publications prepared by the American National Standards Institute.

Much of the necessary subject matter is hinted at in Chapter 11, on engineering design. Also, Appendix H contains a brief summary of most of the standard conventions that are used in drawings. However, Appendix H does not include any treatment of descriptive geometry, because that would increase the size of this textbook enormously. Fortunately, most engineering curricula, at least for mechanical and civil engineers, include formal courses on both descriptive geometry and conventional drawing practices, bearing such titles as *engineering graphics* or *engineering drawing*. Sometimes students (and some professors, too) regard such courses merely as drafting courses, beneath the level of an engineer's professional work. But engineers must understand the graphical language that is being used, because engineers are usually the interpreters of the drawings and are frequently the supervisors of the drafting technicians who make them.

Exercises

10.1 Prepare your own résumé, using Figure 10.4 as a guide.

10.2 Imagine that you have been part of a group that carried out the opinion poll referred to in the letter of transmittal in Figure 10.1. Make up some results—whatever you think would be the likely outcome. Then write a complete report on your imaginary project and its results, following the organizational outline in Section 10.2.

10.3 Look up some technical journals in your college library. (The library will probably have a bewildering variety of them, but the *Proceedings of the IEEE* or the *Journal of Applied Mechanics* might be good ones to start with.) Examine a few of the technical papers in these journals to see if they generally follow the organizational outline recommended herein. If some of them do not, do there appear to be good reasons for the ways in which they are organized?

10.4 Examine the abstracts in some technical journals to see if they do a good job of getting across the three main points recommended in Section 10.2. In particular, see if they give you some clues as to the actual results, instead of principally describing the procedure that was used.

10.5 Select an article from a journal and write your own abstract for it. Compare your abstract with the one that actually appeared with the article.

10.6 Make a list of technical jargon terms that you commonly employ, which may not be readily understood by others who are not "in the know."

10.7 See if you can make a list of bad speaking habits you may have. Listen to yourself as you talk, and see if any "and-uhs" or "you knows" creep in, or if you sometimes do not speak clearly. (Warning: These are among the hardest of habits to detect, and you will have to make a special point of reminding yourself to listen to your own speech.)

10.8 Prepare an oral presentation on one of the following topics or on one approved by your instructor. Research the topic in your library and work from an outline, following the suggestions in this chapter.

The pros and cons of nuclear energy

Technological competition between the United States and Japan

The pros and cons of graduate study

Should professors do research in addition to teaching?

Going into management

The pros and cons of metrication

Sales as a career objective for engineers

Social implications of computer networking and computer security

10.9 See if you can list some words, in addition to those in Section 10.1, that are commonly misused or misspelled. Pay special attention to those that could cause confusion of meaning.

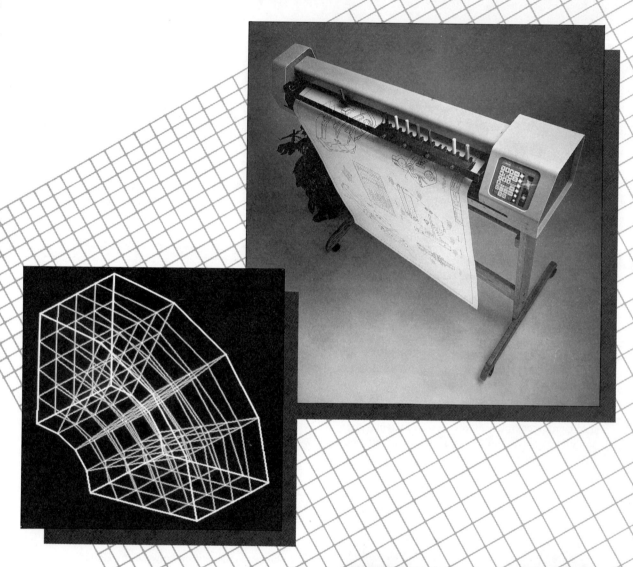

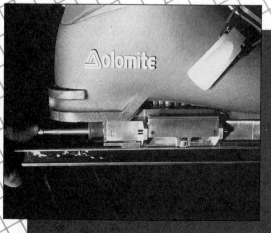

Photographs courtesy of Ioline Corporation (top), Applicon Incorporated (center), and University of California, Davis (bottom right).

11

Engineering Design

The word *design* is very broad and includes all of the activities of engineers as they go about their jobs of creating devices, systems, and structures to satisfy human desires and needs. It is the function of design that separates the work of an engineer from that of a scientist. A scientist's job is to gain new knowledge; an engineer's job is to use knowledge—both old and new—to create new things. It is this job of creation that we call "design."

A chemical engineer who is working on the process for a new chemical plant is doing design. So is the electronics engineer who is laying out a new microchip on a computer console. And so is the civil engineer who is designing a new skyscraper, a new freeway, or a new sewage treatment plant. Thus, we speak familiarly of such things as process design, circuit design, structural design, and system design.

■ 11.1 SOME DEFINITIONS

Human beings have a tendency to inflate their use of terms. Thus, the entire activity of drafting is called "design" by some, and the term "development" is used frequently to cover many of the design activities just listed. Also, there is a tendency in many branches of industry to call all of their engineering activities "research," whereas some of the companies that engage in this practice in truth conduct no research whatever. Let us look at some definitions:

design: Engineering design is the process of applying the various techniques and scientific principles for the purpose of defining a device, a process, or a system in sufficient detail to permit its physical realization. . . . Design may be simple or enormously complex, easy or difficult, mathematical or nonmathematical; it may involve a trivial problem or one of great importance. (Massachusetts Institute of Technology, Committee on Engineering Design)

development: . . . technical activity concerned with *nonroutine* problems which are encountered in translating research findings or other general scientific knowledge into products or processes. . . . The engineering activity required to advance the design of a product or a process to the point where it meets specific functional and economic requirements and can be turned over to manufacturing units. (National Science Foundation)

applied research (sometimes called "developmental research" or "engineering research"): . . . Investigation directed to discovery of new scientific knowledge and which [has] specific commercial objectives with respect to either products or processes. (National Science Foundation)

basic research (sometimes called "fundamental research"): . . . original investigation for the advancement of scientific knowledge and which [does] not have specific commercial objectives. (National Science Foundation)

These definitions have reasonably wide acceptance. It can be seen that the terms *design* and *development* tend to run together, with the former having to do mostly with the physical realization—i.e., fabrication—of the device, and the latter dealing more with pinning down unknown factors that may determine whether or not the device will work as intended. It is clear, however, that the word *research* applies to the discovery of new knowledge, the only utility of the words *applied* and *basic* being to distinguish whether or not the researcher is motivated by the knowledge of potential applications.

■ 11.2 THE DESIGN PROCESS

To give an idea of the way in which design problems are usually handled, it is useful to describe the details of the design process. Usually, we consider the process to consist of five phases:

1. Problem definition
2. Invention
3. Analysis
4. Decision
5. Implementation

Problem Definition

Problem definition is one of the steps an inexperienced engineer is likely to skip entirely. Having been subjected to intensive schooling during which the problems were almost always given in clearly defined form, the new engineer is uncertain of what to do when faced with a problem that is vague and largely unstructured. Because

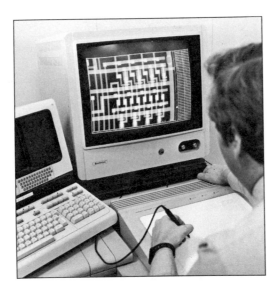

Figure 11.1
It is essential to use a computer system in designing integrated circuits. (Courtesy of Hewlett-Packard Company)

engineering education relies heavily upon problem solving, i.e., *analysis*, the analysis phase is reasonably familiar to the new engineer; thus, there is a strong temptation to get into the analysis phase as rapidly as possible. As a result, analysis may proceed at a lively pace before the determination has been made that the correct problem is being analyzed. The risk is that the analysis will produce an answer to a problem nobody is interested in. Therefore, the first task of the engineer is to *find out what the problem really is*.

In learning more about the problem to be solved, the engineer should articulate the basic needs that are to be satisfied. Generally, if the engineer's company is in the manufacturing business, this means learning something about the ways in which the potential customers will use the device. If the engineering project is in the realm of public works—a bridge, let us say—then an extensive analysis of traffic patterns, shifts in population, and the like will be needed. In any event, a clear vision of the potential need is necessary. Then, when the problem is defined, it will probably be the right problem.

It should be recognized, in contrast to the foregoing, that many useful new products arise in spite of the fact that the public need cannot be clearly identified. Such was the case, for example, in the case of the microprocessor. When the microprocessor was first invented, it was by no means clear that a large market for such a device existed. However, the inventor of the microprocessor had a clear view in mind of the problem to be solved, which was to compress the circuitry of a computer onto a few microchips. He obviously had faith that such a development would be attractive in the marketplace, but even he did not foresee just how explosively large that market would become. (See "Milestones" in Chapter 18.)

An important aspect of problem definition that is frequently overlooked is *human factors*. Consideration must be given to physical limitations imposed by human capabilities. This is a large subject in itself, and numerous books have addressed it, especially with reference to such items as average human body dimensions, reach, and

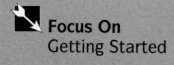

Focus On
Getting Started

A rather frightening experience occurs to virtually every young engineer shortly after graduation who is given a design task to perform. This is when a large blank piece of paper appears, and the engineer is told to put something on it. The "something" is a new machine, device, or structure, and the question is, "How do you begin?" How do you convert that blank piece of paper into lines and figures that will be convertible into something useful and tangible?

One mechanical engineer told of his own first design experience, and how he got past that awful blank sheet of paper:

The first thing I should tell you is that I never thought I would turn out to be a mechanical designer. Originally, I thought I would be an electrical engineer, but then I started taking courses in thermodynamics, and thought maybe I would be a thermodynamicist. After graduation, I couldn't convince anybody to hire me as a thermodynamicist, but I got an offer for a position as a mechanical designer, so that's what I became.

My first assignment was to design a cargo-dropping device, a gadget that worked something like a parachute, but which had wings that autorotated like a helicopter. It was to be filled with cargo, pushed out of an airplane, and would sail gently down to earth, with the wings autorotating.

I was frightened to death. I knew nothing of airplane wings, and didn't know where to begin. But I didn't give my boss enough credit. He didn't really expect me—a green, inexperienced graduate—to start right off to design this thing from scratch. He was an aeronautical engineer, and had already worked out the design of the wings. He gave me lots of guidance, and I soon started learning very fast.

The first thing I learned was this: even if you don't know where to begin, you nevertheless have to make a beginning somehow. If your beginning turns out to be a bad one—and there's a high probability it will—you can always back up and start over. You may have to make several such beginnings, but begin you must. In my case, I started out by trying to put some lines on that piece of paper to show where the structural members might go, how big the cargo compartments might be, how the external skin would be fastened on, where the wings would attach and so on. Pretty soon I found I had to figure out how big some of the parts had to be, to carry the load, and had to make some simple stress calculations. At this point I learned something else important: there really were some useful things in the textbooks I had used in my engineering courses, and I could apply them directly to the job at hand.

Before long, the job had subdivided itself naturally into small sub-tasks I could tackle one by one, and—with my boss' guidance, of course—I was on my way. The blank piece of paper no longer looked so frightening, and I had learned one of the most valuable lessons of my life: the way to get started on a project is—to start.

strength; visual acuity; average manual dexterity; sensitivity to noise, shock, and vibration; and tolerance to environmental factors such as temperature, humidity, and acceleration.

Invention

Engineering, by definition, is concerned with new things. That is precisely what invention is: coming up with new ideas. A new idea may involve a combination of old components, but if a new and useful effect results from this combination, then invention has taken place. Questions of patentability will be excluded from this discussion, and we will concentrate upon the creative act of conceiving an idea for hardware that may solve a particular problem. This potential assemblage of hardware has sometimes been referred to as a "scheme" for solving the problem at hand. At this stage, one cannot be sure that the scheme in hand will actually solve the problem. The purpose of the next phase of design—analysis—is to shed some light upon the probability of success. Obviously, at least one scheme has to be generated; otherwise, there is nothing to analyze.

As an example of what is meant by a scheme, suppose that a supervisor instructs an engineer to design an electric accelerometer. The instrument is intended to measure the acceleration of an automobile and is to be mounted on a dashboard for visual reading, although it must also be possible to use the output of the accelerometer to make a record on paper. It is hoped that the engineer will come up with many systems of hardware that, in principle at least, can perform the desired function, but for the purposes of this book, only two of the possible systems will be described:

1. Mount a mass to be as free from friction as possible, and provide a spring so that, as the automobile accelerates, the inertial resistance of the mass will cause the spring to be compressed. Connect the mass to an iron core mounted within a coil so that displacement of the mass (and core) will change the inductance of the coil and, thus, give an electrical indication of acceleration. (See Chapter 17 for a discussion of inductance.)
2. Gear a DC generator to the drive shaft of the automobile and hook it in series with a capacitor so that the current in the circuit is proportional to the acceleration. (Chapter 17 also includes a discussion of capacitors and capacitance.)

The typical reader of this book will not yet have had sufficient exposure to mathematics or to the subjects of dynamics and electric circuits to understand how the analysis of the two schemes should proceed. But that is the purpose of an engineering education—to equip an engineering student with the mathematical and scientific tools to be able to analyze systems like the foregoing.

In the two schemes just described, many questions of feasibility have been left unsettled, but that is what the analysis phase is for—to answer such questions. The important thing is that the engineer now has something physical and concrete to analyze.

One of the most important aspects of scheme selection is the posing of a textbook-type problem to be solved. For either of the foregoing examples, a physical structure has been described, and the question is asked, "How does the current vary with car speed?" Nobody gave the engineer the problem in this form. It was necessary to pose the problem from scratch, and this posing is one of the most important elements in the design process.[1]

---

[1] See D. W. Ver Planck and B. R. Teare, Jr., *Engineering Analysis, An Introduction to Professional Method* (New York: Wiley, 1954).

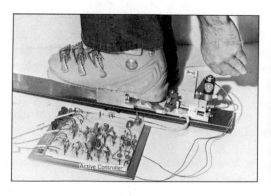 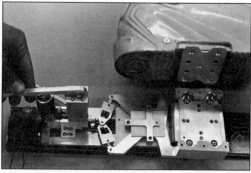

Figure 11.2

An experimental computer-controlled ski release unit. The active controller contains two analog computer circuits, which continuously provide solutions for the differential equations that represent torsion and bending. When a critical condition of impending leg injury is reached, an electrical signal releases the ski. (Courtesy of University of California, Davis)

Analysis

As already stated, it is primarily because of the analysis phase that engineers go to college. All of the other functions—problem definition, invention, decision, and implementation—can be carried out, to a very large extent, without the benefit of a college education. (This helps to account for the remarkable success in design achieved by many non-college graduates in the past.) Even today, much design can be performed without recourse to the analysis step: The designer proceeds directly from invention to decision, on the strength of experience and intuition.

However, the basic rationale for engineering is that a *better* job of design can be done with the intelligent application of science and mathematics than without. In fact, some of today's more difficult design tasks can be accomplished only with the assistance of advanced mathematics and scientific know-how. It was once popular to say that the engineer can do, for one dollar, what the untrained person requires two dollars to do. But an untrained person could not design a jet plane or a computer even with an unlimited amount of money.

The basic analytical tools used by the engineer are mathematics and a collection of scientific laws. Applying a knowledge of science, the engineer constructs a set of equations called a *mathematical model* and then, by means of mathematical manipulation, extracts from this model the information needed. The model should be reasonably representative of the physical system and obviously should be no more complex than is absolutely necessary to produce the required information. Herein lie two common errors of inexperienced engineers. (1) Frequently, the model chosen only slightly represents the physical system but has been chosen primarily because it is one the engineer knows how to analyze or because it is elegant. (2) Far too often, the model is more complex than it has to be: A lengthy and involved analysis could be a complete waste—for example, if the purpose of the analysis were to produce order-of-magnitude figures for a comparison with competing schemes.

If the engineer possesses insufficient scientific information to construct a good mathematical model, it may be necessary to initiate a research program to get the needed information. Therefore, the analysis phase of design may include research. In fact, *all* of the first three phases of design may require research in order to properly define a problem, test a scheme, and thus arrive at an invention or obtain the scientific information necessary for analysis. Because of the frequent intermingling of research with the design (or development) process, a corporation generally refers to its entire technical activity as "research and development," usually abbreviated R & D.

The most useful thing about the analysis phase of design is its production of quantitative information that can be used as a basis for scheme selection, as opposed to the purely qualitative nature of the invention phase. Thus, at this point it is necessary to get *numbers* and to become specific about the hardware components and their interconnection. Through mathematical analysis, the influential system parameters can be identified and optimum values selected. After this has been done, at least in a preliminary way, the decision phase can follow.

Decision

Even after mathematical analysis, individual judgment is necessary. For one thing, economic considerations enter the picture at this stage. The product must be produced at a low enough cost and at sufficient volume to recover all of the development expenses and produce a profit.

Selection of a scheme depends upon which of those offered appears the most favorable as a result of *optimization*. Optimization is finding the best combination of certain variables that will maximize the desired results, such as appearance, weight, durability, selling price, serviceability, quietness, or sensitivity. If all these can be placed together on a value scale, it is possible to pick out an optimum combination and to make a single selection from among the competing alternatives.

Implementation

Before the design process may properly be considered complete, detailed manufacturing instructions must be prepared so that the device, structure, or system can be produced. Historically, the medium for such instructions is the working manufacturing drawing, although computer output data, in the form of punched or magnetic tapes, are also becoming widely used.

However, even before working drawings (or tape) can be produced, much detail design, involving spatial considerations, strength, weight, economy of manufacture, and the like, is necessary. Although detail design is only one of the many phases of design, it is an important one that must not be treated lightly by the engineer. The responsibility for a design extends to the last detail. Many of the engineering technology educational programs described in Chapter 6 are aimed at the implementation stage of design, where manufacturability becomes a major consideration.

Iteration

Lest it be expected that a design project should proceed neatly and directly through all the phases in order, special mention must be made of *iteration*. What this elegant

Figure 11.3
Scale models are essential tools in the design and construction of refineries and other process facilities involving piping. (Courtesy of the Parsons Corporation)

word means is that the designer may frequently need to back up and do something all over again. New data may be discovered, a new idea may be generated, mistakes may be found, or things simply may not work as expected. The last point, *things may not work as expected*, is especially important: Careful analysis must be followed by careful testing. Failures on the test bench may even require complete abandonment of a given scheme.

A couple of examples may help illuminate the design process.

■ 11.3 DESIGN EXAMPLE: PROJECTOR SUPPORT CARRIAGE

Often, in embarking on the design of a new product, it is necessary to conceive a brand-new device that does not yet exist on the market. Under these circumstances, steps 1 and 2 of the design process (problem definition and invention) are probably the most difficult and the most critical. This is because the design engineer has to start with *nothing* and, from that nothing, make something. Furthermore, it has to be something that will find a place in the market and can be sold at a profit. More often, however, the design task is not the conception of a total new product, but instead the design of a portion of a product. Perhaps the product already exists, and some new functions are to be added. Or it may be that a development contract has been signed with a customer, in which case the general nature of the product has been established by the terms of the contract. It is still necessary, under these conditions, to generate a *scheme* (or several schemes) that will ultimately produce a successful design, but the first two steps are less intimidating than when one must start with a "vacuum." Also, it is much more likely that an inexperienced engineer will be given just a portion of a product to work on than an instruction that in effect says, "We need a brand-new product. Go think up one."

So let us assume that you, the new engineer, are working for a company that is in the business of designing and manufacturing precision optical measuring devices. In consultation with a potential customer, one of the company's sales engineers has come up with specifications for a measuring system that is somewhat different from the existing product line but bears a strong family resemblance to it. The company management has signed a contract with the customer, and the design of the new product has been assigned to a small group of engineers. You are a member of this group, and you have been assigned the responsibility of designing the portion of the product that is called the "projector support carriage."

First you must know more about the product. Figure 11.4 is a side view of the product in schematic form. The overall unit is about 5 feet tall. A film projector is mounted in the upper portion of the unit with the lens pointing straight down, as shown. As can be seen in the figure, the projector is mounted on the unit for which you are responsible, the projector support carriage—but more about that later. The projected image is reflected off a mirror, and is focused on a glass screen, the back of which has a finely ground surface so it can capture the projected image. The operator sits in front of this screen and sees an image of the film magnified about 15 times.

The film being used is standard 35-mm movie film, and the pictures being taken are of a rocket in flight. Cross hairs are visible on each frame, and these are known to be aligned with the central axis of the movie camera. Also on each frame is a picture of the rocket being tracked. (See Figure 11.5.) However, it is not possible for the camera operator always to keep the cross hairs exactly on the rocket, so there is usually a considerable distance between the rocket image and the cross hairs.

During filming, a computer system automatically records the direction in which the camera is pointed, meaning the direction represented by the cross hairs. Measurements must be made on each frame to correct the difference between the direction represented by the cross hairs and the rocket's real location. All of these things can

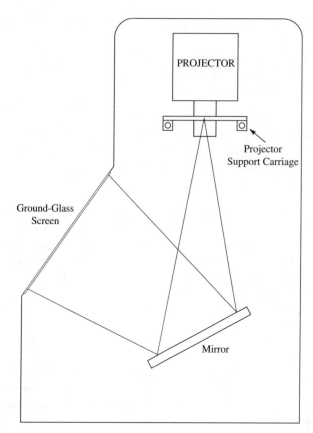

Figure 11.4
Schematic diagram of a precision optical measuring device.

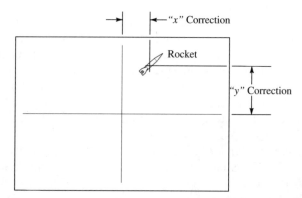

Figure 11.5
Diagram of a 35-mm frame, showing cross hairs, the rocket image, and x and y corrections.

be seen in Figure 11.5. The cross hairs are shown, and these may be thought of as the x and y axes. The intersection of these two axes is, of course, the origin. The horizontal distance of the rocket image from the origin is the x correction, and the vertical distance of the rocket from the origin is the y correction. These two corrections are measured by the use of fine steel cross wires, which are mounted directly behind the ground-glass screen and can be controlled by the operator. The motion of the cross wires is transmitted electronically to a computer so that the x correction and y correction can be recorded automatically and correlated with the direction in which the camera was pointed. The computer uses the x and y corrections to calculate the true location of the rocket for each frame of the film. With a film record consisting of many such frames, the rocket's trajectory can be calculated. But such calculations are not part of your assignment. You are supposed to design the projector support carriage.

You work up some thoughts on what a projector support carriage might look like, and prepare a sketch of your ideas (top and front views) as shown in Figure 11.6. Actually, only a portion of the support carriage is shown (x-direction motion only), in order to keep our example within reasonable bounds. In the top view, a 2.50-inch-diameter hole is visible. The projector lens sticks downward through this

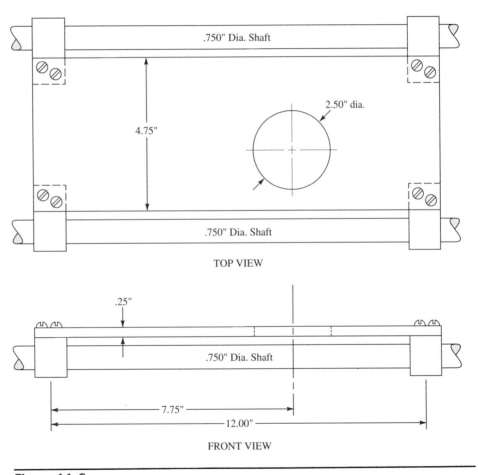

Figure 11.6
The initial design of the projector support carriage (x-direction motion only).

hole and is mounted there (by some means with which we will not concern ourselves at the moment). Suffice it to say that it must be possible for the operator to make a rotational adjustment of the projector within this hole, so that the cross-hairs of any projected image can be rotated to line up squarely with the mechanical cross wires behind the ground-glass screen. Also, it must be possible to move the entire projector ± 0.50 inches in the x direction and ± 0.50 inches in the y direction. This is because each frame of the film may not necessarily line up exactly with the preceding frame. The operator, before making any measurements, has to tell the computer which position of the mechanical cross wires represents zero, and each frame may have to be adjusted slightly to match this initial point.

Only the portion of the projector support carriage that provides the motion in the x direction is shown in Figure 11.6. (There would have to be a similar structure to provide for the y direction, but that is not shown.) The projector, mounted in the 2.50-inch hole, is supported by an aluminum plate 0.25 inches thick, 4.75 inches wide, and approximately 12.00 inches long. The plate is fastened with screws to four bearing blocks that slide back and forth in the x direction on shafts 0.750 inches in diameter. (The mechanical means for causing that motion is not shown.) At least, this is the way in which you initially conceive of the support structure, its size, and the manner in which it will be put together.

However, you are aware of the fact that the 0.25-inch plate and the four bearing blocks could possibly all be cast of aluminum and thus be only one piece instead of five (actually, 13 pieces, if you count the eight screws). Of course, the aluminum casting would not furnish a satisfactory sliding surface on the 0.750-inch steel shafts, so you would have to provide bearing inserts of some suitable bearing material, such as bronze, at each of the four sliding surfaces. Furthermore, if you decide on cast aluminum, a fairly large initial expense will be incurred, because the first step will be to make a wooden pattern, and patterns are costly. If a large number of units ultimately are to be manufactured, the cost of the pattern will be justified, because each casting made from it will be less expensive than machining the 0.25-inch plates and bearing blocks separately and then screwing them together. However, at this point, the contract calls for the fabrication of only one unit, and a vague future possibility of a total of five or ten units at a sales price of $85,000 each has been mentioned. Under these circumstances, it is not clear whether the cost of a pattern would be justified.

All of these thoughts are whirling through your head. Obviously, it is necessary for you to know something about methods of fabrication, such as casting and machining. You also need to know something about suitable materials and their relative costs. You probably will have had some exposure to these topics in college through courses in engineering materials and manufacturing processes, but most of what you need to know will come from experience; from other, more experienced engineers; and from your willingness to ask questions when you lack vital information.

You have now thought of two possible ways to fabricate your support plate (assembling it from machined parts or casting it in one piece), but you have a sneaking worry that there might be yet other, better, ways to go about the whole thing, if you could just think of them. Still, you do have two possible ways, so you are going to go ahead with step 3 of the design process: analysis.

The reason you have decided to do some analysis at this point is that there is a great big worry on your mind. This is a precision optical measuring device you are working on. You know that the whole system, and especially the optical image, must be as free from distortion as possible. You know, for example, that the mirror mounted directly below the projector is made from heavy plate glass and is ground to a precision flat surface. Everything possible is being done to minimize optical distortion, and you certainly do not want your support carriage to be a source of distortion. You know, for example, that any structure will deform slightly under load. The deformation is seldom visible to the naked eye, but in a precision optical system it might be disastrous. As a graduate engineer, you have had courses in mechanics of materials, and you have an idea of how to approach this problem, based on what you learned in those courses.

Note that the projector is mounted on a plate of aluminum that is supported at the corners. This plate can be regarded schematically as a beam 0.25 inches thick, 4.75 inches wide, and 12.00 inches long, supported at the ends, and with a transverse load (the projector) applied 7.75 inches from the left-hand end. Figure 11.7 is a simplified schematic diagram of this condition, in which everything has been reduced to the bare essentials. The load is shown as if it were applied at a single distinct point (whereas it actually is distributed across the width of the 2.5-inch hole), and the same thing has been done with the bearings at the ends. In fact, the bearings themselves have been omitted, and the forces that the bearings apply to the ends of the support plate are shown as arrows directed upward. Forces like these, which are the result of an applied load, are called *reactions*. Also, the schematic diagram is drawn with the assumption that the reactions of the two bearings on the left can be lumped together and treated as one force, and likewise with the reactions of the two bearings on the right.

Figure 11.7 is called a *free-body diagram*. We will have a great deal to say about free-body diagrams in Chapter 14, but let us point out some of their advantages here. First, we should note that our diagram is only an approximation of the actual structure. But its simplicity is also its advantage. We have eliminated all the details that will only have minor effects on our solution. If we tried to include every little detail in our analysis, it would become so cumbersome that it would be time-consuming to analyze and would likely not be worth the effort. One of the tricks of analysis, and one that can be learned only through experience, is to be able to detect which details are important enough to be included, and which are not. It is a waste of valuable time (and an engineer's time is very valuable) to make time-consuming calculations of things which do not turn out to be important.

The second thing we should note about the free-body diagram is that it isolates the particular part we wish to analyze, showing all the forces that act on that part.

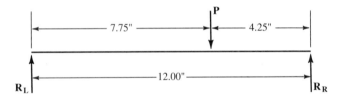

Figure 11.7

Free-body diagram of a projector support system. The symbol **P** represents the weight of the projector.

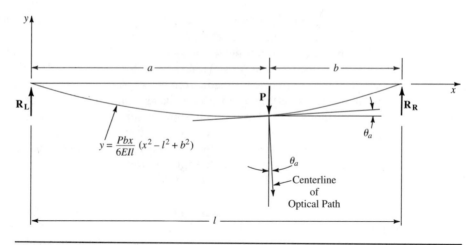

Figure 11.8

Idealized diagram of the deflection curve of a simple beam with a single, concentrated load **P** not in the center. (The amount of deflection is greatly exaggerated.)

All parts that exert forces on the part we are analyzing have been removed, and have been replaced by the forces which those parts exert on the part in question. This is the great trick of free-body diagrams, and this is also their great advantage. We started with a rather complex-looking part and reduced it to a straight line, representing the beam, and the three forces that act on the beam. The forces are: **P**, the load of the projector; $\mathbf{R}_L$, the lumped reactions that the two bearing blocks on the left exert upward on the beam; and $\mathbf{R}_R$, the lumped reactions that the two bearing blocks on the right exert upward on the beam. Note that $\mathbf{R}_L$ and $\mathbf{R}_R$, when added together, must exactly balance the load **P** exerted by the projector. This last statement is an expression of the *law of equilibrium*, about which we will have a great deal more to say in Chapter 14.

The topics we have just been discussing usually arise in courses on *statics*, which is the study of forces acting on stationary structures. (The name given to the study of moving forces is called *dynamics*.) Engineering students usually study statics in their second year. The study of beams, and especially the deflection of beams, usually arises in either the second or third year, under the heading *mechanics of materials*. But you, the young graduate engineer working on this problem, have already taken such courses, and you know how to figure out the deflection of the beam under this load, and from that the likely effect on optical distortion.

Figure 11.8 shows the beam as it deflects under load. The deflection has been greatly exaggerated so we can see what is going on. Note that the x and y axes have been marked on the diagram,[2] and the equation for the curve of the deflected beam

---

[2] These x and y axes are not to be confused with those mentioned in connection with the 35-mm frame of Figure 11.5. The x and y axes here are used solely to describe the equation for the curve of the deflected beam in Figure 11.8.

between the left-hand end and the location of the load is

$$y = \frac{Pbx}{6EIl}(x^2 - l^2 + b^2)$$ (11.1)

where

P is the applied load.

b is the distance from the applied load to the right-hand end of the beam.

l is the overall length of the beam.

E is the *modulus of elasticity* of the material we are using. ($E = 30,000,000$ lb/in.^{2} for steel and $12,000,000$ lb/in.^{2} for aluminum. These are approximate values. Sometimes the value of E for steel is given as $29,000,000$ lb/in.^{2}. More information on the modulus of elasticity is given in Chapter 15.)

I is known as the *moment of inertia* of the beam and is a geometrical property of its cross section. (We will calculate the moments of inertia for a couple of cross sections.)

Equation (11.1) is the deflection curve for a so-called *simple beam with a single concentrated load not in the center*. (It applies only to the section of the beam from the left-hand end to the load.) For a different kind of beam—one with the ends rigidly fixed at the ends, for example—the curve would be different. It also would be different for a different kind of load—one distributed uniformly along the length of the beam, for example.

In the foregoing, we simply presented Equation (11.1) without saying where it came from. In the study of mechanics of materials, you will learn how to derive the equations for the deflected shapes of beams under varying circumstances, using the *double-integration method*. The name of the method implies that integral calculus is used, and that is correct. Integral calculus is also used to derive the moment-of-inertia formulas for various geometrical cross sections. The typical reader of this book will not have been exposed to the techniques of integral calculus, although that exposure lies just over the horizon. However, *you*, the new graduate engineer of this design example, know all about these techniques. In fact, you could derive them yourself, if you had to. But you do not have to, because the formulas for the deflection curves for most typical beam situations are listed in handbooks, and so are the formulas for the moments of inertia for many typical cross sections. We will get the formulas we need from the handbooks because we know that you, the new engineer, are thoroughly familiar with what we are doing.

In Figure 11.8, a tangent has been drawn to the curve at the point of the applied load, and the angle that this tangent makes with a horizontal line, i.e., with the x axis, is marked θ_a. At a point slightly to the left of the location of the load P, a tangent could be drawn that would be exactly parallel to the x axis. The deflection of the beam is greatest at this point. We call it the *maximum deflection*. Again, by application of integral calculus, we can find the x coordinate of this point. But you know that handbooks also list maximum deflections for many types of beams, so

you look this one up. It is

$$\text{max. deflection} = \frac{Pab\,(2a + b)\sqrt{3b\,(2a + b)}}{27EIl} \tag{11.2}$$

You will substitute into this equation your actual numerical values, but first you must calculate a value for I, the moment of inertia for your beam's cross section. Your beam has a rectangular cross section, so you look in your handbook and find the following formula for rectangular cross sections:

$$I = \frac{wh^3}{12} \tag{11.3}$$

where w and h are defined as in Figure 11.9.

You visualize the cross section of the beam you are working with, and see that it is 4.75 inches wide and 0.25 inches thick. But you suddenly realize that your beam has a hole in it 2.50 inches in diameter, so you decide you had better reduce the width of the beam in your calculations so that it is $4.75 - 2.50 = 2.25$ inches. Substituting $w = 2.25$ inches and $h = 0.25$ inches into Equation (11.3) gives the following result:

$$I = \frac{wh^3}{12} = \frac{(2.25\ \text{in.})(0.25\ \text{in.})^3}{12} = 2.930 \times 10^{-3}\ \text{in.}^4$$

You know that the projector you are going to use weighs 25 pounds. (It is a heavy-duty, commercial 35-mm projector of the type used in movie theaters, chosen because it has a precision film-advance mechanism that causes each frame to come into close alignment with the position of the previous frame.) You can now list all the values you need, for substitution into Equation (11.2):

$P = 25$ lb

$a = 7.75$ in.

$b = 4.25$ in.

$l = 12.00$ in.

$I = 2.930 \times 10^{-3}$ in.^{4}

$E = 12,000,000$ lb/in.^{2}

Figure 11.9

A generalized rectangular cross section. The moment of inertia of a rectangular cross section is given by the formula $I = wh^3/12$.

The value of 12,000,000 lb/in.^{2} is used for E because, as already noted, it is the modulus of elasticity of aluminum, and your material is aluminum. We can now substitute these values into Equation (11.2). We will do it in successive steps, because we are less likely to make errors that way:

$$ab = (7.75 \text{ in.})(4.25 \text{ in.}) = 32.94 \text{ in.}^2$$

$$(2a + b) = 2(7.75 \text{ in.}) + 4.25 \text{ in.} = 19.75 \text{ in.}$$

$$\sqrt{3b(2a + b)} = \sqrt{3(4.25 \text{ in.})(19.75 \text{ in.})} = 15.87 \text{ in.}$$

$$\text{max. defl.} = \frac{(25.0 \text{ lb})(32.94 \text{ in.}^2)(19.75 \text{ in.})(15.87 \text{ in.})}{(27)(12.0 \times 10^6 \text{ lb/in.}^2)(2.930 \times 10^{-3} \text{ in.}^4)(12 \text{ in.})}$$

$$= 0.2719 \text{ in.}$$

This result is astonishing. It simply does not seem possible that a 25-pound weight could deflect a 0.25-by-2.25-inch plate of aluminum more than a quarter of an inch. Your instinct of reasonableness has been violated. Suddenly, you remember a professor you had in college who insisted that every solution to a problem be examined for reasonableness. The professor was outraged when his students accepted whatever answers emerged from their calculators as correct, no matter how impossible they might be. You were somewhat annoyed with the professor at the time, but the professor's viewpoint must have stuck in your subconscious. You are now looking at your answer with a suspicious eye. The obvious thing to do is to repeat the calculation, so you run it through your calculator again, being especially careful this time. Your answer is

max. defl. = 0.02266 in.   (We will round this off to 0.0227 in.)

Looking for the cause of this markedly different result, you realize that, in your first calculation, you left out the last entry of 12 inches in the denominator of your equation. You double-check this by the following:

$$0.02266 \times 12 = 0.2719$$

That little calculation shows that your two answers differ by exactly the missing factor of 12. You are satisfied that you have found your error and are thankful that *you* found it instead of your boss. However, you are still suspicious of your answer because, even though it looks small, it is actually rather large under the circumstances. You are aware that beam deflections are usually so small as to be invisible to the naked eye, but this deflection is almost $\frac{1}{32}$ of an inch, and that amount of deflection would be highly visible. You are reasonably confident that your answer does not contain any more computational errors like the one you made before, but maybe there are other errors you are overlooking.

So you decide to make an entirely independent check. You note, for example, that the location of the load P is not really very far off center, so you decide to calculate the deflection of a beam that actually has the load *in* the center. The maximum deflection for such a beam ought to be of about the same order of magnitude as that for your beam, and it is order-of-magnitude errors you are worried about.

You also know that the formula for such a case is very simple; thus, calculation errors are not likely. So you look up the formula in your handbook:

$$\text{max. defl. (simple beam, load in center)} = \frac{Pl^3}{48EI}$$

Before substituting anything into this equation, you recheck your calculations for I (you decide they are correct) and then make the substitutions:

$$\frac{Pl^3}{48EI} = \frac{(25 \text{ lb})(12 \text{ in.})^3}{(48)(12 \times 10^6 \text{ lb/in.}^2)(0.00293 \text{ in.}^4)} = 0.02560 \text{ in.}$$

Ah. This is the same order of magnitude as your answer of 0.0227 inches, so you are now ready to proceed. However, you are still worried that the deflection is so large. In itself the deflection is not serious, provided it remains rock-hard at that level during the period in which precision measurements are being made. If it does remain rock-hard, it can be treated as a constant, and the adjustment mechanisms provided in the system can compensate for it. But the image of a 25-pound projector causing your support member to sag almost $\frac{1}{32}$ inch is very unsettling. You doubt that your support member is rigid enough. If just placing a 25-pound weight on your support member can make it sag that much, perhaps a truck passing on a nearby street, or even somebody walking across the room, could cause the unit to vibrate so much that the film image would not remain stationary. But you resolve to come back to that point, because you have another troubling matter to investigate.

Earlier, in discussing Figure 11.8, we mentioned that a line drawn tangent to the deflection curve at the point where the load **P** is applied would make an angle with the x axis that we called θ_a. This angle is the *slope* of the beam at that particular point. The thing that worries you about the slope of the beam is that it means the projector (which of course is fastened to the beam) will be tilted away from the vertical direction by the angle θ_a. As a result, the lens will not point in a direction perpendicular to the ground-glass screen, but will point off the perpendicular by exactly the value of θ_a.

The situation we are talking about can be seen clearly in Figures 11.10(a) and 11.10(b). In Figure 11.10(a), the optical axis points directly at the screen, and the image is rectangular, as it should be. But in Figure 11.10(b), the optical axis is tilted at the angle θ_a, with the result that the image is no longer a rectangle, but is trapezoidal. (Note that everything has been greatly exaggerated in the figure so we can see what is going on. The actual magnitude of θ_a is very small, and the degree of distortion probably could not be detected by the unaided eye.)

The first thing we must do is calculate θ_a. The equation that gives the slope of the beam at any point between the left-hand end and the location of the load is

$$\frac{dy}{dx} = \frac{Pb}{6EIl}(3x^2 - l^2 + b^2) \tag{11.4}$$

This equation is not listed in many handbooks. Therefore, it is likely that you would have derived it yourself, by differentiating Equation (11.1). If the equation for the deflection curve of a beam is differentiated once, the result is an equation that

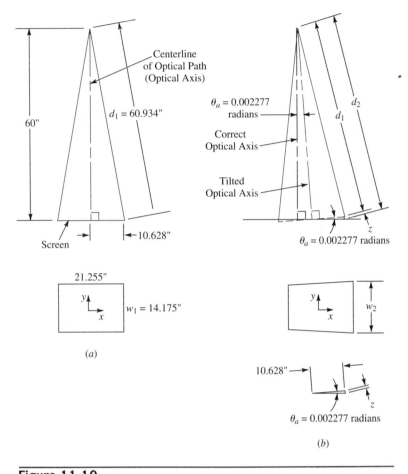

Figure 11.10
Schematic diagrams showing the tilting of the optical axis of the projector because of the deflection of the projector support plate.

gives the slope of a beam, usually designated either by the symbol θ or by the symbol dy/dx. *Differentiation* is a topic covered under differential calculus, which is usually included in the first year of an engineering program. We will not go into the details here.

We must now evaluate Equation (11.4) for the point we are interested in, which is where $x = 7.75$ inches. This is the location of the projector load and of the optical axis. Hence, we substitute the following numerical values for the variables in Equation (11.4):

$P = 25$ lb

$x = 7.75$ in.

$b = 4.25$ in.

$l = 12.00$ in.

$I = 2.930 \times 10^{-3}$ in.^{4}

$E = 12,000,000$ lb/in.^{2}

We will not go through all these numerical calculations the way we did on page 297, but simply give the result (θ_a means the value of θ at the point where $x = a = 7.75$ inches):

$\theta_a = 0.002277$ radians

It is important to note that all the unit symbols involved in this substitution have cancelled each other, so the answer is in *radians*. The value we just calculated for θ_a has been placed at the appropriate locations in Figure 11.10(b).

The dimensions of a standard frame on 35-mm film are 24 mm × 36 mm. When converted to inches, these become 0.945 inches × 1.417 inches and, when enlarged 15 times by our optical system, 14.175 inches × 21.255 inches, as shown in Figure 11.10(a). One-half the width of the enlarged frame is thus 10.628 inches, as shown in the figure, and we can compute the dimension d_1 by the use of the Pythagorean theorem. You have been told that the "centerline distance" of the optical path is 60 inches, so $d_1 = 60.934$ inches. (If you are worried about the fact that the mirror shown in Figure 11.4 has been omitted in Figure 11.10(a), do not be. The optical path can be treated as if the mirror were not there, provided the mirror has an optically flat surface. Such mirrors are customarily ground to be flat within extremely close tolerances and have the "silvering" on the front surface, so that the reflected light does not have to travel through the glass.)

What we want to do now is compute a numerical value for the distance d_2, as shown in Figure 11.10(b). At the outer edge of the enlarged frame, the light must travel a greater distance than it did before the optical axis was tilted. In other words, it must now travel the distance d_2 instead of the distance d_1. This is precisely what causes the image to assume a trapezoidal shape, because the light must travel a greater distance at one side of the image than at the other, and thus produces a greater magnification at that side. In the distorted trapezoidal image, the right-hand edge of the (supposedly) rectangular image is w_2 instead of w_1, as shown in the figure.

We can calculate w_2 as follows. First we must calculate a value for z, which is the amount by which d_1 has been lengthened by the tilting to become d_2. The small triangle containing z has been repeated at the bottom of Figure 11.10(b) for clarity. It is not actually a right triangle, but θ_a is such a tiny value that it doesn't matter. For such small angles, the magnitude of the angle *in radians* is almost the same as the sine of the angle; i.e., θ_a and $\sin \theta_a$ can be treated interchangeably. Therefore, we can compute a value for z as follows. The relationship between an angle θ (in radians), its radius l, and its arc Δ, are given in the sketch below of a sector of a circle (see also Appendix B).

$\Delta = \theta \cdot l$ (11.5)

For our case, the angle is 0.002277 radians, and the radius is 10.628 in. Remember, the angle θ_a is so tiny that we can treat the triangle shown at the bottom of Figure 11.10(b) as if it consisted of a sector of a circle. For an angle as small as ours, the error would not show up until the fifth decimal place (z takes on the role of Δ in the previous sketch). Our calculation is

$$z = (0.00227 \text{ radians})(10.628 \text{ in.})$$
$$= 0.0242 \text{ in.}$$

and,

$$d_2 = d_1 + z = 60.934 \text{ in.} + 0.0242 \text{ in.}$$
$$= 60.959 \text{ in.}$$

The degree by which w_1 is lengthened to w_2 (i.e., the degree of distortion) is directly proportional to the ratio d_2/d_1, which is

$$\frac{d_2}{d_1} = \frac{60.959 \text{ in.}}{60.934 \text{ in.}} = 1.0004103$$

so

$$w_2 = 1.0004103(w_1) = 14.181 \text{ in.}$$

Thus, the increase (distortion) in the length of w_1 is

$$w_2 - w_1 = 14.181 \text{ in.} - 14.175 \text{ in.} = 0.006 \text{ in.}$$

Finally, this is what we are after. With the optical axis tilted this much, if a measurement were made at the extreme edge of the screen that extended all the way from the top to the bottom of the frame, it would be in error by 0.006 inches. But would this be serious? Actually, it would not. From your knowledge of the entire system, you know that the digitizing capability of the measuring system is expected to be 0.010 inches, and our distortion is less than this. (The value of 0.010 was selected because this is about the finest discrimination of which the human eye is capable.) So, after all this calculation, you finally decide that the distortion resulting from the tilted axis can be tolerated, so you need not worry about the deflection of the projector support system on this account, at least.

Now, even though this issue is laid to rest, you begin to worry about the general flexibility of the system you discovered through your beginning calculations. You can visualize the possibility that the image might vibrate whenever anybody walks across the room, so you decide the support system had better be stiff enough to prevent this. Remember, you were already considering the possibility of making the support plate from an aluminum casting in order to save the cost of machining and assembling the bearing blocks. If you do use a casting, you could include stiffening ribs at little cost, and they would provide a very stiff support structure. These considerations are just enough to tip your judgment in favor of an aluminum casting.

But before going ahead, you need to decide just how much additional stiffness you need. There is no sense in using more material than necessary; unnecessary material adds not only cost, but weight.

To calculate the deflection for this new case, we can still use Equation (11.2), but the equation for moment of inertia, Equation (11.3), no longer applies. It was for a rectangular cross section. Our new cross section is going to resemble an inverted shallow "U," a shape that is sometimes called a "channel." The pictorial view in Figure 11.11(a) shows what our cast support plate is going to look like. It is still a thin plate with a big hole in it, but the four blocks on the corners are now to be cast integrally with the plate. Also, a thin stiffening rib will connect the two front blocks, and another one will connect the two rear blocks. One result of this arrangement is that the plate will now be 7.25 inches wide instead of 4.75 inches. This is because the flat-plate portion will extend beyond the outer edges of the four support blocks and merge with the stiffening ribs. Of course, we must reduce this width by the diameter of the 2.50-inch hole as we did before, so the effective width of the support plate is 7.25 − 2.50 = 4.75 inches. (Before, it was only 2.25 inches.) In between the support blocks, the cross section will look like the crosshatched portion of Figure 11.11(a), which is repeated in Figure 11.11(b). If the cross section were turned over, it would look like the cross section of a channel that could carry water.

A generalized cross section of a channel is shown in Figure 11.12. The moment of inertia I of such a cross section is calculated in two steps. First we must calculate the value of x_b, which is the distance from the upper surface of the channel to the so-called "neutral axis." (The subject of the neutral axis and its significance is covered in courses on mechanics of materials.) Then, once x_b is known, the moment of inertia

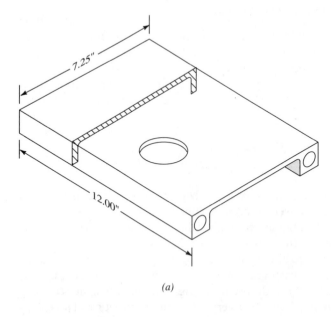

(a)

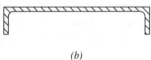

(b)

Figure 11.11
(a) The redesigned projector support plate.
(b) Cross section of the support plate, displaying a shape often called a "channel."

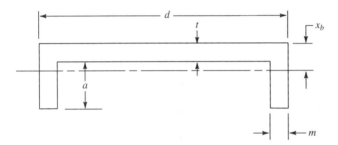

Figure 11.12
Generalized cross section of a channel.

of this cross section can be calculated. The formulas can be found in handbooks, although graduate engineers know how to use calculus to derive these formulas, should the need arise. The formulas are

$$x_b = \frac{\dfrac{t^2 d}{2} + 2am\left(t + \dfrac{a}{2}\right)}{td + 2am} \tag{11.6}$$

$$I = \frac{dx_b^3 - d(x_b - t)^3 + 2m(b - x_b)^3}{3} \tag{11.7}$$

The values from our example that are to be substituted into these formulas are

$d = 4.75$ in. (Remember, we reduced the 7.25-in. width by 2.500 in. to get this value.)

$t = 0.25$ in.

$a = 1.00$ in.

$m = 0.25$ in.

The value for t is taken directly from our first set of calculations, in which we were using an aluminum plate 0.25 inches thick. We retain the same thickness, at least for now. The values for a and m are the result of value judgments made by you, the design engineer. The choice of a is controlled by your judgment concerning how large the blocks must be in order to contain bronze bushings for the support shafts, and the choice of $m = 0.25$ inches is probably controlled by the fact that $t = 0.25$ inches.

If the above values are substituted into Equations (11.6) and (11.7), the following results are obtained. (We will not go through all the calculation steps here, but merely give the results.)

$x_b = 0.3101$ in.

$I = 0.1852$ in.^{4}

With the previous design, I was only 2.930×10^{-3} in.^{4}. Thus, the new design is 63 times as stiff as the previous one, mostly as a result of adding the stiffening ribs. Since I appears to the first power in the denominator of Equation (11.2), this means that the maximum deflection is only $\frac{1}{63}$ the value we calculated before, or 0.0227 inches/63 = 0.00036 inches. This is indeed a small deflection, and your judgment tells you that the new design will be stiff enough so that the image will not vibrate—at least not from this cause. To be certain of this, you make one further calculation: You check to see just how much the center of the image on the screen would shift if the optical axis were tilted. Figure 11.10(*b*) shows the situation nicely. The optical path is 60 inches long. The shift in the point where the optical axis hits the screen can be computed by using Equation (11.5), wherein $\theta = 0.002277$ radians and $l = 60$ inches. The result is

$$\Delta = (0.002277 \text{ radians})(60 \text{ in.}) = 0.1366 \text{ in.}$$

This is more than an eighth of an inch, and you shudder at the prospect of the image vibrating back and forth by this amount. In truth, it probably would not vibrate this much just from someone walking about the room, but it still seems excessive to you. With the stiffer support plate, however, the value of I will be 63 times as great as in your original design. Since I appears in the denominator of the equation for the slope of the beam, just as it does in the equation for deflection, this means the amount by which the optical axis shifts will be only $\frac{1}{63}$ as great as in the first design. A quick calculation shows that this will now be only 0.0022 inches instead of 0.1366 inches. It is unlikely that the eye could even *see* something as small as 0.0022 inches. Besides, our system cannot measure anything less than 0.010 inches, so you are satisfied.

Before making your final judgment, you would probably want to experiment with Equations (11.6) and (11.7) to see what would happen if you made t and m thinner—only 0.20 inches, say, instead of 0.25 inches. You could save some material (and a little weight) this way, but obviously you would also come up with a somewhat smaller value of I, which in turn would allow a larger deflection to occur. Also, if you make these sections thinner, you might be making it more difficult for the molten metal to flow into all corners of the casting. (You will probably talk to the people in the foundry about this.) How far you go with this kind of thing must remain a matter of judgment. Obviously, you can never reduce the deflection absolutely to zero. Real materials in the real world always deflect a bit. But you can indeed make the deflection small enough so that it does not create a problem.

Perhaps, as a relatively inexperienced design engineer, you wish someone had a way of specifying an absolute standard for a matter like this, but such standards are usually the product of extensive experimentation and experience. Those things do not exist for a new design. You are the engineer, and it is your job to make the decisions. You have to use judgment, probably erring on the side of conservatism—providing, of course, that your conservatism does not come at an exorbitant price. If you are very new on the job, your supervisor will probably make the difficult calls. Although it is important to realize that such uncertainties are the essence of design decisions, it is also important to realize, as in this example, that you were able to use your theoretical knowledge to a considerable degree to help guide you through the uncertainties.

Obviously, a great deal more would need to be done in this example to carry out the full design of the projector support carriage. We have dealt with only the support plate for the *x*-direction movement. There must be a *y*-direction movement also, and the plates must interact. There must also be a means of controlling the movements with handles on the control panel of the instrument. Similarly, there has to be a means of rotating the projector. You have to evaluate and make judgments about every little detail, so that you or your drafting assistant can make detail drawings for the shop. Then the people in the shop will have to be able to use these drawings to construct the device exactly the way you intended, with nothing left to assumptions. And—most frightening of all—the device, when constructed, must work.

Design is a big job. It is what engineers are paid for.

■ 11.4 DESIGN EXAMPLE: DIGITIZING READING HEAD

In the previous example, every step of the design process was included, although we terminated the example just as we came to the last one—implementation. The first step, problem definition, was quite clear and straightforward: The problem was to mount the projector on a support system that could provide the necessary adjustment motions, and to do so without introducing unacceptable distortion into the system. Other elements of problem definition were present also, but were taken for granted. These were that the system should function easily, should be manufacturable with minimum expense, and should provide for easy maintenance. The second step—invention—probably was hardly recognizable, because the only invention consisted of figuring out a suitable manner of arranging the support system. Nevertheless, something had to be thought up—the process we previously described as coming up with a "scheme"—before we could embark on the analysis.

That example was dominated by steps 3 (analysis) and 4 (decision). The analysis guided us toward our final act, which was to decide exactly what size and shape our support structure would have and how it would be fabricated. We didn't talk much about implementation. But now we present an example that demonstrates every step of the process. Whereas our previous example involved a combined mechanical-structural device, this example involves an electromechanical device.

Art is an electronics engineer; Ed is a mechanical engineer. They work for a small engineering and manufacturing firm. In fact, it is the firm of our previous example, and the device we are about to discuss is a part of the precision optical measuring instrument already described in part.[3] Art is the supervisor of electrical design for this firm, and Ed is the supervisor of mechanical design. Both are graduates of a nearby state university, and each has about five years of experience.

The company's existing line of products includes a device attached to each cross-wire measuring system that converts the motion of the wires into electrical pulses, which can be transmitted to a computer. The device is termed an *analog-to-digital* converter because it converts a linear function (the cross-wire motion) into digital form. The company's engineers call it a *digitizing reading head*. The operation of this unit is simple. A sheet-metal disk is attached directly to the end of each of the shafts

---

[3] This account is based on a true-life case. The device being described was a successful product. The inventor is A. J. Winter of Los Angeles.

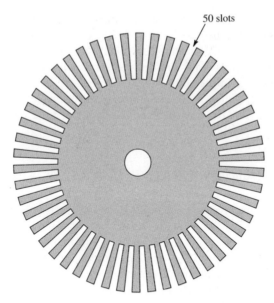

50 slots

Figure 11.13
Layout of a slotted sheet-metal disk with 50 slots.

that carry the cross wires. (See Figure 11.13.) The disk has 50 slots in it, and it is located in the light path between a lamp bulb and a photoelectric cell. (See Figure 11.14.) One full revolution of the shaft causes the light to be interrupted 50 times; thus, 50 electrical pulses are generated and transmitted to the computer. Two revolutions produce 100 pulses, and so on. The computer counts the pulses and thus knows the distance the cross wire has traveled.

There are two cross wires, one for motion in the x direction and the other for motion in the y direction. Each cross wire is mounted on two parallel shafts, as shown in Figure 11.15. Each shaft has machined in it a fine groove just large enough to receive the wire, and the groove has been made into a helical spiral somewhat like a screw thread. The so-called lead of the screw is 0.500 inches, which means that as the shaft is turned one revolution, the cross wire moves laterally 0.500 inches.

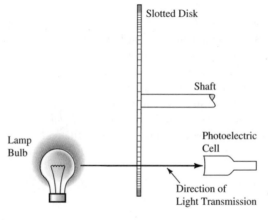

Slotted Disk

Shaft

Lamp
Bulb

Photoelectric
Cell

Direction of
Light Transmission

Figure 11.14
A schematic view showing how the slotted disk interrupts light transmission as the shaft rotates.

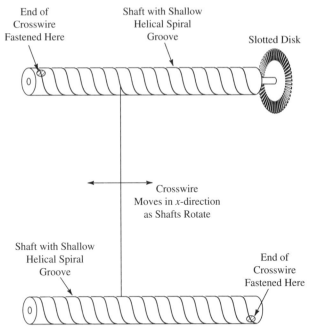

End of
Crosswire
Fastened Here

Shaft with Shallow
Helical Spiral
Groove

Slotted Disk

Crosswire
Moves in *x*-direction
as Shafts Rotate

Shaft with Shallow
Helical Spiral
Groove

End of
Crosswire
Fastened Here

Figure 11.15
A cross-wire system using shafts with shallow helical spiral grooves.

As we said previously, one revolution of the shaft will cause 50 electrical pulses to be generated. This means there will be one pulse for each 0.010 inches of motion of the cross wire, and this is the measuring precision of the system.

Now, the trouble with this system is that it is cumbersome for the operator. In the *x* direction, more than 20 inches of cross-wire travel is required to traverse the entire ground-glass screen. If each revolution of the cross-wire shaft causes 0.500 inches of travel, then more than 40 revolutions are necessary. The operator rotates a handle with a crank attached to move the cross wire. Operators have been observed in action, and they do an enormous amount of cranking, using both hands (remember, there is a *y* axis, too). The effort required is regarded as a negative factor that might affect future sales. Therefore, it is an important problem, and the sales department wishes somebody would do something about it. Step 1, *problem definition*, has taken place: Redesign the cross-wire system so it is faster and easier to use. Specifically, the system is to be designed so that fewer revolutions of the operating handles are needed in order to move the wires across the screen.

Since this is a mechanical system, it is perceived as a mechanical problem, and Ed, the supervisor of mechanical engineering, has been thinking about it a lot. He has an idea: The cross wire might be connected to two steel bands that would be wrapped around steel wheels fastened to shafts, as shown in Figure 11.16. (The cross-wire system for the *x* direction is shown; a similar system for the *y* direction is also needed.) A control wheel is connected to one of the shafts, and an analog-to-digital converter is connected to the other. Each steel band is wrapped around two steel wheels that are, say, 4.00 inches in circumference. The steel bands are under tension, and they apply sufficient force to the steel wheels so that they act as friction drives. That is,

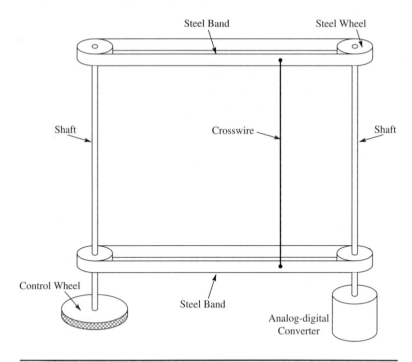

Figure 11.16
A schematic diagram showing the proposed new cross-wire system.

as the wheels rotate, they carry the bands along with them, without slippage. (Nothing is said yet about how the ends of the steel bands are to be fastened together to form loops, or about how the cross wires are to be fastened to the steel bands. The manner in which these problems are handled could make or break the whole project.) For each revolution of the shafts, the cross wire moves a distance equal to the circumference of the wheels: 4.00 inches. Thus, only five revolutions of a shaft and its control wheel, instead of 40, are needed to move all the way across the 20-inch screen. Furthermore, no crank handles are needed. Each control wheel is knurled on its periphery; it is only necessary for the operator to spin a wheel by using the palm of the hand to push the knurled edge. This is an enormous improvement, but there is a catch: In order to preserve the measurement precision—0.010 inches—there will have to be 400 electrical pulses for each rotation of a shaft, instead of 50. What to do?

We are now in the middle of the *invention* phase. Ed's idea is terrific, but a major new idea—how to get 400 pulses per revolution—is needed. So far, Ed is still thinking only of the mechanical system. He assumes that any redesign of the system will continue to use the existing digitizing head. One way to get 400 counts would be to connect the digitizing head to the shaft by means of gears with a ratio of 1:8. Then, for each rotation of the shaft, the digitizing head would rotate eight times and generate $8 \times 50 = 400$ pulses for each rotation of the cross-wire shaft. But Ed, who has had some experience with gears, knows there is a serious problem here. A gear ratio of 8:1, if it is a *reduction* in speed, is no problem. But if it is 1:8 *increase* in speed, then the torque required is also increased by eight. At first glance this may appear to be acceptable because the torque required to turn the digitizing head is small, but when operators spin the wheels for many hours a day, working against this load could become quite disagreeable.

Ed knows there is an even more serious defect in this idea. In the existing system, the photoelectric cell is on the borderline of not reacting fast enough as the slots whirl in front of it. If the rotation of the slots were speeded up by a factor of eight, the photoelectric cell would essentially see nothing but a gray blur spinning in front of it. It could not react properly to such a high rate of input. Ed is aware of the fact that the photoelectric system is touchy and has been the source of many system breakdowns. In the back of his mind he senses that in any redesign of the measuring system, its reliability has to be improved. A new element has unconsciously been added to the problem definition: Make the digitizing head more reliable.

Ed now shifts his attention away from the slotted-disk system and focuses on getting a new idea for the digitizing head. He likes his idea for the new kind of cross-wire system, and if he could just do something so the digitizing head would produce more pulses per revolution, everything would be dandy. The problem definition has again been expanded: Improve the sensitivity of the digitizing head so it produces more pulses per revolution.

Ed briefly considers the possibility of using a helical potentiometer. A potentiometer can produce a voltage that is proportional to the angular position of its input shaft, and helical potentiometers are available that allow a range of ten complete revolutions. The linearly varying voltage would then be converted to digital pulses by standard electronic circuitry. Ed would have to provide a gear ratio of 1:2 in order to convert the five revolutions of the cross-wire shaft (for 20 inches of travel on the screen) to take advantage of the ten revolutions of the potentiometer, but a gear ratio of 1:2 is no problem.

But the system is still in trouble. Ed knows that helical potentiometers are accurate only to one part in a thousand. Thus, if his full travel has to be 20 inches, then 0.001×20 inches $= 0.020$ inches is the measurement error his system would produce. But he has to have an accuracy of 0.010 inches. The helical potentiometer idea is out.

So far, Art, the head of electrical engineering, has not become involved in this problem. But Ed and Art make a practice of discussing the technical activities of the company regularly, and this problem of the analog-to-digital converter comes up. During their initial discussion, Art does not have anything to contribute, but the next day he comes back with an idea.

Art is thinking of using transformers. Ed immediately objects because he does not see how a transformer could help. To him, a transformer is a device for converting a voltage from, say, 220 volts to 110 volts. But Art explains his idea, and sure enough, it depends for its operation on the same kind of principle that transformers use.

Art has made a rough sketch, shown in Figure 11.17. A rotor and a stator face each other across a very narrow gap, perhaps 0.005 inches. (In the sketch the gap is exaggerated so it can be seen.) The rotor and stator both have "printed circuit" patterns on them, made by photographic etching of a thin copper layer, backed up by

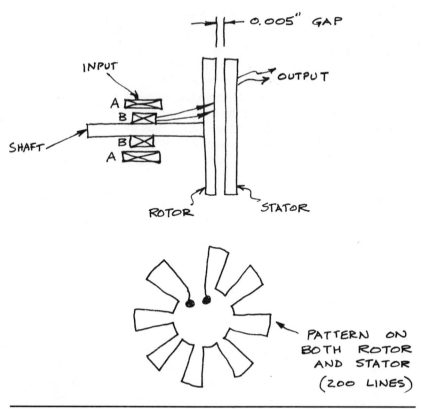

Figure 11.17
Sketch of a new digitizing head.

nonconducting plastic. The patterns look something like the sketch shown in the lower part of Figure 11.17, but contain 200 lines instead of the small number shown in the sketch. (See Figure 11.8.) A doughnut-shaped coil of fine wire is wrapped around the rotor shaft, shown as "B" on the sketch. Another doughnut-shaped coil, marked "A," surrounds coil B. There is an air space between the two coils. Coil A is fixed to the housing of the device (as is the stator), and coil B rotates with the rotor.

The device works as follows. A high-frequency electric current is fed to stationary coil A. This is marked "input." The current induces a current in rotating coil B. (You will learn in physics that an alternating current in a conductor creates an alternating electromagnetic field around the conductor. If another conductor, such as coil B, is placed within the influence of this field, an electric current will be induced in the second conductor. Transformers are based on this principle.) Coil B is connected to the large, round black spots shown on the printed circuit pattern belonging to the rotor (this connection is suggested by the two arrows Art has put on his sketch, leading from coil B to the rotor), so the induced high-frequency current in coil B passes through all the ray-like, radially extending portions of the pattern.

The printed circuit pattern on the rotor and the identical printed circuit pattern on the stator face each other across a small air gap. The high-frequency current in the pattern on the rotor induces a current in the pattern on the stator, again by "transformer action." This current is labeled "output."

What has been accomplished? Just this: When the two facing patterns are aligned, and all of the ray-like conductors "look at" conductors just like them across the narrow air gap, the coupling is good, and the current in the output wires is at a maximum. But when the shaft is rotated a slight amount so that the two patterns are misaligned by one-half of the space between two rays, then the coupling is poor. Each ray-like conducting portion of the pattern on the rotor now "looks" across the narrow air gap at the blank plastic area between two of the conducting portions of the pattern on the stator. The current in the output leads is now at a minimum because of the poor coupling. When the rotor rotates one-half space more, then the patterns are aligned again, and the output current goes back up to a maximum. Thus, as the shaft rotates, the current in the output leads varies up and down, going through 200 such cycles for a complete rotation of the shaft. In the electronic circuits to which the output leads are connected, these variations can be detected and converted to a series of discrete pulses, thus "digitizing" the output.

"But that's only 200 pulses," Ed objects.

Art has a solution for that. "For each line on the pattern, the voltage rises and then falls as the shaft rotates. We'll put a 'trigger' circuit that will trigger a pulse as the voltage rises through a certain level and trigger a pulse again as it comes back down through the same level. We'll get two pulses per line for a total of 400, but only have to put 200 lines on the pattern."

Ed assigns the drawing of the pattern (Figure 11.18) to a member of the drafting pool and assigns the preparation of a design layout of the unit to a young graduate

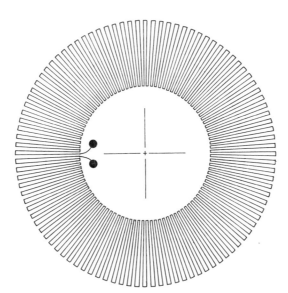

Figure 11.18

Layout of a pattern for rotor and stator, with 200 lines.

mechanical engineer who has been with the firm for about two years. Thus, even though the decision phase has not been wrapped up, the implementation phase has begun.

But what about the analysis phase? Where are all the equations and calculations that we customarily associate with analysis? Art actually plans to skip elaborate calculations and settle questions of technical feasibility with a test model. He knows that developing an accurate mathematical model of the electromagnetic behavior of his device might be difficult, and even if he succeeded, he would finally have to settle matters by using a test model anyway. So he will go directly to the model. Actually, he plans to use his knowledge of theory to make many rough calculations to guide the testing as he goes. Such calculations are often referred to as "back-of-the-envelope" calculations, to emphasize that they are informal.

As soon as the drawing of the pattern is finished, it is sent out to a supplier to have the printed circuits made. The drawing is originally 20 inches in diameter so it can be made with precision, but it is photographically reduced so that the final patterns are only 2.5 inches in diameter. A test rotor and stator are fabricated in the experimental shop, the printed circuit patterns are cemented to them, and everything is set up on the bench for test. During test, many questions of current strength, voltage levels, and component tolerances come up, and decisions are made concerning them. In particular, every component and every voltage is examined to see what the effect on operation would be in case the components and/or voltages drift from their nominal values. Such an examination is known as *tolerance analysis*. Art knows from bitter experience that components can vary as much as 10 or 20 percent from their nominal values, and that voltages can do the same. Since he wants this device to be reliable, he selects components and establishes voltage levels in such a way that the circuit will still operate correctly even in the face of 20 percent variations. Gradually, he satisfies himself that the device will work reliably as planned, and crystallizes the circuitry design in the process.

But, from the first moment he sees the printed circuit as it comes from the supplier, Art is troubled. When the drawing of the pattern was made, it was 20 inches in diameter, and the pattern looked very robust. But now it is the size it actually will be in production—2.5 inches. The lines look awfully thin and close together. He makes a back-of-the-envelope calculation and discovers that the lines are only 0.010 inches wide, separated by spaces that are also only 0.010 inches. He fears that such thin lines of copper will be excessively vulnerable to damage. He could double the size of the unit and thus double the thickness of the lines. But a big unit would be more difficult to incorporate into the overall design of the cross-wire system than a small one. In fact, Ed and Art examined the space limitations carefully before selecting the diameter of 2.5 inches. Another way to double the thickness of the lines would be to have only 100 of them instead of 200. If this were done, a set of gears giving a ratio of 1:2 would be necessary. Although this is operationally feasible, it would, of course, increase the cost. But the reliability also would increase. Art feels embarrassed because he did not make this little calculation right in the beginning, but it did not occur to him how *small* the lines on the actual production unit would be until he saw them. He discusses these issues with Ed, and both are convinced that reliability is vital, even at increased cost. Ed, too, is embarrassed because he did not investigate this seemingly small matter before. He assigns a new drawing, this time with 100 lines, to the person from the drafting pool, and everything starts over again. Finally, the design is set.

■ 11.5 DESIGN LAYOUT

Meantime, the *design layout* of the digitizing reading head discussed in the previous section has been proceeding. (See Figure 11.19.) The layout shows that there is a central shaft, 0.250 inches in diameter and 4.500 inches long, supported by two ball bearings mounted in a housing that is 3.500 inches in its largest outside diameter. The housing has an extended "neck" 1.375 inches in diameter. The neck is to be used to mount the unit by means of a clamp. The rotor and stator are visible in the figure. The rotating shaft assembly, including the rotor and rotor coil, is the only moving part.

The layout of Figure 11.19 is worth some detailed attention, because a layout is a vital design tool. The design of the device in this example involves many phases. Obviously, the most important is the original invention, or basic idea. However, much

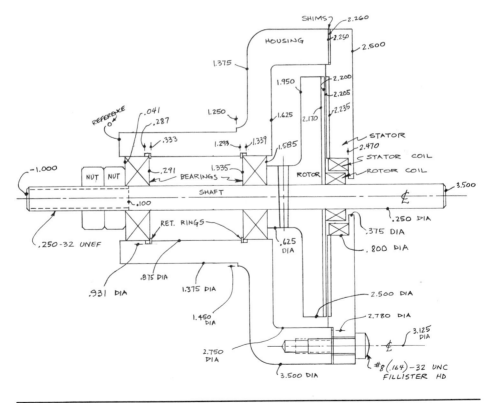

Figure 11.19
Design layout of a digitizing reading head.

design beyond the original idea is necessary before any units can be successfully manufactured and sold. One phase of the design is the theoretical analysis and subsequent
testing of the electromagnetics of the coils and of the patterns on the rotor and stator.
If the electromagnetic coupling of these parts is too weak, the device is worthless.
Another phase is the design of the electrical power supply, which provides the high-
frequency input current, and of the electronic circuits, which detect the current variations in the output and convert them to discrete digital pulses. And finally there is
the phase of the design that establishes the exact geometrical relationships of all of
the parts in the device. These relationships fix the dimensions of the parts and provide the information that is subsequently supplied to the shop (even in large manufacturing companies, the factory is often called "the shop") so the parts can be
manufactured.

The layout in Figure 11.19 resembles an assembly drawing, and in a sense it is
one. But a true assembly drawing (such as the one in Figure 11.28) provides only
such instructions to the shop as are necessary to put the device together. The layout
of Figure 11.19 shows a great deal more than this: It shows all the relationships that
are needed to determine the dimensions to go on the working drawings.

A layout is carefully drawn, as close to exact scale as possible, using sharp, clear
pencil lines. Obviously, such a drawing is not done freehand. Layouts are often drawn
larger than full size to provide greater clarity in the details.

A special kind of dimensional referencing system is shown in Figure 11.19. It is
called *base-line referencing*, and it finds favor with many engineers. Most layouts require three base lines for referencing, which might be thought of as the familiar x, y,
and z axes of analytic geometry. Since our reading head consists almost entirely of
cylindrical surfaces, only two base lines are necessary in this case. One is the centerline of the shaft (marked $\mathcal{C}$). The other is the flat surface at the left-hand end of the
housing, marked "REFERENCE" and given the dimension "zero." All horizontal distances (i.e., parallel to the shaft) are measured with reference to this zero plane, positive to the right, negative to the left. The distances from the zero plane are marked
directly on the layout, with a small leader extending to the surface in question. These
reference dimensions are used later to establish the actual manufacturing dimensions
on the working drawings, as will be seen.

In making the layout, the engineer has a general mental picture of the desired
relationships among the parts. But, as any layout develops, inevitable conflicts and
other problems occur. It may turn out that two surfaces interfere with each other,
or a supporting surface is too flimsy, or inadequate material has been left to provide
a firm supporting surface, or any of countless other things. It also may become apparent that the device will be hard to assemble or repair or that the arrangement
may entail unnecessarily costly manufacturing procedures. Awareness of these problems develops as the layout takes shape, and changes to the layout must be made
frequently. As the changes are made, the accompanying reference dimensions are also
meticulously kept up to date so that they provide an accurate record of dimensional
relationships. Thus, the layout is simultaneously a design tool to discover problems
and a record of the decisions that are made.

To illustrate the use of base-line referencing, consider the location of the first
bearing to the right of the zero reference surface, which is to be a standard ball

bearing. The bearing manufacturer's catalog tells us this bearing is 0.250 inches thick and 0.875 inches in outside diameter. The lines defining the 0.875-inch diameter are drawn, and one of them is marked .875 DIA., as shown. This is, of course, the diameter of the hole through the inside of the housing into which the bearings will be inserted.

An arbitrary decision is now made concerning the lateral location of the bearing. The engineer decides that the reference dimensions of the two sides of the bearing will be .041 and .291, as shown on the drawing. (The "inches" is omitted, it being understood that all dimensions are in inches. Also, the zero in front of fractional dimensions on working drawings is customarily omitted.) The difference between .041 and .291 is .250, which is correct for the thickness of the bearing.

Each bearing is to be held in place by a thin metal piece 0.042 inches thick, called a *retaining ring*. Outlines of typical retaining rings are shown in Figure 11.20. The one on the left is intended for internal use in housings, as is the case here. The one on the right is intended for external use on a shaft, but we are not using any external rings in our design. Retaining rings are made of spring steel and are put in place with a special tool, which fits into the two small holes shown. An internal ring of the type shown here has a diameter slightly greater than that of the hole through which it is to be inserted. When the tool is used, it causes the two free ends of the ring to be pulled toward each other, which reduces the diameter just enough to enable it to be slipped into the housing. When the ring comes into alignment with a groove that has been prepared for it, it is released and snaps into the groove, preparing a shoulder for the bearing to rest against.

The engineer draws the groove on the layout with the ring in place, exaggerating the drawing just enough so the details can be seen, and marks the appropriate reference dimensions. Since the ring is 0.042 inches thick, the .291 location dimension is increased by .042 to give .333, which is the right-hand edge of the groove. The engineer also learns from the retaining ring catalog that the groove should be 0.046 inches wide, so .046 is subtracted from .333 to give the left-hand edge of the groove, .287. Later, when it comes time to place dimensions on the working drawing of the housing, the engineer (or, more likely, a drafter working for the engineer) can pick off the dimension .333 directly from the layout as the distance to be prescribed from the end of the housing to the right-hand edge of the groove.

The engineer (or the drafter) also knows, by looking at the layout, that the right-hand edge of the groove is the one that determines the location of the bearing. This is because when the shaft assembly is inserted through the two bearings, the shoulder on the hub of the rotor will press against the right-hand bearing, pushing it as far as

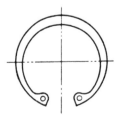

Figure 11.20
Outlines of typical retaining rings. The ring on the left is used inside a housing and is called an internal ring. The one on the right is used on a shaft and is called an external ring. (Courtesy of Waldes Kohinoor, Inc.)

it will go to the left. A nut is placed on the left-hand end of the shaft, which is threaded, and tightened gently until all the internal slack is taken up. This causes the two bearings to move toward each other until stopped by the retaining rings in the grooves. A second nut is then placed on the shaft and tightened against the first nut, to hold it in place.

Perhaps this is enough description to get across the purpose of base-line referencing. Only one other detail will be described, because it pertains to the most critical element of the design: the small air gap between the printed circuit patterns on the rotor and stator. Reference to this part of the layout will reveal the two thin pieces of plastic, each 0.030 inches thick, which carry the printed circuit patterns. The reference dimensions of 2.170 and 2.200 define the 0.030-inch thickness of the pattern fastened on the rotor, and the reference dimensions of 2.205 and 2.235 define the 0.030-inch thickness of the pattern on the stator. The reference dimensions 2.200 and 2.205 locate the facing surfaces of the two patterns and show that a nominal air gap of 0.005 inches has been provided. However, parts cannot be manufactured to the kind of precision necessary to ensure that the gap will be exactly 0.005 inches after assembly, on all of the many units that are to be made, without unreasonable expense. Therefore, the engineer has provided a space of 0.010 inches between the mating surfaces of the housing and stator, as defined by the reference dimensions 2.250 and 2.260. Upon assembly, shims (ultrathin pieces of metal, often made of brass) will be inserted in the 0.010-inch space as needed, until the 0.005-inch air gap is achieved.

The engineer must consider other problems in the production of this device, involving concentricity, parallelism of the two facing patterns, and the like, but they are too complicated to discuss here; they come under the heading of manufacturing tolerances. (We will briefly address the subject of tolerances later, when we discuss how snugly the shaft should fit inside the hole in the rotor.)

Before proceeding to a discussion of working sketches, it should be noted that it is possible to produce a layout like the one in Figure 11.19 directly on the screen of a computer terminal, using a modern computer-aided drafting system. The principles we have discussed here apply whether a computer-aided drafting system or pencil and paper are employed. In either case, we get a layout resembling the one in Figure 11.19. It should be mentioned, however, that in a computer-aided drafting system, the computer automatically keeps track of the locations of all the lines you draw on the screen and, on command, computes the dimension between any two surfaces. It even places that dimension on a drawing for you, if that is your desire.

Exercises

11.5.1 What is the nominal dimension, taken from Figure 11.19, between the edge of the retaining ring groove marked 1.293 and the surface marked 2.250? Why would this dimension be considered important? Examine the housing sketch in Figure 11.22 and locate the dimension in question.

Solution

$$2.250 - 1.293 = \textbf{0.957} \qquad \text{Answer}$$

When the rotor assembly is inserted, this dimension determines the location of printed circuit surface on rotor.

11.5.2 In Figure 11.19, what is the nominal distance from the contact surface between the right-hand bearing and rotor hub (marked 1.585) to the face of the rotor to which the printed circuit is to be cemented? Why is this dimension important? Locate the dimension in Figure 11.25.

11.5.3 In Figure 11.19, locate the two dimensions defining the outside diameter of the rotor and the inside diameter of the housing. What are these dimensions? What is the nominal clearance between the rotor and the housing?

11.5.4 In Figure 11.19, how far is the surface of the left-hand bearing depressed below the zero reference surface? How much of the threaded portion of the shaft (designated by dotted lines and marked ".250-32 UNEF") extends into the bearing?

■ 11.6 SKETCHES FOR THE SHOP

One of the last stages of the implementation step is preparing *working sketches* for the experimental shop. It is necessary to say *one* of the last stages, because the implementation step is characterized by the need to back up many times and make revisions as problems are encountered. The very last step is the preparation of *working drawings* for production, once everything has been ironed out. But, even after a project has entered production, new problems continually crop up, requiring that additional changes be made. Furthermore, a manufacturing company continues to have a major degree of responsibility for its products as long as they are being used by its customers. Thus, a project is never truly finished until every unit that has ever been built has reached the end of its useful life and has been retired from service.

A distinction was just made between working *sketches* and working *drawings*. A further distinction needs to be made between the kind of sketch two engineers may use while discussing a design (we will call this a *design* sketch) and what is called a *working* sketch. A design sketch may indeed be quite sketchy, because its maker is right there to explain it. (See Art's sketch in Figure 11.17.) But a *working* sketch has to have every last bit of information on it so that a person somewhere else who must be guided only by the sketch can make the part correctly. On the other hand, the only real difference between working sketches and working drawings is that the former are made by freehand drawing techniques, and the latter are crisp and perfect, often produced by computer-aided drawing systems. Usually it is faster to make sketches for the experimental shop freehand than to prepare them on a computer, especially if they are made by the engineer who prepared the layout. Under these circumstances, there is no necessity to communicate the information to another human (the drafter), or to a machine (the computer). (More information on computer-aided drafting systems is provided in Chapter 18.)

In the case of the digitizing reading head we discussed in the last two sections, the engineer who made the design layout is going to make the working sketches for the experimental shop, even though the company employs a number of *drafters*, who are very skilled at making good production drawings, and also has a computer-aided drafting system. (As a general rule, the company's engineers are used at the "front end" of the design process, for the steps of problem definition, invention, analysis,

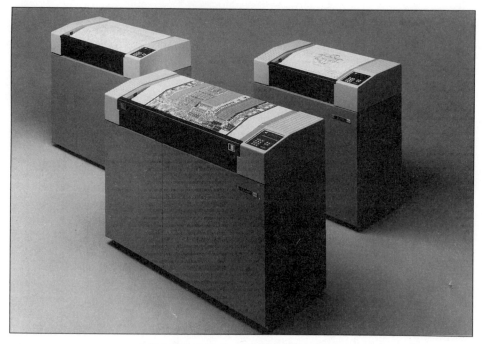

CalComp 5700 Series Electrostatic Plotting Systems

Figure 11.21

In companies equipped with computer-aided drafting systems, high-quality prints can be made directly from computer output. (Courtesy of CalComp)

and decision; in addition, engineers supervise the implementation step. The drawings themselves are usually made by drafting technicians. But in this particular case, the engineer who made the layout will make the sketches.)

In the previous chapter, the point was emphasized that engineers must be intimately familiar with both orthographic projection and conventional drafting practices, because they are generally the supervisors of drafting technicians and are responsible for the contents of the drawings. A couple of additional points need to be made.

First, there is sometimes a tendency to assume that a computer-aided drafting system will automatically take care of such things as orthographic projection and drafting conventions. This is a false assumption. The computer system can do many useful things, but the operator still must be thoroughly familiar with both the theory of projection and with conventional practices, or the computer might produce what amounts to nonsense. The computer can do only what it is told to do; it is the human being who must understand the system.

Second, even with a computer system, an engineer often finds it faster and more cost-effective to make freehand sketches for the shop than to make them with a computer. When full-bore production drawings are required, the computer system is usually the method of choice. But when sketches for the experimental shop are to be made, it is usually faster to sketch them by hand. Hence, a certain amount of sketching skill is desirable for almost all engineers.

So, for the digitizing reading head, the design engineer has prepared a separate sketch of each part, as in Figures 11.22, 11.23, 11.24, and 11.25. These are of the housing, stator, shaft, and rotor. Each sketch gives all the information necessary for making the part. (For brevity, sketches have not been included here for making the rotor coils or the printed circuit patterns.) In the sketches a straightedge has been used in a few places, such as the centerlines, because here it is faster than using freehand, and circles smaller than 1 inch, which are most easily made by use of a template. But all the other lines have been done freehand. (Figure 11.26 is a photograph of two templates, one for making circles and the other for making ellipses. Such templates can usually be purchased at college bookstores.) Large circles are made with a compass, and the lines are "heavied up" with a pencil. Obviously, a certain amount of care has been taken to make these sketches neat, even though they are done freehand. The sketches are made on vellum (transparent paper) with 10 × 10-to-the-inch markings. These markings serve as guidelines to help keep the freehand work neat.

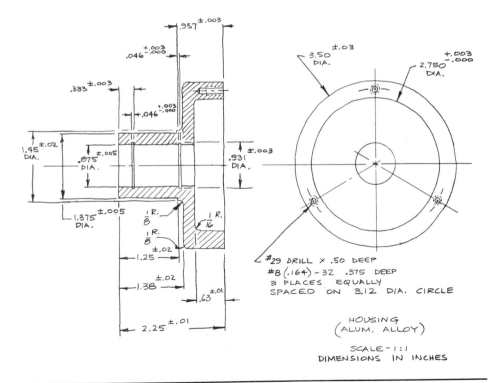

Figure 11.22
Working sketch of housing.

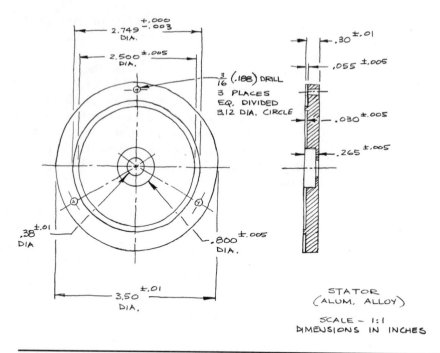

Figure 11.23
Working sketch of stator.

The left-hand view of the housing in Figure 11.22 has been drawn as a cross section (or *sectional view*, as it is called). The view we see here is as if the housing has been cut in half, and we are looking at the inside. The cross-hatching represents the cut surface of the metal that is exposed by the imaginary cutting action. Similarly, the right-hand view in Figure 11.23 is a cross section of the stator.

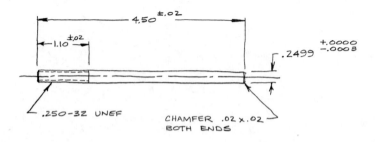

Figure 11.24
Working sketch of shaft.

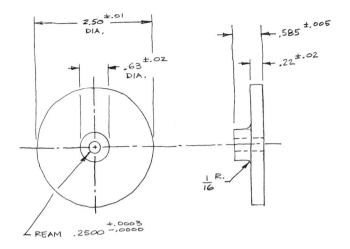

ROTOR
(ALUM. ALLOY)

SCALE – 1:1
DIMENSIONS IN INCHES

Figure 11.25

Working sketch of rotor.

The specifics of the proper handling and placement of dimensions, and other matters relating to the preparation of engineering drawings, are usually covered in required courses in an engineering curriculum. We will not go into them here. However, the following points should be noted:

The dimensions are spread out with lots of space left between them. This helps clarity.

Generally, the dimensions have been placed outside the boundary of the object, again for clarity.

Each dimension has a tolerance. In our example, we have placed a tolerance on almost every dimension. However, it is common practice to have a general note

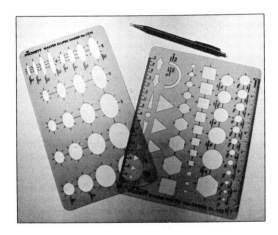

Figure 11.26

Typical templates for making ellipses, circles, and other shapes rapidly.

on each drawing stating that a standard tolerance, say, ± 0.2, is understood to apply to every dimension unless stated otherwise.

Certain conventions have been used. For example, the dashed lines at the left end of the shaft in Figure 11.24 indicate that this end of the shaft is threaded. The associated note, .250-32 UNEF, means that the outside of the threaded portion has a nominal diameter of 0.250 inches, that there are 32 threads to the inch, and that the thread shape is defined by a national standard called the Unified National Extra Fine (UNEF) thread series.

The dimensions on the drawings are taken from the layout. For example, the dimension defining the depth of the large cavity in the housing (Figure 11.22) is found to be 0.63 inches by taking the difference between the reference dimensions 1.625 and 2.250 on the layout. The actual difference is 0.625, but we round this to two figures, because the tolerance we are going to place on this dimension is $\pm .01$. (See Chapter 7 for a discussion of significant figures.)

At every point, the engineer must worry about just how closely controlled each dimension must be—in other words, how small the tolerances should be. Mostly, such judgments are based on experience, but when it comes to the closeness of the fit of a shaft in its hole, design manuals are available to give guidance. In the case of the fit between the shaft and the rotor, we can see that the diameter of the shaft (Figure 11.24) is 0.2499 inches, plus zero and minus 0.0003 inches. Thus, the largest that any shaft can be, and be acceptable, is 0.2499 inches; the smallest is 0.2496 inches. We sometimes express this in the form of so-called *limit dimensions*:

.2499
.2496

If we look at the tolerances on the hole in the rotor (Figure 11.25) we can see that the limit dimensions are

.2503
.2500

Thus, if we happen by chance to select a large shaft (.2499) and a rotor with a small hole (.2500), the clearance will be 0.0001 inches. It will probably take a bit of pressure to make the shaft go into the hole under these circumstances. At the other extreme, with the smallest possible shaft (.2496) and the largest possible hole (.2503), the clearance will be 0.0007 inches. This rotor will wobble just slightly on the shaft, and the engineer will have to decide whether this much wobble is permissible. Since the rotor carries one of the printed circuit patterns, the function of the unit could be affected adversely if the wobble were too great.

Two more sketches are shown in Figures 11.27 and 11.28. Both are so-called assembly drawings, showing how to put parts together once they have been fabricated. One is the final assembly drawing (Figure 11.28), and the other is called a subassembly drawing (Figure 11.27). The subassembly drawing shows how to put the rotor, rotor pattern, rotor coil, and shaft together. The proper location of the rotor on the shaft is shown by the dimension 2.59 inches, found from the layout by taking the difference between the -1.000 and 1.585 reference dimensions and rounding to two significant

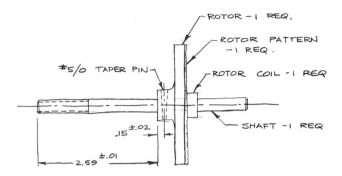

ROTOR - 1 REQ,

ROTOR PATTERN
-1 REQ.

ROTOR COIL - 1 REQ

#5/0 TAPER PIN

SHAFT - 1 REQ

.15 ±.02

2.59 ±.01

ROTOR   SUB-ASSEMBLY
SCALE - 1:1
DIMENSIONS IN INCHES

Figure 11.27
Working sketch of rotor subassembly.

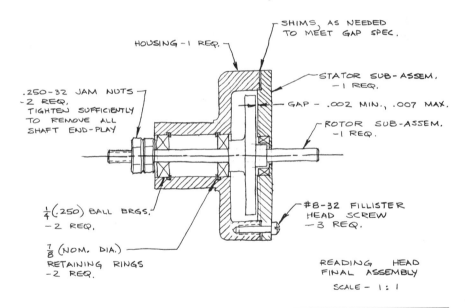

SHIMS, AS NEEDED
TO MEET GAP SPEC.

HOUSING - 1 REQ.

STATOR SUB-ASSEM.
-1 REQ.

.250-32 JAM NUTS
-2 REQ,
TIGHTEN SUFFICIENTLY
TO REMOVE ALL
SHAFT END-PLAY

GAP - .002 MIN., .007 MAX.

ROTOR SUB-ASSEM.
-1 REQ.

$\frac{1}{4}$(.250) BALL BRGS.
-2 REQ.

#8-32 FILLISTER
HEAD SCREW
-3 REQ.

$\frac{7}{8}$ (NOM. DIA.)
RETAINING RINGS
-2 REQ.

READING   HEAD
FINAL ASSEMBLY
SCALE - 1:1

Figure 11.28
Working sketch of final assembly.

figures. (Since the −1.000 dimension is negative, the two are added algebraically to give 2.585.) The two parts are held together by a taper pin. Actually, a subassembly drawing like this one would give further detailed instructions on how to drill and ream a hole for the taper pin. Detailed instructions would also be given on how the rotor pattern and rotor coil are to be held in place, and how to connect the two by soldering. These details are omitted here for simplicity. Our purpose in these few

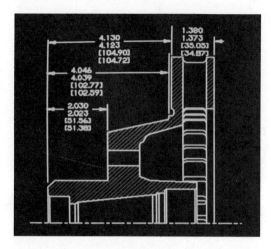

Figure 11.29
In computer-aided drafting, if the location of a surface is changed, such as the shoulder in this casting, the computer not only shows the surface in its new location, but also instantly updates all the dimensions relating to that surface. (Courtesy of Applicon Incorporated)

pages is not to give a complete course on manufacturing processes, but to give some insight into the organization and nature of working drawings.

The sketch of the final assembly in Figure 11.28 shows all the parts needed and their correct relationships. In this case, the only "manufacturing" instruction relates to the use of shims to achieve the desired air gap.

The fundamental points to be drawn from this example of working sketches are the following:

1. A design layout serves as a master control over the working drawings.
2. Each part is drawn on a separate piece of paper, with full manufacturing instructions. All the dimensions necessary to define the shape are given, with tolerances on the dimensions either explicitly shown or implied by a general note.
3. Assembly drawings and subassembly drawings are made, with the necessary instructions for putting the parts together.
4. Perfectly drawn mechanical drawings are not necessary during the prototype stages of a project. Freehand sketches will do, but they should still be prepared carefully and with an eye to clarity.

Exercises

11.6.1 Figures 11.22 and 11.23 show the housing and the stator. The diameter marked $2.749^{+.000}_{-.003}$ on the stator fits inside the diameter marked $2.750^{+.003}_{-.000}$ on the housing. What are the "limit dimensions" on these surfaces? What is the minimum looseness possible? The maximum looseness?

11.6.2 In the housing sketch of Figure 11.22, the overall length of the housing can vary between 2.24 and 2.26. The inner (locating) surface for the left-hand groove can vary between 0.330 and 0.336. The inner (locating) surface of the right-hand groove can vary between 0.954 and 0.960. If the two retaining rings (see Figure 11.19) are each exactly 0.042 thick, what is the smallest dimension that can separate the two inner surfaces of the bearings if manufacturing tolerances vary to their permissible extremes?

■ 11.7 MANAGING THE DESIGN PROCESS

It would seem that the management of the design process is something to be left to the boss and is not the concern of the individual designer. That point of view is incorrect for two reasons: (1) Each individual must manage his or her own activities; the boss is sure to ask, sooner or later, "How long is your design going to take?" (2) You may be the boss sooner than you think, and the scheduling of design activities will become one of your major concerns.

As an example, let us prepare a tentative development schedule for the precision optical measuring device shown in Figure 11.4. Ed is the supervisor of mechanical design, and has the overall responsibility for preparing the schedule. At about the same time as he assigns you the responsibility for the design of the projector support carriage (Section 11.3), he also asks you for your estimate of how long it will take you. After some discussion, you and Ed together decide it might take ten weeks, with another eight weeks to construct the parts, assemble them, and test their operation. Then, acting mostly on the basis of Ed's experience, he adds another five weeks for redesign and four more weeks for testing. This gives an aggregate elapsed time of 27 weeks. Ed quickly realizes this is probably the phase that will take the longest and therefore is the critical element in the overall scheduling (see Figure 11.30). Ed originally planned to assign the design of the support carriage controls to you, also. This is estimated to require five weeks for the design and three weeks for construction, totaling eight weeks. If the carriage and the controls were done sequentially by one person (you), the overall elapsed time would be 35 weeks. Ed believes another competent designer already employed by the company might soon be finished with her current project, and he tentatively plans to assign the support carriage controls to her, so her phase and yours can run in parallel.

At this point, Ed has not yet worked out the details of the cross-wire system that was described in Section 11.4, but he has some ideas. For now, at least, he plans to assign the design of this system to himself. His estimate is that it will take 25 weeks overall (see Figure 11.30), so it is not yet looked upon as the critical element. However, its estimated development time is almost as long as that of the projector support carriage, and it could easily become the critical element if things go badly. Above all, these two critical elements must be undertaken in parallel, or else the overall development time will be the sum of 27 weeks and 25 weeks, which is exactly one year—provided there is no schedule slippage.

The design of the optical system is not expected to take long (see Figure 11.30). The optics of this new model are similar to those in existing models being manufactured by the company, and no serious problems are expected. Similarly, the design of the electronic counters and the design of the power supply, including wiring, are not expected to be troublesome, since these are merely adaptations of existing units being manufactured by the company.

At this point, namely week 1, Ed is unaware that any scheduling is going to be necessary for the digitizing reading head. This is because he has just begun to consider the ramifications of his ideas for the cross-wire system, and he does not know that a design of a totally new kind of head—the one using the arrangement of Art's sketch in Figure 11.17—is going to be required. (The sequence of events that led to Art's

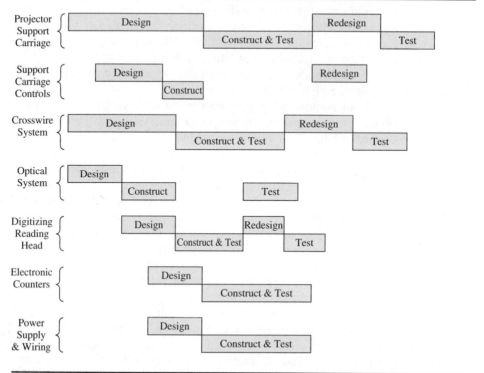

Figure 11.30

An estimated development schedule for a precision optical measuring device.

sketch have not taken place yet.) But, about three or four weeks into the program, Art has the idea for what ultimately turns out to be an important new product, and provision is made for the design, construction, and test of this unit in the schedule.

So a schedule is developed, and everyone goes to work. Will schedule slippage occur? Probably. However, as it becomes apparent that slippage is going to take place, Ed is immediately informed. He then has to re-examine the schedule to see if it is possible to assign additional people to the project (assuming they are available) or if overall slippage, affecting the unit's delivery date, is inevitable. Most especially, he will look to see if some other element of the project has now become the critical element, in view of the new information. By always being aware of which elements are the critical ones, he can be sure that the attention of the group will be applied where it will do the most good.

The project we have described is a relatively manageable one because it does not have very many phases. Hence, it may appear that some of the explanation given is self-evident. But many projects are much larger than the one in our example, and

Figure 11.31
The last stage in the design process is a careful checking and verification of the results. This civil engineer is checking the drawings for the construction of a large industrial building.

their development time may cover many months, or even years, instead of the 27 weeks of our example. In such a case, the scheduling of phases and the recognition of which ones are the most critical become of first-rank importance. A master schedule board is almost sure to be maintained and updated regularly through weekly conferences, to make sure that everyone is on top of the progress of the project. These considerations have assumed such major proportions that a special name has been given to them: *critical path management*.

Exercises

11.1 Make a list of examples (say, ten) of useful engineering accomplishments in which "high technology" played an essential part. In each case, write a brief description of the field (or fields) of high technology that was (were) vital in each case. Then, using your college catalog of courses, list the topics a person should study in order to be able to participate in the design and development work associated with each item. (Do not worry if you feel short on experience; do the best you can, based on your experience to date.) Now make another list of examples, also of useful engineering accomplishments, where "high technology" was *not* involved, but where, instead, good judgment, experience, and bright ideas were the key ingredients. Which list was easier for you to produce? Overall, which list of items has had the greatest impact on humanity, for good or bad? What conclusions do you draw from this exercise?

11.2 The word *design* is employed by professionals in many fields to describe their activities. Write down as many such fields as you can, and consider both the similarities and the differences between meanings of the word in those fields and in engineering.

11.3 Another kind of design, different from that discussed in this chapter, has to do with the attractiveness of an object—its appeal to the eye. Such design is sometimes called "styling" and has become of overriding importance in some fields—the auto industry, for example. See if you can come up with any examples from your experience in which you believe "styling" has gotten in the way of function or, at the very least, seems to have been done for its own sake and to no clear purpose. Include such familiar examples as autos, household appliances, furniture, TV sets, electronic instruments, and hand tools. See, in each such case, if you can figure out what the stylist was trying to accomplish. Could there have been considerations in the stylist's mind beyond functional utility, including such things as product recognition, attempting to attract the customer's attention, or misguided attempts at trendsetting?

11.4 From the literature, generate a list of examples of engineering design in which the use of a computer has become essential to the design process. Generate another list of examples in which the use of a computer seems to have little or no role.

11.5 From your experience, list at least three items in common use in which human factors are important to the design. For each, describe briefly the part that human factors play, and suggest ways in which the utility of these items could be improved.

11.6 Investigate five of the following systems, and describe briefly the "scheme" upon which each operates. Some possible references: Q. Deane (ed.), *Inside Modern Technology: How It Works* (New York: Greenwich House, 1982); *The Way Things Work* (New York: Simon and Schuster; vol. I, 1967; vol. II, 1971).

TV camera
Color TV
Computer memory
Hybrid electric car
AC–DC rectifier
Ammeter
Ski safety binding
Liquid crystal display
DC motor
Electric resistance strain gage
Artificial kidney
Laser
Pressure gage
Distillation
Ship stabilizer
Variable speed control unit
Computerized tomography
Powder metallurgy
Speedometer
Electrocardiograph
Xerography
Heat pump
Gas flow meter
Quartz clock
Automatic exposure control

11.7 Select five systems from the list in Exercise 11.6 that would probably require some sort of research, either in the literature or in the laboratory, in order to complete the design. Assume these systems do not yet exist and it is your job to design them. State your belief concerning the kind of research projects that would be needed and how you would get started. (Do not worry if you have had no experience with research; the purpose of this exercise is to get you thinking about how one goes about learning something when one knows little or nothing about it.)

11.8 For each of the topics you selected in Exercise 11.7, write a short discussion of the possible trouble spots that could cause you to have to back up and start over again.

11.9 See if you can come up with a list of things that you wish "somebody would invent." Some examples are improvements in your car's performance, comfort, or safety; stacking arrangements and methods of use for a stereo system; TV controls; the heating and ventilating system for your home or apartment; seating arrangements in your classrooms, including visibility; the transportation system on your campus; garbage pickup and resource cycling; more "user-friendly" computer operations; an improved postal system. Make up your own list; then try your hand at proposing solutions.

11.10 List several situations in society that you believe are unsafe or are unacceptable for environmental or economic reasons or because of their social injustice. Propose ways in which new engineering solutions might help in these cases. Be specific. Give enough attention to details so that others can understand how your solutions would operate.

11.11 Propose two schemes for converting shaft rotations to digital electrical pulses.

11.12 Come up with a preliminary design for tying newspapers in bundles automatically.

11.13 Design a device for holding a book open to a particular page, freeing the hands.

11.14 Propose a scheme for a wheelchair that can climb stairs.

11.15 Make a preliminary design for a power-driven branch trimmer with a "reach" of 15 feet.

11.16 Design a locking system to prevent bike theft.

11.17 Propose a parking system for your campus that eliminates autos from the central part of the campus, yet minimizes walking distances to academic buildings and offices for students and employees.

11.18 Make a preliminary design for a device that will drive screws and is attachable to a power hand drill.

11.19 Propose a collection system for using discarded newspapers as an energy source.

11.20 Design a device for attaching a hand calculator to a desk, to prevent theft. No holes should be drilled in the desk, at least not where they can be seen.

11.21 Design a security system for your home that will prevent break-ins, yet permit easy exit in case of fire.

11.22 Propose a system for handling heads of lettuce that will result in the least possible number of handling steps between field and market.

11.23 Design a simple bicycle rack for carrying a package that is too large to fit in a knapsack.

11.24 Design an improved instrument panel for your auto.

11.25 Make a preliminary design for a traffic interchange for bicycles and pedestrians, where autos are not involved.

11.26 Make a preliminary design for a battery-operated electric car. Make some estimates on how large the motor(s) should be, how many batteries will be needed, and how long the car can be operated between chargings.

11.27 Propose a scheme for a sorting system that will separate refuse into the following three components: steel pieces such as "tin" cans; aluminum pieces; and paper, plastic, and other assorted garbage.

11.28 Design the perfect hot–cold water controller for a shower.

11.29 Design a rack that can be attached to a car and will carry two bicycles.

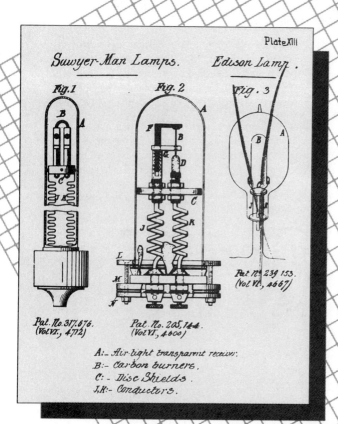

Plate XIII

Sawyer-Man Lamps.   Edison Lamp.

Fig. 1   Fig. 2   Fig. 3

Pat. No. 317,676.   Pat. No. 205,144.   Pat. No. 239,153.
(Vol. VI., 4712)   (Vol. VI., 4600)   (Vol. VI., 4667)

A:- Air-tight transparent receivers.
B:- Carbon burners.
C:- Disc Shields.
J,K:- Conductors.

Photographs courtesy of General Electric Corporation (center) and Parsons Brinckerhoff (bottom). Drawing (top right) from Siren, *Leonardo da Vinci, the Artist and the Man*, Paris, 1916.

12

Creativity

In the previous chapter we discussed invention and creativity, giving the clear impression that engineers deal intimately with these things. And so they do, but to varying degrees, depending upon their assignments. For a few engineers, invention may be the daily order of business: They are expected to come up regularly with new ideas, some of them perhaps patentable. For most engineers, however, the job is not necessarily one of coming up with new inventions every day, but of solving the problems that arise in swarms in any design or development project. Even so, each problem requires that someone come up with an idea for a solution, and the process of generating such ideas is creative, too. Thus, engineers, in varying degrees, all participate in being creative, so it is worthwhile to expand on the subject here.

The word *creativity* is in a rather curious state: Nearly everybody can recognize creativity instantly, yet nobody can define the word satisfactorily. A companion word, *invention*, causes even more trouble. Every layperson is instantly confident of the word, knows exactly what it means, and can think of countless inventions: the electric light, the safety pin, the phonograph, even the atomic bomb. It is only Patent Office examiners and U.S. Supreme Court justices who believe there is a problem in defining what an invention is.

For the purposes of this book, an invention will be regarded as something that is clearly possessed of novelty and usefulness; has identity as a distinct device, process, or system; and has been "reduced to practice." (Even though an idea may have been reduced to practice, it still may not be commercially practical. Reduction to practice may consist of nothing more than a

Famous Engineers
Edison Did Not Invent the Electric Light

In the United States, Thomas Edison is revered as the inventor of the electric light. In England, that honor is accorded to Joseph Swan. In 1860, preceding Edison by 19 years, Swan brought a piece of carbonized paper to incandescence in a vacuum. However, the light lasted for only a short time and Swan lost interest. Many others performed similar experiments during the next 20 years. When Edison was issued his first electric light patent in 1879, 31 other patents on the same type of device already existed.

Why, then, did Edison come to such prominence? There were two basic reasons. First, he devised a lamp that possessed a long enough life to be practical. Second, he perceived that it was vital for each lamp to draw a small current, or the costs of the distribution system would be prohibitive. In both endeavors the collaboration of a university-trained engineer, Francis Upton, was essential for success. Yet Edison's name is known by almost everyone and Upton's by almost no one. Achievement of long lamp life turned out to depend upon achievement of an ultrahigh vacuum. Upton designed a pump that produced a vacuum of 10^{-6} atmosphere, far beyond anything that had been achieved previously. With respect to the workable distribution system, Upton designed a new generator that was 80 percent efficient, as opposed to the best efficiency then available—55 percent. The two men formed a team in which each was indispensable to the other. Edison provided the personal drive and the intuition; Upton contributed the mathematical insight that led Edison's intuition in productive directions.

The creation of an economical electrical distribution system was of paramount importance. Gas lighting was widespread and was very economical. Electric lighting would have to compete with gas, or it would not receive acceptance. The key was, in the parlance of the times, to "subdivide the lighting" into small units suitable for indoor use. Existing lamps had low resistances, on the order of $\frac{1}{3}$ ohm. With a supply voltage of 100 volts, such a lamp would draw 300 amperes. A thousand lamps would draw 300,000 amperes—an impossibility. In England, a group of scientists had in fact studied the problem and pronounced it impossible to solve. However, Edison's intuition told him that if the lamp's resistance could be made very high, say, 1000 ohms, then the necessary current would be correspondingly low. He was right.

Edison had been experimenting with platinum because of its high melting point. He had previously considered carbon and had abandoned it. But carbon caught his interest once

written description in a patent or a series of laboratory demonstrations that succeed only under carefully controlled conditions. The long, and usually painful, task of refining an idea until it reaches the point of commercial realization is what we have previously referred to as *development*.)

It is almost impossible to talk about creativity without also talking about patents. The aforementioned layperson, who had no trouble with the term *invention*, would probably also assume that creativity, invention, and patent all refer to essentially the same thing. A slightly better-informed person would know this is wrong, for it can

more when he read in a magazine that Swan, after almost two decades of inactivity on the electric light, had resumed his experiments and was using carbon. After months of feverish activity, Edison and his associates were able to make a lamp with a carbon filament that burned for 16 hours. Within a few weeks they made lamps that burned for hundreds of hours, as they learned how to make more durable filaments. The key to the use of carbon was the existence of Upton's high vacuum, because carbon heated in the presence of air consumed itself. The high resistance that was necessary was achieved by making the carbon very thin in cross section. However, the resistance was on the order of 100 ohms instead of the desired 1000. Nevertheless, a resistance even of this level made a practical lighting system possible. In subsequent patent infringement lawsuits, it was precisely the thinning down of the carbon, producing a high resistance, that was held to be the key to success. On the basis of this fact, Edison's patent was judged valid and judged not to infringe on earlier patents even though some of them were based on the use of carbon.

In contrast to other important inventions, many of which were initially ignored, the electric light was eagerly anticipated by the public. Edison's success was instantaneous, and he became a wealthy man because of the electric light and other inventions. Lighting systems, mostly powered by coal-fired steam engines, were installed widely in subsequent years. However, the systems were based on direct current (DC), and this limited their use. In order to transmit electricity farther than a very short distance, high voltages are necessary to reduce the amperage. At that time, direct current could not be stepped up to high voltages for transmission purposes, and then stepped back down again for use in the home or office. Alternating current (AC) could do these things by the use of transformers, and within a few years AC systems began to displace the Edison DC systems. Edison, never one to admire an idea not his own, stubbornly refused to accept AC power and even tried to prevent its adoption by claiming it was more dangerous than DC.

In recognizing Edison's genius, it is probably more important to concentrate on his conception of a "subdivided" electric system than upon the single fact that he and his associates developed a long-life lamp. Also, Edison clearly foresaw the importance of electric distribution systems in providing power for motors and for heating systems. In fact, the latter two uses account for more electric power consumption today than does lighting. Without exaggeration, the work of Edison and his associates—especially Upton—revolutionized the world.

Reference
R. Conot, *A Streak of Luck* (New York: Seaview Books, 1979).

immediately be pointed out that Einstein could not have patented his theory of relativity, even though the highest possible order of creativity was involved in bringing it into being. Nor would Einstein's theories normally be termed inventions. As another example, the Golden Gate Bridge represents creative civil engineering of the first magnitude, but it is not an invention, nor is it patentable, because suspension bridges have long been known. The legal meanings attached to the words *inventions* and *patents* will be discussed in greater detail in the next chapter.

Figure 12.1
An engineer checks out a new kind of high-performance gyro for airplane inertial reference systems that costs half as much as existing gyros and is twice as reliable. The gyro features a central glass block in which two laser beams move in opposite directions around a triangular path. (Courtesy of Honeywell, Inc.)

■ 12.1 THE CREATIVE PERSON

Everybody is creative to some degree; creativity is not an all-or-nothing function, so that one is either a thoroughly creative person or else completely lacking in this quality. Creativity is distributed by degrees among humanity and is very likely to be closely associated with intelligence.

Psychologists have shown considerable interest in this last point, and some research actually seems to indicate that there is no correlation between intelligence and creativity. In such cases, however, "intelligence" is likely to mean intelligence *as measured by an IQ test*, and IQ tests measure only certain kinds of intelligence. As Getzels and Jackson put it: "IQ tests tend to reward 'convergent thinking,' since they are largely framed in terms of 'acceptable' answers." Because IQ tests emphasize reading comprehension, vocabulary, and recognition patterns, it can readily be seen that those qualities regarded as creative generally are bypassed.[1]

There is some psychological evidence that our educational system may tend to inhibit the development of creativity in individuals. Anne Roe of Harvard University suggests there is a "subtle something" about the way in which elementary subjects are taught that may have a stultifying effect upon original thinking. She says, "Teaching these subjects in terms of 'right' and 'wrong' answers carries a strong moral connotation of considerable significance." Unconventional answers from children may be sweepingly denounced by teachers as "wild" or "silly" but may actually be indications of creative potential. In addition to this, children usually impose sanctions against members of their group who are "different." As a result, potentially creative children often are subjected to severe repressions and frequently develop into isolates.[2]

Some people appear to be natural creators; that is, ideas seem to come forth almost automatically from them. Others have to engage in deliberate, conscious effort to develop their creative potential. Obviously, there is a limit to each person's ability,

---

[1] J. W. Getzels and Philip W. Jackson, *Creativity and Intelligence* (New York: Wiley, 1962).
[2] Anne Roe, "Personal Problems and Science," in *Scientific Creativity: Its Recognition and Development*, C. W. Taylor and F. Barron (eds.) (New York: Wiley, 1963), pp. 133–134.

Figure 12.2
A completed microprocessor. The invention of the microprocessor was a major creative event. (Courtesy of Intel Corporation)

but one should take heart from the findings of a California psychologist that "the actual creative productivity of almost every individual falls far short of his own level of creative capacity."[3] An executive of Hewlett-Packard underscores the point that its engineers are expected to be creative and that the company does not rely upon isolated individuals for invention. "All engineers are resident inventors," he says.[4]

Clearly, the ideal would be to find ways to induce all individuals to use more of their natural potential for creativity. Certain techniques have been suggested for this and will be mentioned later, but the first step is to realize that everyone has far more creative capacity than is generally used. Simple awareness of this can open the door to more effective creative activity. There is no more poetic advice to the would-be creator than that of Ralph Waldo Emerson, who says, in his essay *Self-Reliance*:

> A man should learn to detect and watch for that gleam of light which flashes across his mind from within, more than the lustre of the firmament of bards and sages. Yet he dismisses without notice his thought, because it is his. In every work of genius we recognize our own rejected thoughts; they come back to us with a certain alienated majesty. Great works of art have no more affecting lesson for us than this. They teach us to abide by our spontaneous impression with good-humored inflexibility the most when the whole cry of voices is on the other side. Else tomorrow a stranger will say with masterly good sense precisely what we have thought and felt all the time, and we shall be forced to take with shame our own opinion from another.

■ 12.2 ENHANCING CREATIVITY

Much effort has been expended in trying to develop deliberate methods for enhancing creativity. One method that has attained a degree of popularity is called *brainstorming*. In this method, the problem to be addressed is announced to a group of people sitting around a table. Then the members of the group are asked to come up with as

---

[3] R. S. Crutchfield, "The Creative Process," in *The Creative Person* (Berkeley: Institute of Personality Assessment and Research, University of California, 1961), p. VI-2.
[4] "Whatever Happened to 'Participative' Management?", *IEEE Spectrum*, February 1979, pp. 60–68.

many ideas as they can that will solve the problem, as fast as they can, without any attempt to judge whether the ideas are good, bad, or plain silly. One member of the group is assigned to write down all the ideas as fast as they come out. Evaluation of the ideas is postponed until later. The expectation is that each group member's imagination will be stimulated by others' ideas and that large numbers of new ideas will be generated as a result.

For a few years after brainstorming was invented, the idea caught fire. It was fun, and marvelous results were claimed. Everybody sat around the table and gushed forth ideas without restraint. (Indeed, the less restraint exercised, the better.) Then a reaction against brainstorming set in that was almost as extreme as the enthusiasm had been; heated debate arose as to whether brainstorming was or was not a useful creative tool.

Attempts have been made to evaluate the effectiveness of brainstorming by means of controlled tests; unfortunately, the tests have not been conclusive. Donald W. Taylor of Yale University conducted one such study and decided that brainstorming "*inhibits* creative thinking." In his research, individuals working alone produced more ideas, and more *high-quality* ideas, than the same number of individuals working as a group under brainstorming conditions.[5]

In another test case, two engineers spent more than a month conceiving 27 possible solutions to a given problem. Then 11 young engineers having no prior acquaintance with the problem came up with all 27 of these ideas, plus many more, in a 25-minute brainstorming session.[6]

In still a third instance, a pair of psychologists repeated Taylor's tests on different subjects and this time found no significant differences between the performance of groups and that of individuals.[7] The major finding of this study was that the key influential factor is the employment of *deferred* judgment as opposed to *concurrent* judgment, whether used by brainstorming groups or by individuals. In the exercise of deferred judgment, ideas are generated without any attempt at evaluation until later; with concurrent judgment, ideas are evaluated as they occur.

This, then, is the lesson of value that brainstorming can teach us about creativity: In any deliberate attempt to be creative, the focus should first be on the generation of a large number of ideas, including even some superficially foolish ones. Later, after the flow of ideas has ceased, judgment can be employed as each idea is carefully examined for new leads and hidden possibilities.

Perspiration and Inspiration

Thomas Edison is often quoted as having said, "Genius is 99 percent perspiration and 1 percent inspiration." There is another maxim, that "inspiration most often strikes those who are hard at work."

---

[5] D. W. Taylor, "Environment and Creativity," in *The Creative Person* (loc. cit.), p. VIII-5.

[6] E. K. Von Fange, *Professional Creativity* (Englewood Cliffs, N.J.: Prentice-Hall, 1959), p. 51.

[7] S. J. Parnes and A. Meadow, "Development of Individual Creative Talent," in *Scientific Creativity: Its Recognition and Development* (loc. cit.), p. 318.

Outstanding creative persons are almost always noted for their great energy and drive. Nevertheless, many people still believe that inspiration comes unbidden, at idle moments, and can strike only those who are blessed with a mysterious gift of some sort. At one time, the U.S. Supreme Court even clothed this notion with official dignity by insisting that each potential patent be tested to determine whether the invention resulted from a "flash of genius." Fortunately, by act of Congress, this idea has been discarded, and inventions are no longer tested on the basis of the manner in which they were made, but on their intrinsic nature. Even so, investigations made by psychologists show that there actually is something called inspiration, although it most certainly does not come unbidden. Instead, it requires a strenuous preparation period.

This preparation (the perspiration part) consists of an intense period of study and search, in which one learns everything possible about the subject, followed by a period of extreme concentration and effort during which the creator makes repeated attempts to solve the problem. A long-time associate of Edison told of coming in on the great inventor one night to find him surrounded by piles of books he had ordered. Edison studied the books night and day for six weeks, making 2000 experiments during the same period. In the end, he produced a solution.[8]

Another example of such diligence is the discovery of vulcanization by Goodyear: the U.S. Commissioner of Patents declared in 1858 that Goodyear had made himself such a master of the subject of rubber that nothing could escape his attention.[9] Thus, the "accidental" discovery of vulcanization was preceded by an intense period of preparation.

W. H. Easton, in "Creative Thinking and How to Develop It," gives this graphic description concerning the occurrence of inspiration (called "illumination"):

> In this case, the thinker encounters a problem of great difficulty; but, as he has no way of knowing this in advance, he proceeds as usual, expecting to clear up the matter without much trouble. This, however, he fails to do. The problem resists all of his initial efforts to solve it, and before long, he discovers he has run into a serious obstacle.
>
> This is a critical point in his work. If he were like most people, he would stop here, giving up the problem as hopeless. But, being a creative thinker, he refuses to accept defeat, so he works on.
>
> But no amount of deliberate thinking gets him anywhere. He develops and applies every promising method of solving his problem he can imagine, but all prove failures.
>
> After struggling for hours, he runs out of ideas. Further thinking is useless, but his intense interest in the matter prevents him from stopping. Yet all he can do is to mill old ideas around in his mind to no purpose. Finally, frustrated and utterly disgusted with himself, he throws the work aside and spends the rest of his day in misery.
>
> Next morning he wakes oppressed. His problem is still on his mind and he thinks about it gloomily. But as the fog of sleep clears from his brain, the tenor of his thoughts changes.

---

[8] M. Josephson, *Edison* (New York: McGraw-Hill, 1959), p. 94.

[9] J. Jewkes, D. Sawers, and R. Stillerman, *The Sources of Invention*, 2nd ed. (New York: Norton, 1969), p. 50.

Figure 12.3
As bicycle use increases in the United States, so do muscle injuries resulting from overexertion. The pedal dynamometer shown here was devised for a research project to record loading data between foot and pedal during simulated bicycling. (Courtesy of University of California, Davis)

If, now, nothing distracts his attention, he soon finds that exactly those ideas he strove so hard to grasp the day before are now flowing through his mind as smoothly and easily as a stream flows through a level meadow. This is illumination.[10]

Blocks to Creativity

Probably the most frustrating block to creativity is what psychologists call "persistence of a misleading set." This means that one solution to a problem is already known, and try as one might, one cannot get past that particular solution to see what other (and perhaps better) solutions might exist. The inventor says determinedly, "All right, here we go for a solution of a totally different sort," but, though strenuous efforts are made, the inventor's mind circles about and comes to rest directly upon the old, familiar solution. Successful creative people say they sometimes get around this kind of block by forcing themselves to adopt extravagantly unorthodox viewpoints of their problem, such as, "Suppose I completely inverted this structure and made the output into the input?" or "Now that I have a mechanical solution to this problem, suppose I deliberately try to make one that is completely electronic?"

Closely allied to the block of a misleading set is the block called "functional fixedness." Here, a potential new use for a familiar object is obscured by its present

---

[10] *Creative Engineering* (New York: American Society of Mechanical Engineers), pp. 6–7.

use. An excellent illustration of this kind of block is given by Harold Buhl:

> Some students were once given the task of removing a ping-pong ball from a rusty pipe that had been bolted upright to the floor. In the room with the pipe, students found hammers, pliers, soda straws, strings, pine, and an old bucket of dirty wash water. After fishing vainly with the various tools most of the students finally saw a solution; they poured dirty water into the cylinder and floated the ball to the top. Then the experiment was repeated on other students with one important change; instead of the bucket, there was a crystal pitcher of fresh ice water surrounded by shining goblets on the table with a gleaming white cloth. Not one student solved the problem because no one could connect the beautiful pitcher and its clean water with the rusty pipe.[11]

Premature criticism has caused many an idea to be stillborn. It is better for the potential creator to adopt an attitude of optimistic reserve, always expecting the most favorable results from new avenues of thought. This, of course, is nothing but the application of deferred judgment. Eventually a choice must be made from among the various alternatives, but a possibility should not be eliminated too early by someone's saying, "Oh, that's ridiculous!" For it might not be so absurd as it seems.

Last, and most pernicious of all, is the block of fear: primarily fear of social disapproval or, perhaps, of supervisorial disapproval. Undoubtedly, fear is at the root of human tendencies to conform. It should suffice to say that anything really new is a departure from past practice and, therefore, represents some individual's nonconforming. If a person is so afraid of failure as to be unwilling to depart from tradition, it is likely that nothing creative will be forthcoming.

■ 12.3 A FORMULA FOR CREATIVITY

Besides the exhortation to work hard, other factors may help to stimulate one's creative capacities. To begin with, a simple *awareness* of the different phases of the creative process[12] may be valuable:

1. *Preparation.* It is essential to obtain every scrap of knowledge concerning the specific problem that one can. In opposition to this, some people point out that many technical bottlenecks have been broken by people who were novices in the fields of their accomplishments. Such successes may be attributable to these individuals' freedom from functional fixedness and from misleading sets. Even so, most engineers will be more successful at being creative if they first go to the trouble of making themselves knowledgeable. Edison, one of the most prolific inventors the world has ever seen, generally avoided scientists, but made it a rule to gain access to every bit of *scientific knowledge* he possibly could. Whenever he moved into a field with which he was unfamiliar, he first

---

[11] H. R, Buhl, *Creative Engineering Design* (Ames: Iowa State University Press, 1960), p. 55.
[12] D. W. MacKinnon is the source for the names of the five phases of creativity. See "The Study of Creativity," in *The Creative Person* (loc. cit.), p. I-1.

Figure 12.4
Vacuum casting is a new technique that produces superior products at lowered cost.
(Courtesy of CWC Castings Division of Textron Inc.)

collected all of the published material on the subject that he could and then digested it in an orgy of reading.

2. *Concentration.* One way of getting started at being creative is to sit down with the deliberate intention of being creative. At first, there may be no discernible result, but each step is a necessary precursor of those that follow. One way or another, one must get to thinking, long and hard, about possible solutions to the problem and developing as many promising leads as possible. Concentration is the hardest part of being creative, but it is also the most characteristic.

3. *Incubation.* Incubation is defined as a temporary withdrawal of the conscious mind from the problem while the subconscious continues to work on it. Some psychologists question the necessity for an incubation period. However, it is probably not accidental that the term "sleep on it" has become common in our language.

4. *Inspiration.* As previously described, inspiration is the sudden appearance of new insight, accompanied by exhilaration and elation. Sometimes the elation is premature.

Figure 12.5
The oldest automatic guided vehicle believed to exist, of those which actually went into production, manufactured in 1956. It was in continuous service for 19 years, during which time it traveled in excess of 50,000 driverless miles. (Courtesy of Mannesmann Demag Material Handling Systems Division)

5. *Verification.* This final period requires steady nerves and enormous determination as one "proves out" an idea, both to stave off despair as obstacles are encountered and to prevent oversights that might result from too much mental intoxication carrying over from phase 4.

Serious attempts have been made to teach the creative problem-solving process. Parnes and Meadow report that research on the effectiveness of such courses has been conducted at the State University of New York at Buffalo, with very encouraging results. In the teaching of creative problem solving, blocks to creative thinking are first discussed. These include difficulty in isolating the problem, rigidity of narrow viewpoints, trouble identifying fundamental attributes, conformity, excessive faith in logic, fear, self-satisfaction, perfectionism, negativism, and reliance on authority.[13]

After the blocks have been identified, the principle of deferred judgment is introduced, together with practice in "attribute listing." In attribute listing, the student is taught to look for fundamental attributes of an object, rather than to focus on its known functions. For example, in considering a piece of paper, a student might discover potential new applications for paper by studying such fundamental attributes as its whiteness, its square corners, its straight edges, or its translucence. This exercise is expected to help avoid functional fixedness, an evil that might easily occur if one were to focus prematurely on the known function of paper as a material for writing.

Students are also taught to keep notes on all ideas that come to them and to allocate definite periods for deliberate idea production. They are urged to list all conceivable facts that might relate to their problems, together with lists of questions and possible sources of answers. Potential sources of answers are then researched. Many people have reported amazement at the number of answers that can be obtained from a library. (For some perverse reason, libraries are often among the last sources tapped for answers.)

In the Buffalo research on the effectiveness of such courses in creativity, individuals were tested on their creative abilities both before and after taking creative problem-solving courses. The results were then compared with those of control groups. The findings were that a significant increase in creative ability was produced by taking a problem-solving course; follow-up research showed that the effect was a lasting one.

---

[13] S. J. Parnes and A. Meadow, in *Scientific Creativity: Its Recognition and Development* (loc. cit.), pp. 311–320.

Exercises

12.1 As an exercise in deliberate creativity, see what you can do with a method called "forced-relationship techniques." (See Eugene K. Von Fange, *Professional Creativity*, Prentice-Hall, Englewood Cliffs, N.J., 1959. Reprinted by permission of Prentice-Hall, Inc.) In this technique, first take a couple of inventions that you have read or heard about and then, using the following checklist, see if you can generate new ideas about them involving new applications, new ways of embodying the inventions, or—perhaps—entirely new inventions with new applications.

1. Put to other uses?
2. Adapt? Copy good ideas from other objects?
3. Modify? Change color, sound, odor, shape?
4. Magnify? Thicker, heavier, multiply components, exaggerate?

5. Minify? Subtract, condense, lighten, streamline?
6. Substitute? Other ingredients, processes, or approaches?
7. Rearrange? Interchange, change sequence, transpose cause and effect?
8. Reverse? Backward, upside down, transpose positive and negative?
9. Combine? Blend, or produce assortment?

12.2 Psychological research has been focused upon the attempt to identify the characteristics of outstandingly creative people, especially in the context of science. A common and consistent core of characteristics has been discovered through these investigations. (See C. W. Taylor and F. Barron, eds., *Scientific Creativity: Its Recognition and Development*, Wiley,

New York, 1963, pp. 385–386.) Briefly, as compiled from studies by Roe, McClelland, Barron, Saunders, MacCurdy, Knapp, and Cattell, the typical traits of the productive *scientist* are:

1. Self-sufficiency and capacity for self-direction
2. Preference for mental challenges; detached attitude in social matters
3. A high ego
4. Preference for exactness
5. Preference for isolation as a defense mechanism
6. High personal dominance, but a dislike for personal controversy
7. High self-control, even overcontrol; little impulsiveness
8. A liking for abstractness
9. Independent thinking; rejection of group pressures
10. Superior intelligence
11. An early interest in intellectual matters
12. Comprehensiveness and elegance in explanation
13. An enjoyment in pitting oneself against uncertain circumstances

Do you believe, from what you know of engineers, that this list applies to engineers as well as scientists? If not, do parts of it apply? Which parts? Apply the list to yourself, and see whether your own characteristics match those of the outstandingly creative scientist. If the match is not very good, do you think this means you are doomed never to get an idea? Or do you think, instead, that there might be many kinds of creativity besides those considered by the makers of the list? If so, what are they, and what characteristics are associated with creative people in those fields?

12.3 Try the technique of brainstorming with a few of your friends. (The technique probably will not work well with fewer than four or five.) First, think of some nice kind of invention to have that, as far as you know, does not exist. Then, setting a time limit, begin mentioning ideas that might be used to solve the problem. Appoint one member of the group to write down the ideas as fast as they come. Do not evaluate or criticize (or even make a face at) any of the ideas until the time period is up. Do not worry if there are protracted silent periods at times; those are normal. At the end, discuss your list to see if you think there is a promising idea on it, or if you have just wasted your time. (If you cannot think of anything to brainstorm about, look over the exercises in Chapter 11.)

12.4 List as many ideas as you can for each of the following, within a 3-minute time limit:

> New uses for Scotch tape
>
> Uses for white typing paper
>
> Blocks to creativity not already mentioned in this text
>
> New products that would be nice to have
>
> Possible adverse social consequences of having the new products you just listed
>
> New applications for computers
>
> A new brand name for a personal computer that will have strong sales appeal
>
> Uses for waste newspaper

12.5 After coming up with as many ideas as you can for a new invention, either by brainstorming or by other means, lay out for yourself the next steps you would have to go through in order to select the three best ideas with which to proceed further. (In other words, define for yourself what you mean by "best.")

12.6 Meet with a professional engineer and learn how creativity is encouraged and rewarded by that engineer's employer.

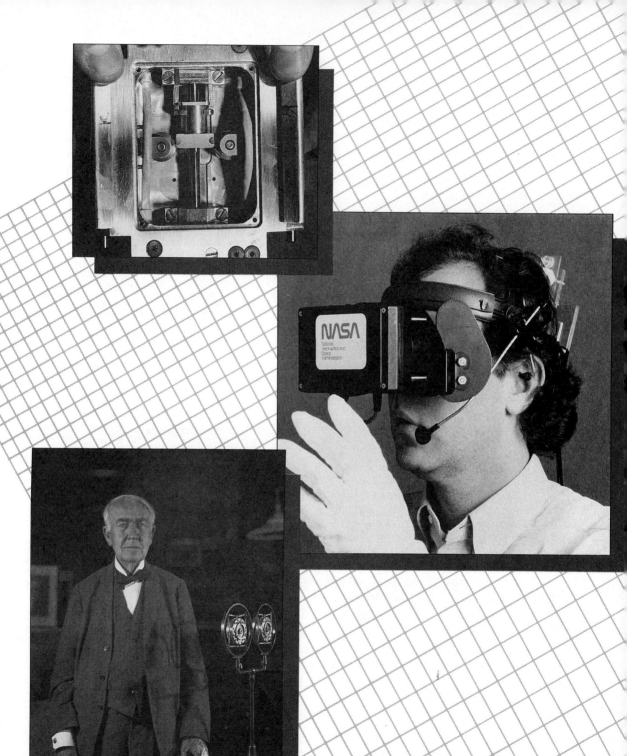

Photographs courtesy of University of California, Davis (top), NASA (center), and General Electric Corporation (bottom).

13

Patents

The distinguished judge Learned Hand is reported to have said, "I am very little certain about anything, but least of all about patents." Since the judge has been considered one of the nation's outstanding authorities on patent law, it seems highly likely that there is more to the subject than first appears. He meant that there is scarcely any subject more complex than patents—or more controversial. In spite of this, most Americans reach adulthood with clearly developed, and usually favorable, attitudes toward patents.

Unfortunately, many patents have only marginal value, and some even skirt the borderline of trivia. The U.S. Supreme Court has spoken harshly of the "list of incredible patents which the Patent Office has spawned" and has struck hard at the patenting of gadgets. The court has said, "It is not enough that an article is new and useful. The Constitution never sanctioned the patenting of gadgets. Patents serve a higher end—the advancement of science." Some people, however, believe the Supreme Court has gone too far in its high number of patent invalidations. One of these was Justice Jackson of the Supreme Court, who said acidly, in a 1949 dissenting opinion,

> It would not be difficult to cite many instances of patents that have been granted, improperly I think, and without adequate tests of invention by the Patent Office. But I doubt that the remedy for such Patent Office passion for granting patents is an equally strong passion in this Court for striking them down so that the only patent that is valid is one which this Court has not been able to get its hands on.

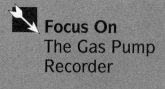

**Focus On
The Gas Pump
Recorder**

An interesting story involving patents and their value (in this case, their lack of value) is told by a mechanical engineer:

I had just changed jobs and had gone to work for a new company where I was more or less in line for an executive position in their engineering division. Not knowing what else to do with me while they "tried me out," so to speak, they stuck me in the Patent Department, where I was put in charge of evaluating new product ideas submitted by outsiders.

I knew next to nothing about patents and wondered what I would do with my time in case there weren't any new product submissions to be "in charge" of. But, lo, there was a submission almost immediately. In came an outside inventor, equipped with a working model of what he called a "gasoline pump recorder," and an impressive-looking patent he was willing to license to us, for a fee.

The idea behind the machine was that a customer would insert a credit card into a slot on the side of the gasoline pump. After the gas was delivered, the gallonage, the price per gallon, and the total dollar value would all be printed on the credit slip simultaneously, along with the customer's name, address, and credit card number.

I wasn't terribly impressed with the model, which frequently malfunctioned, but that was okay because I knew that the first models of most de-

vices are subject to such ills. I also wasn't very optimistic that prospective customers would pay very much for a machine that only saved them a small amount of time, because writing down gallonage and price only takes about five seconds. Beyond that, I really didn't know what to do next. Did we have a potentially good product on our hands, or not? I knew that many terrific products had emerged from unpromising-looking origins. On the other hand, I didn't want to suggest committing a lot of money to a lemon.

The fellow in the office next to mine solved my problem. He was an experienced patent attorney, and he said, "Let's send for the 'prior art' from the U.S. Patent Office." This we did, and before long there came back a package containing, among other things, about six issued patents which to me looked exactly like the one the outside inventor was offering us. Furthermore, these patents had all been granted on much earlier dates than the one we were considering. I couldn't make much out of them and was especially distressed by the "claims," which were listed at the end of each of the patents. To me, they all seemed to be pretty much alike, including the ones in "our" patent.

My patent attorney friend said, "Let me analyze the claims," In a day or two he came back and said, "The only thing in the submitted patent which is different from the others is the fact that the credit slips are contained in *a removable box.* If you eliminate the removable box feature, not only does the patent give zero protection, but there is still the question of the coverage claimed by the earlier patents. If we want to go into this field, we will have to obtain the rights to all the patents. The one with the least value is the one we are considering."

I had absolutely no trouble writing my report after that.

■ 13.1 THE VALUE OF PATENTS

It is almost impossible to find a restrained, dispassionate statement about patents. Those who defend the practice often take the position that America's great technological progress can be directly attributed to the influence of the patent system. They declare that industrial firms will not undertake expensive development programs

Figure 13.1
The first transistor (1947), a point-contact type, amplified electrical signals by passing them through a solid semiconductor material, basically the same operation performed by present-day junction transistors. (Courtesy of Bell Laboratories)

unless they can be sure the results can be protected by patents. Executives of some corporations have declared that they would be driven to cut their research and development expenditures drastically if the patent system were abandoned. Yet, others have stated that their innovative activities would not change at all if there suddenly were no patent system.

Practically no one among the critics of the patent system recommends such an extreme course as its outright abolition. One possible past exception was Thomas Edison, who wrote with some bitterness, during long, drawn-out litigation over his electric light patents, that he didn't care if the whole patent system was "squelched."[1] Edison's bitterness is understandable. Although he ultimately won his electric light suit, the victory cost his company $2 million and is said to have made his adversary, Westinghouse, almost insolvent.

The possibility of such calamitous litigation is one of the patent system's chief evils and at the same time one of its chief advantages. It is bad because of the social waste, advantageous because it makes potential infringers wary of the patents held by others. A prospective litigant would be very foolish to set about breaking another's patent through costly court actions unless virtually certain, in advance, of the outcome.

A patent confers a limited monopoly upon its owner in return for making known something that did not exist before. Supporters of the patent system hold that the granting of a limited monopoly to an inventor is an incentive that is necessary to increase the flow of inventions and thereby benefit society.

Critics of the patent system generally focus their criticism on monopoly. Most often the critics are economists, some of whom maintain that the provision of a financial incentive to the inventor may come at too high a cost to the rest of society. A

---

[1] M. Josephson, *Edison* (New York: McGraw-Hill, 1959), p. 355.

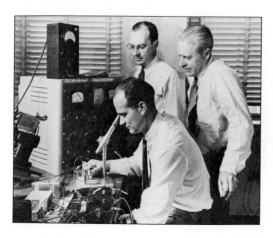

Figure 13.2
With this apparatus at Bell Laboratories, some of the first investigations leading to the invention of the transistor were made. The coinventors of the transistor were Dr. William Shockley (seated), Dr. John Bardeen (standing, left), and Dr. Walter H. Brattain (standing, right). (Courtesy of Bell Laboratories)

dramatization of this point, attributable to the Princeton economist Fritz Machlup, follows:

> Assume that twenty corporations are engaged in selling a particular product and that 100 million units are sold annually at $1 each. One firm patents an improvement on the product that results in a ten-cent saving on each unit. If all firms were free to use the improvement, the public would be able to acquire the product at ninety cents each; obviously this would benefit the public. However, free use is impossible, because of the patent. Instead, the product continues to sell for $1, and the innovating firm makes the extra profit, which represents its incentive. If the firm continues to corner one-twentieth of the total market for seventeen years (the life of the patent), it will realize $8,500,000 extra profit, which is presumably enough to cover its research costs and give a return, besides. But during the same period, the public has been required to forgo the opportunity to purchase the 100 million units per year at a reduction of ten cents each: The total cost to society, in lost savings, is $170 million.[2]

Admittedly, the example is oversimplified and extreme. In actual practice, the other firms would also introduce innovations that, in time, would tend to drive the price down. In addition, there is no valid reason to believe the price would remain at $1, even in the absence of other innovations. The original firm would probably try to increase its share of the market by lowering its price, and this would force the other firms to do the same thing. Still, the example graphically displays a principal ground upon which many economists base their recommendations for revision of the patent system. Some believe the 17-year period is too long; they suggest that 10 years, for example, might be sufficient to reap the rewards. Defenders of the system oppose such proposals as disastrous to the cause of innovation and social progress.

In spite of all the criticisms leveled against the patent system, the net judgment of investigators appears to be that the system substantially fulfills the purposes

---

[2] *The Role of Patents in Research–Part II* (Washington, D.C.: 1962), pp. 198–200. (Proceedings of a symposium sponsored by the National Academy of Sciences, National Research Council.)

for which it was intended: the stimulation of invention and the dissemination of knowledge.

In a sincere effort to get to the heart of the subject, the National Academy of Sciences (NAS) in 1960 appointed a distinguished committee to examine the role of patents in research. The committee held a symposium to which they invited some of the people best informed on patent matters in the United States; the purpose of the symposium was to give full play to all sides of the question.[3] The committee's conclusions were substantially as follows:

1. The patent system does stimulate research and development.
2. The sheltered competitive position offered by patents is especially beneficial to small companies; it diminishes the advantages possessed by large companies because of their size.
3. One of the greatest benefits of the patent system is its encouragement of the publication and dissemination of information. Knowing that it is protected by a patent application on file, a company is thereby willing to release information concerning its development. The alternative would be industrial secrecy, which would be deleterious to progress.

■ 13.2 INVENTION

The mere conception of an idea is insufficient to constitute invention; the invention must be "reduced to practice." Ordinarily, this means that an actual working model of the invention has been constructed and has functioned successfully. (Very seldom does the Patent Office require that an actual working model be sent to Washington to support the application.) However, the simple act of filing an application that discloses an operable, patentable structure is called *constructive reduction to practice* and is usually accepted by the Patent Office. The patentee should realize that if such a patent later becomes involved in litigation, it will be vulnerable to attack; a patent based upon actual working models is much stronger.

As was mentioned before, the term *invention* is difficult to define, and this very difficulty has led to strife between the U.S. Supreme Court and the Patent Office. Dictionary definitions of the word are not very helpful; one can never be certain an invention is truly patentable until the patent has been judged valid by the Supreme Court. Nevertheless, some general rules for patentability are used by the Patent Office. They are listed below. (It should be borne in mind that such a brief treatment cannot reveal all the complexities of patent matters. The purpose here is to impart an awareness to the engineer that the patent system is like an iceberg, with 90 percent of its substance hidden beneath the surface.)

---

[3] The committee members quickly discovered that they were investigating not the effects of the patent system on *research*, but its effects on *development* (in other words, on engineering). In fact, they decided that the patent system and research are scarcely relevant.

✳ Milestones
The Xerox Story

The story of the dramatic growth of Xerox from a small specialty company to a multibillion-dollar corporation is a favorite of financial investors. Also, it is considered good sport to make fun of companies that missed a chance to get in on a good thing in the early years—giants such as IBM, Lockheed, Kodak, and others—and thus failed to reap a bonanza.

But the true story is one of frustration, risk, faith, and endless hard work. The story began in 1938, when Chester Carlson first demonstrated in a crude way that the process later to be called *xerography* could be made to work. Carlson was employed during the day as a patent attorney and did his xerographic work in the evenings at home, in his kitchen. But after he almost set the kitchen afire, his wife suggested he find another location for his experiments, so he moved his equipment into an empty apartment behind a beauty parlor.

The first demonstration involved a laborious process. First, it was necessary to rub a sulfur-coated metal plate with a cloth to give it a static charge. Then the plate, with a glass slide (bearing an inscription of the date) placed over it, was exposed to a bright light. Next the charged metal plate was dusted with powder, and the excess was blown away, leaving powder adhering to the locations on the plate that had *not* been reached by the light, meaning that the powder now reproduced the inscription on the glass slide. Finally, Carlson pressed a piece of paper against the metal plate, and the inscription was transferred to the paper. The basic principle of xerography had been demonstrated. But the results were blurry, and the process was tedious. The world yawned.

Carlson did not stumble on xerography by luck. He had a college degree in physics from Cal Tech. Because of this background, he was able to base his invention on the fundamental principles of electrostatics. He patented his device, but could not interest a company in picking up the patent and developing it into a commercial product. In fact, it was not until 22 years had passed that Carlson's ideas became fully embodied in the Model 914 copier that took the market by storm. But the intervening years had not been spent in idleness. They had been filled with hard work, endless disappointments, and the expenditure of millions of dollars.

To be patentable, an invention must be new. In addition, several rules of invention are commonly applied:

1. A device is not an invention if it is obvious to one "skilled in the art." Naturally, a big source of contention is the fact that something that appears obvious in retrospect may not have been so obvious in advance. Some of the best inventions appear to be the simplest with the improved vision afforded by hindsight.
2. A combination of old elements is not an invention if no new result is achieved; such a combination is called an *aggregation*. However, if there is a new and unexpected result because of the combination, then the combination constitutes a patentable invention. The vast majority of patents that are issued are actually combinations of old elements.

The organization that carried out this part of the process was Haloid, a small company based in Rochester, New York, that specialized in photographic papers. Haloid changed its name in 1958 to Haloid-Xerox and later to Xerox. Before Haloid came into the picture, Batelle Memorial Institute had picked up Carlson's patents and had begun further research on the process. This was in 1944, six years after Carlson's first demonstration. A year later, a magazine article describing the work caught the eye of John Dessauer, who headed the research and engineering division of Haloid. Dessauer was regularly scrutinizing every publication he could find, searching for ideas for new products. He became excited about the possibilities of "electrophotography," as it was then called, and convinced his company to take it on. An agreement with Batelle was not reached for another year, but ultimately Haloid gained exclusive rights to xerography, with Batelle acquiring 50,000 shares of Haloid stock. These shares multiplied in value in later years and were instrumental in transforming Batelle into an enormous international research firm.

But from 1946 to 1960 Haloid lived on a precipice, forced to find ever more resources to pour into a device that seemed determined not to be turned into a reliable, saleable product.

Much basic research ground had to be plowed before Carlson's blurred image gradually took on the superb quality we take for granted today. A great deal of clever engineering was necessary to arrive at a compact machine that made copies automatically instead of by the original laborious, step-by-step manual method. Millions had to be poured into factory construction, and more millions into the creation of a marketing staff. In 1960, Model 914 was finally announced, and Xerox commenced its rise into the financial stratosphere. It was fortunate that Xerox had meticulously taken out patents on all the improvements it had made over the years, because in 1957, three years before the 914 was released, Carlson's original patent expired.

Carlson, who had lived most of his life financially hard pressed, became a multimillionaire in the 1960s. It is claimed that his total earnings from his invention amounted to more than $150 million, two-thirds of which he gave away to a variety of worthy causes, including his alma mater, Cal Tech.

Reference
J. H. Dessauer, *My Years with Xerox—The Billions Nobody Wanted* (New York: Manor Books, 1975).

3. A mere substitution of obvious equivalents or a substitution of materials is not an invention unless some surprising result is thereby obtained. Also, a mere change in size does not constitute invention unless there is a new result. A famous example is Edison's electric light. Others had built operable incandescent lamps before Edison, and it was claimed that the only difference in his lamp was a reduction in the size of the carbon filament. However, Edison's patent was upheld on the basis that it was this reduction in size that caused his lamp to be practical, all previous models having burned for only short periods of time before going out. (See boxed item in Chapter 12.)

In order to establish, in contested cases, that an invention has met all the legal requirements to be patentable, it is important to keep complete, accurate records. For a variety of reasons, it may later be necessary to establish in court the earliest

Figure 13.3
Chester Carlson, the inventor of the xerographic process, shown with an early model of his machine as developed by Battelle Memorial Institute. (Permission, Battelle Memorial Institute)

date upon which conception of an idea took place. Further, it may be necessary to prove that the engineer's company exercised diligence in pursuing an idea. A patent may easily be lost unless diligence can be proved. In such cases, the engineer's notebook almost invariably constitutes the legal evidence that must be produced in court.

In addition to the date of conception, it may be necessary to prove the date on which reduction to practice took place. Therefore, continuing records of the work must be maintained, especially photographic evidence of model construction and operation, properly signed and witnessed. Such records will also be useful in proving diligence. If any substantial gaps appear in the record, they may be sufficient to constitute abandonment, and the patent may go to a more diligent person, even with a later conception date.

Some additional items must be borne in mind by patent applicants:

1. The true inventor(s) must be identified. Affixing any person's name (such as the inventor's superior, for example) who is not legally and truly the inventor or a co-inventor is illegal.
2. The invention cannot have been patented or publicly described before.
3. If the invention is placed in public use, offered for sale, or described in a printed publication, the inventor must file an application within one year or he or she will forfeit any rights to the patent.
4. Laws of nature, systems of logic, methods of doing business, mental processes, theories, and systems of mathematics are not patentable.
5. There are three basic kinds of patents:
    a. *Utility patents*, which may be granted on processes, machines, manufactured articles, or compositions of matter and are good for 17 years.
    b. *Plant patents*, which protect the developers of new strains of plants, such as special kinds of roses.
    c. *Design patents*, which cover external appearance features and may be taken out for $3\frac{1}{2}$, 7, or 12 years at the applicant's option, the only difference being the size of the fee.

6. Patents are issued to individuals, not to corporations. However, the inventor may *assign* a patent to a corporation. In many cases, the assignment is simultaneous with the filing of the application.

■ 13.3 ANATOMY OF A PATENT

Patents are very imposing-looking documents. Many run to scores of pages, some to hundreds. Characteristically, they have page upon page of complicated drawings followed by pages of even more complicated text and, finally, at the end, the most complicated part of all: the claims. There is a natural tendency to assume that the entire structure disclosed is covered by the patent. Such is seldom the case, however; the only part of the patent that has any legal force is that specified in the claims.

First come the *illustrations*, with every part identified by a key numeral that is referred to in the descriptive material. Shading, crosshatching, and the use of line weights are all rigidly prescribed.

The *specification* is a description of the device, complete in every detail so that a competent worker could construct an operable model from this part of the document. Generally, the beginning of the specification presents a statement of the objectives plus a brief résumé showing how the invention achieves those objectives and how it is an improvement over the prior art (previously issued patents). Much "art" is taught by the description, and from the viewpoint of the dissemination of knowledge, this is the most important part of the patent.

The *claims* form the only part of the patent that gives any legal protection to the inventor. Even though a patentable structure may be described in the specification, it can be seizèd upon by competing companies unless it has been described by at least one of the claims.

The claims are usually very upsetting to the newcomer to patents because, often, they all sound very much alike, extending on and on into tedium. There are excellent legal reasons for writing them this way, however, and the future value of the patent depends on the skill, knowledge, and ingenuity of the claims-writer. Even for a basic invention, much of the potential protection can be lost if the claims are poorly drawn.

Generally speaking, the first claim describes the invention in the broadest possible way, and the succeeding ones become progressively more specific. Bear in mind that the more specific a claim is, the narrower it is and the greater the probability that someone will be able to get around it.

Another factor in favor of multiple claims is as follows: If a patent has only one claim and if that claim is ever invalidated, the whole patent is gone. But if there are many claims, graded in their degrees of specificity, it is possible that adverse judicial action may not invalidate all of them. In this way, some of the coverage will still be intact. Many patents have large numbers of claims (frequently as many as 40 or 50). The all-time record holder was issued in 1915 and had 797 claims. At the time it had been pending for 26 years.

Even with careful reading of a patent, it is virtually impossible to tell, with any precision, just what coverage it offers. It is necessary to review the prior art to discover the exact significance of the claims. It is also necessary to obtain a copy of the "file

Figure 13.4
An artificial hip joint, shown here, must have an extremely accurate form and an ultra-high-quality surface finish, to ensure a safe and pain-free seating. The joint is being measured on an airborne precision rotary table that has an accuracy of 0.05 micron. (Courtesy of FAG Kugelfischer)

wrapper" of the patent, which contains the complete record of all the transactions between the inventor's attorney and the Patent Office. File wrappers are open to public inspection and are very revealing documents. They include all the amendments that had to be made to the original claims before any of the claims were finally allowed. The limitations on the issued claims thus become apparent. It is often surprising to discover how exceedingly narrow the claims of an otherwise impressive-appearing patent may actually be. As a general rule, the longer a claim is, the narrower it is likely to be.

Exercises

13.1 Do you believe that innovation and creativity would fail to flourish if there were no patent system? Prepare an analysis and justification for your views, whether pro or con.

13.2 The author of this text once had the responsibility in a company for receiving and reviewing ideas suggested by persons outside the company. The company manufactured small office computers, among other things. One suggestion received from an outsider was that "the computations should be done by nuclear energy." That was it; no other details were forthcoming. Discuss this idea in the context of patentability.

13.3 Scan some issues of the *Official Gazette of the United States Patent and Trademark Office* and see if you can find any examples of patents in which the usefulness or novelty seems to be slight.

13.4 Through careful examination of some issues of the *Official Gazette*, see if you can get ideas regarding some useful new products. List your ideas, showing for each one how you hope it will avoid infringement of any of the patents you found in the *Gazette*.

13.5 The committee that investigated the role of patents in research decided that the patent system and research are scarcely relevant (see page 349). Discuss the meaning of this statement and why it is or is not true.

13.6 Meet with a professional engineer, and find out how patents and patentable ideas are handled by the engineer's employer. Is the engineer required to keep a notebook? Is there any special reward or recognition in the engineer's firm for issued patents? Does the engineer you selected have any issued patents? Does his or her job call for the creation of potentially patentable ideas, or do the responsibilities of the job lead in other directions?

13.7 Assuming you go on to become a professional engineer, you will probably be asked at some time or other to sign an invention-assignment agreement. Discuss your views of such agreements, the rights that you believe should belong to you, and those that should belong to your employer.

Photographs courtesy of Limi-
torque Corporation (top left),
NASA (center right), and International Paper Company (bottom;
photographer: Cheryl Rossum).

14

Statics and Dynamics

Statics and dynamics fall under the general topic of *engineering mechanics*, which is the science dealing with the action of forces on bodies. *Statics* is concerned with bodies at rest; *dynamics* is concerned with bodies in motion. Both statics and dynamics have to do with forces that are external to the bodies, and the bodies are considered to be rigid. (On the other hand, when we take up the topic of *mechanics of materials* in the next chapter, we will deal with internal forces, and the bodies will be considered to be flexible.) First, in order to handle later problems, we must take up the fundamental topics of forces, vectors, and moments.

■ 14.1 FORCES

Most of us have an instinctive feel for the concept of a *force*. In Chapter 9, "SI and Other Unit Systems," we spoke a great deal about forces and discussed the difference between mass and force. Now we need to point out a difference between the quantities that are considered to be scalars and those that are considered to be vectors. Quantities such as mass, length, and area can be described simply by numbers that represent magnitudes; these quantities are called *scalars*. But quantities such as force, velocity, and acceleration not only have magnitude, they also have direction; they are called *vectors*. For example, the direction of the force we call "weight" is toward the center of the Earth. As another example, when we push on a car, we must specify the direction of our push; it makes a great deal of difference whether we push upward, downward, or sideways.

Figure 14.1
A Pratt and Whitney F-100 military jet engine installed for testing in an altitude facility at Lewis Research Center. (Courtesy of Lewis Research Center, National Aeronautics and Space Administration)

Figure 14.2 shows a couple of examples of forces represented by vectors. On the left of Figure 14.2(*a*), a cart is shown being pulled by a man; on the right is shown the force exerted by the man, in the form of an arrow with magnitude F and having the direction of the arrow. On the left of Figure 14.2(*b*), a weight W is suspended from two cables fastened on opposing walls, and on the right the weight is represented by a vector directed downward with magnitude W; the forces in the cables are represented by vectors with magnitudes F_1 and F_2 and having the directions of the arrows. (In what we do subsequently, it will often prove useful to draw vectors with lengths that are proportional to their magnitudes.)

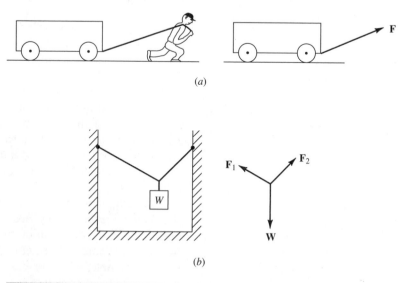

Figure 14.2
Examples of representation of forces by use of vectors.

■ 14.2 VECTORS

In this book, whenever we show a vector we will employ boldface, as in **A**. If one writes on paper or on a blackboard, an arrow may be placed over the top, as in $\vec{A}$. Some people prefer to underline the character representing the vector, as in A̲. In this text, if we are referring to the *magnitude* of the vector **A** we will use an italic character, as in *A*.

When we add two scalars, we simply take the sum of their magnitudes, but when we add two vectors, we must do it in accordance with the rule of vector addition. This rule can be stated most simply in the form of the *parallelogram rule*, as demonstrated in Figure 14.3. In Figure 14.3(*a*), if we add two vectors **A** and **B**, we get a resultant **R**. **A** and **B** can be thought of as two sides of a parallelogram, and **R** the diagonal. (We have drawn **A** and **B** as solid arrows; their resultant **R** is shown with a dashed line.) A portion of the parallelogram is redrawn in Figure 14.3(*b*) in the form of a triangle. If we apply the *law of cosines* (see reference formulas on inside back cover) to this triangle, we obtain

$$R^2 = A^2 + B^2 - 2AB \cos(\pi - \theta) \tag{14.1}$$

From trigonometry, we know that $\cos(\pi - \theta) = -\cos\theta$. Therefore, we can write Equation (14.1) as

$$R^2 = A^2 + B^2 + 2AB \cos\theta \tag{14.2}$$

It is also possible to go in the opposite direction, as shown in Figure 14.4. Here we resolve vector **R** into its components **A** and **B**. (In this case, the original vector **R** is shown as a solid arrow, and it is the components **A** and **B** that are shown dashed.)

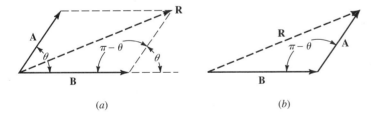

(*a*)   (*b*)

Figure 14.3
The parallelogram rule of vector addition.

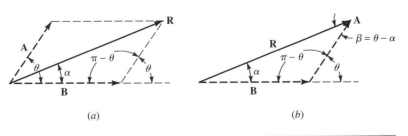

(*a*)   (*b*)

Figure 14.4
Resolution of vector **R** into two components, **A** and **B**.

To do this, we must know the directions we wish the components to have, which are defined by angles α and θ. Again, vectors **R**, **A**, and **B** must obey the parallelogram rule, as shown. Proceeding as in the previous example, we redraw a portion of the parallelogram as shown in Figure 14.4(*b*). Angle α has been specified as the direction we wish one of the components to have, and the angle $(\pi - \theta)$ is the direction of the other. We can solve for two of the sides of this triangle, provided we know at least one of the sides and all three of the angles, by using the *law of sines* (see inside back cover). We know **R** and we know two of the angles, α and $(\pi - \theta)$. We have designated the third angle as β, and we can find a value for it in terms of α and θ by remembering that the three angles of a triangle must add up to π. Thus, we can write

$$\alpha + (\pi - \theta) + \beta = \pi \tag{14.3}$$

and solve this equation for β, getting

$$\beta = \theta - \alpha \tag{14.4}$$

Finally, we can apply the law of sines to the triangle in Figure 14.4(*b*):

$$\frac{R}{\sin (\pi - \theta)} = \frac{A}{\sin \alpha} = \frac{B}{\sin (\theta - \alpha)} \tag{14.5}$$

Solving for the magnitudes A and B, we get

$$A = R \frac{\sin \alpha}{\sin (\pi - \theta)} \tag{14.6}$$

$$B = R \frac{\sin (\theta - \alpha)}{\sin (\pi - \theta)} \tag{14.7}$$

There is another way of representing vectors that is often useful. As we just saw, if we can add two vectors and get a resultant, we can also do the reverse and resolve a vector into its components. An especially convenient pair of component vectors are those directed along the x and y axes. For example, Figure 14.6(*a*) shows a vector **A** that makes an angle α with the x axis. We can break **A** into two vectors, one with magnitude A_x and directed along the x axis, and the other with magnitude A_y and

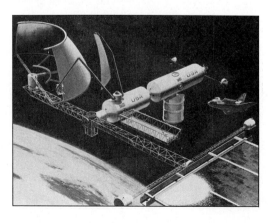

Figure 14.5
A proposed space station. Among other things, a space station would require a great deal of structural engineering. (Courtesy of Jet Propulsion Laboratory)

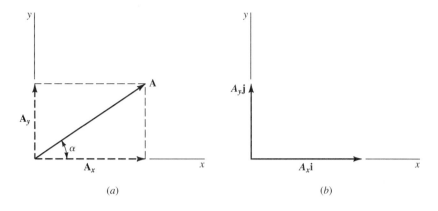

Figure 14.6
Rectangular components.

directed along the y axis. Since the vectors $\mathbf{A_x}$, $\mathbf{A_y}$, and $\mathbf{A}$ all bear the proper relationships to each other, which is required by the parallelogram rule (here, of course, the parallelogram is actually a rectangle), we see that $\mathbf{A}$ can also be considered the resultant of $\mathbf{A_x} + \mathbf{A_y}$. Because of their rectangular relationship, we call $\mathbf{A_x}$ and $\mathbf{A_y}$ the *rectangular components* of $\mathbf{A}$. From trigonometry, we can calculate their magnitudes as

$$A_x = A \cos \alpha \quad \text{and} \quad A_y = A \sin \alpha \tag{14.8}$$

The directions of the x and y axes are arbitrary and should be chosen to make the problem most tractable. Some of the examples to be given later will clarify this point. With practice, the use of rectangular coordinates becomes almost automatic.

We can represent the vector $\mathbf{A}$ in an especially convenient form through the use of unit vectors. A *unit vector* has a magnitude of unity and is directed along one of the three coordinate axes x, y, and z. The symbol $\mathbf{i}$ represents the unit vector along the x axis, the symbol $\mathbf{j}$ represents the unit vector along the y axis, and the symbol $\mathbf{k}$ represents the unit vector along the z axis (see Figure 14.7). In longhand, the symbols $\hat{\imath}$, $\hat{\jmath}$, and $\hat{k}$ would be used. (In this text, we are considering only two-dimensional cases,

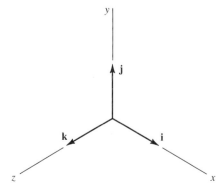

Figure 14.7
The three unit vectors, **i**, **j**, and **k**.

so the symbol **k** will not be needed.) Thus, we can write

$$\mathbf{A_x} = A_x\mathbf{i} \tag{14.9}$$

and

$$\mathbf{A_y} = A_y\mathbf{j} \tag{14.10}$$

Because of the relationship shown in Figure 14.6(a), we can write

$$\mathbf{A} = \mathbf{A_x} + \mathbf{A_y} \tag{14.11}$$

or, equivalently, through use of Equations (14.9) and (14.10), we can write

$$\mathbf{A} = A_x\mathbf{i} + A_y\mathbf{j} \tag{14.12}$$

Equation (14.12) means that we can replace any vector with its rectangular components. This has been done in Figure 14.6(b): **A** has been replaced by its rectangular components, $A_x\mathbf{i}$ and $A_y\mathbf{j}$. These two can now be treated independently as vectors in their own right, if we wish.

As an example, suppose we have a vector **A**, with a magnitude of five units, directed at an angle of 30° with the x axis, as shown in Figure 14.8. By virtue of Equations (14.8) we know that the x component is (5 units)(cos 30°) = 4.33 units, and the y component is (5 units)(sin 30°) = 2.50 units. Using the unit vectors, we can write the x component as 4.33**i** and the y component as 2.50**j**. In Figure 14.8(b), **A** has been replaced by its rectangular components.

We can, of course, simply write vector **A** as

$$\mathbf{A} = 4.33\mathbf{i} + 2.50\mathbf{j}$$

and we will often find this useful.

In Figure 14.9(a), note that a vector with a direction like that shown for **B** would be written

$$\mathbf{B} = -2.50\mathbf{i} + 4.33\mathbf{j}$$

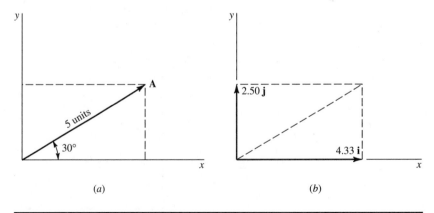

(a)   (b)

Figure 14.8

Vector **A** has been replaced by its rectangular components. We can write
$\mathbf{A} = 4.33\mathbf{i} + 2.50\mathbf{j}$.

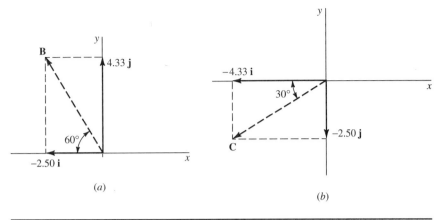

Figure 14.9
Representation of vectors **B** and **C** by use of unit vectors.

and in Figure 14.9(*b*), a vector with a direction like that shown for **C** would be written

$$C = -4.33i - 2.50j$$

If we wished to find the resultant of **B** and **C**, we would add them, collecting all the coefficients of **i** together and all the coefficients of **j** together, thus:

$$B + C = (-2.50i + 4.33j) + (-4.33i - 2.50j)$$
$$= (-2.50 - 4.33)i + (4.33 - 2.50)j$$
$$= -6.83i + 1.83j$$

In like manner, we could combine any number of vectors into a single resultant.

Exercises

14.2.1 What single force can replace the 8-N and 10-N forces in the figure? In other words, what is the *resultant* of these two forces? At what angle does it act?

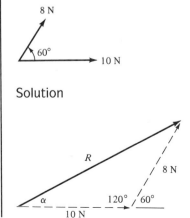

Solution

Law of cosines:

$$R^2 = A^2 + B^2 + 2AB \cos \theta$$
$$= (10 \text{ N})^2 + (8 \text{ N})^2 + 2 \cdot (10 \text{ N})(8 \text{ N}) \cdot \cos 60°$$
$$= (100 + 64 + 80) \text{ N}^2$$
$$= 244 \text{ N}^2$$
$$R = \mathbf{15.6 \text{ N}} \qquad\qquad\qquad \text{Answer}$$

Law of sines:

$$\frac{R}{\sin 120°} = \frac{8 \text{ N}}{\sin \alpha}$$

$$\frac{15.6 \text{ N}}{0.866} = \frac{8 \text{ N}}{\sin \alpha}$$

$$\sin \alpha = \frac{8 \text{ N}}{15.6 \text{ N}} \cdot (0.866)$$

$$= 0.444$$

$$\alpha = \mathbf{26.4°} \qquad\qquad\qquad \text{Answer}$$

14.2.2 In the figure of Exercise 14.2.1, (a) find the vertical and horizontal components of the 8-N force. (b) Then find the vertical and horizontal forces that can replace the 8-N and 10-N forces. (c) Finally, combine these components to find their resultant.

Solution

(a) $V = 8 \text{ N} \cdot \sin 60°$

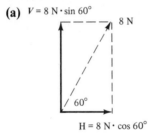

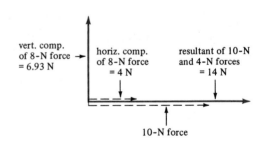

$$\text{vert. component} = 8 \text{ N} \cdot \sin 60°$$
$$= 8 \text{ N} \cdot (0.866)$$
$$= 6.93 \text{ N} = \mathbf{6.93j \text{ N}} \qquad\qquad \text{Answer}$$

$$\text{horiz. component} = 8 \text{ N} \cdot \cos 60°$$
$$= 8 \text{ N} \cdot (0.500)$$
$$= 4 \text{ N} = \mathbf{4i \text{ N}} \qquad\qquad \text{Answer}$$

(b) $8\text{-N vector} = (4\mathbf{i} + 6.93\mathbf{j}) \text{ N}$

$10\text{-N vector} = 10\mathbf{i} \text{ N}$

vertical forces = **6.93j** N   Answer

horizontal forces = (4i + 10i) N

= **14i** N   Answer

(c)

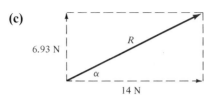

6.93 N

R

α

14 N

$$R^2 = (6.93 \text{ N})^2 + (14 \text{ N})^2$$

$$= 244.02 \text{ N}^2$$

$$R = 15.6 \text{ N}$$

$$\sin \alpha = \frac{6.93 \text{ N}}{15.6 \text{ N}} = 0.444$$

$$\alpha = 26.4°$$

This result can be obtained more directly by simply adding the vectors. Calling the 8-N vector **A** and the 10-N vector **B**, we can write

A = (4i + 6.93j) N

B = 10i N

R = **A** + **B**

= (4i + 6.93j) N + (10i) N

= (4i + 10i) N + 6.93j N

= **(14i + 6.93j)** N   Answer

We will usually keep the answer in this form, but if we want the magnitude and direction of the resultant **R**, we can find them as we did above.

14.2.3 Express each of the following vectors in the form of its rectangular coordinates, using unit vectors.

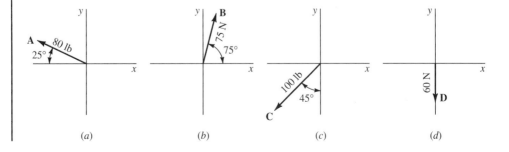

A 80 lb
25°
(a)

B
75 N
75°
(b)

100 lb
45°
C
(c)

60 N
D
(d)

Solution

(a)

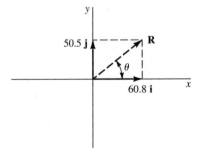

$$\mathbf{A} = (-80 \cdot \cos 25° \, \mathbf{i} + 80 \cdot \sin 25° \, \mathbf{j}) \, \text{lb}$$

$$= (-72.5\mathbf{i} + 33.8\mathbf{j}) \, \text{lb}$$ Answer

14.2.4 Add the following vectors. In each case, find the resultant in the form of its rectangular coordinates, using unit vectors. Then find the magnitude of the resultant, and specify its angle with the x axis.

(a) $\mathbf{A} = (48.2\mathbf{i} - 22.0\mathbf{j}) \, \text{lb}$
  $\mathbf{B} = (12.6\mathbf{i} + 72.5\mathbf{j}) \, \text{lb}$

Solution

$$\mathbf{A} + \mathbf{B} = [(48.2 + 12.6)\mathbf{i} + (-22.0 + 72.5)\mathbf{j}] \, \text{lb}$$

$$= (60.8\mathbf{i} + 50.5\mathbf{j}) \, \text{lb}$$ Answer

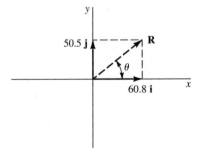

$$R = \sqrt{(60.8)^2 + (50.5)^2}$$

$$= 79.0 \, \text{lb}$$ Answer

$$\theta = \tan^{-1} \frac{50.5}{60.8}$$

$$= 39.7°$$ Answer

(b) $\mathbf{C} = (-32\mathbf{i} - 46\mathbf{j}) \, \text{N}$
  $\mathbf{D} = (46\mathbf{i} + 32\mathbf{j}) \, \text{N}$

(c) $\mathbf{R}_1 = (40\mathbf{i} + 40\mathbf{j}) \, \text{lb}$
  $\mathbf{R}_2 = (-40\mathbf{i} + 40\mathbf{j}) \, \text{lb}$
  $\mathbf{R}_3 = -80\mathbf{j} \, \text{lb}$

14.2.5 Find the components of the 6-N force along the x and y axes.

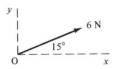

14.2.6 Find the resultant of the 3-N and 6-N forces (a) by law of cosines and (b) by using rectangular coordinates. Do not forget to find the angle with the 6-N force at which the resultant acts.

14.2.7 Find the components of the 10-N force in the $O\text{-}A$ and $O\text{-}B$ directions.

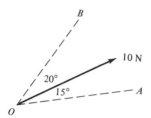

14.2.8 Find the components of the 13-N force in the $O\text{-}A$ and $O\text{-}B$ directions by use of law of sines. Check your answer by computing the rectangular components of the force. (Use $O\text{-}A$ and $O\text{-}B$ as your x and y axes.)

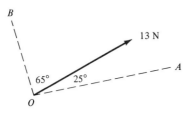

14.2.9 Find the x and y components belonging to the resultant of the 35-lb and 60-lb forces.

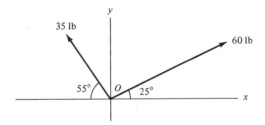

14.2.10 The force of 100 N acts vertically on the bracket, as shown in the figure. Find the rectangular components of the force in the x and y directions.

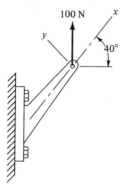

14.2.11 Instead of acting vertically, as in Exercise 14.2.10, the 100-N force now acts at the angle shown below. Find the components of the force in the direction of the bracket arm and parallel to the wall. (The former force would produce what is called *direct tension* on the arm, and the latter force would produce what is called the *shear force* along the wall.)

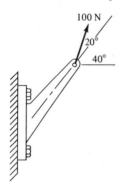

14.2.12 Find the x and y components of the 80-lb force in the figure.

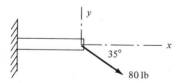

14.2.13 Find the components of the weight of the force in the figure that are perpendicular to the inclined plane, and parallel to it.

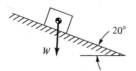

14.2.14 A boom and a cable support a weight, as shown in the figure. The tension in the cable is 100 N. Find the components of the cable force that are perpendicular to the boom, and parallel to it. Does the component along the boom tend to stretch the boom or compress it?

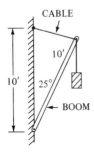

14.2.15 A group of forces $\mathbf{F}_1$, $\mathbf{F}_2$, and $\mathbf{F}_3$ are acting on a common point as shown. Find their resultant $\mathbf{R}$. Express $\mathbf{R}$ in terms of its unit vectors. Then find its magnitude and its direction relative to the x axis.

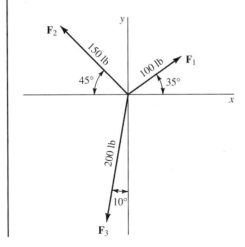

■ 14.3 MOMENTS

The subject of *moments* is very important in engineering mechanics because moments occur everywhere in structural systems. But, to lead up to our discussion of moments, we should first mention two special kinds of cases involving forces.

In the first special case, it should be recognized that a structure containing a cable or other flexible member can effectively support only a pulling force, or a *tension force*, as it is called. If we try to push on the end of a cable, or apply what is called a *compression force*, the cable merely deflects. The same is true if we push sideways on a cable or try to bend it. So, in any system containing a cable, there can only be a tension force in the cable, and the tension force will be directly aligned with the cable. This knowledge often helps to simplify a problem.

The second special case is *two-force members*. One is shown in Figure 14.10(*a*). A two-force member is assumed to be rigid, so it is capable of supporting a compressive force as well as a tensile force. The member is connected to a stationary structure with "frictionless" pins at the ends. Such *pinned-end members* occur often in real systems. With this kind of member, the forces are applied only at the pins. If any force is applied to the member at a point between the pins, it is no longer a two-force member, but a *beam*.

If a member is a part of a stationary structure, then any force applied to one end of the member must be resisted by an equal and opposite force at the other end. Otherwise, the member would move, and we have said it is stationary. In this statement we introduce the principle of *equilibrium*. It will be stated formally in a later section.

We have been leading up to the important feature of a two-force member: *In a member upon which only two forces are acting, the forces must be equal, opposite, and colinear.* In Figure 14.10(*a*), it can be seen that, if the two forces were not colinear, they would produce an unbalanced counterclockwise twisting force, which would set the member spinning in space. But this cannot be so, because the body is stationary. Thus, in a pin-connected structure, where the forces are all applied at the pin joints and where the joints are assumed frictionless, all of the forces are directed along the members of the structure. [See Figure 14.10(*b*)]. Of course, if any forces are applied on a member *between* the joints, that member is no longer a two-force member, and the rule does not apply. The requirement that the pins be frictionless is also important. If they were not frictionless, they would apply twisting forces to the joints, and it would be as though we had gripped the two ends of the member in our hands and bent it. That would be a different matter altogether, because we would not be dealing with a two-force member under such conditions.

Several times we have referred to "twisting" and "bending." This leads us to the very important concept of a *moment*, or *torque*. When we apply a wrench to a nut,

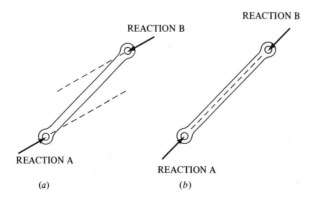

Figure 14.10
A two-force member.

Figure 14.11
The concepts of moment and moment arm are familiar ones from our everyday lives. The *moment* is the twisting force we place on an object, as we are doing here by applying the adjustable end wrench to the nut. We know that we need to apply less force on a long wrench than on a short one to overcome a given resistance. The length of the wrench is the *moment arm*.

as in Figure 14.11, we are applying a moment. The magnitude of a moment is equal to the product of a force and the perpendicular distance from the force to the pivot point about which the twisting action occurs. This perpendicular distance is called the *moment arm*, and the pivot point is called the *moment center*. The units associated with a *force* would typically be pounds, tons, or newtons. But when we are dealing with a *moment*, the units are force times distance—typically pound-inches or newton-meters. Thus, if we have a moment of 12 lb-in. applied to a nut, this could be the result of a 4-lb force acting on the end of a 3-in.-long wrench or a 2-lb force acting on a 6-in. wrench. The resulting moment is the same.

Exercises

14.3.1 Find the moment of the 10-lb force about point *A*.

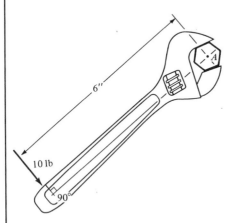

Solution

moment about $A = M_A = 10$ lb $\cdot$ 6 in.

$$= \textbf{60 lb-in.} \quad \textbf{(counterclockwise)} \qquad \text{Answer}$$

14.3.2 Find the moment of the 10-lb force about point A.

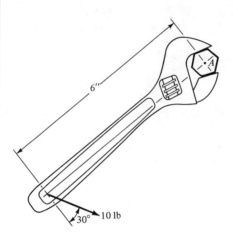

Solution

Note: The moment arm is the *perpendicular* distance to the force. See the next figure.

6 in. $\cdot$ (sin 60°) = 6 in. $\cdot$ (0.866) = 5.196 in.

M_A = 10 lb $\cdot$ 5.196 in.

= **51.96 lb-in.   (counterclockwise)**   Answer

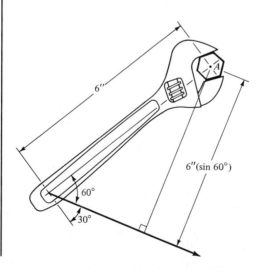

14.3.3 Find the moment(s) the two 6-N forces exert about point A; about point B.

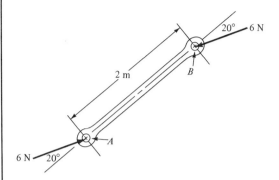

Solution

The force at A passes directly through A. Therefore, it exerts no moment on A. The force at B has a perpendicular moment arm with respect to A of magnitude $2 \text{ m} \cdot \sin 20° = 2 \text{ m} \cdot (0.342) = 0.68 \text{ m}$.

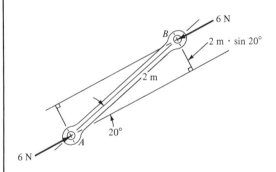

Therefore, the moment about A is

$$M_A = 6 \text{ N} \cdot 0.68 \text{ m}$$
$$= 4.1 \text{ N·m} \quad \textbf{(counterclockwise)} \qquad \text{Answer}$$

A similar line of reasoning prevails for point B. The force acting directly on B exerts no moment at B, but the force acting at A exerts a moment of 4.1 N·m about B.

Two equal, parallel, opposite but noncolinear forces like those shown here are referred to as a *couple*.

14.3.4 A wheelbarrow carries a 100-N load as shown. What force F must be exerted vertically to support the wheelbarrow?

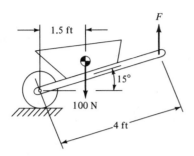

Solution

Two observations should be made: First, the unit systems must be made compatible; second, the *clockwise* moment exerted by the 100-N force must be exactly balanced by the *counterclockwise* moment exerted by the force F.

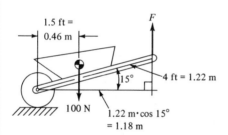

The moment exerted by the 100-N load about the wheelbarrow axle is $100 \text{ N} \cdot 0.46 \text{ m} = 46 \text{ N} \cdot \text{m}$. Therefore, the moment exerted in the opposite direction must be $F \cdot 1.18 \text{ m} = 46 \text{ N} \cdot \text{m}$.

Solving for F, we get

$$F = \frac{46 \text{ N} \cdot \text{m}}{1.18 \text{ m}} = \textbf{39 N}$$ Answer

In the foregoing, we set the two moments equal to each other, and, since they were in opposite directions, the net moment was zero. This is a manifestation of the *principle of equilibrium*, which will be treated in more detail later. In general, moments acting about the same point can be added algebraically.

14.3.5 A brake with length L, sitting at the 30° angle shown, has a force F applied to it in the horizontal direction. Calculate the moment M_A acting about the pivot A.

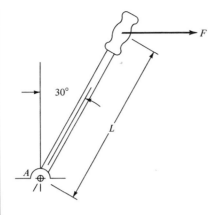

14.3.6 What is the net moment about *A*? About *B*?

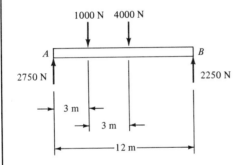

14.3.7 A force of 100 N applied to the nail is necessary to pull it out. What force *F* is necessary on the end of the hammer?

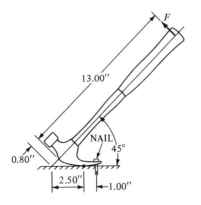

■ 14.4 FREE-BODY DIAGRAMS

One of the most important steps in setting up a mathematical model of a mechanical system is the selection of an appropriate *free-body diagram*. In a machine or structure, we usually want to know the forces and moments that are acting on the components of the structural system, because this information is needed to decide how large and strong the members must be. When we make a free-body diagram, we isolate the portion of the system we are interested in and show all the forces acting on this isolated portion, which is called the *free body*. Setting up a free-body diagram that is both correct and complete is an absolutely vital step in the analysis process. Yet students often tend to not take it seriously enough, and it is the source of many errors.

Disasters
The Kansas City
Hotel Disaster

On July 17, 1981, two overhead walkways collapsed at the Hyatt Regency Hotel in Kansas City, killing 111 people and injuring 188. It was one of the worst hotel disasters in history, and national attention was immediately focused upon the engineering design and construction of the walkways.

The two walkways were crowded with people at the time, and it is estimated that 1000 to 2000 people were jammed into the lobby below to watch and participate in a dance contest. Those on the walkways were dancing also, and whether the dancing produced dynamic loads that the walks could not sustain is unknown. In replacing the walks later, however, a design "live load" of 200 lb per square foot (psf) was used instead of the 100 psf employed in the design of the original walkways.

There were other complicating factors. One had to do with the basic design, wherein all of the weight of the walkways came to bear directly on some weld seams where the structural members had been welded together. The structure did fail at precisely those locations, but whether this was the crucial factor cannot be determined. Another factor is the unconfirmed report that some of the essential load-bearing washers may have been missing.

But there is one exceedingly important factor in this case from the viewpoint of the student of free-body diagrams. It becomes apparent through examination of the accompanying illustration. The original design called for the walkways to be suspended from the ceiling on long rods $1\frac{1}{4}$ inches in diameter, as shown in (*a*). The rods would pass through the fourth-level walkway and continue downward to support the second-level walkway. The loads were transmitted from the walks to the rods by nuts and washers. However, the design was changed later to that shown in (*b*). Instead of being supported on continuous rods, the second-level walkway was now to be supported on separate rods. The failures occurred at the points shown in (*b*). The rods themselves did not break, but the nuts punched through the structural members at their bearing points, leaving the upper rods still attached to the ceiling. The redesign caused the loads at these points to be double those in the original design.

We can best illustrate the process with some examples. In Figure 14.12(a) a boom is shown pivoted at the wall at B and held in place by a cable fastened to the wall at C and to the boom at A. A weight W hangs from the end of the boom. An obvious free-body diagram is to isolate the weight W, as in Figure 14.12(b), making an imaginary cut through the cable that holds it and replacing the part of the cable we cut away with the tensile force T exerted by the cable on the free body we retained. All we can learn from this free-body diagram, however, is that the cable tension T is equal to the weight W, which we knew instinctively.

A more useful free-body diagram is the one shown in Figure 14.12(c). Here the connection at A has been isolated by making imaginary cuts in the two cables and in the boom. The cable forces must necessarily be tensile forces and must coincide with the axes of the cables. We label the downward cable force T and the horizontal one S. The force applied on the imaginary cut end of the boom by the part that was cut away could be either tension or compression, but it must be directed along the

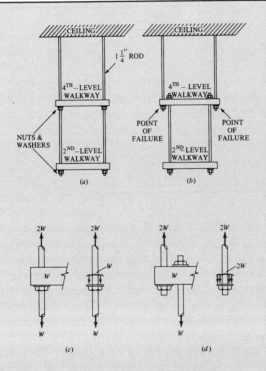

(a)   (b)

(c)   (d)

the load W of the second-level walkway passes directly through the continuous rod, and the nut-and-washer combination has to support only the load of its own walkway, W. In the diagram in (d), the load of the second-level walkway is transmitted through the fourth-level walkway, and this load is added to the load of the fourth-level walkway itself, thus applying a load of 2W to the nut-and-washer combination at the bottom of the upper rod. This is exactly where the structural failure occurred. Whether this doubling of the load actually caused the collapse has not been determined because of the other factors, such as the unanticipated dynamic load, the welding seam that took the load, and the possible omission of some of the load-bearing washers.

A look at a free-body diagram of the failure point shows why the load was doubled in the new design. In the original design, shown in (c),

References

"Column-supported Walk to Replace Hyatt Bridges," *Engineering News-Record*, August 20, 1981, p. 14.

"Hotel Disaster Triggers Probes," *Engineering News-Record*, July 23, 1981, pp. 10–11.

"Hyatt Walkway Design Switched," *Engineering News-Record*, July 30, 1981, pp. 11–13.

"NBS Reproduces Hyatt Walkways," *Engineering News-Record*, October 22, 1981, pp. 12–13.

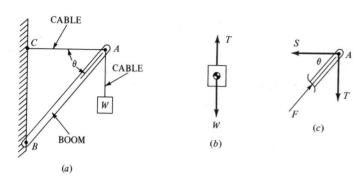

Figure 14.12
Free-body diagrams.

axis of the boom, because this is a two-force member. If you are troubled by the fact that there are *three* forces, *F*, *S*, and *T*, acting on this member, remember that any two forces can be combined into a single resultant. If *S* and *T* are combined into a single resultant, it will be found that the resultant is aligned with the boom's axis, directly balancing the force *F*. Having drawn the free-body diagram of Figure 14.12(*c*), typically we would then replace it with the equivalent force diagram of Figure 14.12(*d*) before proceeding further with the analysis.

A block sitting on an inclined surface, as in Figure 14.14(*a*), provides another example. If we take away the inclined surface and replace it with the forces that the surface exerts on the block, we get the diagram in Figure 14.14(*b*). The force exerted by gravity is the weight *W*, directed vertically downward. The force exerted by the inclined surface is shown resolved into two components: *N*, which acts perpendicular to the surface, and F_f, which acts parallel to it. The force *N* is referred to as the *normal force* ("normal" means "perpendicular"), and the force F_f is the *friction force*, which keeps the block from sliding down the plane. We call these forces that act on the block *reactions* because they react to the applied force *W*.

We can make our free-body diagram a little "cleaner" if we shove the gravity force *W* upward and show it pushing down on the block as in Figure 14.14(*c*). This demonstrates an important point, which is that one can slide a force anywhere along its line of action without altering the outcome of the force analysis. This is true provided, of course, that we do not try to do something we know we cannot, such as pushing on the end of a cable.

Whenever we construct a free-body diagram and show a force or reaction acting upon it, exerted by the part of the system that was cut away, we should realize there

Figure 14.13
Ball bearings are not only used for rotary motion. The longitudinal and cross slides of the numerically controlled sheet-metal working center shown here are supported on *linear* ball bearings. (Courtesy of FAG Kugelfischer and Rasking SA)

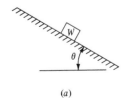

(a)   (b)

(c)

Figure 14.14
Free-body diagram of a block sitting on an inclined plane.

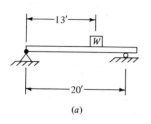

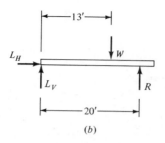

(a)

(b)

Figure 14.15

A simple beam and its free-body diagram.

is an equal and opposite force applied by the free body we keep, acting upon the part we cut away. Thus, if the inclined surface presses upward on the block in Figure 14.14(c) with normal force N, then the block presses down on the inclined surface with the same force N. And, if the inclined surface exerts a frictional force F_f up and to the left on the block, then the block exerts an equal force F_f down and to the right on the inclined surface. This principle is known as Newton's third law: *for every action, there is an equal and opposite reaction.*

An example of a *simple beam* is shown in Figure 14.15(a). The beam is 20' long, and a weight W rests on the beam 13' from one end. The weight of the beam itself is neglected. The left-hand end is fastened with a pinned connection, and the right-hand end rests on a "frictionless" roller. Such an arrangement is often used to permit a beam to expand on a hot day without causing it to buckle. The diagram has been simplified to its essentials, of course, but real-life structures often can be represented by the type of diagram shown here.

A free-body diagram of the beam is shown in Figure 14.15(b). The reaction acting on the left-hand end is represented by its rectangular components L_H and L_V. The weight W is shown acting downward, and we have placed a reaction R, acting upward, at the right-hand roller. We have taken the roller away and replaced it with the force it exerted on the underside of the beam. We know there is no horizontal force acting at the roller because if such a force tried to come into existence, the roller could not resist it, but would just move until the force disappeared. This is an important feature of a frictionless roller: The force exerted by the roller must be at right angles, or normal to the surface.

An interesting thing can be seen by looking at Figure 14.15(b). All the forces that act on the beam are shown, and they all have to balance each other, or else the beam would be moving through space, which it is not. We can instinctively see that the vertically acting forces will somehow work out so they balance each other, but what is going to balance the single horizontal force L_H? The answer is: nothing. Therefore, L_H must be zero, and we have only vertical forces. However, we are getting ahead of ourselves, because the analysis we just made belongs to the subject of *equilibrium*, which we will take up shortly. At this stage of the game, we would go ahead and put both L_H and L_V on the free-body diagram, letting the later step of equilibrium analysis take care of L_H automatically.

Figure 14.16
As a part of the development of a new airplane, a load test may be conducted upon an entire aircraft as a unit, as in the static test fixture shown in the rear of this photograph. (Courtesy of Grumman Aerospace Corporation)

One final example of a free-body diagram, for the case of what is called *simple shear*, is presented in Figure 14.17. We have already discussed two kinds of forces: tension and compression, which act either to elongate or to compress a member. A *shear force* is one that cuts across a member, as a pair of scissors would. In Figure 14.17, we have two straps of material, held by a single bolt, being pulled apart by forces P. Obviously, one force P is exactly balanced by the other. This resembles a two-force member, yet the forces are not aligned exactly. However, the misalignment is small because the straps are relatively thin, so will be ignored in what follows. On the other hand, if the forces became very large, they would begin to warp the bolted connection in the clockwise direction, producing some bending of the plates. For our purposes here, we will assume the bending does not occur.

In Figure 14.17(*b*) an imaginary cut has been made through the bolt that separates the connection into two free-body diagrams. On each diagram, the force exerted by one part of the bolt on the other part is shown by the unknown force F. Of course, since each free-body diagram is a two-force member, we know that $F = P$. Figure 14.17(*b*) re-emphasizes the fact that if we make an imaginary cut in a structural member, the force exerted by the cut-away part on the part we keep is matched by an equal and opposite force exerted by the part we keep on the part we cut away.

A summary of some of the ways in which forces and reactions may be represented in free-body diagrams is provided in Figure 14.18.

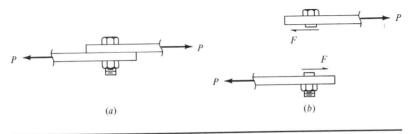

(a)   (b)

Figure 14.17
A case of simple shear, with free-body diagrams.

Representation of forces and reactions in free-body diagrams
(two-dimensional cases only)

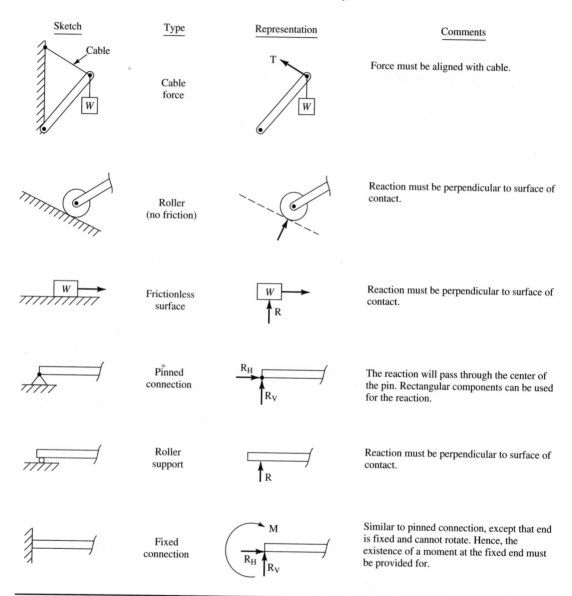

Sketch	Type	Representation	Comments
	Cable force		Force must be aligned with cable.
	Roller (no friction)		Reaction must be perpendicular to surface of contact.
	Frictionless surface		Reaction must be perpendicular to surface of contact.
	Pinned connection		The reaction will pass through the center of the pin. Rectangular components can be used for the reaction.
	Roller support		Reaction must be perpendicular to surface of contact.
	Fixed connection		Similar to pinned connection, except that end is fixed and cannot rotate. Hence, the existence of a moment at the fixed end must be provided for.

Figure 14.18

Summary of the ways in which forces and reactions may be represented in free-body diagrams.

14.4.1 In Figure 14.12(a), imagine that the upper cable is severed at point *C* and that *C* is replaced by the force $\mathbf{T}_1$ that it exerts on the cable. Similarly, imagine that the boom is severed at point *B* and that *B* is replaced by the horizontal and vertical components of the force that *B* exerts on the boom. Call these $\mathbf{B}_H$ and $\mathbf{B}_V$. Finally, imagine that the cable that supports *W* is also severed. Draw the free-body diagram for this case.

Solution

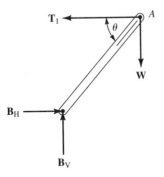

14.4.2 In the diagram of the wheelbarrow in Exercise 14.3.4, imagine that the wheel has been removed and replaced by the horizontal and vertical components of the force exerted by the wheel's axle. Draw the free-body diagram for this case.

14.4.3 In Exercise 14.3.7, imagine that the nail has been removed and replaced by the force it exerts on the hammer, and that the same has been done with the surface on which the hammer rests. (Assume this surface is frictionless.) Draw the free-body diagram.

14.4.4 Draw a free-body diagram for the beam in the following figure. Assume the cable is severed at point *C*.

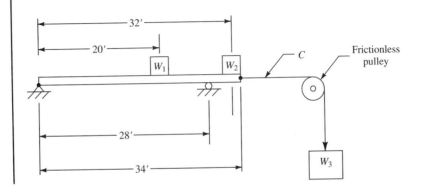

14.4.5 Draw a free-body diagram for the truss in the following figure. Take the entire truss for your free body, severing the cable at point C and removing the supports at A and B. Make your drawing approximately to scale.

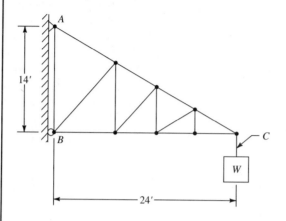

14.4.6 A heavy roller of weight W is held on an inclined plane by means of a cable, as shown. Draw a free-body diagram for the roller.

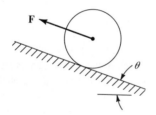

■ 14.5 EQUILIBRIUM

The concept of *equilibrium* is a very powerful one that appears repeatedly in engineering analysis. It simply says that if a body is at rest, then the net sum of all the forces and moments acting on the body must be zero. This is formalized by writing the equations of equilibrium:

$$\sum H = 0 \tag{14.13}$$

$$\sum V = 0 \tag{14.14}$$

$$\sum M = 0 \tag{14.15}$$

meaning that the sums of the horizontal forces, of the vertical forces, and of the moments must all be zero or else the object is not in equilibrium. (The symbol Σ signifies "sum.")

In Equations (14.13), (14.14), and (14.15), we have stated the essentials of the mathematical model we are going to use in several succeeding examples. However, it is important to remember that some of the other essential features of any model we set up are the simplifying assumptions we make, the correct choice of a free-body diagram, and the representation of the applied forces and reactions.

As an example of the use of these equations, let us go back to Figure 14.15(b). Horizontal forces to the right will be taken as positive, and those to the left as negative. Vertical forces upward·will be positive, and the downward ones will be negative. Moments that produce clockwise motion will be positive, and those that produce counterclockwise motion will be negative. When we write the equations of equilibrium, we will provide certain symbols to remind us of the directions we have chosen as positive. If forces to the right are positive, we will use the symbol $\rightarrow$. If upward forces are positive, we will use $\updownarrow$. If clockwise moments are positive, we will use $\curvearrowright$. The directions we choose as positive are entirely arbitrary, but once we pick a direction we must be consistent henceforth.

Another point is worth emphasizing. The direction assigned to the applied force W must be shown to be downward, of course, because it is a known force. However, the reactions L_H, L_V, and R are all unknowns. We can assign them any directions we wish. We could, for example, assign the reverse of the directions we chose. L_H, will turn out to be zero, as we already know, so what we are about to say does not affect it. But if we assigned L_V and R both to be in the downward direction, our solutions for L_V and R would simply turn out to have minus signs attached to them, which means that their true directions would be the reverse of the directions we assigned them. This means that, as a matter of general procedure, we can assign arbitrary directions to any unknown reactions, and the solutions will tell us, by being positive or negative, whether our arbitrarily assigned directions are the correct ones.

If we proceed to use the equations of equilibrium, we will start with $\Sigma H = 0$, putting all the horizontal forces and reactions on the left-hand side of the equals sign, with their correct signs attached, and equating their sum to zero. We get.

$$\Sigma H: \quad +L_H = 0 \quad \rightarrow \tag{14.16}$$

This formally tells us what we already knew: there is no horizontal reaction at the left-hand support, because there are no horizontal forces applied to the free body.

Proceeding to $\Sigma V = 0$, we sum up the vertical forces algebraically and get

$$\Sigma V: \quad +L_V - W + R = 0 \quad \updownarrow \tag{14.17}$$

We cannot do anything with this equation yet because it has two unknowns, L_V and R. Therefore, we set up the moment equation $\Sigma M = 0$, choosing to compute all the moments as if the left-hand support were the center of rotation. This point is called the *moment center* and turns out to be an especially convenient point because the reactions L_H and L_V pass directly through it. This means they both have moment arms of zero, so their moments are zero. The only forces that create moments about this particular moment center are W and R. The former has a moment arm of 13 feet and creates a clockwise moment. The latter has an arm of 20 feet and creates a counterclockwise moment. Therefore, the moment equation $\Sigma M = 0$ is

$$\Sigma M: \quad +13 \cdot W - 20 \cdot R = 0 \quad \curvearrowright \tag{14.18}$$

Equations (14.17) and (14.18) give us two equations in the two unknowns L_V and R, W being a known weight. The solution is

$$L_V = \tfrac{7}{20}W \qquad R = \tfrac{13}{20}W$$

We can substitute these values back into Equation (14.17) to see if it checks:

$$\tfrac{7}{20}W - W + \tfrac{13}{20}W = 0$$

and we can see it does.

Example: Ladder Resting Against a Wall

As a simple example of setting up free-body diagrams and using the equations of equilibrium, let us take the case of a ladder resting against a wall with a man of weight W on it, as in Figure 14.19(a). The questions to be settled are two: (1) Is the ladder resting at a safe angle? (2) If the angle is initially safe, does the situation become unsafe after the man has climbed a certain distance up the ladder?

We conceivably could set up the problem by removing the ladder from the picture and replacing it with the forces F_1 and F_2 exerted by the ladder on the ground and the wall, as has been done in Figure 14.19(b). This is a free-body diagram, but it happens we have not chosen a useful one. The element that has been isolated in Figure 14.19(b) is in fact the wall, the Earth, and maybe the entire universe. If we

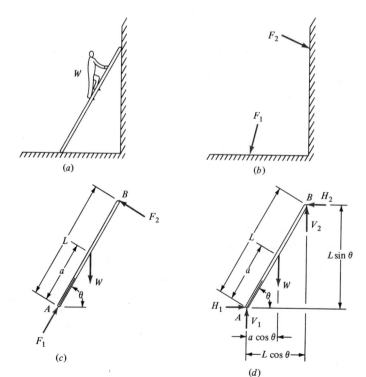

Figure 14.19
A safety analysis for a man climbing a ladder.

instead select the ladder as the element to be isolated, we get a useful free-body diagram, as in Figure 14.19(c).

The forces F_1 and F_2 are unknown in magnitude and also in direction. We could, of course, deal with the angles, as well as the forces, as unknown quantities, but it is more convenient to represent these forces with their rectangular components. Thus, in Figure 14.19(d) the force F_1 has been replaced by V_1 and H_1, which are its vertical and horizontal components, and F_2 has been replaced by V_2 and H_2.

We are now ready to set up our mathematical equations. To do this, we will use some knowledge of friction from physics (after all, it is only friction that holds the ladder in place) and will then use the principle of equilibrium. In the case of friction, physics tells us that the rectangular components of a force acting against a surface, just before slippage begins to occur, are related by the formula

$$F_f = \mu N \tag{14.19}$$

where F_f is the component parallel to the friction surface, N the component perpendicular to the surface, and μ the *coefficient of friction*. Coefficients of friction for most dry surfaces range from 0.2 to 0.5. For lubricated surfaces, μ may be as small as 0.05. F_f will always act in a direction to oppose incipient motion. Thus, in Figure 14.19(d) note that the frictional forces have indeed been selected in their correct directions, for the ladder, when it slips, will move downward and outward. (Note, at the ground, we have assigned the symbol H_1 to the frictional force, whereas at the wall the frictional force is V_2, and H_2 is the perpendicular reaction at the wall.)

For Figure 14.19(d), then, applying Equations (14.13), (14.14), and (14.15), we write

$$\sum H: \qquad\qquad H_1 - H_2 = 0 \quad \leftrightarrow \tag{14.20}$$

$$\sum V: \qquad\qquad V_1 + V_2 - W = 0 \quad \updownarrow \tag{14.21}$$

$$\sum M: \quad W \cdot a \cos\theta - H_2 L \sin\theta - V_2 L \cos\theta = 0 \quad \curvearrowright \tag{14.22}$$

which are the equilibrium equations. The friction equations for the floor and the wall,

Figure 14.20
A chain saw mounted in special testing fixture for measurement of vibration characteristics. (Courtesy of University of California, Davis)

using Equation (14.19), are

$$H_1 = \mu_1 V_1 \tag{14.23}$$

$$V_2 = \mu_2 H_2 \tag{14.24}$$

where μ_1 is the coefficient of friction at the ground and μ_2 is the coefficient of friction at the wall. Forces horizontally to the right have been taken as positive, as have forces vertically upward. Clockwise moments have been taken as positive, and the bottom end of the ladder has been selected as the moment center. In the moment equation, forces H_1 and V_1 drop out because they pass directly through the moment center. The applied force W causes a clockwise moment with a moment arm of $a \cos \theta$. The forces H_2 and V_2 both produce counterclockwise moments, the first with moment arm $L \sin \theta$, and the other with moment arm $L \cos \theta$. (You should verify that the moment arms are correctly derived from the trigonometry of the triangle formed by the ladder and its relationship to the wall and floor.)

If this were a preliminary analysis, we could make an assumption that $\mu_2 = 0$, which would simplify the problem. Taking $\mu_2 = 0$, we immediately get the result from Equation (14.24) that $V_2 = 0$.

Figure 14.21
The engineering science of dynamics often has unusual applications. Here, a student athlete is engaged in a research program employing a mechanical "force plate" that is installed in the floor. By measuring the athlete's dynamic reactions with the floor, correlations can be made with the subject's movements for the purpose of improving athletic performance and preventing injuries. (Courtesy of University of California, Davis)

We now have three unknowns, H_1, H_2, and V_1, in the three Equations [(14.20) through (14.22)], which permits us to solve for them, obtaining the solutions

$$V_1 = W \tag{14.25}$$

$$H_1 = H_2 = \frac{Wa}{L \tan \theta} \tag{14.26}$$

However, we have a fourth equation, Equation (14.23), that places a constraint on the solution. If we substitute Equation (14.23) into Equation (14.26) and use Equation (14.25), we obtain

$$\tan \theta = \frac{a}{L\mu_1} \tag{14.27}$$

Equation (14.27) establishes a relationship between the angle θ and the coefficient of friction μ_1. Note, also, that the weight of the man, W, does not affect the result.

If we assume (conservatively) a coefficient of friction of $\mu_1 = 0.2$, and that the man is halfway up the ladder ($a/L = 0.5$), then θ has the value 68.2°. With the ladder set at that angle, if the man climbs higher than the halfway point, then Equation (14.26) shows that H_2 (which equals H_1) increases in magnitude because a is increasing. But Equation (14.23) shows that H_1, the friction force, cannot exceed $0.2W$. (Remember, $V_1 = W$.) Hence, if H_2 increases, it exceeds H_1; our condition of equilibrium is violated, for we now have an unbalanced force; and the ladder slips.

To be completely safe, the ladder must not slip even when the man has climbed all the way to the top—that is, when $a/L = 1$. Solving Equation (14.27) for this condition gives a value for θ of 78.7°, which is getting uncomfortably close to vertical and introduces a new problem in that the man may inadvertently tip over backward.

The engineer has now gleaned just about everything possible from this mathematical model. It is obvious that something will have to be done about the coefficient of friction at the floor if safety is to be improved. If, for example, the ladder is equipped with rubber feet and used on a rough concrete surface, a coefficient of friction of 0.5

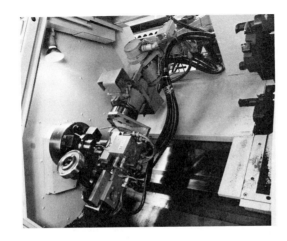

Figure 14.22
Entering a numerically controlled turning center through the rear of the machine, a computer-controlled industrial robot prepares to remove a finished part from the machine's chuck. A second part, shown in the front gripper, will then be loaded into the chuck for machining, and the robot will transfer the finished part to a gauging station. If the part is within tolerances, it will be placed on a delivery pallet, while the robot picks up a new rough part to be machined. (Courtesy of Cincinnati Milacron)

might be assumed. Then, Equation (14.27) would give a value for θ of 63.4°, again assuming that $a/L = 1$. This is certainly a more reasonable angle than 78.7°.

Preliminary Design Example

We will now use the methods we have developed to make a preliminary analysis of the device shown in Figure 14.23. The knowledge gained during such an analysis could then be used to change the design of the device in case any problems become apparent.

Figure 14.23 shows a simple load-carrying device that weighs 500 lb when fully loaded. It is guided in its vertical travel by four rollers running along a 4″ × 6″ wooden post. (Note that a nominal 4″ × 6″ post actually has the dimensions 3.5″ × 5.5″.) When the unit has been raised to its topmost position, it is to be latched in place as shown, by a safety latch. For reasons of economy and avoidance of extra structure, it is planned to pivot the latch hook around a point in the center of the wooden post, which will obviously place a wrenching force on the unit that the rollers must resist in order to keep the unit in place. As a preliminary design question, the engineer wishes to know how much force the rollers will apply to the wooden post. Knowing this, some simple tests can then be made on wood samples to see if the rollers are likely to produce indentations in the wood—a condition to be avoided.

Figure 14.24 is a free-body diagram of the load-carrying device. The first thing we have noted and incorporated into the diagram is that the latch hook can be treated

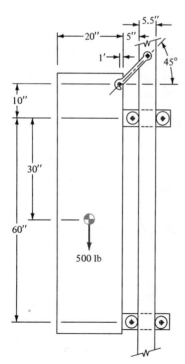

Figure 14.23
An example involving a load-carrying device.

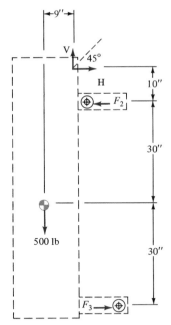

Figure 14.24
Free-body diagram of a load-carrying device.

as a two force member. As a result, the force applied at the point where the latch hooks onto the body must be aligned in the direction of the latch. We can say this because the pivot about which the latch swings probably can be treated as frictionless, at least for a preliminary design. Similarly, the point at which the latch hook engages the body probably also cannot exert much friction. From this, then, we note that the vertical and horizontal components, V and H, of the force applied by the latch must be equal because of the 45° angle. We have assumed arbitrary directions for the forces on the rollers, F_2 and F_3, but note that they must be applied perpendicular to the wooden post because they *are* rollers and can be assumed frictionless (i.e., there cannot be any friction forces parallel to the surface of the post). Only one roller is shown at the top, and one at the bottom, even though there are actually two rollers at each location. This is because the loads will cause one roller of each pair to be pushed against the post while the roller on the opposite side is lifted slightly, so it is no longer in contact with the post. Since the roller is no longer in contact, it can exert no force, so we ignore it in the free-body diagram. Thus, we have established our model and can write the equations of equilibrium:

$$\sum V: \qquad\qquad\qquad V - 500\ \text{lb} = 0 \quad \updownarrow \qquad\qquad (14.28)$$

$$\sum H: \qquad\qquad\qquad H - F_2 + F_3 = 0 \quad \leftrightarrow \qquad\qquad (14.29)$$

$$\sum M: \quad -500\ \text{lb}(9\ \text{in.}) + F_2(10\ \text{in.}) - F_3(70\ \text{in.}) = 0 \quad \curvearrowright \qquad (14.30)$$

In the preceding, the moment center has been chosen to coincide with the point where V and H intersect so as to eliminate these reactions from the moment equation.

Solving these three equations and remembering that $V = H$, we get

$$V = H = 500 \text{ lb}$$

$$F_2 = 508 \text{ lb}$$

$$F_3 = 8 \text{ lb}$$

This is a rather startling discovery, that the force with which the upper roller presses into the wood actually exceeds the weight of the load carrier. Whether such a load would create an undesirable indentation in wood could be determined by a simple experiment. If the results were unsatisfactory, a redesign would be in order, probably involving a better direction of action for the latch. The mathematical model has told us a great deal about our tentative design, even though it contains many simplifying assumptions.

Exercises

14.5.1 A 50-lb lamp is supported by a boom and cable as shown. The boom is connected by a pin joint at point A. Find the tensile force in the cable.

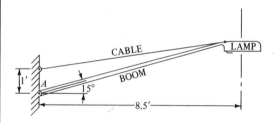

Solution

Assume cable, boom, and vertical force of the lamp coincide in a point, with only a small loss in accuracy.

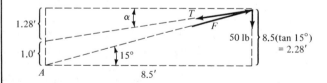

Trigonometry gives angle α:

$$\tan \alpha = \frac{1.28}{8.50} = 0.1506$$

$$\alpha = 8.56°$$

We convert all the force vectors to rectangular components, using unit vectors:

$$\mathbf{F} = F \cos 15° \, \mathbf{i} + F \sin 15° \, \mathbf{j} \tag{14.31}$$

$$\mathbf{T} = -T\cos 8.56° \, \mathbf{i} - T\sin 8.56° \, \mathbf{j} \tag{14.32}$$

$$\text{force vector for weight of lamp} = -50\mathbf{j} \, \text{lb} \tag{14.33}$$

We have only the two unknowns F and T, so we use the first two equations of equilibrium, Equations (14.13) and (14.14). For Equation (14.13), we sum all the horizontal components by adding up all the coefficients belonging to the unit vector $\mathbf{i}$ in Equations (14.31), (14.32), and (14.33). We get

$$F\cos 15° - T\cos 8.56° = 0 \quad \rightarrowtail \tag{14.34}$$

For Equation (14.14), we add up all the coefficients belonging to the unit vector $\mathbf{j}$:

$$F\sin 15° - T\sin 8.56° - 50 \, \text{lb} = 0 \quad \updownarrow \tag{14.35}$$

If we solve Equations (14.34) and (14.35) simultaneously, we get

$$F = \textbf{441 lb} \qquad\qquad\qquad \text{Answer}$$

$$T = \textbf{431 lb} \qquad\qquad\qquad \text{Answer}$$

14.5.2 We have the same 50-lb lamp as in Exercise 14.5.1, but now it is supported by two bolts, 1 ft apart, in a bracket in the base. There is no cable. Assuming the weight of the lamp causes the whole assembly to try to pivot about point A, then what is the force in the upper bolt?

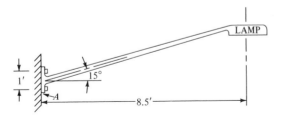

Solution

We assume that point A and the lower bolt coincide. We use point A as the moment center and apply $\Sigma M = 0$:

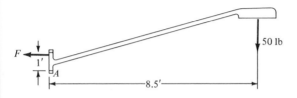

$$-F \cdot 1' + 50 \, \text{lb} \cdot 8.5' = 0 \quad \circlearrowright$$

$$F = \textbf{425 lb} \qquad\qquad\qquad \text{Answer}$$

14.5.3 A beam that weighs 20 N/m is supported as shown and also is subjected to a concentrated load of 100 N. Find the reactions at *A* and *B*.

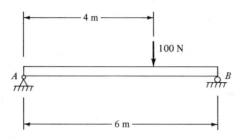

Solution

As far as the reactions are concerned, the weight of the beam can be treated as a concentrated load acting at the center of gravity of the beam. Therefore, our free-body diagram is

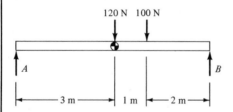

weight of beam = 20 N/m × 6 m = 120 N

$\sum V$: $A + B - 120 \text{ N} - 100 \text{ N} = 0$ $\updownarrow$

$\sum M_A$: $120 \text{ N}(3 \text{ m}) + 100 \text{ N}(4 \text{ m}) - B(6 \text{ m}) = 0$ $\leftrightarrow$

$A = \textbf{93.3 N}$ Answer

$B = \textbf{126.7 N}$ Answer

14.5.4 The beam in the figure overhangs on the right-hand end. Find the reactions at *A* and *B*.

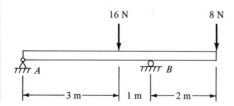

14.5.5 A cable passes over a frictionless pulley as shown, supporting a weight W. Determine the horizontal and vertical reactions at the pulley support shaft A.

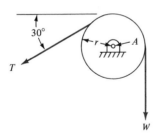

14.5.6 A beam 10 ft long balances on a pivot as shown when a 100-lb weight is suspended from one end. How much does the beam weigh?

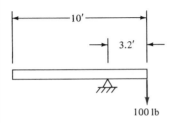

14.5.7 A force T is being exerted on a roller of weight W that is 24″ in diameter. The roller has just encountered a 4″ step. What force T, in terms of W, is necessary in order to just get the roller started over the step?

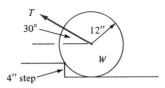

14.5.8 At what angle θ will the block of weight W just start to slide? The coefficient of friction is $\mu = 0.3$.

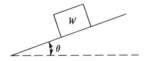

14.5.9 A pickup truck is loaded so that its front-axle load is 1400 lb and its rear-axle load is 1600 lb without trailer. A trailer is added that weighs 3000 lb, with dimensions as shown. The pickup's load limit for the front axle is 1660 lb and for the rear axle is 2080 lb. What are the front and rear axle loads after the trailer is hooked on? Are the limits exceeded?

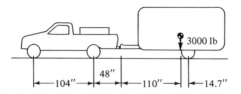

14.5.10 Solve for the force **F** that is required in a cable to keep a heavy roller of weight W from rolling down an inclined plane, as shown.

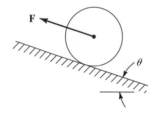

■ 14.6 A SIMPLE BRIDGE TRUSS

At first glance, the bridge truss in Figure 14.25 may look too complicated to be analyzed by the techniques we have discussed so far. But such is not the case, because the three equations of equilibrium are all we need in order to determine the forces in the members. (Of course, once the forces have been determined, the next step will be to figure out how large the members must be in order to withstand those forces. That will be the subject of the next chapter.)

The truss in Figure 14.25 has been simplified to its basic essentials. All the joints are to be treated as if they were connected together with frictionless pin joints. Some bridges are in fact constructed with actual pin joints, and others have joints that are near enough to being pin joints to justify the choice. The joints are never truly frictionless, of course, but such an assumption is close enough to reality to be valid.

A load of $W = 2000$ lb is applied at joint G. The weight of the structure itself is neglected in this example. (In a real bridge, the weight of the structure would be an important element to include in the analysis.) In Figure 14.25(a), note that the support at A is shown fixed by a pinned connection to the Earth, but the one at H is shown supported by a roller. As we saw before, two important results of using the roller type of support are that (1) thermal expansion is allowed for, and (2) there can be no horizontal reaction at such a support.

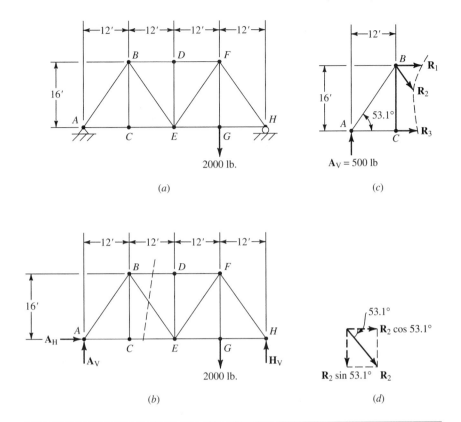

Figure 14.25
A simple truss.

Figure 14.25(b) is a free-body diagram of the whole truss taken as one unit. Horizontal and vertical reactions are shown at joint A and designated as A_H and A_V. The vertical reaction shown at joint H is designated as H_V. If we apply the three equations of equilibrium to this free-body diagram, choosing joint A as the moment center (we can always choose the moment center anywhere we want to), we get

$\sum H$: $\qquad\qquad\qquad\qquad A_H = 0 \quad \rightarrow$ (14.36)

$\sum V$: $\qquad\qquad A_V + H_V - 2000 \text{ lb} = 0 \quad \uparrow$ (14.37)

$\sum M$: $\qquad (2000 \text{ lb})(36 \text{ ft}) - H_V (48 \text{ ft}) = 0 \quad \circlearrowleft$ (14.38)

We can solve these equations for A_V and H_V, getting

$H_V = 1500 \text{ lb}$

$A_V = 500 \text{ lb}$

Now that we know the magnitude of the reactions on the ends of the truss, we can solve for the force in any member of the truss by use of the *method of sections*.

Suppose we wish to find out what force is being supported by the member at the top of the truss between joints B and D. In Figure 14.25(b), if we make an imaginary cut along the dashed line through the entire truss, cutting members BD, BE, and CE, and discarding the portion to the right of the imaginary cut we will have left a section that looks like the one in Figure 14.25(c). We have shown the unknown forces $\mathbf{R_1}$, $\mathbf{R_2}$, and $\mathbf{R_3}$ that exist in the three members we cut. We are entitled to show them as extending in exact alignment with the directions of the three cut members because these are all pin-jointed members, which we previously designated as *two-force members*. A force in a two-force member must be aligned with the member.

We can compute the rectangular components of $\mathbf{R_2}$ in the horizontal and vertical directions, as shown in Figure 14.25(d), and then apply the three equations of equilibrium to the free-body diagram in Figure 14.25(c), but replacing $\mathbf{R_2}$ with its rectangular components. We get (using joint B as a moment center, thereby causing $\mathbf{R_1}$ and $\mathbf{R_2}$ to drop out of the ΣM equation)

$$\Sigma H: \qquad R_1 + R_2 \cos 53.1° + R_3 = 0 \quad \rightarrowtail \tag{14.39}$$

$$\Sigma V: \qquad 500 \text{ lb} - R_2 \sin 53.1° = 0 \quad \updownarrow \tag{14.40}$$

$$\Sigma M: \qquad (500 \text{ lb})(12 \text{ ft}) - R_3 (16 \text{ ft}) = 0 \quad \curvearrowright \tag{14.41}$$

If we solve these three equations for the three unknowns, we get

$$R_1 = -750 \text{ lb}$$

$$R_2 = 625 \text{ lb}$$

$$R_3 = 375 \text{ lb}$$

The minus sign on R_1 means that the true direction of the force is opposite to the direction we assumed in Figure 14.25(c). R_2 and R_3 have positive signs, so this means their true directions coincide with the directions of the arrows in Figure 14.25(c). An arrow directed *away* from the joint means that the member is undergoing *tension*. An arrow directed *toward* the joint means that the member is undergoing *compression*. Therefore, member BD has a compressive force of 750 lb in it, member BE has a tensile force of 625 lb in it, and member CE has a tensile force of 375 lb in it.

Figure 14.26
This continuous steel truss carrying a superhighway over the Brandywine River (Delaware) was chosen to give a clean, aesthetically pleasing appearance. Its design utilized the original granite piers of the bridge it replaced (constructed in 1885). The bridge was designed through computer analysis. It is made of a modern "weathering" steel, which automatically develops a protective layer when exposed to the atmosphere and eliminates the need for paint. (Courtesy of Bethlehem Steel Corporation)

In the foregoing, we were actually after only the force in the member between joints B and D, but by the method of sections we got all three forces in the three members we cut. If we wished, we could continue to apply this method repeatedly at different locations in the truss until we solved for every force in every member.

Exercises

14.6.1 A simple truss is loaded as shown in the following figure. (a) Find the horizontal and vertical reactions at the truss supports A and F. (b) Find the forces that act in all the members, and tell whether these are tensile or compressive forces.

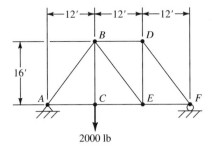

2000 lb

14.6.2 Find all the forces that act in all the members of the truss shown in the following figure, and tell whether they are tensile or compressive forces.

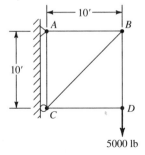

5000 lb

14.6.3 Find the reactions and the forces in the members for the truss in the following figure.

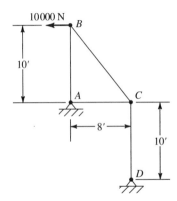

14.6.4 A truss with two loads is shown in the following figure. (a) Find the reactions at A and N. (b) Find the forces in DE and JK. Are these members in tension or in compression?

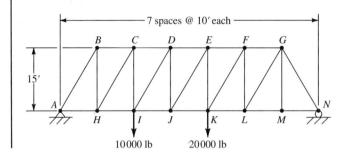

10000 lb   20000 lb

■ 14.7 DYNAMICS

From physics, we know that Newton's laws are the following:

First law. A body at rest will remain at rest, and a body in motion will move at a constant speed in a straight line, unless acted upon by an unbalanced force.

Second law. A body acted upon by an unbalanced force will accelerate in the direction of the force according to the relationship $a = kF/m$, where a is the acceleration, F the unbalanced force, m the mass of the body, and k a constant of proportionality.

Third law. For every action, there is an equal and opposite reaction.

We have already encountered the first and third laws. The first law, in fact, is a statement of the principle of equilibrium. Let us now look at the second law, upon which the important subject of *dynamics* is largely built. (Incidentally, the foregoing "laws" are not really laws at all, because bodies moving at very high speeds do not obey Newton's second "law," but behave in accordance with Einstein's relativity equations. However, in the vast majority of cases encountered by engineers, Newton's second law serves very well indeed to describe what happens, within our normal ability to measure physical qualities. Sir Isaac Newton deduced his second law by correlating the results of many physical observations, and his encapsulation of physical behavior into such a simple mathematical expression was a scientific achievement of the first magnitude.)

The foregoing statement of the second law included the equation $a = kF/m$, but in use, the constant of proportionality k is generally made equal to unity by an appropriate selection of units, and the equation is usually written in its more familiar form:

$$\mathbf{F} = m\mathbf{a} \tag{14.42}$$

where we have written **F** and **a** in boldface to remind us that force and acceleration are vectors. (Velocity is a vector also, and is generally written as **v**.)

Units

Some reminders regarding the proper units to use in Equation (14.42) may be in order here. From Chapter 9, we know that if we use SI units, the magnitude of **F** must be in newtons, m must be in kilograms, and **a** must be in meters per second per second, customarily written m/s². In the English gravitational system, **F** is in lbf (pounds force), **a** is in ft/sec², and m has the unusual units lbf-sec²/ft, commonly called the *slug*. It may help, in remembering these units, to think of Equation (14.42) with the units substituted in place of the variables:

$$\mathbf{F} = m\mathbf{a} \tag{14.42}$$

$$\text{lbf} = \left(\frac{\text{lbf-sec}^2}{\text{ft}}\right) \cdot \left(\frac{\text{ft}}{\text{sec}^2}\right) \tag{14.43}$$

If we cancel units in the numerator and denominator of the right-hand side of Equation (14.43), we get consistency on both sides of the equals sign, winding up with

$$\text{lbf} = \text{lbf}$$

Frequently, in the gravitational system, the magnitude of the weight W will be given, rather than its mass. To compute the mass, we can rewrite Equation (14.42) so it is in terms of the magnitudes of its vectors,

$$F = ma$$

and then substitute $F = W$ and $a = g$ (remembering that W is the force resulting on an object of mass m under the influence of the acceleration of gravity g), getting

$$W = mg \tag{14.44}$$

Figure 14.27

This enormous centrifuge, one of the largest in the world, has a 30-foot radius. The capsule at the end was designed to carry an astronaut. When the centrifuge rotates, heavy acceleration loads—many times the acceleration of gravity g—are experienced by the astronaut. The purpose of the device was to determine the ability of humans to perform certain tasks at high g loads. Note at the left the assembly of solid steel blocks to provide a counterweight. (Courtesy of Ames Research Center, National Aeronautics and Space Administration)

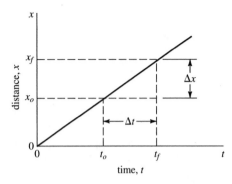

Figure 14.28

A distance-versus-time diagram, showing a case of constant velocity.

Equation (14.44) is an extremely useful relationship and is worth remembering. Using it, we can compute the mass m, if we know the weight W, as

$$m = \frac{W}{g}$$

Velocity

In more advanced courses, you will learn how to handle cases in which accelerations and forces are not constant, but vary with time. Such problems require the use of differential equations. In this book we will deal only with cases of constant accelerations and constant forces. But to discuss acceleration, we need first to discuss velocity.

Figure 14.28 is a graphical representation of constant velocity. Distance is represented by x and is plotted along the vertical axis. Time t is plotted along the horizontal axis. The definition of velocity is that it is equal to a change in distance Δx divided by the corresponding change in time Δt. In Figure 14.28, we assume an object is located at a beginning point, which we label x_0, and moves steadily to a final point, which we label x_f. We have marked this change of distance as Δx. When the object is at its beginning point, the corresponding time is marked as t_0, and when the object is at its final point, the corresponding time is marked as t_f. We have marked this change of time as Δt. The magnitude of the velocity (which we are assuming to be constant) is

$$v = \frac{\Delta x}{\Delta t} = \frac{x_f - x_0}{t_f - t_0} \qquad (14.45)$$

Acceleration

In Figure 14.29 we show the case of constant acceleration. The initial velocity and time are v_0 and t_0, and the final velocity and time are v_f and t_f. Between the initial and final points the velocity changes uniformly, as can be seen in Figure 14.29. The magnitude of the acceleration is

$$a = \frac{\Delta v}{\Delta t} = \frac{v_f - v_0}{t_f - t_0} \qquad (14.46)$$

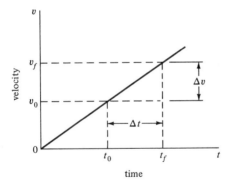

Figure 14.29
A velocity-versus-time diagram, showing a case of constant acceleration.

If we write Equation (14.46) in the form $\mathbf{a} = (\mathbf{v}_f - \mathbf{v}_0)/\Delta t$, then we can solve for $\mathbf{v}_f$ and get the useful equation

$$\mathbf{v}_f = \mathbf{v}_0 + \mathbf{a} \cdot \Delta t \tag{14.47}$$

Under conditions of constant acceleration, the distance Δx traveled by an object in time interval Δt is the average velocity v_{av} times the time interval Δt:

$$\Delta x = v_{av} \cdot \Delta t$$

$$= \frac{v_f + v_0}{2} \cdot \Delta t$$

If we substitute for v_f from Equation (14.47), we get a useful equation for finding Δx:

$$\Delta x = \left(\frac{v_0 + a \cdot \Delta t + v_0}{2} \right) \Delta t$$

$$= v_0 \cdot \Delta t + \tfrac{1}{2}a(\Delta t)^2 \tag{14.48}$$

If an object is falling freely under the influence of gravity g, Equations (14.47) and (14.48) can be written

$$\mathbf{v}_f = \mathbf{v}_0 + \mathbf{g} \cdot \Delta t \tag{14.49}$$

$$\Delta x = v_0 \cdot \Delta t + \tfrac{1}{2}g(\Delta t)^2 \tag{14.50}$$

Exercises

14.7.1 An automobile starts from rest and reaches 60 miles per hour in 45 seconds. Calculate v_0, v_f, t_0, t_f, Δv, Δt, and a.

Solution

$$v_0 = 0 \qquad\qquad\qquad\qquad \text{Answer}$$

$$v_f = 60\,\frac{\text{mi}}{\text{hr}} \left(5280\,\frac{\text{ft}}{\text{mi}} \right) \left(\frac{\text{hr}}{3600\ \text{sec}} \right)$$

$$= 88\ \text{ft/sec} \qquad\qquad\qquad\qquad \text{Answer}$$

$t_0 = 0$ Answer

$t_f = 45$ sec   Answer

$\Delta v = v_f - v_0 = (88 - 0)$ ft/sec

$\qquad\qquad = 88$ ft/sec   Answer

$\Delta t = t_f - t_0 = (45 - 0)$ sec

$\qquad\qquad = 45$ sec   Answer

$$a = \frac{\Delta v}{\Delta t} = \frac{88 \text{ ft/sec}}{45 \text{ sec}}$$

$\qquad\qquad = 1.96$ ft/sec^2 Answer

14.7.2 If a 1-kg mass is dropped from rest at a height of 2 m, how long will it take to hit the floor? How fast will it be going when it hits? (Use $g = 9.8$ m/s^2.)

Solution

$$\Delta x = v_0 \cdot \Delta t + \tfrac{1}{2}g(\Delta t)^2$$

$$2 \text{ m} = 0 \cdot \Delta t + \frac{9.8 \text{ m/s}^2}{2}(\Delta t)^2$$

$$(\Delta t)^2 = \frac{2(2 \text{ m})}{9.8 \text{ m/s}^2} = 0.408 \text{ s}^2$$

$$\Delta t = \mathbf{0.64} \text{ s} \qquad\qquad\qquad\qquad\qquad \text{Answer}$$

$$v_f = v_0 + g \cdot \Delta t$$

$$\qquad = 0 + 9.8 \, \frac{\text{m}}{\text{s}^2} \cdot 0.64 \text{ s}$$

$$\qquad = \mathbf{6.3} \text{ m/s} \qquad\qquad\qquad\qquad \text{Answer}$$

Note that the magnitude of the mass does not affect the result.

14.7.3 In Exercise 14.7.1, how far has the automobile gone in the first 45 seconds? If it keeps on accelerating at the same rate, how far will it go in the next 45 seconds? How fast will it be going?

14.7.4 An automobile assembly line travels at the rate of 0.05 ft/sec. The distance between two assembly stations is 25 ft. How long do the workers have between the time an auto arrives at their station and the time it arrives at the next station?

14.7.5 An elevator starts from rest and moves with a constant acceleration $a = 0.1g$ until it reaches its maximum velocity of 5 ft/sec. How long does it take to reach this velocity?

14.7.6 An airplane is heading due north at 200 miles per hour. (a) It is flying directly into a head wind of 50 miles per hour. What is its net forward velocity? (b) What is the magnitude of its net velocity if it is flying into a 50-mph wind from the northwest? In what direction is this velocity?

Solution

(a)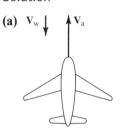

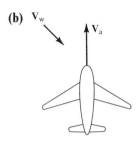

$$v_a \text{ (of airplane)} = 200\mathbf{j} \text{ mph}$$
$$v_w \text{ (of wind)} = -50\mathbf{j} \text{ mph}$$

net velocity $\mathbf{v} = \mathbf{v}_a + \mathbf{v}_w = (200 - 50)\mathbf{j}$ mph

$$= 150\mathbf{j} \text{ mph} \qquad \text{Answer}$$

(b)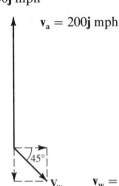

$$v_a = 200\mathbf{j} \text{ mph}$$

$$v_w = [(.707)50\mathbf{i} - (.707)50\mathbf{j}] \text{ mph}$$
$$= (35.3\mathbf{i} - 35.3\mathbf{j}) \text{ mph}$$

net velocity $= \mathbf{v}_a + \mathbf{v}_w$

$$= [35.3\mathbf{i} + (200 - 35.3)\mathbf{j}] \text{ mph}$$
$$= (35.3\mathbf{i} + 164.7\mathbf{j}) \text{ mph} \qquad \text{Answer}$$

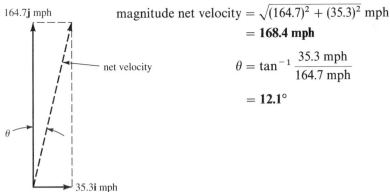

magnitude net velocity $= \sqrt{(164.7)^2 + (35.3)^2}$ mph

$$= 168.4 \text{ mph} \qquad \text{Answer}$$

$$\theta = \tan^{-1} \frac{35.3 \text{ mph}}{164.7 \text{ mph}}$$

$$= 12.1° \qquad \text{Answer}$$

14.7.7 The airplane of Exercise 14.7.6 now has a tail wind from the southeast of 80 mph. What is the magnitude of its net velocity, and what is its direction?

14.7.8 A car passes a given station with a velocity of 20 mi/hr, but is accelerating at a constant rate. Two minutes later it is going 50 mi/hr, and it maintains the 50-mi/hr velocity for 10 minutes longer. (a) What is its acceleration during the first 2 minutes? (b) How far has it traveled after 12 minutes?

■ 14.8 APPLICATION OF NEWTON'S SECOND LAW

As mentioned earlier, Newton's second law is expressed by the equation

$$\mathbf{F} = m\mathbf{a} \tag{14.42}$$

This equation not only demonstrates the proportionality that must exist between $\mathbf{F}$ and $\mathbf{a}$, it also states that $\mathbf{F}$ and $\mathbf{a}$ must have the same direction.

As an application of Equation (14.42), let us suppose that a rocket is being propelled vertically upward with a constant thrust of 10 000 N. (See Figure 14.30.) At the beginning of a measured interval of 10 s, the rocket is moving with a velocity of 5 m/s. We want to know the following: (a) How fast is the rocket moving at the end of the 10-s interval? (b) How far did it go during the interval? We will assume that the rocket is near the Earth's surface, so that $g = 9.81$ m/s^2 can be taken as constant, and we will also assume that its mass is constant. (If the rocket continued to consume fuel over an appreciable period of time, we would probably have to take into account the fact that the rocket is using up fuel and that its overall mass is therefore decreasing.)

Figure 14.30 shows that there is an upward force of 10 000 N acting on the rocket, and that its weight of $F_w = mg = 500$ kg$(9.81$ m/s$^2) = 4905$ N is acting downward. Thus, there is an unbalanced upward force acting on the rocket of 5095 N.

$$\mathbf{F} = 5095\mathbf{j} \text{ N} = 5095\mathbf{j} \frac{\text{kg} \cdot \text{m}}{\text{s}^2}$$

We can now apply Newton's second law:

$$\mathbf{F} = m\mathbf{a}$$

$$5095\mathbf{j} \frac{\text{kg} \cdot \text{m}}{\text{s}^2} = 500 \text{ kg } \mathbf{a}$$

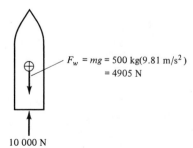

$F_w = mg = 500$ kg$(9.81$ m/s$^2)$
$= 4905$ N

10 000 N

Figure 14.30
A 500-kg rocket propelled upward by a 10 000-N force.

We can solve this equation for the acceleration **a** and get

$$\mathbf{a} = 10.19\mathbf{j} \text{ m/s}^2$$

The final velocity after the 10-s interval is given by Equation (14.47).

$$\mathbf{v_f} = \mathbf{v_0} + \mathbf{a} \cdot \Delta t$$
$$= 5\mathbf{j} \text{ m/s} + (10.19\mathbf{j} \text{ m/s}^2)(10 \text{ s})$$
$$= \mathbf{106.9j} \textbf{ m/s} \qquad\qquad\qquad \text{Answer}$$

To answer question (b) (How far did the rocket go during the interval?), we can use Equation (14.48):

$$\Delta x = v_0 \cdot \Delta t + \tfrac{1}{2}a(\Delta t)^2$$
$$= 5\frac{\text{m}}{\text{s}} \cdot 10 \text{ s} + \frac{1}{2}\left(10.19 \frac{\text{m}}{\text{s}^2}\right)(10 \text{ s})^2$$
$$= 50 \text{ m} + 509.5 \text{ m} = \mathbf{559.5} \textbf{ m} \qquad\qquad \text{Answer}$$

As another example, we will consider the two-mass system shown in Figure 14.31(a). The masses m_1 and m_2 are connected by a string, which runs over a pulley. The mass m_1 rests on a smooth table, which is considered to be frictionless. The pulley is considered to be both massless and frictionless, and the string's mass is neglected because it is so tiny. (These idealizations would probably be justified if we were making a preliminary analysis of the system, to get an overall idea of magnitudes.)

The mass m_2 is subject to the pull of gravity, and what we want to know is, how far will it move in the first interval of time, Δt, after it is released from rest and is allowed to fall?

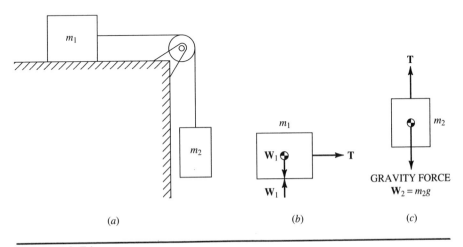

(a)   (b)   (c)

Figure 14.31
Two connected masses influenced by gravity.

We see immediately that the masses m_1 and m_2 always move together because of the string. We also see that the tension everywhere in the string will be uniform because the pulley cannot exert any forces on the system, having been considered both massless and frictionless.

Figures 14.31(b) and (c) are free-body diagrams. They differ substantially from those we drew previously, for their free bodies are *dynamic*, whereas those we dealt with earlier were *static*. Nevertheless, free bodies they are, and we can apply Newton's second law individually to each one. In doing so, we will not bother to write $F = ma$ in vector form, because we know that $\mathbf{F}$ and $\mathbf{a}$ will have the same direction.

The free body in Figure 14.31(b) has only one unbalanced force acting upon it: the force applied by the string, which we designate as T. The free body also has its own weight acting upon it, but this is exactly balanced by the force that the table exerts upward. There is a horizontal friction force exerted by the table on the body, but we are ignoring friction in our preliminary analysis.

The free body in Figure 14.31(c) has two forces acting on it: the string tension T and the gravity force m_2g. (Remember, weight and mass have the relationship $W = mg$.)

We can write two equations embodying Newton's second law, one for each free body:

$$T = m_1 a \tag{14.51}$$

$$m_2 g - T = m_2 a \tag{14.52}$$

Both masses move together, since they are connected by a common string, and we can solve for their common acceleration a by eliminating T from the foregoing two equations. We do this by substituting Equation (14.51) into Equation (14.52):

$$m_2 g - m_1 a = m_2 a$$

$$a = \frac{m_2}{m_1 + m_2} g \tag{14.53}$$

The bodies are starting from rest, so we can find Δx after an interval of time Δt by using Equation (14.48) and setting $v_0 = 0$. We get

$$\Delta x = \frac{1}{2} \cdot \frac{m_2}{m_1 + m_2} g(\Delta t)^2 \qquad\qquad \text{Answer}$$

Exercises

14.8.1 An object with mass of 4 kg is allowed to drop in free fall under the action of gravity. What is the force exerted on the object by gravity? What is the acceleration of the object? (Assume conditions at the Earth's surface.)

14.8.2 Repeat Exercise 14.8.1 in U.S. Customary (English) units. What is the weight of the object in lbf? The mass in slugs? The acceleration?

14.8.3 An automobile (see the following figure) accelerates with a constant acceleration of 0.5 ft/sec² during a specified interval. The auto weighs 3500 lb. What force is developed between the tires and the ground to produce this acceleration? (Ignore any air drag.)

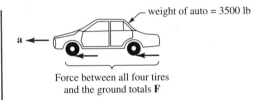

weight of auto = 3500 lb

a ←

Force between all four tires
and the ground totals **F**

14.8.4 A rocket starts from the Earth's surface and moves vertically upward for 5 s under a constant thrust of 1000 N. The rocket's mass is 100 kg. At the end of the 5-s interval, the thrust ceases. How far does the rocket rise before it starts to fall back to the Earth? (See the following figure.) Assume the mass of the rocket remains constant.

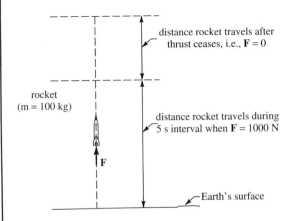

distance rocket travels after
thrust ceases, i.e., **F** = 0

rocket
(m = 100 kg)

distance rocket travels during
5 s interval when **F** = 1000 N

F

Earth's surface

14.8.5 A weight W is supported by a rope passing over a pulley, as shown in the figure. The acceleration g is the acceleration of gravity at the Earth's surface. The pulley is frictionless.

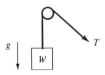

g

W

T

(a) If the weight is stationary, what is the force T? What is the net force (meaning the net sum of all the forces, positive and negative) acting on the weight?

(b) If the rope is suddenly released, what is the force T that is now present in the rope? What is the net force acting on the weight?

(c) Assume T is increased sufficiently so that the weight moves upward with a constant acceleration of 9.81 m/s^2. What force T_1 is necessary to accomplish this result? What is the net force now acting on the weight?

14.8.6 A man of weight 200 lb is standing in an elevator, as shown in the figure. The acceleration g is the acceleration of gravity at the Earth's surface.

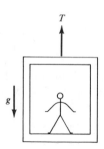

(a) Assume the elevator is stationary. Draw a free-body diagram of the man and show the forces acting on him. What is the force (call it P) between his feet and the floor of the elevator?

(b) If the elevator is moving upward at constant velocity, what is P?

(c) If the elevator is moving downward at constant velocity, what is P?

(d) If the supporting cable breaks, so that the elevator goes into free fall, what is P?

(e) If the elevator accelerates upward with a constant acceleration of $g/2$, what is P? Draw a free-body diagram for this case.

14.8.7 An elevator moves upward with constant acceleration $\mathbf{a} = 2\ \text{ft/sec}^2$ for a short interval of time. The elevator weighs 5000 lb and is supported by a cable, as shown in the following figure. (a) Find the tensile force $\mathbf{T}$ in the cable during this time. (b) Find the tensile force $\mathbf{T}$ after the acceleration ceases and the elevator's velocity becomes constant.

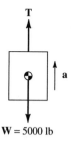

W = 5000 lb

14.8.8 The arrangement of the two-mass system in Figure 14.31 is modified as shown in part (a) of the following figure. The surface on which m_1 rests is considered to be frictionless. Note that the weight W_1 of the mass m_1 has been broken into two components, one parallel to the surface and the other perpendicular to it. A free-body diagram for m_1 is shown in (b), and a free-body diagram for m_2 is shown in (c). What is the acceleration a of this system?

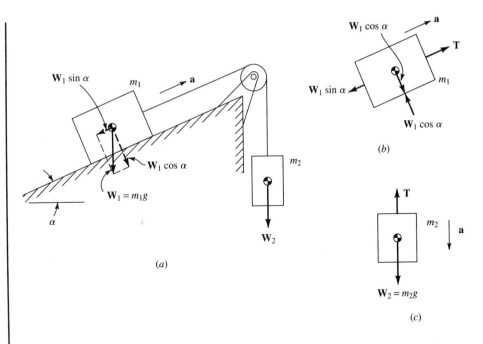

(b)

(a)

(c)

14.8.9 Two masses m_1 and m_2 are connected by a string, as shown in the following figure. A constant acceleration **a** is given to m_1. What is the tensile force **S** in the string? (Assume the surface is frictionless.)

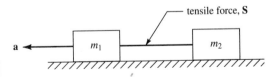

tensile force, **S**

14.8.10 A block of mass 10 kg sits on an inclined surface, as shown in the following figure. It is released from rest. What will be its velocity **v** after 4 seconds have elapsed? (Assume the surface is frictionless.)

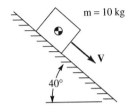

m = 10 kg

V

40°

Photographs courtesy of Aerospace Corporation (top right), FAG Kugelfischer (center), and DMJM (bottom).

15

Mechanics of Materials

Chapter 14 was concerned with finding the external forces that act upon a device or structure. In order to design the device or structure, an engineer also needs to know what the *internal* forces are. Usually, the determination of internal forces requires knowledge about the behavior of materials. Materials under load deform, and their deformation is directly related to a very important quality: strength. All these matters come under the subject called *mechanics of materials*.

■ 15.1 STRESS AND STRAIN

Once the forces acting upon a structural element are known, the next steps are usually to find out if the member is (1) strong enough and (2) stiff enough. The factor relating to the question of strength is called *stress*; that relating to stiffness is called *strain*. Stress and strain are in turn related by a property of the material being used, the *modulus of elasticity*. These concepts are simple, but extremely important and useful. For most structural materials, they are related by the equation

$$E = \frac{\sigma}{\varepsilon} \tag{15.1}$$

where E is the modulus of elasticity, given typically in units of newtons per square meter (N/m^2) or pounds per square inch $(lb/in.^2$, or simply psi); σ is the stress, given in the same units as E; and ε is the strain, which is dimensionless.

Stress is defined as force per unit area, and strain as a change in length of a structural ele-

Famous Engineers
Stephen Timoshenko

In a library containing engineering textbooks, one will find a large number of books bearing the name of Stephen Timoshenko. His name is little known among today's engineering students, even though they have been profoundly influenced by his work.

Timoshenko was born in 1880 in Kiev, in the Ukraine. He was educated in St. Petersburg and served on the structural mechanics faculties of St. Petersburg Polytechnic Institute and the Kiev Polytechnic Institute. He became a noted professor at a very young age but was fired in 1911 for political reasons. His "period of disgrace" lasted until 1913, when, back in St. Petersburg, he was appointed professor once more. At the start of the Russian Revolution, he left St. Petersburg and returned to Kiev, where the situation appeared more stable. But the Russian Civil War surged back and forth through Kiev, and during this time Timoshenko became convinced he was on the Bolshevik "wanted" list. He fled, first to Yugoslavia and then in 1922 to America. During a brief occupation of Kiev by the Polish Army in 1920, he was able to get his family out, just ahead of the Bolsheviks. He was not to return to Russia for almost 40 years, convinced he was still on the Bolshevik "wanted" list.

While he was in eastern Europe, Timoshenko published several books and many technical papers and was famous there among structural engineers, although he was virtually unknown in western Europe and the United States. He was startled by one of his first impressions upon arriving in America—the metal structures of New York's elevated railways. In his own words, "their technical design revealed a staggering ignorance.... When trains went by ... the swaying of those structures reached utterly inadmissible limits."

Timoshenko became a technical consultant for Westinghouse and remained there five years. Always, he was interested in the application of theory to practical problems, and this concern later influenced his teaching and writing. In fact, it was largely his interest in applications that gave rise to the Applied Mechanics Division of the American Society of Mechanical Engineers, which he helped organize.

Timoshenko became a professor at the University of Michigan in 1927, then moved to Stanford University in 1936. At Stanford he was given great freedom, teaching exclusively graduate students and working on his many textbooks. Toward the end of his life, he moved back to Europe, dying in Germany in 1972.

Throughout the 40 years or so that Timoshenko lived in the United States, he remained critical of the relatively low theoretical level of American technical education. When he worked for Westinghouse, for example, he observed that European-trained engineers, because of their better schooling in mathematics and theory, had "an enormous advantage over Americans, especially in the solving of nonstereotyped problems."

Timoshenko's textbooks were to help change all that. Some of his texts on elasticity and elastic stability were firsts in their field, and he influenced so many generations of American engineers that he has sometimes been referred to as "the father of engineering mechanics." His ideas, new at the time, have become the accepted manner of presentation in countless later textbooks. Today, in the welter of texts on engineering mechanics, his name is almost invisible to new generations of students, even though they owe much to his work.

Reference
Stephen P. Timoshenko, *As I Remember* (New York, N.Y.: Van Nostrand, 1968).

Figure 15.1
Simple tension.

ment, compared to its original length. In Figure 15.1, a uniform bar is shown subjected to the tensile force F. The bar has original length L and cross-sectional area A. After the force F is applied, the bar lengthens slightly by length ΔL. (It also contracts slightly in diameter so that A is decreased by a small amount, but the amount is so tiny for the usual structural materials that it is ignored.) The stress in the bar is defined as

$$\sigma = F/A \tag{15.2}$$

and the strain is defined as

$$\varepsilon = \Delta L/L \tag{15.3}$$

Thus, for the tensile case given here, Equation (15.1) becomes

$$E = \frac{F/A}{\Delta L/L} = \frac{FL}{A \cdot \Delta L} \tag{15.4}$$

Equation (15.4) permits us to determine E for any given material by inserting a bar of known dimensions L and A in a testing machine, applying a known force F, and measuring the resulting stretch ΔL. The value of E is computed by inserting the known and measured quantities into Equation (15.4). Thus, E is an experimentally determined property, which is intrinsic to a given material. For a stiff material such as steel, E is typically on the order of 200×10^9 N/m^2 (29 000 000 psi). Since *pascal* (Pa) is the special name given to N/m^2 and because a quantity like 200×10^9 is very large, values for moduli of elasticity may be given in GPa. Thus, a typical value of E for steel is 200 GPa. For a less stiff material, such as aluminum, E would be much smaller, typically 70 GPa (10 000 000 psi).

Note that we have been talking about moduli of *elasticity*. This means that, for the concepts we have been discussing the deflections must be elastic; that is, when the load is removed, the member must return to its original dimensions. If any permanent set remains in the material after load removal, the deflection is said to be *plastic* and our simple concepts do not apply.

Figure 15.2

Engineers frequently conduct strength and fatigue tests on metal specimens, but here a cannon bone from the foreleg of a Thoroughbred race horse is being tested. The bone is subjected to thousands of repetitive cycles of compression loading until it fractures, which gives a measure of the strength of bone as a structural element. This research project was directed toward the improvement of the dynamic characteristics of race track surfaces in order to reduce injuries to race horses. (Courtesy of University of California, Davis)

Figure 15.3 shows a typical stress-strain curve, such as would be obtained in a tensile test. Equation (15.1) applies to stress values up to σ_0, referred to as the *proportional limit*. Beyond this point, the relationship between stress and strain is no

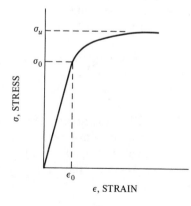

Figure 15.3

A stress-strain curve.

longer linear; and at about this point, for a material like steel, the test specimen no longer returns to its original dimensions when the load is released—that is, it begins to yield and becomes plastic. If the test is continued, the stress continues to increase with increasing load, accompanied by a rapid increase in strain. A maximum value of stress is reached at σ_u, known as the *ultimate strength*; and finally, as the test is continued, the member breaks. Tests of materials show that the relationship in the linear range (*i.e.*, up to σ_0) is not exactly a straight line, but is nearly so for many structural materials. For some other materials (wood and plastic, for example) the experimental curves obtained through testing do not exhibit any linear portions at all and are curved throughout. Thus, the simple linear relations we have been employing do not apply for all materials.

We have been discussing materials in *tension*, but similar relationships apply for materials in *compression*, and E for many materials (steel, for example) has the same value in tension and compression. There is another simple mode of stress known as *shear*. Figure 15.4 shows a case of shear, where two plates have been fastened together with a bolt and subjected to force F. If the bolt has a cross-sectional area A, the shear stress τ is

$$\tau = F/A \tag{15.5}$$

analogous to Equation (15.2).

Figure 15.4
Simple shear.

The stress limits for many materials are given in handbooks. For example, the usual stress limit for "mild" steel in tension is about 240 MPa (35 000 psi). For an actual case, a *factor of safety* would be involved, so a working limit might be 100 MPa (about 15 000 psi). Thus, in utilizing these formulas for design, no tension or compression stresses in excess of 100 MPa would be allowed, and the sizes of the members would be adjusted until this criterion has been met. Similarly, if very low stresses were encountered in some of the members, the members would be redesigned to use less material, because economy in the use of material is important to engineers.

15.1.1 A specimen that is 5.000 in. long between accurately scribed gauging marks is placed in a testing machine. Its cross-sectional area is 0.500 in.². A load of 10 000 lb is applied, and then it is observed that the length has increased by 0.0033 in. What are the stress, the strain, and the modulus of elasticity for this material?

Solution

Stress:
$$\sigma = \frac{P}{A} = \frac{10\ 000\ \text{lb}}{0.500\ \text{in.}^2} = \textbf{20 000 lb/in.}^2 \qquad \text{Answer}$$

Strain:
$$\varepsilon = \frac{\Delta L}{L} = \frac{0.0033\ \text{in.}}{5.000\ \text{in.}} = \textbf{0.000 67} \qquad \text{Answer}$$

Modulus of elasticity:
$$E = \frac{\sigma}{\varepsilon} = \frac{20\ 000\ \text{lb/in.}^2}{0.000\ 67}$$

$$= \textbf{29.8} \times \textbf{10}^6\ \textbf{lb/in.}^2 \qquad \text{Answer}$$

15.1.2 A bracket fastened by two bolts carries a 10 000-N load, as shown. If the working limit for the stress in the bolts is 100 MPa, what is the minimum diameter each bolt may have?

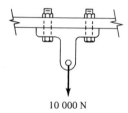

10 000 N

Solution

Let the area of each bolt be A.

$$\sigma = \frac{10\ 000\ \text{N}}{2A} = 100\ \text{MPa} = 100 \times 10^6\ \text{N/m}^2$$

$$A = \frac{10\ 000\ \text{N}}{2 \cdot 100 \times 10^6\ \text{N/m}^2} = 5 \times 10^{-5}\ \text{m}^2$$

$$A = \frac{\pi d^2}{4}$$

$$d^2 = \frac{4}{\pi}(5 \times 10^{-5})\ \text{m}^2 = 6.37 \times 10^{-5}\ \text{m}^2$$

$$d = 0.008\ \text{m} = \textbf{8 mm} \qquad \text{Answer}$$

15.1.3 If a steel member is 3.5 m long and has a cross-sectional area of 25 cm², how far will it stretch under a load of 1000 N? (Remember, E for steel is approximately 200 GPa.)

15.1.4 In Exercise 15.1.3, convert all the units into English units. After you get your answer in English units, convert it back into metric units to see if it checks with the answer you got in Exercise 15.1.3.

15.1.5 If a steel rivet is to bear a load of 1000 lb in so-called *single shear* (meaning the load is applied to only one cross-sectional area undergoing shear), how large a diameter should be used? Use a working limit for shear stress of 10 000 lb/in.^{2}.

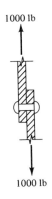

1000 lb

1000 lb

15.1.6 A riveted joint with two rivets supports a load of 10 000 N in shear. If a working shear stress limit of 70 MPa is used, what should the diameters of the rivets be? (Note that each rivet is in so-called *double shear*, meaning that two cross-sectional areas of each rivet must be sheared in order for the joint to fail.)

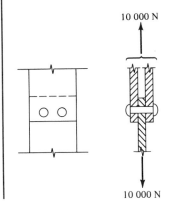

10 000 N

10 000 N

■ 15.2 ELEMENTARY BEAM ANALYSIS

Beams are structural elements that turn up almost everywhere. Obviously, they occur in structures such as buildings and bridges, but they also occur in machines of every conceivable type. Beams are structural members that are acted on by transverse forces, as opposed to the two-force members we discussed in Chapter 14, which were acted on only by forces that were aligned with the members. We will discuss three basic types of beams: (1) a cantilever beam with a concentrated load at the end, (2) a simple beam with a concentrated load in the center, and (3) a simple beam with a uniformly distributed load. The details of the derivations of the stress and deflection formulas belong in more advanced courses. In this book, we will simply write down the formulas we are going to use and emphasize their applications.

Cantilever Beam with a Concentrated Load at the End

A *cantilever beam* has one end rigidly fixed, as if it were embedded in a wall, and the other end free. Figure 15.5(a) shows such a beam with length L and with a concentrated load **P** at the free end. The beam deflects slightly under the load **P**, and this deflection has been greatly exaggerated in the figure so that we can see it. Figure 15.5(b) is a free-body diagram, with the beam represented by a thin line and the deflection designated as Δ. Since the wall has been removed, we must put on the diagram the force and moment that the wall exerted; these cause the beam to be held in equilibrium. The upward force $\mathbf{V_0}$ at the wall balances the downward force **P**. A counterclockwise moment $\mathbf{M_0}$ is exerted by the wall and holds the end of the beam rigidly in place. (To visualize this, imagine you are gripping the end of the beam with your hand and are exerting a twisting moment $\mathbf{M_0}$, which is enough to balance the opposite twisting moment exerted by force **P** acting with moment arm L.) If we apply the equations of equilibrium to this free-body diagram, with the moment center at the wall, we get

$$\sum H: \qquad\qquad 0 = 0$$
$$\sum V: \qquad -P + V_0 = 0 \quad \updownarrow$$
$$\sum M: \qquad -M_0 + PL = 0 \quad \curvearrowright$$

It is easy to solve these and get $V_0 = P$ and $M_0 = PL$.

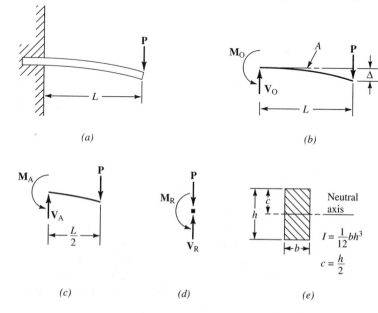

(a)

(b)

(c)

(d)

(e)

Figure 15.5

A *cantilever* beam is fixed at one end and free at the other. (a) A cantilever beam with load **P** at its free end, and length L. (b) Free-body diagram of the entire beam. (c) Free-body diagram of the right-hand portion of the beam. (d) Free-body diagram of the beam tip. (e) Cross section of a rectangular beam.

The load causes the beam to deflect downward so that the fibers on the upper surface are stretched and those on the lower surface are shortened. Thus, we know that the upper fibers are subjected to tension, and the lower fibers to compression. If the load is increased sufficiently, the stress in the upper fibers will become so large that the upper surface of the beam will fail in tension. We know from experience that if we embed a yardstick in a wall and apply a force on the free end, the yardstick will break off at the wall; we infer from this that the stress must be greatest at that point. It will be shown in more advanced courses that the stress at a given point in a beam is directly related to the magnitude of the moment that exists at the point.

In Figure 15.5(b), we took the moment center to be where the beam enters the wall. But if we take an imaginary cut at the midpoint of the beam (point A), we can create a free-body diagram for the portion of the beam remaining, as in Figure 15.5(c). At the left end of this remaining part of the beam, the portion of the beam we removed has been replaced by the force and moment it exerted. This requires an upward force $\mathbf{V_A}$ to balance $\mathbf{P}$ and a counterclockwise moment $\mathbf{M_A}$ to balance the moment of $\mathbf{P}$ times its moment arm $L/2$. The equations of equilibrium (ignoring $\Sigma H = 0$, because there are no horizontal forces, and using point A as the moment center) tell us that

$$\Sigma V: \qquad -P + V_A = 0 \quad \updownarrow$$
$$\Sigma M: \qquad -M_A + P \cdot (L/2) = 0 \quad \circlearrowleft$$

and from these we get $V_A = P$ and $M_A = PL/2$.

Finally, if we take an imaginary cut through the beam an infinitesimal distance to the left of the load $\mathbf{P}$, we get a free-body diagram like the one in Figure 15.5(d). We have put a force $\mathbf{V_R}$ and a moment $\mathbf{M_R}$ on this diagram to represent the force and moment that the removed portion of the beam exerted on the infinitesimal portion remaining. By inspection, we can see that

$$V_R = P$$
$$M_R = 0$$

The result $M_R = 0$ comes about because the moment center is at the point where $\mathbf{P}$ and $\mathbf{V_R}$ meet; hence, any moments created by these two forces must be zero, because their moment arms are infinitesimally small and can be taken equal to zero.

Now we can go back over what we just learned and see how it correlates with what our common sense told us. We learned that the moment (and therefore the stress) is greatest right at the wall and is equal to PL. At the right-hand tip of the beam, the moment is zero; halfway in between, the moment is half as large as it is at the wall. From all this we can deduce that the moment varies linearly in the beam from its maximum value at the wall to zero at the tip. Therefore, the stress varies in the same way and is greatest at the wall, which is what experience also told us. The kind of moment we have been discussing is called *bending moment* because of the manner in which it causes the beam to bend.

As already mentioned, there are tensile stresses in the fibers on the upper surface of this beam and compressive stresses in the fibers on the lower surface. It will be shown in more advanced courses that these tensile and compressive stresses are equal in magnitude, though opposite in sign. (Tension is taken as positive, and compression

Figure 15.6
An unusual bridge design to connect Penang Island and the Malaysian mainland. The bridge is described as a three-span, cable-stayed concrete structure; it has a 738-foot center span and two 353-foot side spans. (Courtesy of Howard Needles Tammen and Beigendorff)

as negative.) Also, it can be shown that the magnitude of the stress in a particular fiber is proportional to its distance from the central axis of the beam, called the *neutral axis*. The beam in our example is assumed to be rectangular in cross section, and its cross section is shown in Figure 15.5(*e*), with width *b* and depth *h*. The neutral axis is shown. For a beam of regular cross section like this one, the neutral axis is at the midpoint of the cross section. The stress at the neutral axis is zero; it increases positively (tension) to its maximum value at the upper surface of the beam, and it increases negatively (compression) to its maximum at the lower surface.

The relationship between the moment and the stress σ in the outer fibers of a beam, given here without derivation, is

$$\sigma = \frac{Mc}{I} \tag{15.6}$$

where M is the bending moment at the particular point in question, c is the distance from the neutral axis to the outer fibers of the beam, and I is the quantity known as the *moment of inertia*. For a beam of rectangular cross section, $I = \frac{1}{12}bh^3$, and $c = h/2$. See Figure 15.5(e).

Sometimes Equation (15.6) is written

$$\sigma = \frac{M}{S} \qquad (15.7)$$

where $S = I/c$. The quantity S is called the *section modulus*.

Equation (15.6) is a useful and powerful formula. As an example, let us suppose we have a cantilever beam 10 ft (120 in.) long, with $b = 3$ in., $h = 6$ in., $c = h/2 = 3$ in., and $P = 2000$ lb. First we compute I

$$I = \frac{1}{12}bh^3 = \frac{1}{12}(3 \text{ in.})(6 \text{ in.})^3 = 54 \text{ in.}^4$$

and substitute this in Equation (15.6), remembering that the maximum value of M is equal to PL, and occurs at the wall:

$$\sigma = \frac{Mc}{I} = \frac{(2000 \text{ lb})(120 \text{ in.})(3 \text{ in.})}{54 \text{ in.}^4}$$

$$= 13\,330 \text{ lb/in.}^2 \quad (91.9 \text{ MPa})$$

On page 417 we said the "allowable stress" that might be permitted in a member, if it were made of mild steel, could be taken as 100 MPa, or about 15 000 psi. The stress in the above example is slightly smaller than that, and this is the maximum stress that occurs anywhere in the beam, so we conclude that the stress level in the beam is acceptable. If the stress in the beam came out to be larger than the allowable, we would have to redesign the beam to make it stronger, by increasing b or h or both. On the other hand, if the stress were significantly smaller than the allowable, we would conclude that the beam has larger dimensions than is necessary, which means we are wasting material. In this case we would redesign the beam to make it smaller.

The maximum deflection for a *cantilever beam with a concentrated load at the end* is given here without derivation:

$$\Delta = \frac{PL^3}{3EI} \qquad (15.8)$$

It is important to realize that Equation (15.8) holds true only for exactly this kind of beam and loading. For other kinds of beams and other loading conditions, the formulas for deflection are different from Equation (15.8).

If we substitute the same values in Equation (15.8) that we used to calculate the stress, we obtain

$$\Delta = \frac{PL^3}{3EI} = \frac{(2000 \text{ lb})(120 \text{ in.})^3}{3(29 \times 10^6 \text{ lb/in.}^2)(54 \text{ in.}^4)}$$

$$= 0.736 \text{ in.}$$

The value of $E = 29 \times 10^6$ lb/in.^{2} for mild steel that was just used was taken from page 415. From our calculations, we can see that the 2000-lb load causes a deflection of only about $\frac{3}{4}$ in. at the end of this 10-ft beam.

Simple Beam with a Concentrated Load in the Center

The following analysis for a simple beam with a concentrated load in the center parallels what we just did for the cantilever beam with a concentrated load at the end. Figure 15.7(a) shows a beam of the type we are going to discuss. A "simple" beam is one that is carried by two supports that are not capable of exerting any moments at the supports; i.e., the moment applied to the beam at each end is zero. However, nonzero moments exist everywhere else in the beam, as will be seen. In fact, the bending moment varies linearly from zero at each end to a maximum at the center, directly under the load.

A free-body diagram of the beam is shown in Figure 15.7(b). The load P is shown, as is the maximum deflection (which will occur at the center point of the beam), designated as Δ. The two end supports have been removed and replaced by the reactions they exerted on the beam, designated as $\mathbf{R_L}$ and $\mathbf{R_R}$. If we write the equations of equilibrium, again leaving out $\Sigma\,H = 0$, and using the left-hand end of the beam as the moment center, we get

$$\Sigma\,V: \qquad R_L + R_R - P = 0 \quad \updownarrow$$

$$\Sigma\,M: \qquad P \cdot L/2 - R_R \cdot L = 0 \quad \curvearrowright$$

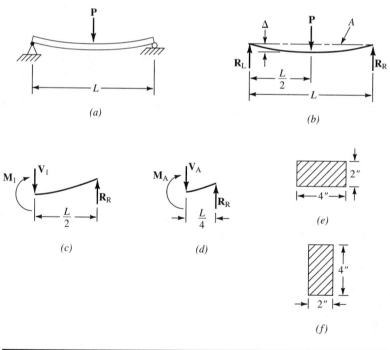

(a)   (b)   (c)   (d)   (e)   (f)

Figure 15.7

(a) A simple beam with a concentrated load at its center. (b) Free-body diagram. (c) Free-body diagram of the right half of the beam. (d) Free-body diagram of the right one-quarter of the beam. (e) Cross section of the beam laid flat. (f) Cross section of the beam set on edge.

We can solve these to get $R_L = R_R = P/2$. Notice, for this beam, that the tensile stress is now in the fibers on the bottom surface of the beam, and the compressive stress is in the fibers on the top surface—the reverse of the case for the cantilever beam.

Let us make a new free-body diagram of just a portion of this beam, by taking an imaginary cut an infinitesimal distance to the right of the load **P**. This produces Figure 15.7(c). The load **P** has been cut away along with the left half of the beam. In place of the left half are the vertical force $\mathbf{V}_1$ and the moment $\mathbf{M}_1$ that the left half of the beam exerted on the right half in order to hold it in equilibrium. If we take the moment center directly under force $\mathbf{V}_1$, we can write the equations of equilibrium as

$$\sum V: \qquad R_R - V_1 = 0 \quad \updownarrow$$

$$\sum M: \qquad M_1 - R_R \cdot L/2 = 0 \quad \rightleftarrows$$

If we use the value $R_R = P/2$ that we obtained above, we get the result that $V_1 = P/2$ and $M_1 = PL/4$.

If we make an imaginary cut at point A in Figure 15.7(b), which is at a distance $L/4$ from the right-hand end, we get the free-body diagram shown in Figure 15.7(d). If we apply the equations of equilibrium, we get

$$\sum V: \qquad R_R - V_A = 0 \quad \updownarrow$$

$$\sum M: \qquad M_A - R_R \cdot L/4 = 0 \quad \rightleftarrows$$

Using $R_R = P/2$, we get the results $V_A = P/2$ and $M_A = PL/8$. We see that the bending moment in the beam at the point halfway between the load **P** and the right-hand support is only half as great as it is at the center of the beam. This affirms what we said at the beginning: The bending moment in a simple beam with a concentrated load in the center is a maximum at the center of the beam, where $M = PL/4$, and varies in a linear fashion to zero at each end.

Now let us apply Equation (15.6) to this beam, with $L = 10$ ft (120 in.), $P = 2000$ lb, $b = 4$ in., $h = 2$ in., and $c = h/2 = 1$ in. Note that the values of b and h are those given in Figure 15.7(e), which are for the beam laid on its flat side. First we compute I as

$$I = \tfrac{1}{12}bh^3 = \tfrac{1}{12}(4 \text{ in.})(2 \text{ in.})^3 = 2.67 \text{ in.}^4$$

and substitute it in Equation (15.6), using the maximum value $M = PL/4$:

$$M = \frac{PL}{4} = \frac{(2000 \text{ lb})(120 \text{ in.})}{4} = 60\,000 \text{ lb-in.}$$

$$\sigma = \frac{Mc}{I} = \frac{(60\,000 \text{ lb-in.})(1 \text{ in.})}{2.67 \text{ in.}^4}$$

$$= 22\,470 \text{ lb/in.}^2 \quad (154.9 \text{ MPa})$$

Consternation! This is too great, if we are committed to a material that permits an allowable stress of only $15\,000$ lb/in.^{2}. But, just as we start to redesign the beam so that its dimensions (and its cost) will be larger, we get the idea that we could do a lot better by taking the same beam and turning it on its edge; as shown in Figure

15.7(f), rather than laying it flat as we did in Figure 15.7(e). So we recalculate, now using $b = 2$ in., $h = 4$ in., and $c = h/2 = 2$ in., and we get

$$I = \tfrac{1}{12}bh^3 = \tfrac{1}{12}(2\text{ in.})(4\text{ in.})^3 = 10.67\text{ in.}^4$$

$$\sigma = \frac{Mc}{I} = \frac{(60\,000\text{ lb-in.})(2\text{ in.})}{10.67\text{ in.}^4}$$

$$= 11\,240\text{ lb/in.}^2 \quad (77.5\text{ MPa})$$

This is much better. In fact, it might be possible to reduce the dimensions of the beam slightly (and use less material) so that the actual stress in the beam is closer to the allowable limit of 15 000 lb/in.^{2}. In this example, it can be seen how sensitive the calculations are to the value of h, which appears to the third power in the formula for I.

Finally, let us compute the value for Δ, using the same values we just did to calculate σ. For a *simple beam with concentrated load in the center*, the formula for Δ (given without derivation) is

$$\Delta = \frac{PL^3}{48\,EI} \tag{15.9}$$

For our beam, then, Δ is

$$\Delta = \frac{PL^3}{48\,EI} = \frac{(2000\text{ lb})(120\text{ in.})^3}{48(29 \times 10^6\text{ lb/in.}^2)(10.67\text{ in.}^4)}$$

$$= 0.233\text{ in.}$$

We can make one final comparison here. In Chapter 9, Equation (9.4), we said that the maximum stress in a simple beam with a concentrated load in the center is

$$\sigma = \frac{PL}{4S}$$

We can get this same formula from Equation (15.6) by remembering that it can be written in the alternative form of Equation (15.7), using $S = I/c$ and noting that we found the maximum moment in a simple beam with a concentrated load in the center to be $M = PL/4$. We can then get

$$\sigma = \frac{Mc}{I} = \frac{M}{S} = \frac{PL}{4} \cdot \frac{1}{S} = \frac{PL}{4S}$$

Simple Beam with a Uniformly Distributed Load

A simple beam with a uniformly distributed load is shown schematically in Figure 15.8(a). A perfect example of a uniformly distributed load is the weight of the beam itself. In previous examples we have ignored the weight of the beam, but now we will have a way of taking it into account. Also, the uniformly distributed load is a good means of handling other sorts of loading conditions that occur frequently, such as the weight of the paving materials that make up the road surface of a bridge. In our

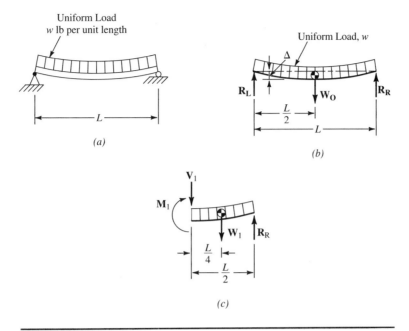

Figure 15.8
(a) A simple beam with a uniform load. (b) Free-body diagram. (c) Free-body diagram of the right half of the beam.

example, the loading condition will be described in terms of pounds per unit length of the beam material (or newtons per meter).

Figure 15.8(b) is a free-body diagram of the beam. The loading is described as w lb per ft, and the maximum deflection Δ is shown. The two end supports have been replaced by R_L and R_R. For the purpose of finding R_L and R_R, we can treat the uniformly distributed load as if it is all concentrated at its center of gravity (which, of course, is at the midpoint of the beam) and has a magnitude of w lb per ft times the beam's length. We have designated this total weight as W_0, and it has magnitude wL.

Now we can write the equations of equilibrium (with the moment center at the left-hand end of the beam) as

$$\sum V: \qquad R_L + R_R - wL = 0 \quad \uparrow$$

$$\sum M: \qquad wL \cdot L/2 - R_R \cdot L = 0 \quad \circlearrowright$$

We can solve these to get $R_L = R_R = wL/2$.

As we did in the previous section, let us make a new free-body diagram of just the right-hand portion of the beam by taking an imaginary cut down through the center. This produces Figure 15.8(c). In place of the left half of the beam are the vertical force V_1 and the moment M_1 that the left half of the beam exerted on the right half. Note that, because we only have half the beam in this diagram (and thus half the uniformly distributed load), the load applied to this portion of the beam is the uniform load w multiplied by half the beam's length, so that $W_1 = wL/2$. For the purpose of finding $\mathbf{V}_1$ and $\mathbf{M}_1$, $\mathbf{W}_1$ can be treated as if it is concentrated at its own center of gravity, which is located at a distance $L/4$ from the midpoint of the beam.

If we take the moment center directly under force $\mathbf{V}_1$, we can write the equations of equilibrium as

$$\sum V: \qquad\qquad R_R - wL/2 - V_1 = 0 \quad \uparrow$$

$$\sum M: \qquad M_1 + (wL/2)(L/4) - R_R \cdot L/2 = 0 \quad \curvearrowright$$

If we solve these two equations, using the value $R_R = wL/2$ that we obtained above, we get $V_1 = 0$ and $M_1 = wL^2/8$. Since this is a simple beam, the moment at each support is zero, and the moment in the center is a maximum. However, in a uniformly loaded simple beam, the moment does not vary *linearly*. We will not pursue the question of exactly how it varies, but leave that subject for more advanced courses.

As we did before, let us apply Equation (15.6) to this beam, with $L = 10$ ft (120 in.), $w = 200$ lb/ft, $b = 2$ in., $h = 4$ in., and $c = h/2 = 2$ in. Note that this loading condition produces a total load on the beam of $wL = (200\text{ lb/ft})(10\text{ ft}) = 2000$ lb, which is the same total load as in the previous example. But in the current example the load is distributed uniformly, so we would expect to get a different result.

$$I = \tfrac{1}{12}bh^3 = \tfrac{1}{12}(2\text{ in.})(4\text{ in.})^3 = 10.67\text{ in.}^4$$

$$M = \frac{wL^2}{8} = \frac{(wL) \cdot L}{8} = \frac{(2000\text{ lb})(120\text{ in.})}{8} = 30\,000\text{ lb-in.}$$

$$\sigma = \frac{Mc}{I} = \frac{(30\,000\text{ lb-in.})(2\text{ in.})}{10.67\text{ in.}^4}$$

$$= 5620\text{ lb/in.}^2 \quad (38.8\text{ MPa})$$

This is exactly half the stress we calculated for the previous example, when we had a 2000-lb load concentrated at the center of a simple beam. Obviously, a distributed load affects a beam less severely than does a concentrated load of the same magnitude.

A comment about units is appropriate here. In the equation $M = wL^2/8$, we had to substitute the length of the beam into the formula. If we proceed, using $L = 10$ ft, we get

$$M = \frac{(200\text{ lb/ft})(10\text{ ft})^2}{8} = 2500\text{ lb-ft}$$

This answer is certainly correct as it stands, but if we substitute this value into the equation for stress, $\sigma = Mc/I$, we get

$$\sigma = \frac{Mc}{I} = \frac{(2500\text{ lb-ft})(2\text{ in.})}{10.67\text{ in.}^4} = 468.6\ \frac{\text{lb-ft}}{\text{in.}^3}$$

Figure 15.9
It is not only bridges and buildings that use structural members such as beams. Most machines also have them, as does this industrial robot at work. (Courtesy of ASEA)

Obviously, we would have to do something with this in order for our units to come out right. We could use the conversion factor 1 ft = 12 in. and get

$$\sigma = 468.6 \, \frac{\text{lb-ft}}{\text{in.}^3} \left(12 \, \frac{\text{in.}}{\text{ft}} \right) = 5620 \, \text{lb/in.}^2$$

which of course is what we need.

Finally, we calculate the value for Δ, using the values we just used to calculate σ. For a *simple beam with uniformly distributed load*, the formula for Δ (given without derivation) is

$$\Delta = \frac{5wL^4}{384EI}$$

For our beam,

$$\Delta = \frac{5(wL)L^3}{384EI} = \frac{5(2000 \, \text{lb})(120 \, \text{in.})^3}{384(29 \times 10^6 \, \text{lb/in.}^2)(10.67 \, \text{in.}^4)}$$

$$= 0.145 \, \text{in.}$$

15.2.1 A mild steel cantilever beam with rectangular cross section is shown in the following figure. $P = 500$ lb, $b = 2$ in., $h = 4$ in., and $L = 8$ ft. Calculate the maximum stress in the beam in psi, and tell where it is located. Calculate the maximum deflection in inches. (Use $E = 29\,000\,000$ psi.)

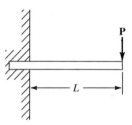

15.2.2 Repeat Exercise 15.2.1 with $P = 3000$ N, $b = 50$ mm, $h = 100$ mm, and $L = 3$ m. Calculate the maximum stress in the beam in MPa. Calculate the maximum deflection in mm. (Use $E = 200$ GPa.)

15.2.3 A mild steel beam with rectangular cross section is shown in the following figure. $P = 800$ lb, $b = 2.25$ in., $h = 4.75$ in., and $L = 15$ ft. Calculate the maximum stress in the beam in psi, and tell where it is located. Calculate the maximum deflection in inches.

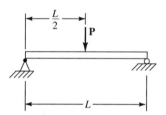

15.2.4 Repeat Exercise 15.2.3 with $P = 2000$ N, $b = 45$ mm, $h = 95$ mm, and $L = 5$ m. Calculate the maximum stress in MPa and the maximum deflection in mm.

15.2.5 A mild steel beam with rectangular cross section is shown in the following figure. $w = 100$ lb/ft, $b = 2$ in., $h = 5$ in., and $L = 10$ ft. Calculate the maximum stress in the beam in psi. Calculate the maximum deflection in inches.

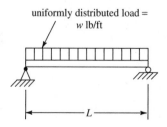

15.2.6 Repeat Exercise 15.2.5 with $w = 1000$ N/m, $b = 30$ mm, $h = 100$ mm, and $L = 4$ m. Calculate the maximum stress in MPa and the maximum deflection in mm.

■ 15.3 EXAMPLES OF APPLICATIONS IN DESIGN

A few examples will help to give you a feel for how the ideas in the foregoing sections might be used in design situations. In Chapter 11 on engineering design, we gave some examples and used some of the same formulas we have been discussing in this chapter. But in Chapter 11 we were concerned mostly with the overall design process, whereas in this section we will focus on selected small elements of design. The primary purpose is to demonstrate ways in which the theoretical ideas of this chapter might be used.

Selection of a Bolt

This example harks back to Exercise 14.5.2, which involved a free-body diagram for a lamp bracket. The configuration of Exercise 14.5.2 is repeated in Figure 15.10. The lamp, which weighs 50 lb, is supported at the end of a bracket 8.5 ft long by two bolts that are fastened into a wall. Earlier, nothing was said about the manner in which the bolts were fastened. Here we will assume that steel anchors were placed and held in their proper positions while the concrete wall was poured, and that the anchors contain steel nuts to receive the bolts.

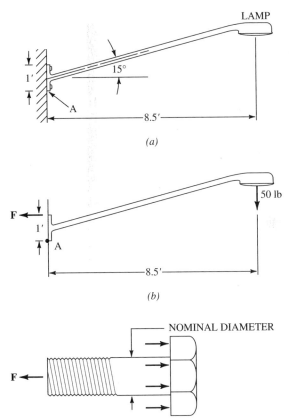

(a)

(b)

(c)

Figure 15.10

(a) A lamp bracket supported by two bolts. (b) Free-body diagram. (c) Enlarged view of the upper bolt.

In Exercise 14.5.2 we assumed that the lamp, under the influence of its own weight, would attempt to rotate clockwise about point A and that point A and the lower bolt coincided approximately at the same point. Using this point as the moment center, we set up a free-body diagram like that in Figure 15.10(b) and found that $\mathbf{F}$, the force in the upper bolt, was equal to 425 lb. Now it would be wise for us to discuss whether the assumptions we made in that exercise are reasonable.

Suppose for a moment that the upper bolt were replaced by a rubber band. It is not hard to visualize that the lamp, rather than merely *tending* to rotate clockwise, really would do so, and that a gap would open at the upper end of the bracket. The lower end of the bracket (point A) would certainly act as the pivot under these circumstances. Furthermore, we can readily imagine that the sharp lower edge of the bracket would dig slightly into the concrete and break small pieces of it away. Soon, as a result of all this, the lower bolt would begin to take over the support of the lamp, and, assuming neither it nor the concrete around it failed completely, the lower bolt would hold the lamp and its supporting arm in a sagging position, with an alarming gap showing at the location of the rubber band.

None of this will occur in practice, of course, but the scenario we just described does illuminate two factors: (1) Almost all of the work of supporting the lamp is done by the upper bolt, and the lower bolt hardly enters at all; (2) if there were any actual rotation of the lamp bracket, the concrete under the sharp lower edge of the bracket at point A would yield slightly, leaving the lower bolt as the actual point about which the lamp and bracket would try to rotate. Furthermore, point A is perhaps 15 in. to 18 in. distant from the upper bolt, whereas the distance between the two bolts is 12 in. In Exercise 14.5.2 we used the 12 in. distance as our moment arm and got $\mathbf{F} = $ 425 lb. If we had used the distance to point A as our moment arm (15 in. to 18 in.), we would have gotten a value for $\mathbf{F}$ less than 425 lb. Therefore, our assumptions were on the conservative side.

This reasoning process demonstrates an important consideration in analyzing forces and stresses in a structure: Assumptions should always err on the side of forces and stresses being larger than they might prove to be in the actual case. This is what we mean by being conservative. We want our structures to be safe.

The next step is to decide how large the upper bolt must be in order to support a load of 425 lb. Figure 15.10(c) is a free-body diagram of a typical bolt. (The head of the bolt is hexagonal in shape; hence, bolts of this type are generally referred to as hex-head bolts.) The tensile load $\mathbf{F}$ is balanced by the forces pressing against the under surface of the bolt head, shown by a group of small arrows. The bolt is assumed to be made of mild steel, and in an earlier discussion we said that a reasonable value for the allowable stress in mild steel is 15 000 psi. However, in this case we will use a much smaller value of allowable stress, 6000 psi. This is appropriate because some factors that we have not yet taken into account could affect our bolt. For example, some unusual tightening stresses may occur when the bolt is assembled on the job. Also, there will most certainly be "stress concentrations" at certain locations in the bolt. What this means is that if a structural member has any sharp angles in it (such as the V angle at the root of a bolt thread), the actual stress at that location can be several times as large as we would expect from our usual stress formulas. There also might be occasional shock loads on our lamp, which would cause the stresses to be larger than we might expect. In a more careful analysis, all of these elements would be pinned down: We might specify a maximum tightening torque that could be used, in order to limit tightening stresses; we might look up the actual magnitude of the stress concentration at the root of a bolt thread (handbooks contain such data); and we might assign an actual value to the shock loads that could occur from, say, an earthquake. One thing we certainly would do in the case of our lamp is to design it to withstand a wind load. Such a load would be expected to be at right angles to the gravity load we are dealing with and would probably require two more bolts, one on each side of the support bracket.

Bearing in mind that a full analysis would have to take all these things into account, we will proceed with our simplified example and calculate how large the bolt must be to hold a load of 425 lb. The formula we will use is Equation (15.2), $\sigma = F/A$, but we will rearrange it so we have $A = F/\sigma$. For σ we will use the allowable stress of 6000 psi and find out how large A must be:

$$A = \frac{F}{\sigma} = \frac{425 \text{ lb}}{6000 \text{ lb/in.}^2} = 0.0708 \text{ in.}^2$$

The formula for the area of a circle is $A = \pi d^2/4$, so

$$d^2 = \frac{4}{\pi} A = \frac{4}{\pi} (0.0708 \text{ in.}^2) = 0.0902 \text{ in.}^2$$

$$d = \sqrt{0.0902 \text{ in.}^2} \cong 0.300 \text{ in.}$$

It is important to realize that the diameter where failure of the bolt will occur is at the so-called "root" of the thread. In Figure 15.10(c), the *nominal diameter* is marked. If we are dealing with a nominal $\frac{1}{2}$ in. bolt, then the nominal diameter is the part of the bolt that is actually $\frac{1}{2}$ in. in diameter. But the bolt will not fail there; it will fail by breaking apart at the bottom of the screw threads, where the diameter is considerably less than the nominal diameter. This dimension is called the *root diameter*.

Our bolt must have a root diameter of 0.300 in. or greater. Tables in handbooks are set up conveniently so that they list the root diameters of bolts. If we look in such a handbook, we will find that the root diameter of a standard $\frac{3}{8}$ in. bolt is 0.294 in., and that of a $\frac{7}{16}$ in. bolt, which is the next larger size, is 0.345 in. The $\frac{3}{8}$ in. bolt is too small, so we will have to use $\frac{7}{16}$ in.

Selection of an I-beam

As another example, we will use the elementary beam formulas developed in Section 15.2 to select a standard I-beam to be subjected to a load as shown in Figure 15.11. I-beams are rolled steel shapes that in cross section look like the letter "I" [see Figure 15.11(b)]. Data concerning the properties of certain standard I-beam shapes are given in tables like Table 15.1, which contains selected data from a manual published by the American Institute of Steel Construction.

The beam configuration in Figure 15.11 is that of a *simple beam with a concentrated load in the center*. We know that the applicable formulas for this particular kind of

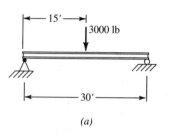

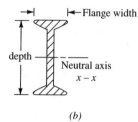

(a)

(b)

Figure 15.11

An example using a steel I-beam.

beam are as follows:

Maximum moment: $M = \dfrac{PL}{4}$

Maximum stress: $\sigma = \dfrac{Mc}{I} = \dfrac{PL}{4S}$

Maximum deflection: $\Delta = \dfrac{PL^3}{48EI}$

Table 15.1   Selected Properties for I-Beams

Nominal Size (in.)	Weight per Foot (lb)	Axis *X-X*	
		I (in.⁴)	*S* (in.³)
24 × 7	79.9	2087.2	173.9
20 × 7	95.0	1599.7	160.0
	85.0	1501.7	150.2
20 × 6¼	75.0	1263.5	126.3
18 × 6	70.0	917.5	101.9
12 × 5	35.0	227.0	37.8
	31.8	215.8	36.0
10 × 4⅝	35.0	145.8	29.2
	25.4	122.1	24.4
8 × 4	23.0	64.2	16.0
	18.4	56.9	14.2
6 × 8⅜	17.25	26.0	8.7
5 × 3	10.0	12.1	4.8
4 × 2⅝	7.7	6.0	3.0

Source: American Institute of Steel Construction.

We do not yet know what I is, of course. That is what we need to find. We will use a maximum allowable stress of 15 000 psi, and we know that E for steel is 29 000 000 psi.

An inspection of Table 15.1 shows us that values of both I and S are given for each size of beam. Axis X-X is the neutral axis for the case in which the beam is set on edge, as in the cross-sectional view in Figure 15.11(b). We already know from experience that this is the most effective orientation for a beam, although we recognize that the beam could be laid flat, in which case the values of I and S would be different.

We will use the form of the stress equation $\sigma = PL/4S$, since values of S are given in the table. Let us solve it for S and then substitute values for $\sigma = 15\,000$ psi, $P = 3000$ lb, and $L = 30$ ft $= 360$ in.

$$S = \frac{PL}{4\sigma} = \frac{(3000 \text{ lb})(360 \text{ in.})}{4(15\,000 \text{ lb/in.}^2)}$$

$$= 18 \text{ in.}^3$$

We now scan the values in Table 15.1 to find the smallest value of S that is larger than $S = 18$ in.^{3}. A beam of nominal size 10 in. $\times$ $4\frac{5}{8}$ in., and weighing 25.4 lb per ft, is our best choice because it has a value of $S = 24.4$ in.^{3}, and the next smaller beam has a value of only $S = 16.0$ in.^{3}, which is too small. Thus, the beam we have selected has a depth of 10 in. and a flange width of $4\frac{5}{8}$ in.

Since this beam has a value of S larger than the minimum of 18 in.^{3} that we require, it is desirable to see what the level of stress actually will be, as follows:

$$\sigma = \frac{PL}{4S} = \frac{(3000 \text{ lb})(360 \text{ in.})}{4(24.4 \text{ in.}^3)}$$

$$\cong 11\,070 \text{ lb/in.}^2$$

So the actual stress in the beam is about 25 percent less than the allowable level.

Having now selected our beam, we begin to worry about its weight, which is 25.4 lb per ft. For a 30-ft beam, this produces a total weight of 762 lb, which might or might not be serious. We decide to use the formulas for a *simple beam with a uniformly distributed load* to see if the additional stress caused by the weight of the beam is serious. The formulas for the uniformly loaded beam are as follows:

Maximum moment: $M = \dfrac{wL^2}{8} = \dfrac{(25.4 \text{ lb/ft})(30 \text{ ft})(360 \text{ in.})}{8}$

$$= 34\,290 \text{ lb-in.}$$

Maximum stress: $\sigma = \dfrac{Mc}{I} = \dfrac{M}{S} = \dfrac{34\,290 \text{ lb-in.}}{24.4 \text{ in.}^3}$

$$= 1405 \text{ psi}$$

We are now going to use a principle we have not encountered before in this book: the *principle of linear superposition*. The kinds of solutions we have been examining fall within a class where the variables behave in a linear fashion. In more advanced courses you will learn when the principle applies and when it does not, but it does apply in the present case. What this means to our problem is this: If we get a solution for a simple beam with a concentrated load, and then get another solution for the same simple beam with a uniformly distributed load, we can add the two solutions together and treat them as the solution for a simple beam that has both loads at once. In our beam, both loads certainly are operating simultaneously, because the 3000-lb load is resting on the beam, and the beam cannot escape being subjected to its own distributed weight of 25.4 lb/ft. Hence, if we add the value of 1405 psi to the one we obtained previously, 11 070 psi, we get a total of 12 475 psi, which is still well within the allowable limit of 15 000 psi. Therefore, we are justified in ignoring the weight. We should recognize, however, that this happy result only comes about because we were forced to select a standard beam with a value for S that is larger than the required value we originally calculated. If we had selected a beam that barely had a value of $S = 18$ in.^{3}, then, when we added the stress from the beam's own weight, the total stress would have exceeded the allowable limit, and we would have been forced to select a larger beam.

As a final step, let us compute the maximum deflection for the beam—first for a simple beam with a concentrated load in the middle, and then for a simple beam with a uniformly distributed load. Since the deflection formulas require the use of I rather than S, we must go back to Table 15.1. Next to the column in which we found $S = 24.4$ in.^{3}, we find the corresponding value of $I = 122.1$ in.^{4}. After we calculate these two deflections separately, we will add them, using the principle of linear superposition.

Concentrated load:

$$\Delta = \frac{PL^3}{48EI} = \frac{(3000 \text{ lb})(360 \text{ in.})^3}{48(29 \times 10^6 \text{ lb/in.}^2)(122.1 \text{ in.}^4)} = 0.824 \text{ in.}$$

Uniformly distributed load:

$$\Delta = \frac{5wL^4}{384EI} = \frac{5(wL)L^3}{384EI}$$

$$= \frac{5[(25.4 \text{ lb/ft})(30 \text{ ft})](360 \text{ in.})^3}{384(29 \times 10^6 \text{ lb/in.}^2)(122.1 \text{ in.}^4)} = 0.131 \text{ in.}$$

Combined:

0.824 in. + 0.131 in. = 0.955 in.

Manufacturing Tolerances

The use of beam formulas can arise in unexpected situations. In the foregoing examples we dealt with typical structural elements. Now we will use the beam formulas to assist us in analyzing manufacturing tolerances.

In Figure 15.12, let us suppose we have two parts of a mechanism that must be assembled under conditions where there is likely to be a misalignment between the part on the left and the slot on the right into which it is to be inserted. Such misalignment can easily result from normal variations in manufacturing. In fact, any design

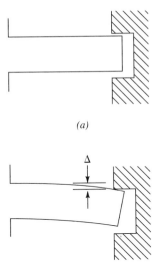

(a)

(b)

Figure 15.12

Misalignment resulting from manufacturing tolerances.

must always take such variations, called *tolerances*, into account. The problem here is to determine how much misalignment, Δ, can be tolerated without overstressing the part on the left.

The part on the left, which we will refer to as the "finger," can be treated as a cantilever beam. During the assembly process, the finger can be inserted into the slot even if there is misalignment, because the parts have not yet been tightened into their final positions. This condition is shown in Figure 15.12(*a*). But, as the parts settle into their final positions (perhaps by tightening down on some bolts), they may assume the relationship shown in Figure 15.12(*b*). (In accordance with our usual practice, we have exaggerated the misalignment Δ. If we drew it to actual scale, it would be invisible to the eye.) The finger is deformed because of the misalignment, and its resemblance to a cantilever beam with a concentrated load at the end is evident.

We use Equations (15.6) and (15.8):

$$\sigma = \frac{Mc}{I}$$

$$\Delta = \frac{PL^3}{3EI}$$

The maximum stress is at the base of the finger, and the moment there is $M = PL$. The value of c is equal to $h/2$, where h is the thickness of the finger. The width of the finger (perpendicular to the plane of the paper) is b, and we know that I, for a beam of rectangular cross section, is $bh^3/12$. Therefore, by making appropriate substitutions and eliminating P between Equations (15.6) and (15.8), we get

$$\Delta = \frac{2\sigma L^2}{3hE} \tag{15.10}$$

In the process of deriving this equation, we discover that b, the width of the finger, does not enter at all, which is a useful thing to know. We also note that I gets cancelled out in the process, which is also useful. If the finger is 1 in. long and 0.125 in. thick, and if we use an allowable limit for σ equal to 15 000 lb/in.^{2}, and $E = 29\ 000\ 000$ lb/in.^{2} (assuming the parts are made of mild steel), we obtain

$$\Delta = \frac{2(15\ 000\ \text{lb/in.}^2)(1\ \text{in.})^2}{3(0.125\ \text{in.})(29 \times 10^6\ \text{lb/in.}^2)}$$

$$= 0.0028\ \text{in.}$$

This is a disturbingly small quantity. It means that if this much misalignment occurs, the finger will be stressed up to the allowable limit. To keep the misalignment this small or smaller, the manufacturing tolerances will have to be controlled very closely, perhaps more closely than is economically justified. In fact, this condition might be so difficult to deal with that you would begin to try to come up with a completely different design to eliminate the misalignment problem.

Exercises

15.3.1 In the cantilever beam example of Section 15.2, we had a beam 10 ft long, with $b = 3$ in., $h = 6$ in., and $P = 2000$ lb. For that beam, we found the magnitude of the stress to be 13 330 lb/in.^{2}. If all of the beam's dimensions are to be kept constant except b, the width, what should b be in order to produce a maximum stress of 15 000 lb/in.^{2}?

15.3.2 In Exercise 15.3.1, assume that b is to be kept constant at 3 in., and h is to be changed until a maximum stress of 15 000 lb/in.^{2} is achieved. What should h be?

15.3.3 In a mild steel cantilever beam with a concentrated load at the end (see the following figure), the deflection must be limited to 10 mm when a load of 1000 N is applied. If $b = 170$ mm, what must h be? (Use $E = 200$ GPa.)

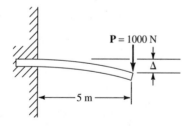

$P = 1000$ N

Δ

5 m

15.3.4 Suppose we have a mild steel cantilever beam with the dimensions and loading shown in the following figure. (a) What is the maximum stress in the beam, and what is the maximum deflection? (b) Suppose the beam is made of an aluminum alloy; the allowable stress for this alloy is the same as that for mild steel, 15 000 psi, but E for the aluminum alloy is 10 000 000 psi. Now what is the maximum stress in the beam? The maximum deflection? Compare your answers with those for part (a).

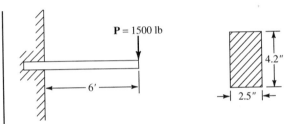

15.3.5 A mild steel cantilever beam supports a weight of 3000 lb, as shown in the following figure. Select an I-beam from Table 15.1 that is the smallest possible, without exceeding the allowable stress limit of 15 000 psi. Ignore the weight of the beam.

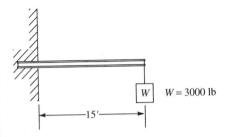

15.3.6 The beam in the following figure, when loaded by nothing but its own weight, must not sag more than 0.125 in. in the center. Select an I-beam from Table 15.1 that will satisfy this requirement. Assume the beam is made of mild steel. (*Hint:* Arbitrarily select a beam and see what deflection results. Then try the next-size beam and repeat until you satisfy the deflection criterion.)

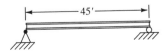

15.3.7 A simple beam is loaded as shown in the following figure. Select an I-beam from Table 15.1 that will have a maximum stress no greater than 100 MPa (which is approximately 15 000 psi). Use $E = 200$ GPa (which is approximately 29 000 000 psi). Ignore the weight of the beam.

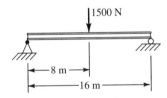

Photographs courtesy of General Atomics (top left), Grumman Aerospace Corporation (center), and United States Steel (bottom).

16

Work, Energy, and Heat

Most of the activities of engineers have something to do with energy. Certainly, the flow of electrons in integrated circuits involves energy, and civil and mechanical engineers use what they call "energy methods" for solving some kinds of structural problems. But in this chapter we will deal with special kinds of energy systems in which thermal energy is converted into some kind of work, for it is through such processes that machines are made to do useful tasks. *Thermal energy*, of course, is the kind of energy a substance acquires when it is heated. There are other kinds of energy, such as *potential energy* and *kinetic energy*, which will be discussed in the next section. But thermal energy is especially important because most of the energy used by humans is produced by burning something—most often oil, natural gas, or coal. A significant share of this thermal energy is used directly to heat houses, fire ore furnaces, provide heat for chemical plants, and the like. But much of it also is used to run engines or to produce steam, which is used to turn turbines, which then turn generators, which produce electricity. In these latter cases, the thermal energy ultimately causes a shaft in a machine to rotate, and for this reason we usually refer to the product of such a process as *shaft work*.

■ 16.1 WORK, ENERGY, AND POWER

In dealing with energy we first must have a precise definition of the term *work*; in other words, we must create a mathematical model. The basic concept is simple: Work is done on

✺ Milestones
The Jet Engine

For more than seven years after Frank Whittle invented the jet engine in 1929, he could get little official attention for his ideas. Whittle had early recognized that high-speed flight would have to occur at high altitudes, where the air density is low. First he considered, then rejected, a rocket propulsion system. Next he considered a jet with a conventional piston engine driving a compressor. But the piston engine would make the system too heavy, and Whittle rejected it. It was then that he conceived the idea of inserting a gas turbine in the expanding air downstream from the combustion chamber and using the turbine to run the compressor. It was this idea that eventually made the jet engine successful, for it resulted in a powerful, compact, lightweight device.

The development of the jet engine began in earnest in 1936, and the first jet flight in England occurred five years later. Those five years were filled with a succession of frustrations and disappointments. Turbine blades had a habit of breaking constantly. Parts became misaligned and destroyed themselves. Unexplained surges in engine speed occurred. Carbon built up and blocked the fuel vaporizing system. Sand got in the lubricating oil. Parts overheated locally and burned. The most troublesome was the vaporizing and combustion system; its problems were not overcome until the vaporizing process was dropped and a liquid spray substituted. The first flight using a Whittle engine occurred in 1941, and by the end of World War II two models of turbojet-powered aircraft were in production; a hundred or so saw actual combat.

About the same time Whittle was working, a parallel project was going on in Germany, pioneered by Hans von Ohain. Neither Whittle nor von Ohain knew of the other's work. Von Ohain came up with essentially the same system as did Whittle, although his ideas appear to have been conceived about four years later than the Englishman's. However, events moved more rapidly in Germany than in England, and as a result the Germans flew the world's first turbojet aircraft almost two years before the English, in 1939. They also moved into production faster, making between 1000 and 2000 jet planes of varying models before the war's end. Many of these saw combat service, causing some consternation on the Allied side.

The Americans entered the picture also, having acquired Whittle's plans from England in early 1941. The General Electric Company was assigned the further development of Whittle's engine. By the end of the war, the United States had three prototype models of jet aircraft. None saw service, although late in the war some prototypes were sent to Europe to raise the morale of American pilots by showing that something was on the way to match the German jets.

In the end, the jets did not influence the course of the war. However, an event that has been called the "turbojet revolution" had taken place, and the stage had been set for the virtual takeover of commercial flights by jet aircraft only 10 or 15 years later. For their creative work, Whittle and von Ohain in 1991 were jointly awarded the Charles Stark Draper Prize, which is considered to be the engineering equivalent of the Nobel Prize.

References

E. W. Constant, *The Origins of the Turbojet Revolution* (Baltimore: Johns Hopkins University Press, 1980).

R. R. Whyte, *Engineering Progress Through Trouble* (London: Institution of Mechanical Engineers, 1975).

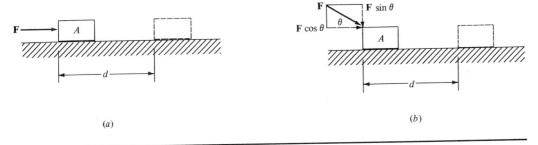

Figure 16.1
Work done on a body by a force **F** acting through a displacement *d*.

a body by a force acting on the body through a displacement in the same direction as that of the force. As shown in Figure 16.1(*a*), the work done on the body *A* in moving through the displacement *d* is

$$W = F \cdot d \tag{16.1}$$

In the case shown, if the force **F** shoves a body along a flat surface, it is likely that the force is merely overcoming the frictional resistance of the surface. In this case, the work done is certainly $W = Fd$, but in most such cases no useful result has occurred, because the work W that was done to overcome friction merely appears as thermal energy.

If the force does not act in the direction of the motion, but is inclined at an angle θ, as shown in Figure 16.1(*b*), then we use the component that is aligned with the direction of movement. In the case shown, the component of interest is $F \cos \theta$, so the work is $W = F \cos \theta \cdot d$.

Since most of the cases of interest cause a shaft to rotate, we should look at Figure 16.2, where we again have a force **F** acting through a displacement *d*, represented as if the force were being exerted by pulling on a rope at the outer rim of a pulley *S* having radius *r*. Instead of the force **F**, we may visualize a twisting moment, or *torque*, acting to cause the shaft to turn. This twisting moment has the value $T = F \cdot r$, by definition. We also know, from geometry, that the displacement $d = r\alpha$,

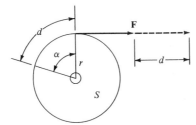

Figure 16.2
Work done in causing a shaft to rotate.

where α is in radians. Therefore, we can write

$$W = F \cdot d$$

$$= \frac{T}{r} \cdot r\alpha$$

$$W = T\alpha \tag{16.2}$$

The units of work are expressed in newton-meters (SI) or foot-pounds (English). Since 1 newton-meter is given the name *joule*, work is ordinarily expressed in SI in joules. In Equation (16.2), since torque is expressed as the product of a force times a distance, the units of work in this equation are also in joules (newton-meters) or foot-pounds. The angle α is expressed in radians and is therefore dimensionless, so work is expressed in the same units whether resulting from a force acting along a straight line or a torque acting through an angle.

Potential Energy and Kinetic Energy

The concept of *energy* is directly derivable from that of work, because the amount of energy possessed by a system is looked upon as its capacity for doing work. That is, energy and work are equatable. In Figure 16.3, if the body A is lifted by a force **F** (**F**, of course, is equal to the weight of the body) a distance d above a reference plane, the work done is $W = Fd$. As long as the body remains poised at the height d, it has the capacity to deliver this work back again if it is allowed to drop. We say, then, that the body has Fd joules (or foot-pounds) of *potential energy*. If the body is allowed to drop, it begins to lose its potential energy as it falls toward the reference plane. But, as it does so, it begins to pick up speed, and we say that its potential energy is being converted into *kinetic energy*. Thus, potential energy is energy that a body possesses by virtue of its position, and kinetic energy is energy that a body possesses by virtue of its motion. When the body has fallen all the way to the reference plane (which might be the surface of the Earth, in a practical example), its potential energy is equal to zero with respect to that reference plane, and all of its potential energy has been converted into kinetic energy.

We can compute how much kinetic energy a body will possess after it has fallen from a height d. From Equation (14.44) in Chapter 14, we know that the force acting

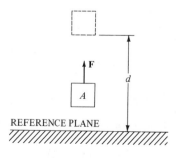

REFERENCE PLANE

Figure 16.3

Potential energy acquired by a body as a result of lifting it to a height d above a reference plane.

on the body, from Newton's second law, is

$$F = W = mg \qquad (16.3)$$

because the acceleration under free fall is g, the acceleration of gravity. If a body starts from rest and moves under constant acceleration a until it reaches a velocity v in time t, then the definition of constant acceleration gives us

$$a = \frac{v}{t} \qquad (16.4)$$

Similarly, from basic definitions, the distance s traveled by a body starting from rest is the average velocity times the time t. Under constant acceleration, the velocity increases uniformly from zero to the velocity v so that the average velocity is $v/2$. Thus, for this situation we can write

$$s = \frac{v}{2} t \qquad (16.5)$$

For our example of a falling body, the acceleration a in Equation (16.4) is g, the acceleration of gravity, so that we have $g = v/t$. If the work done by the body falling from height d is $F \cdot d$ and if all of this work now appears as the kinetic energy of the body, then we can write

$$\text{kinetic energy} = F \cdot d \qquad (16.6)$$

From Equation (16.3) and from Equation (16.5), where s now is the distance d, we can substitute into Equation (16.6) and get

$$\text{kinetic energy} = mg \cdot \frac{v}{2} t \qquad (16.7)$$

Finally we substitute Equation (16.4) into Equation (16.7), noting that here a is the acceleration of gravity g, and get

$$\text{kinetic energy} = m \cdot \frac{v}{t} \cdot \frac{v}{2} t$$

$$= \frac{1}{2} mv^2 \qquad (16.8)$$

Even though Equation (16.8) has been derived under the special conditions of constant acceleration, it is in fact the general expression for kinetic energy. (Proof of this will not be given here, but is readily demonstrated through the use of calculus.)

In Equation (16.8), if m is in kilograms (kg) and v is in meters per second (m/s), then the units of kinetic energy can be computed as $kg \cdot m^2/s^2$. But we know the newton (N) has the units $kg \cdot m/s^2$, so the units of kinetic energy are $N \cdot m$, which is the definition of the joule.

If we are working in the English gravitational system, the mass m must be given in slugs, lb-sec^2/ft. If we substitute this into Equation (16.8) along with the velocity squared, we get

$$\text{kinetic energy} = \frac{\text{lb-sec}^2}{\text{ft}} \cdot \frac{\text{ft}^2}{\text{sec}^2}$$

which, after appropriate cancellations, gives us ft-lb, as it should.

We should note that the equation derived in the foregoing example for the kinetic energy of the falling body applies just prior to the instant that the body arrives at the reference plane. If the reference plane is a solid surface such as the surface of the Earth, all of the kinetic energy is dissipated in the impact and is wasted by merely heating up the surroundings. If the reference plane is not solid, but is elastic like a spring, then some portion of the kinetic energy is stored momentarily in this elastic structure. If the elastic member is restrained after storing this energy, it can give it back later, in a controlled fashion, and do useful work (think of it like a clock spring). Otherwise, the elastic structure immediately delivers its potential energy back to the object and drives it upward again, and the process is repeated with decreasing amplitude of motion until all the energy has been dissipated as heat.

Alternatively, if the falling body strikes a machine structure in a preplanned way, as by striking a lever attached to a crankshaft, it can do useful work. This is exactly what happens in a waterwheel, although in that case the "body" is not a solid object, but a stream of water. Or if the kinetic energy is that belonging to a jet of steam, and the steam is directed against the blades of a turbine, then the steam gives up its kinetic energy by striking these blades and causing the turbine shaft to rotate.

Conservation of Energy

In physics it is shown that energy can be neither created nor destroyed. This fact leads to the *law of conservation of energy*, which states that if energy in one form disappears, it will simply appear in another form. We can state this idea as the *first law of thermodynamics*: For any given system, the net heat supplied to the system must equal the increase in internal energy of the system plus the energy that leaves the system in the form of work. (Here we omit any consideration of the equivalence between mass and energy arising from the theory of relativity.) We have just seen a case of the conversion of potential energy into kinetic energy, and earlier we mentioned how mechanical work could be converted to thermal energy through friction. In the nineteenth century, James Joule designed experiments to measure the equivalence between work and thermal energy, leading ultimately to the following values:

Q (thermal energy)		W (mechanical work)	
1 Btu	=	778	ft-lb
1 calorie	=	4.186 J	

The British thermal unit (Btu) is defined as the amount of thermal energy needed to raise 1 pound of water 1°F. The calorie formerly was defined as the amount of thermal energy needed to raise 1 gram of water 1°C. However, the calorie is no longer approved for use as a unit because the joule is the unit we use in SI for all kinds of energy, whether in the form of thermal energy or of work.

In Joule's original experiments, he used falling weights connected by pulleys and ropes to a paddle wheel, which rotated in a container with water. He calculated the work input from his knowledge of the potential energy loss of the weights and measured the temperature rise of the water resulting from the friction of the paddle wheels. Later experiments have confirmed and refined Joule's early measurements.

Consideration of the matter of temperature rise resulting from the input of heat or work into a substance leads us to the concept of *specific heat*. If we have M lb of a substance, the quantity of heat Q necessary to raise the M lb by a temperature differential ΔT is

$$Q = Mc \cdot \Delta T \tag{16.9}$$

where c is the specific heat of the substance. In the English system, if Q is in Btu, M is in lb, and ΔT is in °F, then c for water is equal to 1 Btu/lb-°F at 63°F. (The temperature must be specified because c is not constant with temperature.) The value of c in SI is 4.186 J/g·K. In an earlier form of the metric system, it was 1 cal/gm-°C, but the calorie is no longer an approved unit. In SI, Q is in joules, M is in grams (if the value $c = 4.186$ J/g·K is used), and ΔT is in kelvin.

The specific heat for a gas depends not only upon its temperature, but also upon the process under which heat is added. Two important processes are those at constant pressure and at constant volume. For air, the specific heat at constant pressure is $c_p = 0.2375$, and that at constant volume is $c_v = 0.169$.

Power is defined as the rate of doing work. In the English system, then, we would typically have units of power such as foot-pounds per second. One *horsepower* is defined as 550 ft-lb/sec. In SI units, a *watt* is defined as 1 joule per second. Thus, it is immediately obvious that there must be an equivalence relation between horsepower and watts. That relation is

$$1 \text{ hp} = 745.7 \text{ watts} \tag{16.10}$$

Exercises

16.1.1 Five gallons of water are stirred, $\frac{1}{2}$ hp of energy being delivered to the stirring process continuously. Assume all the stirring energy appears in the form of a temperature rise of the water. What is the temperature rise after 15 minutes?

Solution

Equation (16.9): $Q = Mc \cdot \Delta T$

$$\Delta T = \frac{Q}{Mc}$$

$$Q = 0.5 \text{ hp}\left(42.41 \frac{\text{Btu/min}}{\text{hp}}\right)(15 \text{ min}) = 318.1 \text{ Btu}$$

$$M = 5 \text{ gal}\left(0.1337 \frac{\text{ft}^3}{\text{gal}}\right)\left(62.4 \frac{\text{lb}}{\text{ft}^3}\right) = 41.7 \text{ lb}$$

(See Appendixes K and L for density and for conversion factors.)

$$c = 1\,\frac{Btu}{lb\text{-}°F}$$

$$\Delta T = \frac{Q}{Mc} = \frac{318.1\ Btu}{(41.7\ lb)(1\ Btu/lb\text{-}°F)}$$

$$= 7.6°F \qquad\qquad \text{Answer}$$

16.1.2 A brake shoe applies a braking force on a brake drum 13 in. in diameter. The force, applied tangentially to the drum, is constant at 20 lb. The drum is brought to a complete stop after 25 revolutions. How much heat was generated by the braking action?

Solution

$$\text{work} = T\alpha$$

$$T = Fr = 20\ lb(6.5\ in.)\left(\frac{1\ ft}{12\ in.}\right)$$

$$= 10.8\ lb\text{-}ft$$

$$\alpha = 25\ rev\left(\frac{2\pi\ rad}{rev}\right) = 157.1\ rad$$

$$\text{work} = T\alpha = 10.8\ lb\text{-}ft(157.1\ rad)$$

$$= 1697\ ft\text{-}lb$$

Appendix L: $1\ ft\text{-}lb = 1.285 \times 10^{-3}\ Btu$

$$1697\ ft\text{-}lb\left(\frac{1.285 \times 10^{-3}\ Btu}{ft\text{-}lb}\right) = 2.18\ Btu \qquad \text{Answer}$$

16.1.3 A 2-kg mass is released from rest and falls 3 m under the action of gravity ($g = 9.81\ m/s^2$) until it strikes the Earth's surface. Using Equations (14.49) and (14.50), calculate the velocity at the instant of impact. Then, using the concepts of potential energy and kinetic energy introduced in this chapter, calculate the velocity at impact. Compare your answers.

Solution

$$\Delta s = v_0 \cdot \Delta t + \frac{1}{2}g(\Delta t)^2$$

$$v_f = v_0 + g \cdot \Delta t$$

$$v_0 = 0 \quad \text{so} \quad \Delta t = \frac{v_f}{g}$$

$$\Delta s = \frac{1}{2}g\left(\frac{v_f}{g}\right)^2 = \frac{v_f^2}{2g}$$

$$v_f^2 = 2g \cdot \Delta s = 2(9.81 \text{ m/s}^2) \cdot 3 \text{ m}$$

$$v_f = \sqrt{58.86 \frac{m^2}{s^2}} = \textbf{7.67 m/s} \qquad \text{Answer}$$

All of the original potential energy (P.E.) is converted into kinetic energy (K.E.) at impact, so

$$(\text{P.E.})_{\text{orig.}} = (\text{K.E.})_{\text{impact}}$$

$$(\text{P.E.})_{\text{orig.}} = F \cdot d \qquad \text{but} \qquad F = mg$$

$$(\text{P.E.})_{\text{orig.}} = mg \cdot d = 2 \text{ kg} \times \left(9.81 \frac{m}{s^2} \right) \times 3\text{m}$$

$$\text{K.E.}_{\text{impact}} = \frac{1}{2} mv^2 = \frac{1}{2} (2 \text{ kg})v^2 = (v^2) \text{ kg}$$

$$(2 \cdot 9.81 \cdot 3) \frac{\text{kg} \cdot \text{m}^2}{s^2} = v^2 \text{ kg}$$

$$v^2 = 2 \cdot 9.81 \cdot 3 \frac{m^2}{s^2}$$

$$v = \textbf{7.67 m/s} \qquad \text{Answer}$$

16.1.4 A vessel containing water is 1 m in diameter and 1 m deep. It is stirred continuously, with 2000 W of energy delivered continuously. After 10 minutes, assuming no energy has been transferred from the water to its surroundings, what is the temperature rise, in kelvin? Convert all values to English units, and recompute your answer.

16.1.5 A vehicle is moving 20 km/h. A brake shoe is suddenly applied to a brake drum 13 inches in diameter and exerts a tangential braking force of 1000 N to the drum. The mass of the vehicle is 2000 kg. The diameter of the vehicle's wheels is 27 in. How many meters will be required to bring the vehicle to a complete stop? Assume the braking force is the only element that retards the vehicle's motion. (*Hint:* Assume all the initial kinetic energy of the vehicle is consumed by braking energy.)

16.1.6 An object that weighs 10 lb falls to the Earth's surface from a height of 100 ft, having started from rest. Calculate its impact velocity using the "energy method" of Exercise 16.1.3, and then calculate its impact velocity using Equations (14.49) and (14.50).

16.1.7 A motor uses 9500 W of electricity, input, and delivers 10 hp in the form of useful output work. Over a 24-hour day, how much energy, in joules, is wasted? If efficiency is defined as

$$\text{efficiency} = \frac{\text{output work}}{\text{input work}}$$

then what is the efficiency of this motor?

■ 16.2 ENERGY CONVERSION

Converting work to heat is easy. For example, we can apply friction, and the available work quickly goes into heating up the surroundings. Unless heating up the surroundings is the effect we want (as in heating a house), we say that the energy has been degraded from a high-quality form to a lower-quality form, because it is less available to produce useful work.

On the other hand, converting heat to work is not so easy. Left to itself, thermal energy usually does not do work at all, but eventually is dissipated throughout the universe, presumably warming it slightly.

Mostly, if heat is converted to useful work, it is done by extremely clever (even though familiar) devices that we call *heat engines*. This is what the engineering science of *thermodynamics* is all about. Mostly, such devices use gaseous media as their so-called working fluids, so we must pause for a moment to look at the behavior of gases.

Ideal Gas Law

A very useful mathematical model of gaseous behavior is provided by an equation known as the *ideal gas law*:

$$PV = MRT \qquad (16.11)$$

where P is the absolute pressure in units of force per unit area, V is the volume, M is the mass of the gas under consideration, T is the absolute temperature, and R is the *gas constant*. The term *absolute pressure* is used in order to distinguish it from *gauge pressure*. Gauge pressure is the pressure measured above atmospheric pressure (14.7 psi, or 101.3 kPa); absolute pressure is measured from true zero. Thus, the two are related as follows:

absolute pressure = gauge pressure + 14.7 psi   (English units)

absolute pressure = gauge pressure + 101.3 kPa   (SI units)

Equation (16.11) holds only for so-called ideal gases (sometimes called perfect gases), and, in fact, Equation (16.11) represents the definition of ideal gas behavior. Unfortunately, many gases do not conform very well to ideal gas behavior, although air is one important gas that comes close (within limits), and so do hydrogen, nitrogen, oxygen, and carbon dioxide. Another important gas—superheated steam—does not come so close to perfect gas behavior as one would like. (The term *superheated* means heated to a temperature above the boiling point of water, which, in turn, depends upon the pressure.)

Nevertheless, Equation (16.11) is useful as a reference frame for the behavior of working gases used in heat engines. For example, we note that, for a given mass of gas, the quantity MR is a constant. Thus, we have

$$\frac{PV}{T} = \text{constant}$$

Then, if we assign the subscript 1 to denote an initial condition (called a *state*, in thermodynamics) of P, V, and T for a given quantity of gas, and the subscript 2 to denote any other state, we can write

$$\frac{P_1 V_1}{T_1} = \frac{P_2 V_2}{T_2} \qquad (16.12)$$

which shows the relationship between pressure and volume of a perfect gas at different states.

Internal Energy

Earlier, we said Joule showed that putting mechanical work into a medium increased its thermal energy. It is obvious that putting *heat* into a medium also increases its thermal energy. The total energy possessed by a medium is referred to as its *internal energy*, to which we usually assign the symbol U. Like potential energy, internal energy is measured with respect to some reference point. In general, we are interested in *changes* of internal energy, and therefore we write an equation relating the change in internal energy ΔU to the amount of heat Q put *into* the system and the amount of shaft work W taken *out of* the system (see Figure 16.4):

$$\Delta U = Q - W \qquad (16.13)$$

This equation is valid for conditions where no flow of the working fluid is taking place and is called the *energy equation for a closed system* (i.e., one with no flow). Obviously, it must be dimensionally consistent. For example, if ΔU and Q are given in Btu, then W must also be converted into its Btu equivalent.

Note that internal energy relates to a state of the medium or system, whereas heat Q and work W are quantities that are put into or taken out of the system. It is not proper to speak of a system as containing a certain amount of heat, just as we would not speak of a system as containing a certain amount of work. We put work (or heat) into or take it out of a system, and the change of state of the system we refer to as its change in internal energy. For an ideal gas, this change of state can be measured by a thermometer, and Joule formulated an important observation, based on experimental evidence, that states that the internal energy *of an ideal gas* depends only upon its temperature.

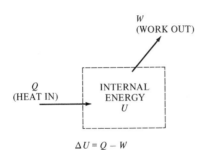

Figure 16.4
The internal energy of a system increases as heat Q is put into the system and decreases as shaft work W is taken out of the system.

Equation (16.13) provides something new beyond what has been said so far about transferring energy from one form into another, because ΔU represents a way in which energy can be stored (or given up) by a substance in case Q and W are not in balance.

Another important law, called the *second law of thermodynamics*, says that heat cannot flow "uphill," i.e., from a cold body to a hot body, without something else happening (such as putting work into the system) and that, although energy in the form of work can be converted entirely to heat, only a portion of the energy in the form of heat can be converted back to work.

Example: Heat Pump

A refrigerator is a system that pumps heat "uphill." Figure 16.5(*a*) is a diagram of a compressor-type refrigeration system, which is commonly used for household refrigerators. The system contains a fluid refrigerant, selected for its special properties, that passes around the system in the direction of the arrows. The fluid enters the *compressor* in the form of a gas and is compressed. When it exits it is still a gas, but under greater pressure. W_{IN} is the amount of work that must be put into the system to operate the compressor. When a gas or a vapor is compressed, it heats up. Therefore,

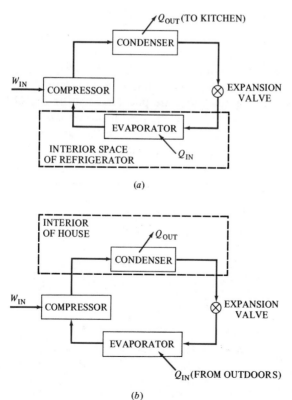

(*a*)

(*b*)

Figure 16.5

Schematic diagrams of a refrigerator (*a*) and a heat pump (*b*).

when the compressed fluid reaches the *condenser*, it is very hot. The condenser is a heat exchanger with its exterior in contact with an air space (in this case, the kitchen) that is cooler than the compressed vapor. Heat Q_{OUT} flows out from the heat exchanger, permitting the vapor to cool off and condense into liquid form. The cooled liquid is forced into the *expansion valve* (the compressor is also acting as a pump) and expands into the lower-pressure area inside the *evaporator*. Expansion causes some of the liquid to turn back into a vapor. If the right working fluid has been chosen, its temperature is reduced by the expansion process to the point where it is cold enough to cool off the interior space of the refrigerator. The evaporator is another heat exchanger, with the inside colder than the space to be cooled, so Q_{IN} flows from the refrigerated space into the working fluid. The working fluid passes on to the compressor, where it is compressed (and heated) once more so that the cycle for the element of fluid we have followed begins anew. The working fluid circulates continuously in the system whenever the compressor is running. An amount of heat Q_{IN} must be constantly removed from the interior of the refrigerator in order to keep it cold, because its insulation is not perfect (and people are regularly opening the door), and heat is always trying to penetrate to the interior of the refrigerator from the outside.

An energy balance on the refrigerator shows that

$$Q_{OUT} = Q_{IN} + W_{IN}$$

which means that the condenser is ejecting more heat into the kitchen than it removes from the inside of the refrigerator. The additional heat, of course, is the heat equivalent of W_{IN}. Thus, a refrigerator heats up the room it is in, an observation that can be confirmed by standing near a refrigerator when it is running and noting the current of warm air that emerges from the back or bottom. In a very real sense, heat has been "pumped" from inside the refrigerator into the kitchen (whereupon it immediately tries to leak back inside the refrigerator through the insulation).

A *heat pump* capitalizes on the phenomenon just described and is, in fact, a refrigerator. Figure 16.5(b) shows a heat pump. It operates exactly like the refrigerator of Figure 16.5(a). The only difference is that the region of interest (the interior of the house) now surrounds the condenser instead of the evaporator. The heat Q_{OUT} removed from the condenser now goes to heat up the interior of the house, and heat Q_{IN} is being transferred from the outdoors into the evaporator heat exchanger. The heat pump is, in fact, refrigerating the outdoors. And, in a very real sense, heat has been pumped from the cold exterior into the house where it is wanted.

Needless to say, the working fluid must be capable of being cooled (by expansion into the evaporator) to a temperature lower than the lowest expected outdoor temperature, or else heat will not flow from the outdoors into the evaporator, and the heat pump will not function as intended. As the outdoor temperature gets lower and lower, the heat pump is less and less effective. Finally, as the outdoor temperature gets to the same value as the temperature of the working fluid in the evaporator, then Q_{IN} becomes zero and Q_{OUT} is merely the heat equivalent of W_{IN}, the work put into the compressor. This observation points up one of the advantages of a heat pump, because it means that, at the planned working temperatures, Q_{OUT} is *greater* than the heat equivalent of W_{IN}. In some practical heat pumps, Q_{OUT} can be four or five times as great as W_{IN}. The amount of heat pumped, Q_{IN}, divided by the input work, W_{IN}, is called the *coefficient of performance*:

$$\text{coefficient of performance} = \frac{Q_{IN}}{W_{IN}}$$

An important feature of the heat pump is that it is capable of getting more heat indoors through the expenditure of work W_{IN} than if W_{IN} were just converted directly into its heat equivalent, as by electric resistance heaters. This explains why heat pumps are more attractive as heating devices than are electric resistance heaters.

Exercises

16.2.1 A volume of ideal gas occupies 1 m³. The gas is heated from 20°C to 60°C at constant pressure. After heating, what volume does it occupy?

Solution

$$\frac{P_1 V_1}{T_1} = \frac{P_2 V_2}{T_2}$$

$$P_1 = P_2 \qquad \text{(these cancel)}$$

$$V_1 = 1 \text{ m}^3$$

$$T_1 = (20 + 273.15) \text{ K} \qquad \text{(absolute temperatures must be used)}$$

$$T_2 = (60 + 273.15) \text{ K}$$

$$\frac{1 \text{ m}^3}{293.15 \text{ K}} = \frac{V_2}{333.15 \text{ K}}$$

$$V_2 = \mathbf{1.14 \text{ m}^3} \qquad\qquad \text{Answer}$$

16.2.2 A system has 15 000 J removed from it in the form of shaft work and has 70 Btu put into it in the form of heat. Its initial internal energy is 500 Btu. What is its final internal energy?

Solution

$$\Delta U = Q - W$$

$$Q = 70 \text{ Btu}$$

$$W = 15\ 000 \text{ J} \left(9.478 \times 10^{-4} \frac{\text{Btu}}{\text{J}} \right)$$

$$= 14.2 \text{ Btu}$$

$$\Delta U = 70 \text{ Btu} - 14.2 \text{ Btu}$$

$$= 55.8 \text{ Btu}$$

$$\text{final internal energy} = 500 \text{ Btu} + 55.8 \text{ Btu}$$

$$= \mathbf{555.8 \text{ Btu}} \qquad\qquad \text{Answer}$$

16.2.3 A volume of ideal gas initially occupies 45.5 ft^3. The gas is cooled from 100°F to 40°F at constant pressure. After cooling, what volume does it occupy?

16.2.4 A system has 100 hp put into it for 5 minutes in the form of shaft work and has 5000 Btu put into it in the form of heat. Its initial internal energy is 10 000 Btu. What is its final internal energy?

16.2.5 A quantity of ideal gas is kept at constant volume while its absolute pressure increases from 150 kPa to 200 kPa. What is the final temperature if the initial temperature is 70°F?

16.2.6 A quantity of ideal gas is compressed from 3 m^3 to 2 m^3, and the absolute pressure increases from 150 kPa to 225 kPa. How much did the temperature change?

16.2.7 A heat pump draws 10 000 Btu per hour from the outdoors, and the compressor requires 1.5-hp input energy. Compute how much energy, in Btu and in joules, is put into the house each hour. What is the coefficient of performance?

■ 16.3 THERMAL CYCLES

Equation (16.11), the equation of the gas law for an ideal gas, is repeated below. (We will use ideal gases for our discussion, because the principles involved will be rendered clearer that way.) Thus, we have

$$PV = MRT$$

For a fixed mass of gas, M is constant, and R by definition is a constant. Therefore, if T is also held constant (that is, isothermal), then Equation (16.11) becomes

$$PV = \text{constant}$$

The preceding equation is that of a hyperbola on the P-V plane. Therefore, if different values of the constant are chosen (i.e., for different values of T), Equation (16.11) defines a family of hyperbolas on the P-V plane, each member of the family being plotted for a given value of T.

In Figure 16.6(a), such a family of hyperbolas has been drawn, each one for a different constant value of temperature $T = T_1$, T_2, T_3, and so on. Curves drawn farther from the origin correspond to larger values of T. Each of the curves is referred to as an *isothermal* because the value of T remains constant along each curve. If the pressure and volume are allowed to change from, say, point a on the T_2 curve to point b so that the temperature is held constant at $T = T_2$, we say that an *isothermal expansion* has taken place. The pressure has decreased from P_1 to P_2, and the volume has increased from V_1 to V_2. The circumstances can be visualized by reference to Figure 16.6(b), which shows a cylinder with a movable piston. At point a, the volume between the piston and the closed end of the cylinder is V_1 and the pressure is P_1. When the piston is moved a distance d to position b, the volume has changed to V_2 and the pressure has dropped to P_2.

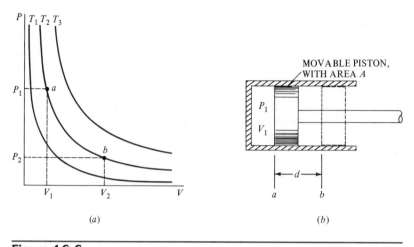

Figure 16.6
A family of hyperbolas for different values $T = T_1$, T_2, T_3, and so on. Each curve is for a constant value of T and is referred to as an *isothermal*.

In moving from point a to point b, it is necessary to put heat into the cylinder to maintain a constant temperature, as can be seen from the following. If we look at Equation (16.13), and recall from Joule's experiments that the internal energy of a perfect gas depends only upon its temperature, then the change in internal energy ΔU must be zero, because the temperature does not change. Thus, for an isothermal process,

$$0 = Q - W \tag{16.14}$$

In Figure 16.6(b), if the piston moves a distance d under the influence of a force F, then we have a case where work is being done—in this case, *removed* from the system. Therefore, for Equation (16.14) to hold true, heat Q must be put into the system.

The foregoing discussion permits us to calculate something very important: the amount of work done during expansion. In Figure 16.7(a) we show an expansion process in a cylinder going from point a to point b. We know that the definition of work is $W = F \cdot d$, and in our case the force $F = PA$, where P is the pressure in the cylinder at any instant, and A is the cross-sectional area of the cylinder. The work, then, is

$$W = PA \cdot d \tag{16.15}$$

where $A \cdot d$ is the change in volume. However, Equation (16.15) holds only if P is constant, which is not the case. In Figure 16.7(b), we have divided up the area under the expansion curve into n narrow segments, each of them with the width ΔV. We have shaded one of them with crosshatching and indicated its height as P_i, the subscript i implying that a different value of P must be used for each narrow segment.

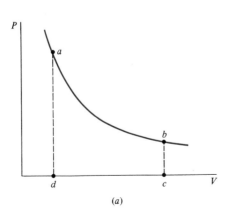

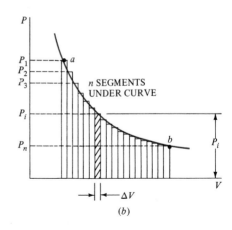

(a)   (b)

Figure 16.7
Work done by an expanding gas.

The area of this small shaded segment represents the amount of work done when the pressure is P_i and the expansion is the width of the tiny segment ΔV, because its area equals $P_i \cdot \Delta V$, and Equation (16.15) says this is the expression for the work. If we choose ΔV to be quite small, calculate the area of each narrow segment, and then add up all the areas so calculated, it can be seen that this procedure will give a close approximation to the area *under the curve ab*, which is the area of the shape labeled *abcd* in Figure 16.7(a). This total area, then, represents the total work done by the expansion from a to b.

The foregoing process can be shown symbolically as

$$W = \sum_{1}^{n} P_i \cdot \Delta V \tag{16.16}$$

where the symbol Σ implies the summation of all the small areas $P_i \cdot \Delta V$. Successive values of $P_1, P_2, \ldots$ are used as the values of the heights of the areas until all n areas have been calculated and summed. The symbology tells us that the subscript i has been allowed to take on all the integer values from 1 to n.

In calculus, we go the next step and write Equation (16.16) as

$$W = \int P \, dV \tag{16.17}$$

where the summation sign has been replaced by the integral sign $\int$ and the finite interval ΔV has been replaced by the infinitesimal interval dV. (See the discussion of calculus in Appendix F.) Equation (16.17) is very important in the study of thermodynamics and will turn up repeatedly as the student proceeds to more advanced studies. We will go no further than this with calculus, but will observe the fact that the work done by this system during the expansion process is equal to the area under the curve ab in Figure 16.7(a). We could also observe that, if the process went from b to a (i.e., a *compression*), then the work done (this time *on* the system) would also be represented by the area under the curve.

Our discussion involved only an isothermal process, but there are many kinds of expansion and compression processes. By a judicious combination of these, it is possible to go through a cycle that starts at a and comes back to a again, wherein the work done during expansion is greater than that done during compression, with *net work being extracted from the system.* Thus, heat has been converted into work.

Exercises

16.3.1 A system like the one in Figure 16.6(b) undergoes an expansion from 3 ft³ to 8 ft³. During this time, the absolute pressure varies according to the equation

$$PV = 50\,000 \text{ ft-lb}$$

How much work was performed by this system during expansion? [Use an approximate method of finding the area under the P-V curve by computing the area of n narrow segments as shown in Figure 16.7(b).]

Solution

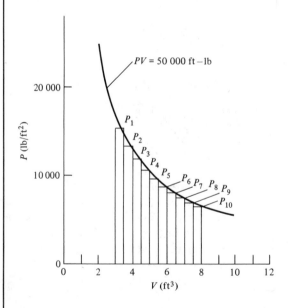

In the figure, a value of 10 has been chosen for n, so we have ten segments to deal with, where $\Delta V = 0.5$ ft³. The ten corresponding values of $P_i = P_1, P_2, P_3, \ldots$ have been scaled off the figure and entered in the table. The areas of the ten segments are then calculated, and their sum computed. Hence, the work done during expansion is

work (area under curve) $= 49.05 \times 10^3$ ft-lb

i	P_i $(\times\,10^3$ lb/ft$^2)$	$P_i \cdot \Delta V$ $(\times\,10^3$ ft-lb$)$
1	15.4	7.70
2	13.3	6.65
3	11.8	5.90
4	10.6	5.30
5	9.6	4.80
6	8.7	4.35
7	8.0	4.00
8	7.5	3.75
9	6.8	3.40
10	6.4	3.20
	$\sum P_i \cdot \Delta V =$	$\overline{49.05} \times 10^3$ ft-lb

Answer

16.3.2 A system like the one in Exercise 16.3.1 undergoes an expansion from 4 m^3 to 10 m^3. During this time, the absolute pressure varies according to the equation

$$PV = 25\,000 \text{ J}$$

How much work was performed by this system during expansion?

16.3.3 A system undergoes an expansion from 5 ft^3 to 10 ft^3. During this time, the absolute pressure varies according to the following equation, where P is given in lb/in.^{2}:

$$P = 200 - 10\,V$$

How much work was performed by the system during expansion?

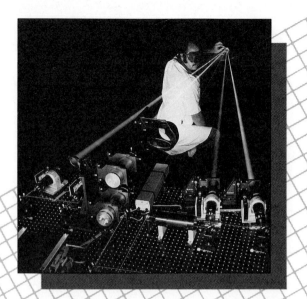

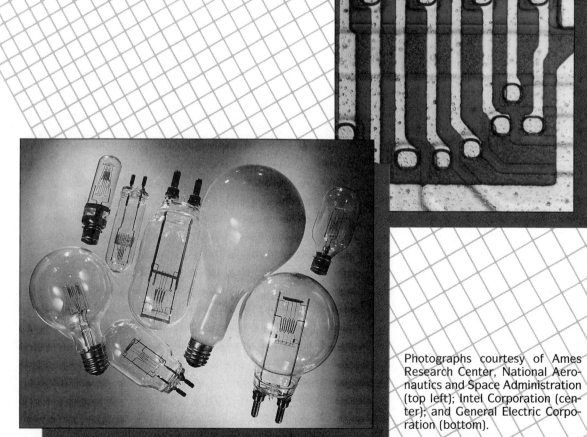

Photographs courtesy of Ames Research Center, National Aeronautics and Space Administration (top left); Intel Corporation (center); and General Electric Corporation (bottom).

17

Electrical Engineering

The term *electrical engineering* once applied solely to the generation and distribution of electrical power. Later the term *electronics* came into being, and it originally was meant to apply to systems wherein electrons moved through evacuated or gaseous media, as in vacuum tubes, or to electric fields generated in the atmosphere, as in radio waves. The older field of electrical engineering, then, presumably dealt with the motion of electrons through solid conductors. But, as semiconductors were developed and largely replaced vacuum tubes, the term *electronics* had to be expanded to include electron motion through these new devices, which of course were solid. This meant that the distinction between *electronics engineering* and *electrical engineering* was increasingly difficult and wound up being mostly arbitrary. Today the field of electrical engineering is generally understood to include electronics engineering, but not the reverse. Electronics engineering includes matters relating to integrated circuits, signal processing, computers, communications, fields, and waves, and in fact includes just about everything from electrical engineering except its original components: generators, motors, transformers, transmission lines, and distribution systems.

In this chapter we will deal with a few of the fundamental ideas of electrical engineering. The purpose is to give some of the flavor of the field, including especially the way in which it uses mathematics.

■ 17.1 FUNDAMENTAL CIRCUIT CONCEPTS

In this section we will explore some of the basic concepts used in the modeling of electrical circuits. As before, a glimpse into more advanced engineering studies will be provided, even though it is recognized that at this point you may not as yet have had sufficient mathematics for a full appreciation of these topics.

As a beginning, let us look at three important elements that turn up repeatedly in electrical (and electronic) circuits: the resistor, the capacitor (sometimes called a condenser), and the inductor (sometimes called a coil). Circuit diagram symbols for these are shown in Figure 17.2. In actual physical situations, a resistor may appear in one of a variety of forms: as a straight wire of resistive material, as a coil of wire, as a cylindrical piece of solid material, or as a "doped" region of an integrated circuit. Real capacitors may be structured in a way to strongly resemble the symbol shown in Figure 17.2 (as in parallel layers in an integrated circuit) or in other forms (such as layers of metallic foil separated by an insulating sheet of paper and rolled up in a cylinder). Inductors may consist of simple coils of wire strongly resembling their circuit symbols, or they may appear in more elaborate forms, such as the windings of electric motors or transformers.

Figure 17.1
An intense beam of light from a laser scribes microscopic lines that connect integrated circuits on the surface of a material. (Courtesy of General Electric Corporation)

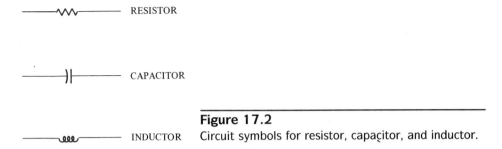

RESISTOR

CAPACITOR

Figure 17.2

INDUCTOR   Circuit symbols for resistor, capaçitor, and inductor.

The mathematical modeling of these elements can be set up as follows. (The physical bases for these mathematical models will be explored in the physics courses taken by all engineering students; they will not be presented in detail here.)

Resistance

For a resistor, the voltage and the current are related by *Ohm's law,*

$$v = iR \tag{17.1}$$

where v is the voltage across the resistor, measured in volts (with symbol V); i is the electric current in the resistor, measured in amperes (with symbol A); and R is the resistance, measured in ohms (with symbol Ω). Current is a fundamentally defined quantity, as was described in Chapter 9. Its unit, the ampere, is equivalent to the rate of flow of electric charge, measured in coulombs per second. The *volt* is defined as the difference in electric potential between two points in a circuit carrying a constant current of 1 ampere, when the power dissipated between these points is equal to 1 watt. The definition for resistance is exactly Equation (17.1) because we can solve it for R, producing the equation

$$R = \frac{v}{i}$$

Equation (17.1) says that a voltage of 1 volt across a resistance of 1 ohm will produce a current of 1 ampere in the resistance.

The validity of Ohm's law rests upon experimental evidence, although it does not hold precisely for all materials. It is as simple an equation as one could ask for, and its usefulness is amazingly broad.

In a resistive circuit, the power dissipated in the circuit is

$$p = vi$$

and is measured in watts (symbol W). Then, since

$$v = iR$$

$$p = iR \cdot i = i^2 R$$

Thus, power lost in the resistance of a circuit (any circuit always has some resistance, however small) is frequently referred to as "*i*-squared-*R*" losses.

Figure 17.3
Using an extremely powerful electron microscope, catalytic reactions at the molecular level are studied as they actually take place. (Courtesy of Exxon Research and Engineering)

Capacitance

For a capacitor, an equation somewhat analogous to Ohm's law can be written

$$v = \frac{q}{C} \tag{17.2}$$

In this equation, v is the voltage measured across the capacitor; q is the amount of electric charge stored in the capacitor, measured in coulombs (with symbol C); and C is called the *capacitance*, measured in *farads* (with symbol F).[1] The volt was defined previously, and the coulomb is defined as the quantity of electric charge transported in 1 second by a current of 1 ampere. Therefore, Equation (17.2) is the defining equation for C, the capacitance. Equation (17.2) says that a voltage of 1 volt placed across a capacitor of 1 farad will cause a charge of 1 coulomb to be stored in the capacitor.

---

[1] Note that the italic letter C is used for capacitance, whereas the roman letter C is used to signify coulombs. Also, the italic letter V is used for voltage, whereas the roman letter V is used to signify volts, as in the expression $V = 100$ V.

As stated before, current is measured in coulombs per second, which means that it is defined in terms of a certain quantity of charge transported by a conductor during a given interval of time. Using the symbol Δq to signify the quantity of charge transported in an interval of time Δt, we can write the current i as

$$i = \frac{\Delta q}{\Delta t}$$

In differential notation (see Appendix F), we write this

$$i = \frac{dq}{dt} \tag{17.3}$$

From calculus, we know that $\Delta q/\Delta t$ is the ratio of a finite change in q divided by a finite interval of time, whereas dq/dt is an instantaneous value of that same ratio, which occurs at a particular instant of time. The two values are usually close to each other, provided Δt is small, and actually become identical as Δt becomes infinitesimally small.

Using these notations, we can alter Equation (17.2) as follows:

$$\frac{\Delta v}{\Delta t} = \frac{1}{C} \cdot \frac{\Delta q}{\Delta t} \tag{17.4}$$

which says that, for a capacitor, if we make a small change Δv in the voltage v during a time interval Δt, then there will be a small change Δq in the charge q during that same time interval Δt, the two changes being related by Equation (17.4). In differential notation, Equation (17.4) would be written

$$\frac{dv}{dt} = \frac{1}{C} \cdot \frac{dq}{dt} \tag{17.5}$$

In the language of calculus, we have *differentiated* both sides of Equation (17.2) with respect to t to obtain Equation (17.5). By use of Equation (17.3), Equation (17.5) becomes

$$i = C \frac{dv}{dt} \tag{17.6}$$

which is a simple mathematical model of the behavior of the voltage (in this case, the *rate of change* of the voltage) and the current of a capacitor. Any time there is a *constant* voltage across a capacitor (meaning $dv/dt = 0$), then no current is flowing in the capacitor.

Inductance

In the case of inductance, we again have a simple equation relating current (this time, the rate of change of current) and voltage:

$$v = L \frac{di}{dt} \tag{17.7}$$

Equation (17.7) is simply the result of experimental evidence and is the defining equation of L, the *inductance* of a coil. L is measured in henries (with symbol H). Experimental evidence has shown that, if there is a *constant* current flowing through a coil, then no voltage will be induced in that coil. However, if there is a *change* of current in the coil, a voltage will be induced in it, governed by Equation (17.7). If a current change of 1 ampere per second in a coil causes a voltage of 1 volt to be induced in the coil, then we say, as a matter of definition, that the coil has an inductance of 1 henry.

These equations have some important consequences. First, Equation (17.1) says that if a voltage v is placed across a resistor R, then a current of i appears instantly. It cannot actually be instantaneous, of course, because all real-world circuits have at least slight amounts of capacitance and inductance, and as we shall see, the action of capacitors and inductors is to oppose instantaneous changes. However, for most practical design purposes, the relationship between voltage and current in a resistor can be considered instantaneous.

Capacitive Circuits

In the case of a capacitor, let us consider the simple circuit in Figure 17.4. The capacitor is initially charged to a voltage v by some external means and is hooked into a circuit with a resistor R, but with the switch open. Of course, as long as the switch remains open, nothing will happen. Intuitively, we expect that as soon as the switch is closed, we will have a complete circuit and some current will flow. What in fact happens is that electric charge flows from one side of the capacitor through the switch and resistor to the other side of the capacitor until it is discharged. The rate of discharge is controlled by how big the resistor is. We instinctively know that if the resistor is small, then the capacitor will discharge very fast, and if the resistor is large, then the rate of discharge will be slow.

We first note that, with the switch closed, if we connect a voltmeter across C, we measure exactly the same voltage as if we had connected the voltmeter across R, because we treat the wires between them as having zero resistance, capacitance, and inductance. Therefore, the wires have no voltage "drops" associated with them. Thus, the voltages across C and R must at all times be equal, and from Equations (17.1) and (17.2) we write

$$iR = \frac{q}{C} \tag{17.8}$$

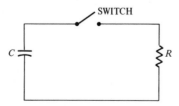

Figure 17.4
A simple circuit for discharging a capacitor through a resistor.

But we know that $i = dq/dt$. However, when we substitute this quantity in Equation (17.8), we write it with a minus sign, because a decrease in the charge in the capacitor, dq, produces an increase in the current i. Thus, we get

$$-R\frac{dq}{dt} = \frac{q}{C}$$

Rearranging, we get

$$\frac{dq}{q} = -\frac{1}{RC}\,dt \tag{17.9}$$

Equation (17.9) is what we call a *differential equation*, and differential equations turn up frequently when we make mathematical models of physical systems. This equation can be solved for q by integrating both sides:

$$\int \frac{dq}{q} = -\frac{1}{RC}\int dt \tag{17.10}$$

However, we will not carry out the details of the integration here. You will learn how to integrate such equations in courses on calculus and physics. For our purposes, it is enough to give the results (although the details of the integration are provided in Appendix G for those who are interested).

The solution to Equation (17.10), which gives us an expression for the charge q as a function of time, is

$$q = Be^{-t/RC} \tag{17.11}$$

where B is a constant that must be evaluated. We have a way of evaluating the constant B, because at the initial time $t = 0$ (just before the switch is closed), we know that q is equal to the initial charge on the capacitor, which we call q_0. Substituting these values into Equation (17.11), we get

$$q_0 = Be^0 \tag{17.12}$$

which means that $B = q_0$, because $e^0 = 1$. Therefore, our complete solution is

$$q = q_0 e^{-t/RC} \tag{17.13}$$

Also, because of Equation (17.2), we can write

$$v = \frac{q_0}{C}\,e^{-t/RC} \tag{17.14}$$

Equation (17.14) is plotted in Figure 17.5 for the particular values of $R = 10^5\ \Omega$ and $C = 10^{-6}$ F. At time $t = 0$, the initial voltage in the capacitor is q_0/C, and the voltage decays exponentially with time. Theoretically, the voltage does not reach zero until infinite time has passed, but as a practical matter, the voltage has become nearly zero after 0.5 s or so.

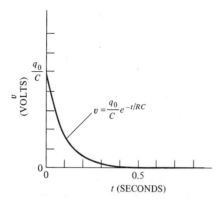

Figure 17.5
Discharge of voltage v in a charged capacitor as a function of time for $C = 10^{-6}$ F and $R = 10^5$ Ω.

The particular value of time for which $t = RC$ is important and is referred to as the *time constant* of the circuit. At $t = RC$, we get

$$v = \frac{q_0}{C} e^{-1}$$

$$= 0.368 \frac{q_0}{C} \qquad (17.15)$$

It is apparent that a circuit with a small time constant discharges very rapidly, and one with a large time constant behaves oppositely. For the circuit in Figure 17.5, $RC = 10^{-1}$ s, or 100 ms. In the period of time equal to one time constant, the voltage on the capacitor is down to approximately 37 percent of its initial level. Time constants and voltage decay rates are very important in electronics, pulse-shaping circuits, and the like.

Inductive Circuits

In the case of inductive circuits, the concept of a time constant comes into play once more. Figure 17.6 shows a simple circuit consisting of a battery with constant voltage V, a resistance R, and an inductor L. Figure 17.6 shows the current i flowing inside

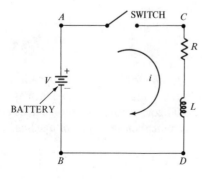

Figure 17.6
A simple circuit with constant voltage source V, containing a resistor and an inductor.

the battery from the negative terminal to the positive one; i.e., it "exits" from the terminal marked +. This is the *positive current* convention. One could argue with some justification that a *negative current* convention would be more proper, since current is actually a flow of electrons, which are negatively charged. However, equally valid results are obtained whether one uses positive current or negative current. It is only necessary to be consistent. In this book, the positive current convention is used, in accordance with common practice.

In Figure 17.6, the question to be dealt with is: When the switch is closed, how does the current in the circuit behave with respect to time? Before the switch is closed, the current i in the circuit is obviously zero. As soon as the switch is closed and the current starts to flow, we know from Equation (17.1) that the voltage across the resistor is iR, and from Equation (17.7) that the voltage across the inductor is $L \cdot di/dt$. As in the case of the RC circuit, we reason that a voltmeter placed across the terminals of the battery, AB, will measure the same voltage value as if it were placed across the terminals CD, because A and C are essentially the same point, and so are B and D—provided the switch is closed, of course. This means that a voltmeter placed across CD will measure the sum of the voltages in the resistor and inductor, which is $iR + L \cdot di/dt$. This is equal to the voltage across the battery V, so we write[2]

$$V = iR + L\frac{di}{dt} \tag{17.16}$$

This equation can be rearranged to give

$$\frac{di}{V/R - i} = \frac{R}{L}\,dt \tag{17.17}$$

As in the case of Equation (17.9), we will not give the details of the integration of Equation (17.17) here (they are included in Appendix G). We will simply give the solution for the current i as a function of time:

$$i = I_f(1 - e^{-Rt/L}) \tag{17.18}$$

In Equation (17.18), I_f is the final value the current reaches after it has stopped changing and settled down. If the circuit had contained only the resistance R, then the current would have jumped "instantaneously" to the value V/R, as required by Ohm's law, Equation (17.1). But there is an inductor in the circuit, and there is a voltage across it only as long as the current is changing, as expressed by Equation (17.7). After the current stops changing, the voltage across the inductor becomes zero, and the circuit behaves as if only the resistor were connected across the battery. Therefore, the final, or *steady-state*, value of the current is $I_f = V/R$.

---

[2] The student should note that we use lowercase letters for v, i, and q when those variables are unknown or are functions of time. When the values are constant, such as for batteries, or are final (steady-state) values, we use capital letters V, I, and Q. It is also customary to use capital letters to indicate the "peak" value of a sinusoidally varying current or voltage, as in $i = I \sin \omega t$ or $v = V \sin \omega t$.

Figure 17.7
A research engineer with precision laser equipment. (Courtesy of University of California, Davis)

Figure 17.8 shows a plot of Equation (17.18), which illustrates the behavior of the current i in an LR circuit after the switch is closed, for the case where $L = 10$ H and $R = 100\ \Omega$. Initially the current is zero, but it rises exponentially, reaching the final value $I_f = V/R$ after infinite time. However, once again we note that i has reached about 99 percent of its final value after 0.5 s.

In the case of an LR circuit, the time constant is $t = L/R$. For this value of t, Equation (17.18) becomes $i = I_f(1 - e^{-1}) = 0.632 I_f$, so we say that the current has reached approximately 63 percent of its final value after a passage of time equal to one time constant.

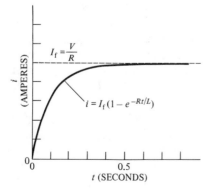

Figure 17.8
The rise in current in an LR circuit as a function of time, for $L = 10$ H and $R = 100\ \Omega$.

Waveform Example

In electronics, it is sometimes necessary to generate an electrical signal that has a sawtooth waveform, such as the one shown in Figure 17.9(a). A circuit that can generate an approximation of the desired waveform is shown in Figure 17.9(b), in which the switch action will be accomplished by an electronic switching circuit that is not shown. The part of the circuit of concern to us here is that containing R and C as it charges and discharges. When the switch (symbolically) is in position b, the capacitor is short-circuited, so for our problem we can assume the discharging of the capacitor is instantaneous, or nearly so, because R for this short-circuit is nearly zero, and therefore so is the time constant RC. During the charging cycle, when the switch is (again, symbolically) in position a, the voltage across the battery also is equal to the sum of the voltages across the resistor and capacitor. The voltage of the battery is V, and the sum of the voltage across the resistor and capacitor is iR, from Equation (17.1), plus q/C, from Equation (17.2). Therefore, we have

$$V = iR + \frac{q}{C} \qquad (17.19)$$

We substitute $i = dq/dt$ from Equation (17.3), noting this time that a positive change in current i produces a positive change in charge dq, so the plus sign is used. Thus, we get

$$V = \frac{dq}{dt}R + \frac{q}{C} \qquad (17.20)$$

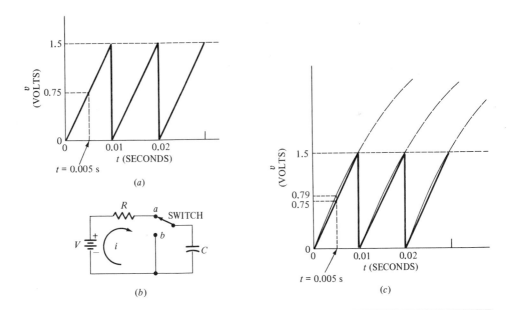

(a)

(b)

(c)

Figure 17.9

A sawtooth waveform. (a) The desired form. (c) An approximation of part (a) generated by the circuit in part (b).

Again we have an equation that must be integrated. (The details of the solution are in Appendix G.) The solution is

$$q = CV(1 - e^{-t/RC}) \tag{17.21}$$

If we substitute Equation (17.2) into Equation (17.21), we get the expression for voltage v across the capacitor:

$$v = V(1 - e^{-t/RC}) \tag{17.22}$$

Equation (17.22) describes an exponential curve resembling the one we plotted in Figure 17.8, but we can only use a short initial segment of the curve if we hope to have it come close to matching one of the desired straight line segments of Figure 17.9(a). We make a first guess that if the desired voltage of 1.5 V, which occurs at $t = 0.01$ s, were reached in one-half time constant, the pulse might be fairly close to linear. So we set the time constant RC to twice 0.01s, or $RC = 0.02$s. We then substitute into

Figure 17.10
A giant electrical generator at Potomac Electric Power Company's Chalk Point installation. This unit is capable of generating 600 MW. (Courtesy of General Electric Corporation)

Equation (17.22) as follows, using $v = 1.5$ V, $t = 0.01$ s, and $RC = 0.02$ s:

$$1.5 \text{ V} = V(1 - e^{-0.01/0.02}) \qquad (17.23)$$

from which we calculate that $V \cong 3.8$ V. We have now tentatively fixed the supply voltage V and the time constant RC. (The actual determination of R and C themselves would finally have to be made, of course, but for our pulse-shaping problem we need to know only the product RC.)

Now we need to check how close we have come to producing a straight line. A quick way to do this is to compare the ideal pulse with the one produced in Equation (17.22) by comparing their midpoint values. The ideal curve has the value of 0.75 V at the midpoint $t = 0.005$ s [see Figure 17.9(a)]. From Equation (17.22) we calculate, for $t = 0.005$ s,

$$v = 3.8(1 - e^{-0.005/0.02}) \text{ V}$$
$$\cong 0.84 \text{ V}$$

This does not look very good, compared to the ideal value of 0.75 V, so we make a second trial with $RC = 0.05$ s. Repeating the previous steps, we get $V = 8.3$ V and v at the midpoint of the pulse equal to 0.79 V. (You should check these calculations to see if they are correct.) This is getting much closer to the ideal desired value of 0.75 V. Whether it is close enough will have to be decided by the designer, based upon the uses to which the waveform is to be put. The result of the second trial is shown in Figure 17.9(c).

Exercises

17.1.1 Three basic electrical measuring instruments are the *voltmeter*, to measure voltage between two points; the *ammeter*, to measure the current (in amperes) flowing in a conductor; and the *ohmmeter*, to measure the resistance (in ohms) of an element. If we have a resistor whose resistance has been measured at 1000 Ω, and it is connected into a circuit that has 0.1 A flowing in it, what is the voltage across the resistor? (We refer to such a voltage as the voltage "drop" across the resistor.) How much power is dissipated in the resistor?

Solution

$$v = iR \qquad \text{(Ohm's law)}$$
$$= 0.1 \text{ A } (1000 \ \Omega)$$
$$= \mathbf{100 \text{ V}} \qquad \qquad \text{Answer}$$

$$\text{power} = vi$$
$$= 100 \text{ V } (0.1 \text{ A}) = \mathbf{10 \text{ W}} \qquad \qquad \text{Answer}$$

or,

$$\text{power} = i^2R$$
$$= (0.1 \text{ A})^2(1000 \ \Omega) = \mathbf{10 \text{ W}} \qquad \qquad \text{Answer}$$

17.1.2 A capacitor of 1 microfarad (1 μF) has a current $i = 0.1$ mA flowing through it at a particular instant. What is the rate of change of voltage, in volts per second, across the capacitor at that instant in time?

Solution

$$i = C \frac{dv}{dt} \qquad \text{[Equation (17.6)]}$$

$$C = 1 \ \mu\text{F} = 10^{-6} \text{ F}$$

$$i = 0.1 \text{ mA} = 10^{-4} \text{ A}$$

$$\frac{dv}{dt} = \frac{i}{C} = \frac{10^{-4} \text{ A}}{10^{-6} \text{ F}} = 10^2 \frac{\text{A}}{\text{F}} = 100 \frac{\text{A}}{\text{F}}$$

From Table 9.3 we see that the farad, F, is equivalent to C/V = A · s/V. Making this substitution, we get

$$\frac{dv}{dt} = 100 \frac{\text{A} \cdot \text{V}}{\text{A} \cdot \text{s}} = \mathbf{100 \frac{\text{V}}{\text{s}}} \qquad \text{Answer}$$

At that particular instant, the voltage across the capacitor is changing at the rate of 100 V/s.

17.1.3 An inductor of 10 H has a voltage across it at a particular instant (we call this an *instantaneous* voltage) of 24 V. What is the instantaneous rate of change of the current in the inductor, in amperes per second?

Solution

$$v = L \frac{di}{dt} \qquad \text{[Equation (17.7)]}$$

$$L = 10 \text{ H} \qquad v = 24 \text{ V}$$

$$\frac{di}{dt} = \frac{v}{L} = \frac{24 \text{ V}}{10 \text{ H}} = 2.4 \text{ V/H}$$

From Table 9.3, we find H = Wb/A = V · s/A, so

$$2.4 \frac{\text{V}}{\text{H}} = 2.4 \frac{\text{V} \cdot \text{A}}{\text{V} \cdot \text{s}} = \mathbf{2.4 \frac{\text{A}}{\text{s}}} \qquad \text{Answer}$$

The instantaneous rate of change of the current in the inductor is 2.4 A/s.

17.1.4 A particular capacitor is observed to have a current of 1.5 mA flowing through it at the same instant that the voltage is observed to be changing at the rate of 55 V/s. How large is the capacitor?

17.1.5 A capacitor is connected to a voltage source, and a chart recorder makes a record of both the current in the capacitor and the voltage across it. Both of these vary with time, of course, until the voltage across the capacitor becomes equal to that of the voltage source, at which time the current ceases flowing. The chart runs through the recorder at constant speed, so measurements lengthwise along the chart are proportional to time; in fact, the chart has marks along it with the space between each pair of marks equal to 0.1 s (see the following figure). Two such marks 0.1 s apart are chosen. The voltage at the first mark is 15.3 V. The voltage at the second mark is 16.4 V. The average current during the interval is 27.3 mA. How large is the capacitor?

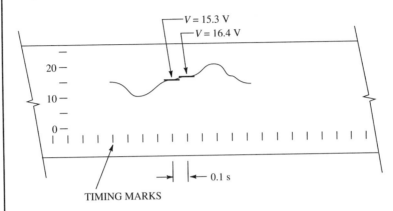

TIMING MARKS

17.1.6 An inductor is observed to have an instantaneous voltage across it of 62.5 V at an instant when the current is changing at the rate of 400 mA/s. What is the size of the inductor?

17.1.7 A resistance of 100 Ω is connected into a circuit with 110 V. What current will flow in the circuit? How much power will the resistor dissipate?

17.1.8 A home lighting circuit is wired for 115 V and has a circuit breaker that will trip when the current in the circuit exceeds 15 A. How many 100-W light bulbs can be connected in the circuit without causing the breaker to trip?

Solution

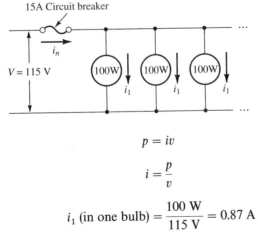

$$p = iv$$

$$i = \frac{p}{v}$$

$$i_1 \text{ (in one bulb)} = \frac{100 \text{ W}}{115 \text{ V}} = 0.87 \text{ A}$$

$$i_n \text{ (in circuit breaker)} = i_1 + i_1 + i_1 + \cdots \text{ for } n \text{ bulbs}$$

$$15 \text{ A} = ni_1$$
$$= n(0.87 \text{ A})$$

$$n = \textbf{17 bulbs} \qquad \qquad \text{Answer}$$

17.1.9 The circuit of Exercise 17.1.8 has a 1500-W electric heater connected in it. How many 100-W light bulbs can be connected to the same circuit without tripping the breaker?

17.1.10 A capacitor of 0.005 F is connected to 100 V. What charge will be stored in the capacitor?

17.1.11 Equation (17.13) states that the charge on a capacitor will discharge through a resistor in accordance with the following:

$$q = q_0 e^{-t/RC}$$

Sketch the decay curve for a circuit like that in Figure 17.4, with $R = 50\,000\ \Omega$ and $C = 5\ \mu\text{F}$. Assume the initial voltage on the capacitor is 5 V.

Solution

$$V_0 = 5 \text{ V} = \frac{q_0}{C} \qquad \left(V = \frac{q}{C} \right)$$

$$q_0 = 5 \text{ V}(5 \times 10^{-6} \text{ F})$$
$$= 25 \times 10^{-6} \text{ C}$$

$$RC = 50\,000\ \Omega(5 \times 10^{-6} \text{ F})$$
$$= 0.25 \text{ s}$$

$$q = 25 \times 10^{-6} e^{-4t} \text{ C}$$

The following table lists values of e^{-4t} and q for values of t.

$t(s)$	e^{-4t}	$q = 25 \times 10^{-6} e^{-4t}$ (coulombs)
0	1.00	25×10^{-6}
0.1	0.67	16.8×10^{-6}
0.2	0.45	11.3×10^{-6}
0.25	0.37	9.3×10^{-6}
0.3	0.30	7.5×10^{-6}
0.4	0.20	5.0×10^{-6}
0.5	0.14	3.5×10^{-6}
0.6	0.09	2.3×10^{-6}
0.7	0.06	1.5×10^{-6}
0.8	0.04	1.0×10^{-6}
0.9	0.03	0.8×10^{-6}
1.0	0.02	0.5×10^{-6}

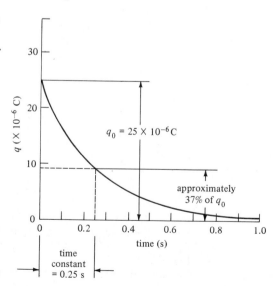

17.1.12 In Exercise 17.1.11, what is the time constant? What is the charge on the capacitor when the time t is equal to the time constant? What percentage of the initial charge q_0 is this? What is the voltage on the capacitor at this point in time? How much (in volts) of the initial voltage has been discharged? How much (in percent) has the initial voltage discharged?

Solution

time constant $= RC = $ **0.25 s**   Answer

q at $t = 0.25$ s $= $ **9.3 × 10^{-6} C**   Answer

percent of $q_0 = $ **37%**   Answer

$$v = \frac{q}{C} = \frac{9.3 \times 10^{-6}\ \text{C}}{5 \times 10^{-6}\ \text{F}} = \textbf{1.86 V}$$ Answer

discharged voltage $= 5\ \text{V} - 1.86\ \text{V}$

$\qquad\qquad\qquad = $ **3.14 V**   Answer

$$\text{discharged voltage (percent)} = \frac{3.14\ \text{V}}{5\ \text{V}} \times 100\% = \textbf{63\%}$$ Answer

17.1.13 In a capacitor-discharge circuit like that of Exercise 17.1.11, assume that $R = 1.25 \times 10^4\ \Omega$, $C = 16\ \mu\text{F}$, and the initial capacitor voltage is 12 V. Sketch the decay curve. What is the time constant?

17.1.14 A capacitor charged to an initial voltage V_0 has a capacitance of 10 μF. A resistor of 1 MΩ is connected across it. How long will it take for the capacitor to be discharged to 37 percent of its initial level? How long before 99 percent of its initial charge is gone?

Solution

37% level: one time constant $= RC$

$\qquad\qquad\qquad\qquad = 10^6\ \Omega(10^{-5}\ \text{F}) = $ **10 s**   Answer

$q = q_0 e^{-t/RC}$

99% gone   when $e^{-t/RC} = 0.01$

$e^{-t/10} = 0.01$

$-0.1t \cdot \log e = \log 0.01$

$$0.1t = -\frac{\log 0.01}{\log e} = -\frac{-2.0000}{0.4343}$$

$\qquad\qquad t = $ **46 s**   Answer

17.1.15 In an inductive circuit like the one in Figure 17.6, the current behaves in accordance with the equation

$$i = \frac{V}{R}(1 - e^{-Rt/L})$$

If the resistance $R = 1.5\ \Omega$ and the inductance L is 3 H, how long will it take the current to reach 95 percent of its final value?

17.1.16 In the first trial of the waveform example of the text, we got a value for the supply voltage $V = 3.8$ V and used a value for the time constant $RC = 0.02$ s. This means the waveform equation for this trial would be

$$v = 3.8\ \text{V} \cdot (1 - e^{-t/0.02})$$

Calculate the values of v for t from zero to $t = 0.04$ s, at 0.005-s intervals. Then sketch the waveform. If this waveform is to be a reasonable approximation of the one shown in Figure 17.9(a), then v should be 0.75 V at $t = 0.005$ s. What value for v does your trial waveform give you? What do you get for $t = 0.01$ s?

17.1.17 Repeat Exercise 17.1.16, using the values from the second trial of the waveform example. First, calculate the value of V that will give 1.5 V when $t = 0.01$ s, as we did in Equation (17.23), but with $RC = 0.05$ s. Then, write out the equation with these values substituted. Finally, plot the waveform for this second trial directly on the same plot you made for Exercise 17.1.16. What values does your trial waveform now give for $t = 0.005$ s and $t = 0.01$ s?

17.1.18 Careful measurements show that a particular light bulb consumes 5880 J in 1 minute when connected to 110 V. How much power does the bulb consume? What is its resistance?

■ 17.2 KIRCHHOFF'S LAWS

Two rules from physics are of great assistance in the analysis of circuits. They are known as *Kirchhoff's laws*.

1. In a loop of a circuit, the sum of all the voltage rises must equal the sum of the voltage drops.
2. At a circuit junction (also called a "node"), the sum of all the currents directed into the junction must be equal to the sum of all the currents directed away from the junction.

As a first example, we will use these rules ("laws") to demonstrate calculations dealing with resistors in series and in parallel.

Figure 17.11(a) shows three resistors connected end to end—or in "series," as it is called—and connected to a battery V. The current i flows through all elements of the circuit. Kirchhoff's first rule (sometimes referred to as "Kirchhoff's voltage law") says the following:

$$V = iR_1 + iR_2 + iR_3 \tag{17.24}$$

There is only one voltage "rise" in the circuit (the battery), and Ohm's law gives the voltage drop in each resistor. We can rewrite Equation (17.24) as follows:

$$V = i(R_1 + R_2 + R_3) \tag{17.25}$$

which shows that the equivalent resistance of several resistors connected in series is simply the sum of the resistances.

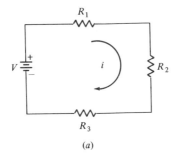

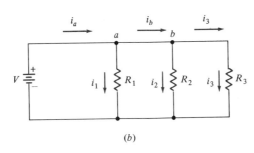

Figure 17.11

Resistors connected (*a*) in series and (*b*) in parallel.

Figure 17.11(*b*) shows the same three resistances connected in parallel so that each resistor has the full voltage V connected across it. We know, then, that

$$V = i_1 R_1 = i_2 R_2 = i_3 R_3 \qquad (17.26)$$

From Kirchhoff's second rule (sometimes called "Kirchhoff's current law"), we can set down certain relationships. For nodes *a* and *b* we can write

$$i_a = i_1 + i_b \qquad (17.27)$$

$$i_b = i_2 + i_3 \qquad (17.28)$$

In each case, we have assigned arbitrary directions for the currents and then applied Kirchhoff's current law, equating the current(s) flowing into a node to the current(s) flowing away from the node. Equations (17.27) and (17.28) can be combined to become

$$i_a = i_1 + i_2 + i_3 \qquad (17.29)$$

But the current through each resistor can be found from Equation (17.26) as

$$i_1 = \frac{V}{R_1} \qquad i_2 = \frac{V}{R_2} \qquad i_3 = \frac{V}{R_3} \qquad (17.30)$$

Substituting Equations (17.30) into Equation (17.29) gives

$$i_a = V\left(\frac{1}{R_1} + \frac{1}{R_2} + \frac{1}{R_3}\right) \qquad (17.31)$$

The expression in parentheses in Equation (17.31) gives us the equivalent value for the three resistors in parallel, if we were to replace them with a single resistor that would give the same current i_a coming from the battery as is produced with the original three resistors. Thus, designating R_{eq} as the equivalent value of the resistance, we have

$$\frac{1}{R_{eq}} = \frac{1}{R_1} + \frac{1}{R_2} + \frac{1}{R_3} + \cdots \qquad (17.32)$$

for any number of resistors in parallel. If we solve Equation (17.32) for R_{eq} in the case involving three resistors, we get

$$R_{eq} = \frac{R_1 R_2 R_3}{R_1 R_2 + R_2 R_3 + R_1 R_3} \qquad (17.33)$$

The foregoing relationships can be used to simplify circuits, as shown in Figure 17.12. The task is to replace the four resistors of Figure 17.12(a) with a single equivalent resistor that will provide the same net resistance between points a and b in both the original and equivalent circuits. In undertaking such a circuit simplification process, one starts at the point most remote from the terminals. The 8-Ω and 4-Ω resistors are in series, so can be replaced by a single 12-Ω resistance that is in parallel with the 10-Ω resistance [see Figure 17.12(b)]. The 10-Ω and 12-Ω resistances of Figure 17.12(b) are in parallel, and the equivalent resistance for these two is found from Equation (17.32):

$$\frac{1}{R_{eq}} = \frac{1}{10} + \frac{1}{12}$$

$$R_{eq} = \frac{120}{10 + 12} = 5.45 \ \Omega$$

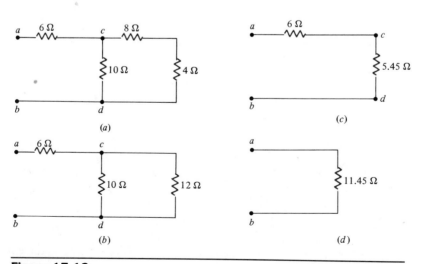

(a)

(b)

(c)

(d)

Figure 17.12

(a) A circuit. (b, c) Steps in the derivation of its equivalent resistance (d).

This gives the result shown in Figure 17.12(c). The final step is to add the 6-Ω and 5.45-Ω resistors, which are in series, to give the final equivalent resistance of 11.45 Ω, shown in Figure 17.12(d).

Wheatstone Bridge Example

A procedure based upon the foregoing that is useful in circuit analysis is called the *mesh analysis method* or *loop current method*. To illustrate this method, we will use the circuit of the *Wheatstone bridge*, shown in Figure 17.13. The Wheatstone bridge circuit is used in instruments to determine unknown resistances and in circuits using devices called *strain gauges*, which are widely employed in structural testing and stress analysis.

In the loop current method, the first step is to assign an arbitrary current to each independent loop of the network, as has been done with currents i_1, i_2, and i_3 in Figure 17.13. A common practice is to arbitrarily show all the loop currents in the clockwise direction, as has been done in the figure. If, in the solution, any of the loop current values turn out to have minus signs, this means the current is actually in the direction opposite to that assumed. For each loop, we can write an equation for Kirchhoff's voltage law as follows:

$$
\left.\begin{array}{ll}
\text{Loop 1:} & V = (i_1 - i_2)R_3 + (i_1 - i_3)R_4 \\
\text{Loop 2:} & 0 = i_2R_1 + (i_2 - i_3)R_0 + (i_2 - i_1)R_3 \\
\text{Loop 3:} & 0 = i_3R_2 + (i_3 - i_1)R_4 + (i_3 - i_2)R_0
\end{array}\right\} \tag{17.34}
$$

In writing these, we are careful to be consistent. We move about each loop in the direction of the assigned loop currents, placing voltage "rises" to the left of the equals sign and voltage "drops" to the right. Having used the "positive current" convention, wherein a positive current is assumed to exit from a voltage source at its positive terminal, we can then assume that such a positive current flowing through a resistor produces a voltage "drop" in the direction of the current. We note, also, that resistor R_1 has only current i_2 flowing in it and R_2 has only current i_3 in it, but R_3 has the *difference* between currents i_1 and i_2 in it, flowing in opposite directions. The net current in R_3, then, is $i_1 - i_2$. Similar situations prevail for resistors R_0 and R_4.

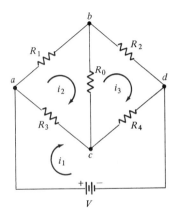

Figure 17.13

A Wheatstone bridge, illustrating the loop circuit method of analysis.

Equations (17.34) can be rearranged as follows:

$$
\left.\begin{array}{l}
(R_3 + R_4)i_1 - R_3i_2 - R_4i_3 = V \\
-R_3i_1 + (R_0 + R_1 + R_3)i_2 - R_0i_3 = 0 \\
-R_4i_1 - R_0i_2 + (R_0 + R_2 + R_4)i_3 = 0
\end{array}\right\} \quad (17.35)
$$

Except for the loop currents i_1, i_2, and i_3, all values in Equations (17.35) are known. Thus, with three equations and three unknowns, we can solve the equations by use of determinants (see Appendix D).

In Figure 17.13, let us assign the following values:

$V = 20$ V $R_0 = 5\,\Omega$

$R_1 = 10\,\Omega$ $R_3 = 5\,\Omega$

$R_2 = 20\,\Omega$ $R_4 = 10\,\Omega$

Then, by substitution and simplification, Equations (17.35) become

$3i_1 - i_2 - 2i_3 = 4$

$i_1 - 4i_2 + i_3 = 0$

$2i_1 + i_2 - 7i_3 = 0$

We can solve for i_1, i_2, and i_3 using the method of determinants. The solution is

$i_1 = 2$ A

$i_2 = \frac{2}{3}$ A

$i_3 = \frac{2}{3}$ A

The current in resistor R_3 is $i_1 - i_2$, or $1\frac{1}{3}$ A, and that in R_4 is $i_1 - i_3$, or $1\frac{1}{3}$ A. We then note that the current in R_0 is $i_2 - i_3$, or zero. Looking back at the assigned values of resistances, we can see that this interesting result is the necessary consequence of the ratios R_1/R_2 and R_3/R_4 being equal. Since the ratio of R_1 to R_2 is the same as that of R_3 to R_4, it means that the voltages across the resistors have to divide the same

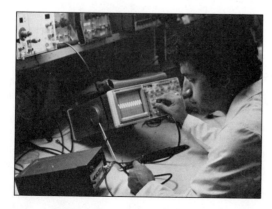

Figure 17.14
Electrical engineers, in addition to performing mathematical analyses of their circuits, often must confirm their results experimentally in the laboratory. Hence, they should be intimately familiar with laboratory instruments.

way. Therefore, the voltage drop from a to b must be the same as from a to c, meaning that points b and c have the same voltage level. As a result, there is no voltage across R_0, and thus no current in it. This is, in fact, the principle of the Wheatstone bridge. In such a bridge, R_3 and R_4 may typically be a pair of fixed resistances of known value. R_2 may be a variable resistor and R_1 a resistor of unknown value that has been connected to the bridge. The resistance R_0 belongs to an instrument called a *galvanometer*, which displays a needle showing how much current is flowing through R_0. The variable resistor R_2 is adjusted until the current flowing through R_0 is zero. At that point we can read the final value of the variable resistor R_2 from a calibrated dial fixed to it and determine the value of R_1 from the relationship

$$\frac{R_1}{R_2} = \frac{R_3}{R_4} \qquad (17.36)$$

Exercises

17.2.1 Four resistances of 5 Ω, 10 Ω, 15 Ω, and 20 Ω are connected in series. What is their equivalent resistance? What is their equivalent resistance if they are connected in parallel?

Solution

Series: $R_{eq} = 5\,\Omega + 10\,\Omega + 15\,\Omega + 20\,\Omega$
          $= \mathbf{50\ \Omega}$ Answer

Parallel: $\dfrac{1}{R_{eq}} = \dfrac{1}{5\,\Omega} + \dfrac{1}{10\,\Omega} + \dfrac{1}{15\,\Omega} + \dfrac{1}{20\,\Omega}$

            $= (0.2 + 0.1 + 0.067 + 0.05) \cdot 1/\Omega$
            $= 0.417 \cdot 1/\Omega$

$$R_{eq} = \frac{1}{0.417}\,\Omega = \mathbf{2.40\ \Omega} \qquad \text{Answer}$$

17.2.2 Reduce the three resistors in series to a single equivalent resistor.

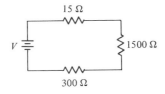

17.2.3 Reduce the three resistors in parallel to a single equivalent resistor, to four significant figures.

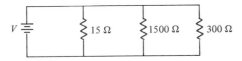

17.2.4 If each resistor R in the figure is 100 Ω, what is the total equivalent resistance?

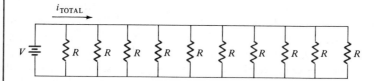

If the voltage V is 100 V, how much current flows in each resistor? How much is the total current, i_{TOTAL}, that the battery must supply?

17.2.5 Reduce the three resistors to a single equivalent resistor.

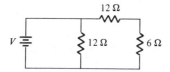

17.2.6 Find the loop currents i_1, i_2, and i_3. What is the current in each resistor?

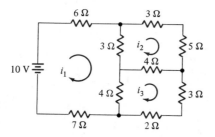

17.2.7 Reduce the resistors in the circuit shown to a single equivalent resistor.

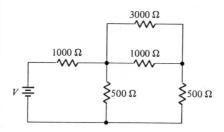

17.2.8 In a Wheatstone bridge [see Equation (17.36)], R_3 and R_4 are both 1000 Ω. An unknown resistance R_1 is connected to the bridge, and R_2 is adjusted until the galvanometer reads zero. The calibrated dial of R_2 gives a reading of 392 Ω. What is the resistance of the unknown resistor R_1?

17.2.9 In the circuit shown in (*a*), the circuit resistance obviously is equal to R. What is the equivalent resistance in (*b*)? From this, what can you conclude as a general rule? In other words, as you add resistances in parallel, does the equivalent resistance increase or decrease?

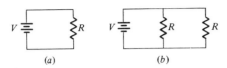

(a)   (b)

17.2.10 Reduce the resistors in the circuit shown to a single equivalent resistor.

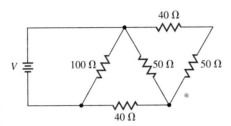

17.2.11 The circuit shown in part (*a*) of the following figure is called a delta circuit, and the one in part (*b*) is called a Y circuit.

(a) What is the equivalent resistance between terminals A and B in the delta circuit? (Assume the third terminal, C, is not connected to anything.)

(b) What is the equivalent resistance between B and C?

(c) What is the equivalent resistance between A and C?

(d) In the Y circuit, assume all three resistances have the same value, *R*. What should *R* be, if you want to have the same equivalent resistance between terminals A and B of the Y circuit as you got for the delta circuit in (*a*)?

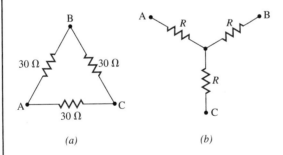

(a)   (b)

17.2.12 What is the value of resistor *R* in the following figure?

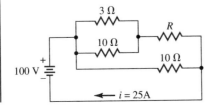

Famous Engineers
Nikola Tesla

Few Americans have heard of Nikola Tesla, although his name has been memorialized in the International System of Units as the unit of magnetic flux density (one tesla equals one $V \cdot s/m^2$). Yet he was a brilliant engineer whose inventions have profoundly affected all our lives. He was a contemporary of Edison, and his AC power generation and transmission system overwhelmed and doomed the Edison DC system. Furthermore, Tesla supporters claim that he, and not Marconi, was the inventor of radio.

Tesla was born in Yugoslavia in 1856. He was educated at Graz, Austria, and at the University of Prague. The idea that was later to revolutionize the electrical power industry occurred to him suddenly while he was walking with a friend in the City Park of Budapest. A mental image came to him of an iron rotor spinning in a magnetic field which itself was rotating, the rotation being caused by two alternating currents that were out of phase. This was the principle of the synchronous motor and of the induction motor.

Tesla's ideas did not come from the blue. He had been mentally wrestling with these matters for four years, ever since he was a student at Graz. There he had observed a DC motor in operation and remarked that the sparking of the brushes might be eliminated by inventing a motor that did not use brushes.

In 1883, Tesla went to work for the Continental Edison Company in Germany. While there, he built his first brushless induction motor. The following year he emigrated to America to work for Thomas Edison, designing DC generators, but soon quit.

Tesla cared for money not as an end in itself, but only because money was needed for his experiments. He had an image in his mind of a complete system for the generation, transmission, and end use of electric power. The generators were constructed in such a way that they could produce separate alternating currents in their windings that were out of phase with each other (see Section 17.3 for a discussion of *phase lead* and *phase lag*)—the forerunner of our modern three-phase AC distribution systems. Being AC, the currents could be stepped up to high voltages with transformers, which brought the cost of transmission down. Tesla began patenting his inventions in 1887 and attracted the attention of George Westinghouse, who bought the patents.

In 1893, Westinghouse signed a contract to generate electrical power at Niagara Falls and transmit it to Buffalo, 22 miles away. Such a transmission distance was beyond the practical reach of DC systems, but Tesla's generation-transmission-utilization system made it possible. By the time the Niagara plant was complete, in 1903, all new generation plants in the United States were being set up using the Tesla system.

Some years after Tesla's death in 1943, a long-standing patent suit was settled that confirmed Tesla's position as a priority inventor of radio, anticipating Marconi. Through most of his life, Tesla lived as a recluse, becoming more eccentric as he grew older. He never married. It is claimed that millions slipped through his hands because of his disinterest in money and his inability to work with others.

References
E. Marshall, "Seeking Redress for Nikola Tesla," *Science*, 30 October 1981, pp. 523–525.
K. M. Swezy, "Nikola Tesla," *Science*, 16 May 1958, pp. 1147–1159.

■ 17.3 AC CIRCUITS

If the voltage source in a circuit has a constant value, as in the case of a battery, the current will vary for a short time, as we have seen, but soon will settle to a steady value. The short-term variation is referred to as a *transient*. Once the current has settled to its final value, it is referred to as *steady-state*. Circuits in which the currents and voltages have constant steady-state values are referred to as *direct current*, or DC, circuits. However, most electrical power circuits supply AC, or *alternating current*.

Typically, the voltage (and current) in an AC circuit have the shape of a sine wave with respect to time and can be expressed mathematically as

$$v = V \sin \theta$$

$$i = I \sin \theta$$

It is customary to make the substitution $\theta = \omega t$, so these equations become

$$v = V \sin \omega t$$

$$i = I \sin \omega t$$

(17.37)

Equations (17.37) have been plotted in Figure 17.15. The quantities v and i are the so-called *instantaneous* values of voltage and current at any given time t, and V and I are the "peak" values of the sine waves.

The reason for the substitution $\theta = \omega t$ will become clearer later, but at this point we will note that θ is an angle, usually given in radians (although radians can always

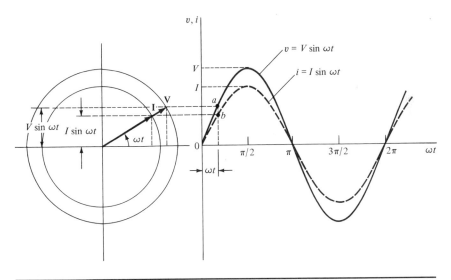

Figure 17.15
AC waveforms for voltage and current in phase.

be converted to degrees).[3] The quantity ω is referred to as the *angular frequency*, given in radians per second, and t is the time, given in seconds. When we multiply ω and t together, the units become

$$\omega t = \theta$$

$$\frac{\text{radians}}{\text{s}} \cdot \text{s} = \text{radians}$$

as they must.

In the situation shown in Figure 17.15, the voltage v and current i reach their peak values V and I at the same instant and are therefore said to be *in phase*. This is what happens in a circuit containing only resistance. The peak values $\pm V$ and $\pm I$ are reached at $\omega t = \pi/2, 3\pi/2, 5\pi/2, \ldots$.

In Figure 17.15, the voltage (or current) has gone through one full cycle when $\omega t = 2\pi$, or 360°, which occurs when $t = 2\pi/\omega$. A full cycle is shown in the right-hand portion of Figure 17.15, wherein both the v and i curves start at zero for $t = 0$ and return to zero for $\omega t = 2\pi$. The same circumstance is displayed graphically in the left-hand portion of the figure, where **V** and **I** are displayed as vectors, in the resistive case here shown aligned, since they are in phase. (Boldface characters are used for vectors.) The two vectors start together at $\omega t = 0$, aligned with the horizontal axis, and rotate together counterclockwise with the angular frequency ω until they return to their starting point after one full rotation, when $\omega t = 2\pi$.

At any instant of time, such as the one depicted, the projections of the vectors on the vertical axis represent the instantaneous values of the functions. The projection of **V** on the vertical axis is, by trigonometry, equal to $V \sin \omega t$, and the projection of **I** on the vertical axis is $I \sin \omega t$. These projections are shown at the left of Figure 17.15 and are projected onto the sine waves at the right to show that the **V** projection strikes the $v = V \sin \omega t$ curve at point a and that the **I** projection strikes the $i = I \sin \omega t$ curve at point b, which is correct for the value of ωt shown.

It can be seen that the projections of **V** and **I** on the vertical axis do indeed trace out the two sinusoidal curves shown with increasing ωt. Functions that behave in the foregoing manner are referred to as *circular functions*. It is customary to speak of such functions as having a frequency f cycles per second, meaning that the waveform is repeated f times in each second. The time for one cycle is thus $t = 1/f$, which occurs at the point marked $\omega t = 2\pi$ in Figure 17.15. Therefore, at this point,

$$t = \frac{1}{f} = \frac{2\pi}{\omega}$$

---

[3] For those unfamiliar with radian measure, we note that a radian is defined as an angle of a circle for which the subtended arc (segment of the circumference) is equal to the radius. Thus, since the circumference of a circle is equal to $2\pi \cdot r$, where r is the radius, it follows that a full circle has 2π radians in it, or 2π radians = 360°. Then, $\pi/2$ radians = 90°, π radians = 180°, $3\pi/2$ radians = 270°, and so forth.

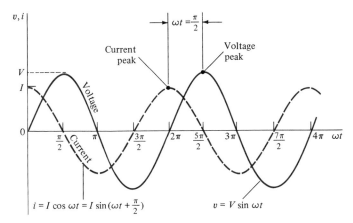

Figure 17.16

AC waveforms for voltage and current in a capacitive circuit. The current *leads* the voltage by $\omega t = \pi/2$, or 90°.

which is easiest to remember in the form

$$\omega = 2\pi f \qquad (17.38)$$

Further examination of AC circuits is beyond the scope of this textbook. However, we will note that in real AC circuits the voltage and current are practically never in phase, because real circuits usually contain capacitance and inductance as well as resistance. If the circuit contains *only* capacitance, the voltage and current are out of phase by exactly 90° ($\pi/2$ radians), with the peak value of the current occurring 90° ahead of the peak value of the voltage (see Figure 17.16). We speak of this as the current *leading* the voltage by 90°.

If the circuit contains only inductance, circumstances are reversed. The voltage and current are still out of phase by 90°, but now the peak value of the current occurs 90° *after* the peak value of the voltage (see Figure 17.17). We speak of this as the current *lagging* the voltage by 90°. In real circuits containing all three properties—resistance, inductance, and capacitance—the current either leads or lags the voltage

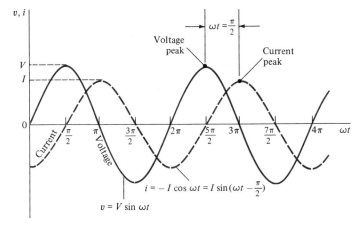

Figure 17.17

AC waveforms for voltage and current in an inductive circuit. The current *lags* the voltage by $\omega t = \pi/2$, or 90°.

depending upon whether the capacitance or inductance dominates. The amount of lead or lag can vary from 0° to 90°, depending upon the relative sizes of the R, L, and C components.

Exercises

17.3.1 Sketch voltage and current curves like those in Figure 17.16, but with the current leading the voltage by 30°. Does capacitance or inductance dominate this circuit?

Solution

(See the following figure.) Note: $30° = \pi/6$.

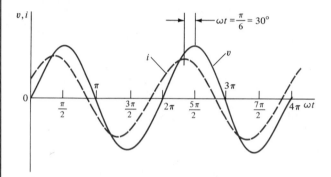

Capacitance dominates because the current is leading the voltage.

17.3.2 Sketch voltage and current curves like those in Figure 17.17, but with the current *lagging* the voltage by 30°. Does capacitance or inductance dominate this current?

17.3.3 Sketch voltage and current curves like those in Figure 17.16, but with the current *leading* the voltage by 270°. Does capacitance or inductance dominate this circuit? (Compare your sketch with Figure 17.17.)

■ 17.4 TRANSISTOR AMPLIFIER

We will deal with the transistor as a device with certain known external characteristics, and will make no attempt here to describe its internal physics. In Figure 17.18, a simple transistor amplifier circuit is shown. The transistor has three terminals: B, the *base*; C, the *collector*; and E, the *emitter*. The design of the transistor is such that the collector current i_C is very nearly equal to the emitter current i_E. That is, $i_E \cong i_C$. The collector and base currents are related by the equation

$$i_C = \beta i_B \tag{17.39}$$

The parameter β is designed into the transistor and is called the *current gain*. It is approximately constant for a given transistor, provided the voltages V_{BB} and V_{CC} in the circuit of Figure 17.18 are maintained within appropriate ranges.

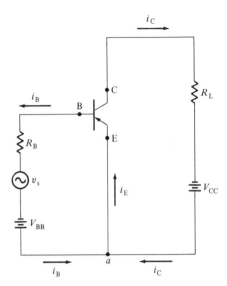

Figure 17.18
A simple transistor amplifier.

Remembering Kirchhoff's current law, you may also note that the conditions at node a require that $i_E = i_B + i_C$. However, the current i_B in a transistor is much smaller than either i_E or i_C, so we can write $i_E \cong i_C$, as stated before.

The job of the circuit in Figure 17.18 is to amplify a varying "signal voltage" v_s, which might be very small, as in a phonograph pickup cartridge. There are *constant* currents in the circuit that are the results of the constant "bias" voltages V_{BB} and V_{CC}, but these are of no interest to us, and we will ignore them. We will focus our attention instead upon the varying currents i_B and i_C, which are superimposed upon the constant currents. (In a practical circuit, means are provided to separate the effects of the variable current i_C from the constant current flowing in the circuit, but such matters are too extensive to be considered here.)

Much of the analysis of circuits such as the one we are considering can be performed by the mathematical models we used for DC circuits, even though we are considering varying voltages and currents such as v_s, i_B, and i_C.

Ignoring the constant voltage V_{BB} and the constant current in the left-hand loop, we can write, from Ohm's law,

$$v_s = i_B(R_B + r_B) \tag{17.40}$$

where r_B is the resistance between the emitter and the base of the transistor.

In the right-hand loop, we write, for the "load resistor" R_L,

$$v_L = i_C R_L \tag{17.41}$$

where the constant voltage V_{CC} and constant current in that loop are of no interest and have been ignored, and where the internal emitter-collector resistance is assumed to be small enough to ignore.

Figure 17.19

A challenging design problem in modern computers has been to pack digital data onto magnetic disks in ever greater concentrations. (Left) Four magnetic read/record heads. Since the magnetic head assembly rides between two disks, two heads float against the upper surface of one disk, and two heads float against the under surface of the next higher disk. (Right) Twenty digital recording tracks can be concentrated in the space between two lines of a thumbprint, which is $\frac{1}{50}$ inch. The resulting information packing density is about 20 million bits of data per square inch. (Courtesy of IBM General Products Division, San Jose, CA)

If we solve for i_B in Equation (17.40) and substitute it in Equation (17.39), we get

$$i_C = \beta \frac{v_s}{R_B + r_B} \tag{17.42}$$

and if we substitute Equation (17.42) into Equation (17.41), we get

$$v_L = \beta \frac{v_s R_L}{R_B + r_B} \tag{17.43}$$

We can now solve for the voltage "gain" between the amplified signal voltage v_L in the load R_L and the original signal voltage v_s as follows:

$$\frac{v_L}{v_s} = \beta \frac{R_L}{R_B + r_B} \tag{17.44}$$

Equation (17.44), giving the voltage gain of the circuit, is one of the important aspects of an amplifier that interest circuit designers. Transistors are available with different values of β, depending upon the circumstances of the circuit to be designed. If a transistor with $\beta = 100$ is selected, and the ratio $R_L/(R_B + r_B)$ is, say, 0.3, then the voltage gain is 30.

Exercises

17.4.1 In a transistor circuit like that of Figure 17.18, if $\beta = 80$, $R_B + r_B = 2000\ \Omega$, and $R_L = 1000\ \Omega$, what is the voltage gain?

17.4.2 In Exercise 17.4.1, if a voltage gain of 50 is desired, what value should the load resistor have?

17.4.3 In Exercise 17.4.1, if the signal voltage is 75 mV, what is the voltage across the load resistor? What is the collector current i_C?

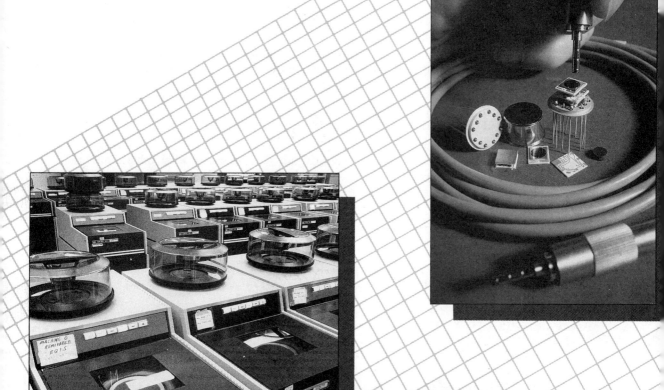

Photographs courtesy of Hewlett-Packard Company (center), Los Alamos National Laboratory (top right), and International Business Machines Corporation (bottom).

18

Computers

■ 18.1 COMPUTER ORGANIZATION

A block diagram of a computer system is shown in Figure 18.1. If the computer is a large one, the different blocks in the diagram will be made up of physically separate items of equipment. For example, the *input/output* (I/O) portion will typically be represented by several input units, such as keyboards or magnetic tape or disk readers, plus one or more output units, such as cathode ray tubes, light-emitting diodes, printers, or magnetic tape or disk recorders. The *central processing unit* (CPU) may be represented by one or more large cabinets full of integrated circuits (ICs). In a really large computer system, the *memory* may be represented by scores of disk drive units, which may be the most strikingly visible components, resembling nothing so much as a room full of washing machines.

On the other hand, most of the blocks in Figure 18.1 may be self-contained in a hand-held calculator. The input is the keyboard, the output the visual display, and the CPU and memory are contained on IC chips inside the case.

The functions of the units are as follows. The *memory* stores information in the form of binary numbers.[1] The CPU consists of three

---

[1] Binary numbers are those that use only the digits 0 and 1 instead of the familiar digits 0 through 9, which we use in the decimal system. No details will be given here regarding the theory or rules of addition for binary numbers; we will merely note that binary numbers are more adaptable to computer operations than are decimal numbers. As an example of a binary number, we will write here the binary equivalent to the decimal number 39, which is 100111. The binary number 100111 may be written out in full as $(1 \times 2^5) + (0 \times 2^4) + (0 \times 2^3) + (1 \times 2^2) + (1 \times 2^1) + (1 \times 2^0) =$ decimal 39.

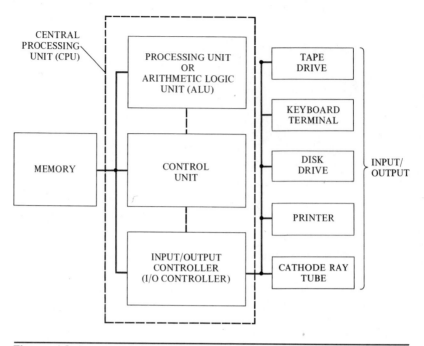

Figure 18.1
Block diagram of a computer system.

subsections: the *control unit*, the *processing unit*, and the *I/O controller*. It is the job of the control unit to interpret the numbers transmitted to it, either as operating instructions or as the data being manipulated. The control unit is the master element: It tells the I/O controller when to accept information or transfer it out; it also supervises the transfer of information to the processing unit, where the arithmetic operations actually occur. (The processing unit is also called the *arithmetic logic unit*, or ALU.)

It is possible to purchase a complete CPU in miniaturized form, known as a *microprocessor*. Some typical microprocessors, each measuring about $\frac{1}{2}$ by 2 inches, are shown in Figure 18.2. Each microprocessor is a tiny IC chip enclosed in a plastic case. The case serves the multiple purposes of protecting the chip, providing a tough frame for ease of handling, and giving a structure for holding the parallel rows of connectors, which are visible in the illustration. Such a microprocessor, plus various memory units possessing a similar appearance (see Figure 18.3), are typically assembled together on printed circuit boards. A printed circuit board is shown in Figure 18.4.

The memory of a computer serves as a place to store the program that will be used to operate the computer and also as a place to store pieces of data. Typical pieces of data would be inputs to be used in calculation, partially computed answers being stored temporarily until needed, and final answers being stored in readiness for final readout.

Famous Engineers
The Invention of the Electronic Computer

If you had never seen or heard of an electronic computer, how would you go about inventing one? When John Mauchly first began thinking about this question in the 1930s, his basic motivation was to improve the process of weather forecasting. He possessed a Ph.D. in physics and was teaching at a small liberal arts college. He knew something about electronics, and knew that vacuum tubes had already been used for simple counting circuits. He also knew that IBM and Remington Rand had used electrical relays to perform addition and multiplication. In his own words, it seemed "obvious" that vacuum tubes could be used for computational purposes and that tubes would be a thousand times faster than relays.

But, as Mauchly learned, going from the conception to the reality was a long, arduous process. While he was searching for the resources to turn his idea into reality, three things happened: World War II broke out, the University of Pennsylvania offered him a job as an engineering professor, and he met a young electronics engineer, J. Presper Eckert.

Eckert and Mauchly found they were both interested in electronic computation and engaged in many discussions about the topic. Eckert had been hired by the University of Pennsylvania to make improvements upon an elaborate electromechanical computer called a "differential analyzer" that was being used to compute artillery ballistics tables for the U.S. Army. Both Eckert and Mauchly thought an electronic computer would be many times better than the mechanical machine, but such a unit would contain 18,000 vacuum tubes. No one but

them believed an electronic computer of that size would be reliable enough to do the job, because even a slight flickering of one of the tubes could cause a wrong answer to be produced.

By 1943 the situation had changed, and Eckert and Mauchly suddenly had a contract from the army to build their machine. They went at it and moved toward their goal in a steady, straightforward way with remarkably few reverses. They made several crucial decisions. First, they would use only standard, off-the-shelf components. Second, they would design every circuit to operate under so-called worst-worst circumstances, assuming that every part would vary to the limit of the expected commercial variations (called tolerances) in the most troublesome possible ways. Third, they operated every component at levels far below its rating: if the "plate voltage" in a tube could be 400 volts, they would use 200: if a resistor was expected to dissipate $\frac{1}{4}$ watt of power, they would use a resistor that could handle four times as much; tube currents would be kept to 25 percent of the rated permissible levels; and so on. As a result, their computer, called the ENIAC, achieved a remarkable 1 in 10^{14} probability of malfunction—enough to keep it operating for 12 error-free hours at a time. It was enough.

Eckert and Mauchly tried to interest Wall Street in the future prospects of the electronic computer, but with no success. Finally, on the basis of a contract from the Bureau of the Census, they formed Eckert-Mauchly Computer Corp. in 1946. In 1950 they were bought out by Remington Rand and launched UNIVAC, the first large commercial computer. The rest is history.

Reference
"Mauchly on the Trials of Building ENIAC," *IEEE Spectrum*, April 1976, pp. 70–76.

Figure 18.2
Finished microprocessors. (Courtesy of Intel Corporation)

Figure 18.3
High-speed 16 384 × 1-bit random access memory (RAM). (Courtesy of Intel Corporation)

Figure 18.4
After the varied kinds of integrated circuits have been sealed in their plastic covers, they are assembled and interconnected on printed circuit boards like the one shown here. (Courtesy of Grumman Aerospace Corporation)

Memories come in many forms. In microprocessor systems, two frequently used memories are the ROM (read-only memory) and the RAM (random access memory). A RAM is shown in Figure 18.3. Externally, ROMs and RAMs may look much alike.

As the name implies, the data in a ROM can only be read out. Its contents cannot be changed after it has been fabricated. Typically, ROMs are used to store fixed programs or data. A typical fixed program in a ROM might be the internally stored subroutines to compute such things as sin x, cos x, e^x or $\sqrt{x}$. A stored piece of data might be a value such as π. A RAM, on the other hand, can have data either entered into it or read out of it and is intended to be used for storing data or programs during computer operation and retrieving, altering, and re-storing them as needed.

A computer typically has an internally stored program that controls the interaction of the units, providing for user programs to be accepted and run, receiving data, transferring data from CPU to memory to output, and the like. Such a program is called an *operating system*. In a large computer, the operating system might be "loaded" into the machine by the computer operator. In a microprocessor-based system, the operating system is stored permanently in a ROM.

Another kind of special internally stored program, a *compiler*, is necessary whenever a high-level programming language such as FORTRAN is used to instruct a computer. The computer does not understand FORTRAN. Instead, it understands instructions given to it in so-called *object language*, which tells it what to do at a given point in time: go to point "A" and accept so many bits of information, place the information in such-and-such a register location, get another set of bits from another location, place it in some other register, retrieve a particular piece of information from a memory register and put it into a particular register in the arithmetic unit called the *accumulator*, retrieve another piece of information from another register

Figure 18.5

A read-only memory (ROM), in this case an *erasable programmable* ROM, or EPROM. Ultraviolet light can be used to erase the information in this ROM; then the ROM can be reprogrammed. (Courtesy of Intel Corporation)

and add it to the one already in the accumulator (that is why it is called an "accumulator"—it accumulates sums), retrieve the sum from the accumulator and store it in
yet another register, and so on. If we give an instruction such as X = Y + Z in a
high-level language, this simple expression must be converted into several instructions in object language. The program that performs the conversion is the compiler.
A compiler, then, is a special program, internally stored in the computer, that converts a program written in a language the programmer understands into another
language the machine understands. Ultimately, the computer must have its instructions given to it in binary number form, because it understands only binary numbers.

■ 18.2 ALGORITHMS

When a key such as $\boxed{\sqrt{}}$ or $\boxed{\text{COS}}$ is pressed on a hand calculator, the answer quickly
appears. But one may wonder just what the calculator is doing inside when such a
key is pressed. The fact that the answer does not appear immediately but is delayed
somewhat may give a hint that the internal process, whatever it is, must be fairly
lengthy, since we know that a calculator performs operations such as multiplication
and division almost instantaneously.

    The fact is that certain elaborate problem-solving procedures, or *algorithms*, are
stored internally in a calculator to compute approximations of the desired functions.
Since these approximations are typically good to eight or more significant figures,
they are usually considered good enough for ordinary purposes. Nonetheless, approximations they are, and it can be instructive to consider what we would do if we were
asked to dream up an internally stored algorithm to compute, say, the square root
of a number. Therefore, we will pretend that we are helping to design a hand-held
calculator, and designing the algorithm for $\boxed{\sqrt{}}$ is our assignment.

    As a first idea, we might remember that $\sqrt{a}$ is also $a^{1/2}$ and that there is an infi-
nite series for calculating the value of a^x, which is

$$a^x = 1 + x \ln a + \frac{(x \ln a)^2}{2!} + \frac{(x \ln a)^3}{3!} + \cdots$$

We might then substitute the value of a (the number whose square root we are
seeking) and $x = 1/2$ into the above series, and start cranking out the successive terms
until we finally come to a term that is very close to zero (i.e., the series converges).
We then take the sum of all these terms, and that is our approximation of $\sqrt{a}$.

    However, infinite series often have a habit of converging at an irritatingly slow
pace, so we might have to compute the values of a very large number of terms. It
might take so long to compute the square roots of some numbers that the potential
buyer of our calculator might think the machine was broken. So we look for something else.

    As a second idea, we might suggest that we just go at the job by trial and error,
but this seems so clumsy, we are embarrassed by the thought. Nevertheless, we work
it out, just to cover all the bases. To do this, let us take an actual number, say, 53 824.

Right away, if we remember any of our arithmetic rules, we know that our answer will contain three digits. A computer procedure can easily be set up to count the digits in a, and thus know how many digits will be in the answer $\sqrt{a}$, so we are confident in this case that we can start with an initial guess of 100, which is the smallest three-digit number there is. We then square 100 and subtract the result from 53 824. If our first guess were a lucky one and in fact were the true square root of the number, the result of our subtraction of course would be zero, and we would be finished. But since the square root of 53 824 is *not* 100, the difference is not zero, and we make a new guess, 200. Each time we do the following: We let x be the number for which the square root is sought, and a any trial value. Then, for each trial value, we take

$$x - a^2 = R \qquad (18.1)$$

and continue with trial values until R is sufficiently close to zero.

Continuing our example, we have the following:

$x = 53\,824$

Trial Value a	a^2	$R = x - a^2$
100	10 000	43 824
200	40 000	13 824
300	90 000	−36 176

Since R changed from positive to negative in the last step, we obviously passed through zero, and the desired square root must lie between 200 and 300. So we back up to 200 and start increasing it by 10 each time until R changes sign again, whereupon we back up and start increasing by 1 each time, and so on. Continuing with the above tabulation, we have

Trial Value a	a^2	$R = x - a^2$
210	44 100	9724
220	48 400	5424
230	52 900	924
240	57 600	−3776
231	53 361	463
232	53 824	0

This seemingly crude method is actually employed in some calculators. It converges on the answer quite rapidly, although in the case of many machines some very clever embellishments have been added to make the process even faster.

Let us describe yet one more way in which the square root of a number may be found, called *Newton's method*. Although it involves a little bit of calculus, even those students who have not yet had calculus should have no difficulty following this explanation, because the only element of calculus we will use is that the derivative dy/dx is equal to the slope of the tangent to a curve. (Appendix F provides simple explanations of the terms *derivative*, *slope*, and *tangent*.)

We begin by letting S be equal to the number for which we wish to find the square root, x. Then, obviously, $x^2 = S$. If we rewrite this as $x^2 - S = 0$, we can see that we have written a function of x, symbolically written $f(x)$:

$$f(x) = x^2 - S \tag{18.2}$$

Now we let y be equal to this function of x, and then make a graph of y against x in the familiar way, as in Figure 18.6(a), for the function

$$y = x^2 - S \tag{18.3}$$

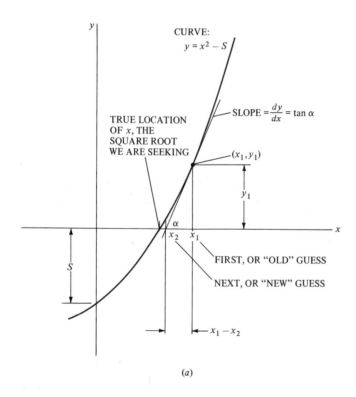

(a)

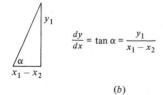

$$\frac{dy}{dx} = \tan \alpha = \frac{y_1}{x_1 - x_2}$$

(b)

Figure 18.6

A plot showing Newton's method for $y = f(x) = x^2 - S$.

The point at which this curve crosses the x axis—i.e., the value of x for which $y = x^2 - S = 0$—is the square root we are looking for, because if $x^2 - S = 0$, then $x = \sqrt{S}$.

We can compute the derivative of the function y (see Appendix F) as

$$\frac{dy}{dx} = f'(x) = 2x \qquad (18.4)$$

We know that $f'(x) = \tan \alpha$ is the slope of the tangent to the curve at a given point. In Figure 18.6(a) we have arbitrarily selected a value $x = x_1$ as our first guess for the square root of S and have drawn a tangent to the curve at point (x_1, y_1). We have projected the tangent downward and to the left until it intersects the x axis at point x_2. We see that x_2 is very much closer to the true point where the curve crosses the x axis than was x_1. Therefore, if we take x_2 as a new guess and repeat the process, we can come even closer. We can keep this up until we get as close to the true location as we wish. We need to find a mathematical expression to give us a value for x_2, knowing x_1.

A method for doing this is provided by the diagram in Figure 18.6(b), which repeats the triangle from Figure 18.6(a). From trigonometry, we can see that

$$\tan \alpha = \frac{y_1}{x_1 - x_2}$$

but we know that the derivative $dy/dx = \tan \alpha$, so we write

$$\frac{dy}{dx} = f'(x) = \frac{y_1}{x_1 - x_2} \qquad (18.5)$$

But for our function, we already found that $f'(x) = 2x$, so at point x_1, the derivative is evaluated as

$$f'(x_1) = 2x_1$$

and Equation (18.5) becomes

$$2x_1 = \frac{y_1}{x_1 - x_2} \qquad (18.6)$$

Now we are all set. All we must do is solve for x_2:

$$x_2 = \frac{2x_1^2 - y_1}{2x_1}$$

If we substitute $y_1 = x_1^2 - S$ [found by substituting $y = y_1$, and $x = x_1$ into Equation (18.3)], we get

$$x_2 = \frac{1}{2}\left(x_1 + \frac{S}{x_1}\right) \qquad (18.7)$$

or

$$\text{new guess} = \frac{1}{2}\left(\text{old guess} + \frac{S}{\text{old guess}}\right) \qquad (18.8)$$

This is an *iterative process*, meaning that it is repetitive, and computers are very good at iterative processes. Shortly, we will show how to program this particular process, or algorithm, for the computer, but first we must talk about *flow charts*.

■ 18.3 FLOW CHARTS

We will use only the four flow chart symbols shown in Figure 18.7. As an example, we will use them to prepare a flow chart for the following computation:

$$a = \sqrt{b^2 + c^2 - 2bc \cos \theta}$$

This flow chart (Figure 18.8) is straightforward. After "start," the first block shows that we input values of b, c, and θ. In the next block we calculate a, in the next we print it (output), and then we look to see if the values of b, c, and θ are the last ones we are going to use to compute a. If the answer to this question is "yes," then we stop; if the answer is "no," we branch off and loop back to the beginning to input new values of b, c, and θ. We keep this up, computing and printing values of a, until we do come to the last set of values for b, c, and θ, whereupon the process stops.

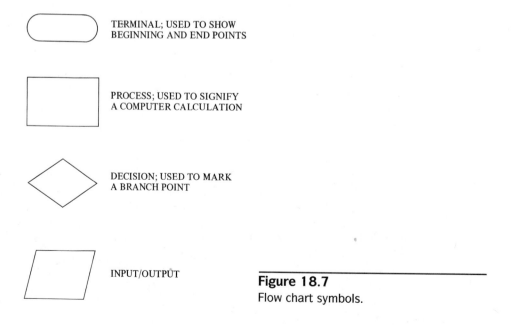

TERMINAL; USED TO SHOW
BEGINNING AND END POINTS

PROCESS; USED TO SIGNIFY
A COMPUTER CALCULATION

DECISION; USED TO MARK
A BRANCH POINT

INPUT/OUTPUT

Figure 18.7
Flow chart symbols.

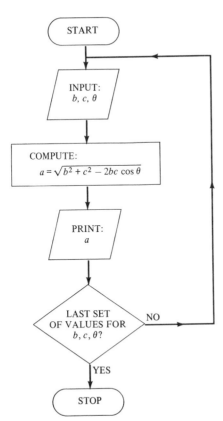

Figure 18.8
A flow chart for computing
$a = \sqrt{b^2 + c^2 - 2bc \cos \theta}$.

Let us now make a flow chart for our Newton's method formula of Equation (18.7), as shown in Figure 18.9. At the outset, we arbitrarily take our first guess for x_1 as 1.0. Next we input the value of S, the number for which the square root is to be found. In the next block we compute x_2, the new guess, according to the formula in Equation (18.7). In the diamond-shaped box we compare the new guess to the old guess by taking their difference, and here we have arbitrarily said that if the absolute value of the difference is more than 0.0001, meaning that the answer to the condition in the diamond is "yes," we will branch off and loop back to the beginning, replacing x_1 with x_2. In other words, we replace the old guess with the new guess, after which we go on to repeat the calculation of x_2 and test the difference $|x_2 - x_1|$ once more to see if it is still greater than 0.0001. We repeat this process, looping around and around, until the condition $|x_2 - x_1| > 0.0001$ no longer holds, meaning that the absolute difference between x_2 and x_1 is now *less* than 0.0001. When this occurs, the answer to the stated condition is now "no," so we proceed immediately to print x_2 as a suitable approximation to the square root of S, and stop.

This should be enough for an introduction to flow charts. Each flow chart gives us an algorithm, i.e., a procedure for solving a problem. In the next section we will convert an algorithm into an operable computer program.

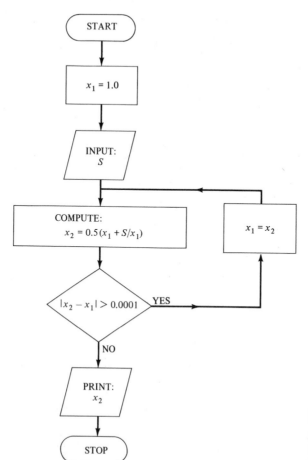

Figure 18.9
A flow chart for Newton's method to find the square root of a number, S.

Exercises

18.3.1 We can solve for the roots of the quadratic equation

$$ax^2 + bx + c = 0$$

by use of the formula

$$x = \frac{-b \pm \sqrt{b^2 - 4ac}}{2a}$$

If

$b^2 - 4ac > 0$, the roots are real and unequal.
$b^2 - 4ac = 0$, the roots are real and equal.
$b^2 - 4ac < 0$, the roots are imaginary.

Draw a flow chart for an algorithm that will accept many successive values of a, b, and c, and solve for the roots. Each time, your algorithm should check to see if $b^2 - 4c < 0$; if so, it should cause a statement "ROOTS IMAGINARY" to be printed, and go to the next set of values of a, b, and c.

Solution

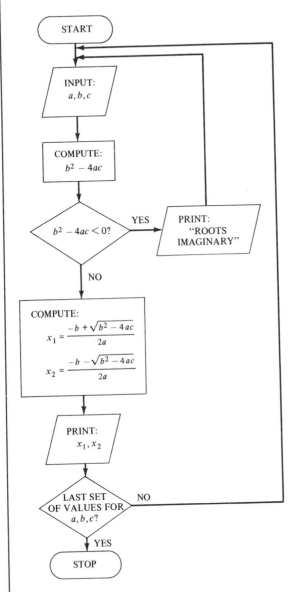

18.3.2 Draw a flow chart for an algorithm to solve the equation

$$c = \sqrt{a^2 + b^2}$$

for many successive values of a and b.

18.3.3 Draw a flow chart for the first three steps of solving Equation (18.1), assuming that x consists of either five or six digits so that our first trial value for a can be 100, as in the example. Compute $R = x - a^2$. If R is positive, add 100 to a, and cause a new calculation of $R = x - a^2$. If R is negative, cause a calculation to occur that will subtract 100 from a, then print a, and stop. If R is zero, print a and stop.

18.3.4 Draw a flow chart that will compute the approximate value of $e^{2.5}$ by computing the first 10 terms of the infinite series

$$e^x = 1 + x + \frac{x^2}{2!} + \frac{x^3}{3!} + \cdots$$

Assume that your computer does not automatically provide for the computation of factorials, so you must provide for this yourself. Calculate the result of your first ten terms and compare to the value of $e^{2.5}$ you get from your hand-held calculator.

■ 18.4 PROGRAMMING IN FORTRAN

The trickiest part of writing a computer program is the one we introduced in the past two sections, which is to reduce the problem at hand to a step-by-step solution procedure. Converting that procedure into computer instructions is the part we will now take up. To do this, we must discuss certain standardized ways of writing instructions; in other words, we must use a *programming language.*

There are many programming languages. Some of them deal with specific instructions to the computer—to tell it where to get a piece of data (from an input register, for example), where to put it (into a register called the *accumulator*, perhaps), where to put a product of multiplying two numbers (into a memory register), and so on. Such a programming language would be called an *object language*, or *machine language*. We will not deal with that kind of programming language here, but will deal with so-called high-level languages. These are constructed in a way that is natural to human beings, using familiar algebraic expressions and instructions written in English. Three languages that are widely used by engineers and scientists are BASIC, FORTRAN, and PASCAL. Of course, the computer must have an interpretive program that converts the high-level language into object language, but we will not concern ourselves with that here. Such a program is called either an *interpreter* or a *compiler*, depending upon how it functions, but in either case we will consider such a program as if it were a part of the machine.

BASIC is an acronym for *Beginner's All-purpose Symbolic Instruction Code*, a language that was developed at Dartmouth College. Do not be misled by the use of the word *beginner's* into thinking BASIC is a trivial language. It is not. It is a surprisingly powerful language, and many engineering problems can be solved through its use. Also, the reader should not suppose there is only one BASIC language. Unfortunately, there are many "dialects" of BASIC, and a computer user will quickly learn that each make of computer may use a slightly different BASIC dialect. Thus, a program that works on one computer may not work on another without some modification.

FORTRAN was developed by IBM in the 1950s. Its name is a contraction of *FORmula TRANslation*, and it has been the most popular programming language among engineers and scientists. Its structure was originally conditioned largely by the fact that it was designed to be key-punched into IBM cards and so was limited to the characters that typically appear on a key-punch keyboard. FORTRAN 77, the newest version of FORTRAN, has a character set somewhat larger than previous versions of FORTRAN, plus an expanded range of capabilities. To the level we will reach in this book, BASIC and FORTRAN are pretty much the same kind of language except that their instructions differ somewhat in appearance. A person who knows FORTRAN picks up BASIC quickly, and vice versa.

PASCAL, named for a famous seventeenth-century mathematician, was designed by Professor Niklaus Wirth of the Federal Institute of Technology in Zurich, Switzerland. It is growing in popularity, largely because its highly structured nature permits the programmer to avoid many of the errors that typically arise in the writing of large and complex programs. The structuring takes the form of blocks, each beginning with the word BEGIN and ending with the word END. Furthermore, each new subordinate block is indented several spaces, as in an outline, so that the blocks stand out clearly with respect to each other. This structure is a substantial aid in troubleshooting the program later. Also, programs in PASCAL exhibit little of the kind of jumping from statement to statement that results from the use of the "go to" statement frequently employed in FORTRAN or BASIC programs.

To the level we shall reach in this book, PASCAL does not appear to differ much in philosophy from BASIC and FORTRAN, although it differs in appearance. PASCAL is preferred by computer scientists, although FORTRAN continues in popularity with many engineers because so much existing software uses it. FORTRAN 77 is the language we will use for the examples of this text.

Algebraic Statements

The first rules in FORTRAN have to do with assigning names to the quantities to be manipulated. In our examples, these quantities will be algebraic variables. The rules for assigning names are:

1. The name must begin with an uppercase letter.
2. The name must contain only uppercase letters or numerical digits.
3. The name must contain one through six characters.

Thus, A, B, X, A1, B1, T435WE, and X91 are all permissible variables, but 2X, 345672, and A4-56 are not.

The next rule has to do with writing algebraic statements. Generally, these are written so they resemble algebraic equations, as in

$$X2 = A + B$$

The expression on the left of the equals sign must be a valid variable name, and the expression on the right may consist of a combination of numbers, previously defined variables, or mathematical functions. However, it is very important to realize that such an expression, even though it resembles an algebraic equation, actually is not to be interpreted as one. In FORTRAN and other similar algebraic languages, a statement such as the foregoing is actually a *replacement* statement. What it means is that the computer is to go to the register named "A" and read what is in that register, then go to the register named "B" and read what is in that register, then add

Figure 18.10
A microprocessor chip magnified to show its intricate structure. (Courtesy of Intel Corporation)

the values from register A and register B together, and then place the sum into the register named "X2," erasing whatever was previously in X2. This explanation permits us to give meaning to such strange-looking expressions as the following, which do not appear to follow the rules of algebra:

X1 = X2

X = X + 1

The first of these expressions, X1 = X2, means that we should take the number we find in the register set aside for the variable X2 and place it in the register set aside for variable X1, erasing the value that was previously in that register. The second expression says that we should take the number in the register set aside for the variable named X, add 1 to it, and put it back into the location for X, erasing the value that was previously in that register.

Valid operation signs for algebraic statements are

Addition	+
Subtraction	−
Multiplication	*
Division	/
Exponentiation	**

When FORTRAN expressions are evaluated by the computer, there are certain rules of priority:

1. All expressions within parentheses are evaluated first. If the expression contains parentheses nested within parentheses, these will be evaluated from the inside out.

2. Exponentiations are evaluated next.
3. * and / operations are evaluated next, working from left to right.
4. + and − operations are evaluated last, working from left to right.

Here are some valid algebraic statements:

X2 = 0.5*(X1 + S/X1)

D1 = D1/10.0

Y = A*X**3 + B*X**2 + C*X + D

The first statement gives a way in which we might write the algebraic expression

$$x_2 = \frac{1}{2}\left(x_1 + \frac{S}{x_1}\right)$$

The second statement says we should take the number we find in D1, divide it by 10, and put the result back in D1. The third statement is the way we would write the algebraic expression

$$y = ax^3 + bx^2 + cx + d$$

Input and Output

We obviously must have a way to get data into the computer, and for this we use READ statements, which are written as in the following examples:

READ *, S

READ *, A, B, C

When the computer encounters such an instruction, it stops and waits until the necessary information has been entered. In the first example above, after the operator enters the desired value of S, the return key is pressed, and the computer proceeds with the program. In the second case, the computer waits until values for A, B, and C have been entered; the return key must be depressed after the entry of each value.

There must also be a way to get information out of the computer, so we have a PRINT instruction:

PRINT *, X1

When the computer encounters this instruction, it will print whatever value is in the X1 register and then go on to the next command. But we might prefer to have the value labeled, in which case we would write instead

PRINT *, 'X1 = ', X1

Then, before printing the value from the X1 register, the computer would print whatever we wrote between the quotation marks. For example, if the X1 register happened to have the number 195 in it, the computer would print

X1 = 195

Figure 18.11
A single-chip microprocessor: a computer on a fingertip. (Courtesy of Intel Corporation)

We combine the PRINT and READ commands whenever we want something to be read into the computer. Otherwise, when a READ command is encountered, the computer will simply stop and wait, and you, the operator, will not know what it wants you to do. Hence, instead of writing only

READ *, S

as we did in the example above, we couple it with a PRINT statement, as follows:

PRINT *, 'Enter value of S'

READ *, S

Now, when the computer gets to the PRINT statement, it will print the information you wrote between the quotation marks, and then it will go to the READ statement and wait for you to enter the desired value for S. To the uninitiated observer, it looks like you are working with a very intelligent machine that knows what questions it should ask you. But you, the programmer, know it is only doing exactly what your program told it to do.

Program: Law of Cosines

Now we are ready to write a simple program for the flow chart of Figure 18.8. The program is shown in Figure 18.12. First it should be given a name, and the first line does this. The name must obey the rules that were listed on page 509; that is, it must contain no more than six uppercase letters or numerical digits and must begin with an uppercase letter. We have named our program "LAWCOS."

The next line involves what is called a *comment*. Note that this line has a "C" at the left margin, which is designated as column 1. If a "C" is placed in this column, it means that the computer is supposed to print whatever is on that line, but otherwise

```
      PROGRAM LAWCOS
C  LAW OF COSINES
      REAL A, B, C, T
 5    PRINT *, 'Enter values of B, C, and T'
      READ *, B, C, T
      A = (B**2 + C**2 - 2.0*B*C*COS(T))**0.5
      PRINT *, 'A=', A
      GOTO 5
      STOP
      END
```

Figure 18.12

A program to implement the flow chart of Figure 18.8.

ignore it. Comments such as this are useful for letting you or perhaps another person know what the program is supposed to be doing. In our case, we are providing for a title that says that this program is concerned with the law of cosines.

The next line is another new item. When we say REAL A, B, C, T, we are telling the computer that all of the quantities A, B, C, and T are going to be real algebraic variables. These are the numerical quantities with which we are familiar, such as 1.0, 0.95768, or 256.75. Note that each quantity contains a decimal point, and this is a requirement for real algebraic variables. Other possible kinds of variables are INTEGER, CHARACTER, and LOGICAL, but because we will not use them in our brief treatment, we will not discuss them further.

On the fourth line of the program is a familiar statement. When the computer comes to this statement, it will print "Enter values of B, C, and T," then it will go to the next line of the program and wait for you to enter the necessary values. Note that the numeral "5" has been entered in column 1 of this statement; this is called a *label*. We will come back to labels a little later.

The sixth line of the program is the FORTRAN equivalent of the algebraic formula

$$a = \sqrt{b^2 + c^2 - 2bc \cos \theta}$$

Note that we have assigned the symbol A to the variable a, B to b, C to c, and T to θ. Note also that we took the square root of $b^2 + c^2 - 2bc \cos \theta$ by enclosing the entire expression in parentheses and raising it to the power $1/2$.

In the computation of the value of A, we wrote COS(T) for $\cos \theta$. Computers have a number of built-in functions, such as the following:

ABS(X)	absolute value of X
SIN(X)	sine of X
COS(X)	cosine of X
TAN(X)	tangent of X
EXP(X)	e^X
ALOG(X)	natural log of X
SQRT(X)	square root of X

As a general rule, the argument X for the trigonometric functions should be entered in radians.

In the next statement (seventh line), we tell the computer to print "A =" and then the value of A.

Now we come to an interesting step in our flow chart (Figure 18.8), which says that, if we have no more values to input for *b*, *c*, and θ, we should stop, but if the opposite is true, we should go back to the input step. For this simple program we will allow the computer operator to make this decision and will automatically send the computer back to the input step every time. We do this by insertion of a statement we have not mentioned before, called the GOTO ("go to") statement. In our case, the statement is

GOTO 5

Now we see why we put the label "5" in column 1 of the fourth line of our program. When the computer encounters a GOTO statement, it will automatically shift to whatever statement bears the label included in the GOTO command. In our case, the label is "5," so the computer shifts to the statement bearing that label. (There is nothing special about the numeral "5." We could have chosen any number between 1 and 99999.) So the computer shifts back to the statement that says

PRINT*, 'Enter values of B, C, and T'

and prints the material between the quotation marks. Then it moves down one line to the "READ*, B, C, T" statement and waits for you to enter the new values for *b*, *c*, and θ. It will keep doing this forever, returning to the input step each time after printing A, until you terminate the process by turning off the computer. This is so because you, the programmer, have put the computer into an *unconditional loop*, and there is no way for it to get out of that loop without figuratively "killing" the computer.

Our program ends with two lines, the statement STOP followed by the statement END. Every program must conclude with these two statements. We have put them in this program even though we know the computer will never get to them because of the unconditional loop. However, shortly we will write some programs that *will* get to these statements, so we put them here as a reminder that we must always do so.

There are some additional restrictions on writing statements. Note that all of the statements in Figure 18.12 except two are indented six spaces. This is because of the following rules. The leftmost column of a page is called column 1, and the rightmost column is column 72. The rules are:

Column	
1	A "C" in this column says that the statement is a *comment*. The statement is to be printed, but otherwise ignored. It can occupy any of the columns from 3 through 72 and may contain lowercase letters.
1–5	May be used for a statement *label*, consisting of any number from 1 to 99999.
7–72	These columns contain the FORTRAN statement, written in uppercase letters. (Statements may occupy more than one line, provided special symbols are used, but we will not discuss that here.)

Branching: The IF statement

The IF statement permits us to branch along different paths in the program, depending upon outcomes. An IF statement takes the following form:

If (condition) THEN

    (statement)

    (statement)

    $\vdots$

ENDIF

If the condition in parentheses which follows the word IF is met, then the computer executes the statements on the following lines. If the condition that follows the word IF is *not* met, the computer skips all the statements on the following lines and goes directly to the instruction immediately below ENDIF.

Some of the conditions that might be specified in IF statements are

equal to	.EQ.
greater than	.GT.
greater than or equal to	.GE.
less than	.LT.
less than or equal to	.LE.
not equal to	.NE.

The following example will clarify the IF statement and the use of these conditions.

Program: Newton's Method

The flow chart in Figure 18.9 was for the use of Newton's method to find the square root of a number. The program for it is shown in Figure 18.13. Let us go through it step by step.

The first line gives the name of the program. We have called it "NEWTON." The second line is a comment that tells us what the program is doing. The computer prints the comment and otherwise ignores it.

The third line states that X1, X2, and S are real algebraic variables.

The fourth line arbitrarily sets X1 = 1.0, as a first guess for the variable x_1.

The fifth line prints the statement "Enter value of number for which square root is desired," and the sixth line causes the computer to wait until you have entered the value of S and have pressed the return key.

The seventh line causes a value of the "new guess" X2 to be calculated according to the formula

$$x_2 = \frac{1}{2}\left(x_1 + \frac{S}{x_1}\right)$$

```
          PROGRAM NEWTON
C   SQUARE ROOT BY NEWTON'S METHOD
          REAL X1, X2, S
          X1 = 1.0
          PRINT *, 'Enter value of number for which square root is desired'
          READ *, S
    1     X2 = 0.5*(X1 + S/X1)
          IF (ABS(X2 - X1) .GT. 0.0001) THEN
          X1 = X2
          GOTO 1
          ENDIF
          PRINT *, 'SQUARE  ROOT=', X2
          STOP
          END
```

Figure 18.13

A program to implement the Newton's method flow chart of Figure 18.9.

The eighth line sets up the condition to be tested in the diamond of Figure 18.9, namely, whether the absolute difference between x_2 and x_1 is greater than 0.0001. If the difference *is* greater than 0.0001 (condition YES), then the computer performs the operation on the next line, which is to replace the value of x_1 with the newly calculated value x_2.

The next line below $x_1 = x_2$ (line 10) tells the computer to go to the statement bearing the label "1," which sends it back to the statement on line 7 of the program, and a new value for x_2 is calculated. The computer continues to follow this loop until the condition $|x_2 - x_1| > 0.0001$ is no longer met, which means the difference is now *less* than 0.0001. When this occurs, then the answer to the condition in the IF statement is NO, so the computer skips over the commands on lines 9, 10, and 11 and comes to the command on the line below ENDIF, which is line 12.

The command on line 12 causes the expression between the quotation marks to be printed and then prints the value in the X2 register as the final estimated value for the square root. After printing, the next two statements tell the computer the program has come to an end, so the computer stops.

Program: Determinants

The purpose of this last example is to illustrate an application involving an *array*. Arrays will be introduced in the context of determinants, which in turn are related to the very important topic of matrices. Matrices and matrix algebra usually come in the second or third year of an engineering curriculum and are vital to every branch of engineering. For example, matrices crop up in circuit analysis, control systems, vibrations, fluid mechanics, heat transfer, and structural analysis.

Let us take a 3 × 3 determinant. (See Appendix D and Chapter 17, where determinants are used to find the currents flowing in a Wheatstone bridge.) We will designate our determinant by the notation

$$D = \begin{vmatrix} a_{11} & a_{12} & a_{13} \\ a_{21} & a_{22} & a_{23} \\ a_{31} & a_{32} & a_{33} \end{vmatrix} \qquad (18.9)$$

Such a determinant can arise from a system of three simultaneous equations:

$$\left. \begin{array}{l} a_{11}x_1 + a_{12}x_2 + a_{13}x_3 = k_1 \\ a_{21}x_1 + a_{22}x_2 + a_{23}x_3 = k_2 \\ a_{31}x_1 + a_{32}x_2 + a_{33}x_3 = k_3 \end{array} \right\} \qquad (18.10)$$

(The algebraic details are in Appendix D.) We know that D is simply a number and can be evaluated as

$$D = a_{11}(a_{22}a_{33} - a_{23}a_{32}) - a_{12}(a_{21}a_{33} - a_{23}a_{31}) + a_{13}(a_{21}a_{32} - a_{22}a_{31}) \qquad (18.11)$$

Also, we know that the solution to Equations (18.10) is

$$x_1 = \frac{D_1}{D} \qquad x_2 = \frac{D_2}{D} \qquad x_3 = \frac{D_3}{D} \qquad (18.12)$$

where

$$D_1 = k_1(a_{22}a_{33} - a_{23}a_{32}) - a_{12}(k_2a_{33} - k_3a_{23}) + a_{13}(k_2a_{32} - k_3a_{22}) \qquad (18.13)$$

$$D_2 = a_{11}(k_2a_{33} - k_3a_{23}) - k_1(a_{21}a_{33} - a_{23}a_{31}) + a_{13}(k_3a_{21} - k_2a_{31}) \qquad (18.14)$$

$$D_3 = a_{11}(k_3a_{22} - k_2a_{32}) - a_{12}(k_3a_{21} - k_2a_{31}) + k_1(a_{21}a_{32} - a_{22}a_{31}) \qquad (18.15)$$

See Equations (D.14), (D.15), and (D.16) in Appendix D.

If we have a lot of determinants, a computer program can save us a lot of time and effort in evaluating these lengthy algebraic expressions. Furthermore, anyone who has had to evaluate determinants knows how easy it is to make errors, so a computer program can also save us a lot of errors (provided the program is correct, of course).

In principle, there is no reason why an expression as long as the ones in Equations (18.11), (18.13), (18.14), and (18.15) cannot be written in a single instruction. But we will break these expressions up into shorter ones by making the following substitutions:

$$G_1 = a_{22}a_{33} - a_{23}a_{32} \qquad G_2 = a_{21}a_{33} - a_{23}a_{31} \qquad G_3 = a_{21}a_{32} - a_{22}a_{31} \qquad (18.16)$$

$$H_1 = k_3a_{21} - k_2a_{31} \qquad H_2 = k_2a_{33} - k_3a_{23} \qquad H_3 = k_2a_{32} - k_3a_{22} \qquad (18.17)$$

With these substitutions, Equations (18.11), (18.13), (18.14), and (18.15) become

$$
\left.
\begin{aligned}
D &= a_{11}G_1 - a_{12}G_2 + a_{13}G_3 \\
D_1 &= k_1G_1 - a_{12}H_2 + a_{13}H_3 \\
D_2 &= a_{11}H_2 - k_1G_2 + a_{13}H_1 \\
D_3 &= a_{11}(-H_3) - a_{12}H_1 + k_1G_3
\end{aligned}
\right\} \tag{18.18}
$$

Before programming these equations, let us look at the nature of a 3×3 array. Such an array has nine elements. Suppose we tell the computer to set up nine registers in its memory and reserve them for later entry of the nine elements. Suppose, also, that we visualize these registers being set up and reserved in the following pattern:

$$
\begin{array}{|c|c|c|}
\hline
a_{11} & a_{12} & a_{13} \\
\hline
a_{21} & a_{22} & a_{23} \\
\hline
a_{31} & a_{32} & a_{33} \\
\hline
\end{array} \tag{18.19}
$$

The resemblance between this array and the determinant Equation (18.9) is intentional, of course, because arrays arise so often in connection with determinants and matrices. However, we should not fool ourselves into thinking the actual physical relationship among the memory locations in the computer is like the one in the array. More likely, all nine registers come one after the other in sequential locations. (All the computer knows is that we have designated nine locations as storage registers for nine variables. It does not know yet what we plan to do with them.) Nevertheless, we are entitled to think of the registers as if they are arranged physically like array Equation (18.19).

To simplify our example, we will treat all of the values $a_{11}, a_{12}, \ldots, k_1, \ldots,$ $G_1, \ldots, H_1, \ldots, D_1, \ldots$ (22 values in all) as separate algebraic variables. We will give them the names A11, A12, ..., K1, and so on. There are more elegant ways to handle arrays, but we will keep our example as simple as possible.

We are now ready to draw a flow chart, as shown in Figure 18.14. The chart is straightforward, and we have already discussed its elements. Only the branch $D \neq 0$ requires discussion. Because D appears in the denominator for computing x_1, x_2, and x_3, a value of $D = 0$ would make the answer indeterminate. Thus, if the condition $D \neq 0$ is NO, this is the same as saying that $D = 0$, and the computer is instructed to print "$D = 0$" and stop. But if the condition $D \neq 0$ is YES, the computer goes on to compute x_1, x_2, and x_3; print them; and then stop.

The complete program is shown in Figure 18.15. The first statement provides the program name, and the second provides a comment. The next two statements declare that all the variables listed are real algebraic variables.

The next four statements provide for the entry of all the A's and all the K's from Equations (18.10). If we use the numerical values from the Wheatstone bridge example in Chapter 17, we proceed as follows (the relevant equations from that chapter are

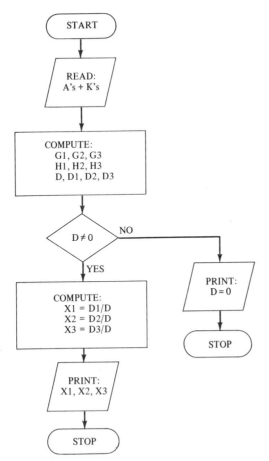

Figure 18.14

A flow chart for evaluating a 3 × 3 determinant.

repeated below, with x's substituted for the i's):

$$3x_1 - x_2 - 2x_3 = 4$$

$$x_1 - 4x_2 + x_3 = 0$$

$$2x_1 + x_2 - 7x_3 = 0$$

The computer prints "Enter values of A11, A12, A13, A21, A22, A23, A31, A32, A33," and we enter the following numbers, taken directly from the above set of equations: 3.0, −1.0, −2.0, 1.0, −4.0, 1.0, 2.0, 1.0, −7.0.

Next, the computer prints "Enter values of K1, K2, K3," and we enter the following numbers: 4.0, 0.0, 0.0.

The next ten statements correspond to Equations (18.16), (18.17), and (18.18).

Following the calculation of D3, we come to an IF statement. All of the commands from IF to ENDIF are shown indented. This indentation is not mandatory,

```
        PROGRAM DETERM
C   CALCULATE VALUE OF THIRD-ORDER DETERMINANT
        REAL X1, X2, X3, A11, A12, A13, A21, A22, A23, A31, A32, A33
        REAL K1, K2, K3, G1, G2, G3, H1, H2, H3, D, D1, D2, D3
        PRINT *, 'Enter values of A11, A12, A13, A21, A22, A23, A31, A32, A33'
        READ *, A11, A12, A13, A21, A22, A23, A31, A32, A33
        PRINT *, 'Enter values of K1, K2, K3'
        READ *, K1, K2, K3
        G1 = A22 * A33 - A23 * A32
        G2 = A21 * A33 - A23 * A31
        G3 = A21 * A32 - A22 * A31
        H1 = K3 * A21 - K2 * A31
        H2 = K2 * A33 - K3 * A23
        H3 = K2 * A32 - K3 * A22
        D = A11 * G1 - A12 * G2 + A13 * G3
        D1 = K1 * G1 - A12 * H2 + A13 * H3
        D2 = A11 * H2 - K1 * G2 + A13 * H1
        D3 = - A11 * H3 - A12 * H1 + K1 * G3
           IF (D .NE. 0.0) THEN
           X1 = D1/D
           X2 = D2/D
           X3 = D3/D
           PRINT *, 'X1 = ', X1
           PRINT *, 'X2 = ', X2
           PRINT *, 'X3 = ', X3
           GOTO 100
           ENDIF
        PRINT *, 'D = 0'
100     STOP
        END
```

Figure 18.15
A program to implement the flow chart of Figure 18.14.

but it is useful because it sets the IF group off by itself, which can help in later checking of the program. The condition associated with the IF statement is set up as D *not equal* to zero. If D is not equal to zero, then the condition is satisfied, and we can proceed with the calculation and printing of the values of X1, X2, and X3. After the printing of X3, the computer encounters the command GOTO 100, so it jumps to that command and stops. On the other hand, if the condition $D \neq 0$ is *not* met, which means that D is zero, the computer will jump all the way to the command that causes "D = 0" to be printed, and then stop.

The foregoing gives the essence of programming and provides a framework upon which to build. Many problems can be programmed with no more than the set of instructions presented here. However, there is a great deal more to programming than this, and most engineering schools offer a series of courses on programming. Also, very little has been said here about how to turn a particular computer on, how to get it into a "program" mode, and how to tell it to begin, or "run." These procedures vary with the model of computer and are covered in instruction manuals.

Exercises

18.4.1 Write a FORTRAN statement for each of the following.

(a) $y = ax^2 + bx + c$ (b) $r = (3x^2 + 2)/(x + 1)$

(c) $t = \dfrac{ax^3 + bx^2}{x + c}$ (d) $A = A + 100$

(e) $c = \sqrt{a^2 + b^2}$ (f) $s = \dfrac{x}{\sqrt{\sin x + 1}}$

18.4.2 Which of these are permissible names for variables in FORTRAN?

BA, AB, 22, 3Y, YY, Y, 4, Y4, X45

18.4.3 Write a FORTRAN "IF" statement for each of the following.
(a) $R > 0$? If yes, go to statement labeled "2."
(b) $R \geqslant 0$? If yes, go to statement 250.
(c) $b^2 - 4ac < 0$? If yes, go to 300.
(d) $|x^2 - a| < 0.001$? If yes, go to 800.

■ 18.5 APPLICATIONS OF COMPUTERS

There is almost no end to the ways in which computers are or might be used. In the examples given in the foregoing sections, we have emphasized the use of computers for numerical calculation, the solution of equations, and the like. Basic to such methods is the field known as *numerical analysis*. Engineering students generally study numerical analysis as an advanced subject, because it depends upon a good knowledge of calculus and differential equations. However, *Newton's method*, which we discussed earlier in this chapter, is a topic that generally arises in courses on numerical analysis, so you have already gained some insight into the subject.

We have presented ways in which an engineer might use a computer, doing all the programming from scratch. Every engineer should know how programming is done, but in actual professional practice, engineers generally use prepackaged programs prepared by others, generally referred to as *software*. Such programs sometimes come in an accessory unit called a ROM (read-only memory) or may come in recorded form on a flexible magnetic diskette frequently referred to as a "floppy disk." (More general names are *diskette* and *minidisk*, since not all such disks are actually floppy.) The disk is inserted into the disk memory unit (Figure 18.16), and the desired program is transferred internally from it to the computer memory. Subsequently, all the operator need do is provide the necessary numerical data, together with any special instructions regarding the output format. The output can be in tabulated form or, with appropriate accessories, in graphical form. Following are some examples of types of preprogrammed packages.

☸ Milestones
From Transistor to Microprocessor

The first transistor was announced in 1948 by Bell Laboratories. As with most inventions in their initial forms, the transistor looked unimpressive. More than anything else, it resembled the old "cat's whisker" detectors used in the crystal radio sets of the 1920s. But it did something new and important: It provided amplification of a signal, and did so by using germanium, a solid-state material.

Most of the world did not take notice, but the scientific research community did. Many small new firms sprang into existence in the next few years, their objective being to manufacture transistors. In 1954, a company that few had heard of before, called Texas Instruments (TI), produced the first silicon transistor. TI ultimately became an industry giant, and silicon later became the most frequently used material in the field. In fact, this material, used as the substrate of transistors, inspired the nickname for the greatest concentration of semiconductor firms in the world—Silicon Valley, in the Santa Clara Valley region of California.

Five years after the birth of the silicon transistor, a new production technique was devised by Fairchild Semiconductor that was to revolutionize transistor technology. This was the planar technique for producing a transistor, which paved the way for integrated circuits and microprocessors. By utilizing intricate masks and the process of photoetching, borrowed from the printing industry, it became possible to produce thousands and finally hundreds of thousands of transistors on a single chip.

In 1968, a few engineers left Fairchild to form Intel Corporation, one of many new companies spawned by former employees of Fairchild—so many that some people referred to them as the *Fairchildren*. Intel was the most successful of all, partly because it came up with a new kind of integrated circuit called a *microprocessor*.

A microprocessor is essentially a computer on a chip. The idea was born in the mind of Ted Hoff, a young engineer with a Ph.D. from Stanford, who had just joined Intel as its twelfth employee. The idea for a completely new way of organizing the circuits came to him, he says, as he looked at a PDP-8 computer sitting near his desk and wondered why some of the operating philosophy of the PDP-8 couldn't be condensed into microcircuitry. The result was a set of four chips that could perform all the operations of a computer. The great advantage was that these were general-purpose units that could be produced in high volume and low cost.

The company was worried that there might be only a small market for the chips. Nevertheless, in 1971, they decided to go ahead, and announced the product. By the 1980s, hundreds of millions of microprocessor chips were being produced annually, and Intel had become a billion-dollar corporation.

It is often the case with new technologies that many people are working on similar ideas at the same moment in time. In the case of the microprocessor, credit for its origination must be shared with Gilbert Hyatt, who was finally granted a patent in 1990 for a "single-chip computer" after a 20-year battle with the patent office. But Intel had also been issued patents on single-chip computers, which raises the specter of extended legal battles as the potentially conflicting claims on the patents are sorted out.

References

The Daily Democrat, August 31, 1990, p. 11.

T. Forester (ed.), *The Microelectronics Revolution* (Oxford, England: Basil Blackwell Publisher, 1980).

Solutions (Santa Clara, Calif.: Intel Corporation, Nov./Dec., 1981).

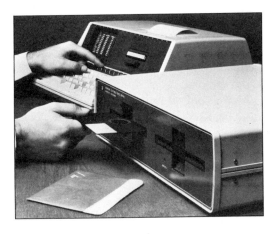

Figure 18.16
Close-up of a dual floppy disk drive, showing the operator inserting a floppy disk, enclosed in a protective cover. (Courtesy of Hewlett-Packard Company)

Text Editing

Many times, computers are used for preparing memos and reports as well as for solving mathematical problems. Software packages for this purpose provide for such editing features as adding or deleting text, moving lines of text from one location to another, and renumbering text lines. Computers that are devoted primarily to this purpose are referred to as *word processors*. In a word processor, the initial draft of a memo or report is retained on a diskette by the machine operator. When revisions are received, the diskette is reinserted in the disk memory unit, and all modifications are made through the computer keyboard to produce a second draft, which replaces the first draft. As many drafts as are needed are made, with the machine operator only inserting the changed items each time, and with the computer memory automatically retaining all the unmodified text. When the last change has been made, the output of the word processor is directed to a printing unit, which types a high-quality copy at very high speed—much faster than a human typist. Also, book publishers are encouraging authors to compose their manuscripts on word processors. The manuscript draft is recorded on disk, and corrections, additions, and deletions are made that continually modify the disk memory storage. At the end, when the manuscript has been perfected, the disks containing the final draft are mailed to the publisher, who can then use them to control the setting of type directly, without having to enter everything again through a keyboard.

Math Solutions

A bewildering variety of prepackaged software exists for math problems. For example, statistics packages perform regression analysis, calculate standard deviations, correlations, statistical distributions, and many other analyses of interest to statisticians. Math software solves sets of simultaneous equations, finds roots of polynomials, evaluates integrals, solves differential equations, and performs many other advanced mathematical operations. As mentioned earlier in this chapter, matrices turn up repeatedly in every branch of engineering, so many software routines are available to carry out a variety of matrix operations.

Finite Element Method

An extremely important numerical analysis technique, called the *finite element method*, has become widespread in recent years and is based upon the use of matrices in computers. The method has been used in fluid mechanics, heat transfer analysis, and structural analysis. It has permitted the analysis of many engineering problems that once were considered unsolvable. For example, the blade of a gas turbine was once considered to be too complicated to analyze by mathematical techniques. In the finite element method, such a part is considered to be made up of a large number of very small, more or less brick-shaped pieces called finite elements, connected only at their corners. If the pieces are taken small enough, and if the mathematical representation of the relationships at the corners is chosen correctly, then the assemblage of interconnected blocks will behave mathematically in a way that closely resembles that of the original part.

Choosing the distribution of the blocks (called generating a finite element *mesh*) and handling the resulting matrix equations require extensive knowledge of structural analysis and of a field called the *theory of elasticity*. The matrices can be very large, involving hundreds or thousands of elements. Hence, their solution would be

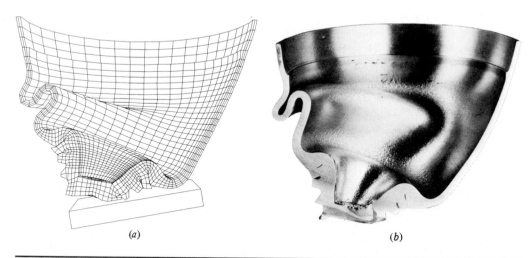

(a)   (b)

Figure 18.17

(a) A computer-generated finite element mesh for a crushable nose cone, which is designed to absorb energy. The finite element mesh is shown in deformed mode after a simulated impact with a surface at an angle. (b) An experimental model after it experienced an actual impact and then was cut in cross section to show its deformed shape. Agreement between the deformation predicted by the computer in part (a) and that of the actual experiment can be seen to be excellent. (Courtesy of University of California Lawrence Livermore National Laboratory and U.S. Department of Energy)

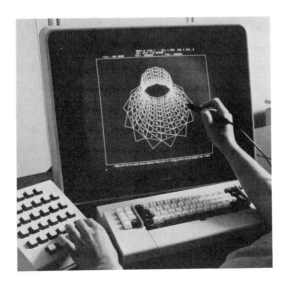

Figure 18.18

An engineer developing a three-dimensional finite element analysis model, using a light pen and a special control panel. (Courtesy of Gibbs and Hill, Inc.)

impossible without large computers. As a result, the development of the finite element method and of digital computers occurred hand in hand. Figure 18.17(*a*) shows a finite element mesh for a nose cone and the deformed shape resulting from a computer analysis, assumed to be the result of an impact on the cone. Figure 18.17(*b*) shows the deformed shape resulting from an actual experiment on a nose cone. As can be seen, the agreement is excellent. Figure 18.18 shows an engineer at an interactive computer terminal, developing a finite element mesh for a structural shell. Many software packages are available that can generate meshes and solve for almost anything that is wanted, including stresses, deflections, temperature distribution, vibrational behavior, and creep, the last named referring to the tendency of some materials to deform steadily under a sustained load.

Business Applications

Some uses of computers have become commonplace. Everyone knows, for example, that banking, credit card sales, and airline reservations are done entirely by computers. Such far-reaching applications have given rise to a field of activity known as *computer security*. This term has two meanings: (1) If something happens to the computer (a power failure, for example), the system must be *fail safe*; that is, it must be designed and operated in such a way that no information is lost. (2) The system must be provided with safeguards against unauthorized access or manipulation of records. These are very important matters and cause computer system designers a great deal of uneasiness.

Much commerce is carried on through *computer networking*, involving not only financial transactions, but also communication of such things as reports, sales orders, and memos. It is possible to hook up to a network through a device referred to as a *modem*—for "*MO*dulator/*DEM*odulator"—which couples a terminal through telephone lines to a computer in a remote location.

Process Management

Increasingly, chemical plants, refineries, and utilities are coming under computer control. The plant operation is typically managed by the computer software, which continually adjusts temperatures and flow rates to maintain the plant at the optimum operating conditions. All of the electrical signals that show what the plant is doing at any given moment are brought into a central control room, where an operator can display graphical images on the screen with whatever detail is desired. The claim has been made that computer control can produce an efficiency improvement of up to 20 percent in a plant, produce greater reliability, and help the plant meet pollution control requirements.

Medical Engineering

Without electronics, the medical profession would be deprived of much of its ability to see, hear, and touch. Enormous advances have been made in medicine because of electronic instrumentation, and almost all such instrumentation today either is, in effect, a computer or contains a built-in microprocessor. Because of this, a whole field of *medical engineering*—or *clinical engineering*, as it is sometimes called—has grown up. A few examples will be given here.

In intensive care units, a microprocessor-based system may monitor heart rate, blood pressure, the respiratory system, the metabolic control system, and the nervous system. All of these vital signals permit assessment of the patient's condition, comparison with established criteria, and display of these at a central control station. In operating room situations, computer-based systems control many channels of data relating to concentration of oxygen and carbon dioxide in the patient's breath, the heart rate, blood pressure, internal temperature, and the like. If a sudden change occurs in any of these, the circumstances are instantaneously analyzed and an error message is flashed, if needed. In electroencephalography (brain waves), a computer can automatically make the analyses of records that formerly were made visually, and detect abnormalities. In prosthetics (the field of medicine dealing with artificial limbs), microprocessors collect signals from the patient's nervous system and use them to control motors, producing coordinated motions of artificial arms or legs. Finally, in a *computed tomography* (CT) scanner, a patient is placed in the middle of a circular array of x-ray detectors located opposite an x-ray source. The source is rotated through a circle, and signals are thus produced relating to a complete cross-sectional "slice" through the patient's body. These signals are processed in a computer, which then reconstructs an image of a cross section of the body (see Figure 18.19).

Spreadsheets

Spreadsheets have been used by engineers (and others) for decades. Generally, a spreadsheet takes the form of columns of data, which are the results of calculations with values from preceding columns. Prior to the advent of the computer, the calcu-

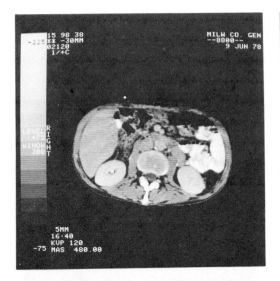 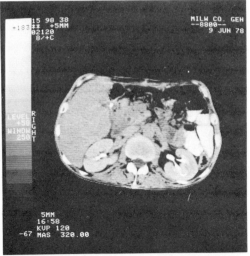

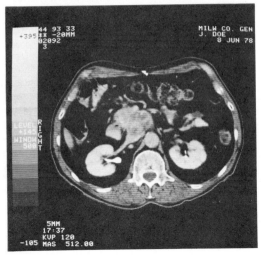 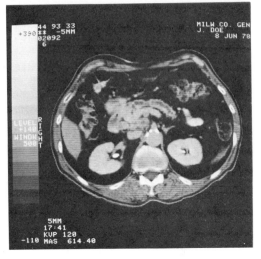

Figure 18.19
Cross-sectional x-ray images of the abdomen, taken on a CT (computed tomography) scanner.
Each section takes only 4.8 seconds to make. (Courtesy of General Electric Company)

lations had to be performed with slide rules or mechanical calculators. As a result,
the process was time-consuming and error-prone. But with a computer the calculation process is nearly instantaneous and free of errors—provided the programming
is done correctly, of course.

Figure 18.21 is an example of a spreadsheet, in this case applied to the repayment
of a loan. In Chapter 19 you will learn how to use a formula that tells how large each
payment must be if a loan is to be repaid in a fixed series of payments. The formula

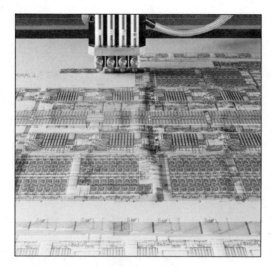

Figure 18.20
An automatic printer reproducing a diagram of a microcircuit.

A	B	C	D	E	F
Period	Principal At Beginning of Period (F)	Payment ($888.49)	Interest (0.01 × B)	Payment Toward Principal (C − D)	Principal At End of Period (B − E)
1	$10,000.00	$888.49	$100.00	$788.49	$9211.51
2	9211.51	888.49	92.12	796.37	8415.14
3	8415.14	888.49	84.15	804.34	7610.80
4	7610.80	888.49	76.11	812.38	6798.42
5	6798.42	888.49	67.98	820.51	5977.91
6	5977.91	888.49	59.78	828.71	5149.20
7	5149.20	888.49	51.49	837.00	4312.20
8	4312.20	888.49	43.12	845.37	3466.83
9	3466.83	888.49	34.67	853.82	2613.01
10	2613.01	888.49	26.13	862.36	1750.65
11	1750.65	888.49	17.51	870.98	879.67
12	879.67	888.49	8.80	879.69	(.02)

Figure 18.21
Example of the spreadsheet method applied to a repayment schedule for a $10,000 loan repaid in 12 monthly installments, at 12 percent annual interest.

is given by Equation (19.8).

$$A = P\left[\frac{i(1 + i)^n}{(1 + i)^n - 1}\right]$$

where A is the amount of each payment, P is the amount of the original loan (called the *principal*), i is the interest per payment period, and n is the number of periods. Thus, if $P = \$10,000$ and is to be repaid in 12 equal monthly payments at an annual interest rate of 12 percent, then the interest per month is $i = 1\%$, and $n = 12$. If we substitute these values in Equation (19.8), we get

$$A = \$10,000 \left[\frac{0.01(1 + 0.01)^{12}}{(1 + 0.01)^{12} - 1} \right]$$

$$= \$888.49$$

In the spreadsheet of Figure 18.21, column A shows the period, and column B shows the amount of principal at the beginning of the period. (The amount of the principal is reduced each month, as will be shown.) Column B begins with $10,000, which is the original value of the loan. In each subsequent period the value in this column is transferred from the entry in column F on the line just above. Thus, the entry of $9211.51 in the second row of column B is taken from the entry in the first row of column F, and so on.

Column C is always the constant value of $888.49, which is the amount of the monthly payment. Column D is the result of multiplying 1% by the amount of principal shown on the same row in column B. Column E is the result of subtracting the amount in column D from the amount in column C—in other words, the result of subtracting the amount of the interest from the $888.49 monthly payment to see how much can be applied to reducing the principal. Column F is the result of subtracting the amount in column E from the amount in column B; in other words, column F shows the amount of principal remaining at the end of the period, when we reduce the principal at the beginning of the period by the amount in column E. The amount remaining at the end of the period is then transferred to column B in the next row and becomes the amount of principal at the beginning of the next period. The calculations begin in the upper left-hand corner and proceed from left to right, a row at a time. The entry shown in parentheses in the heading of each column shows how the amounts in the columns are to be calculated.

A computer is a "natural" for carrying out operations like the foregoing (although the actual programming can be complex and will not be presented here). In fact, a spreadsheet program on a computer would probably also provide for the freedom to vary the quantities P, i, and n. Also, it undoubtedly would have the formula of Equation (19.8) built into it so that the payment amounts in column C could be automatically calculated. Thus, each time you entered a different value of P, i, or n, the spreadsheet values would instantly be recalculated and displayed. In fact, with appropriate programming, the results could also be printed in the form of graphs. Some of the commercially available spreadsheet programs are *Visicalc*, *Supercalc*, and *Lotus 1-2-3*.

The variety of operations that computers perform goes on almost without end. Computers are used to forecast the weather, to simulate river flow and tidal action in estuaries, to make economic analyses, and to simulate the operation of new transportation systems. They turn up in burglar alarms, lighting systems, telephone consoles, traffic signals, and "smart" scientific instruments. But, rather than examine any of these applications, we will turn next to an area that has an enormous potential impact upon productivity—the activity known as CAD/CAM, standing for *Computer-Aided Design* and *Computer-Aided Manufacturing*, together with the closely related field of robotics.

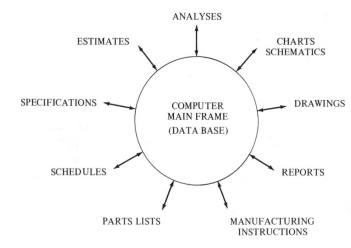

Figure 18.22
Schematic representation of the CAD/ CAM system, visualized as a central data base. (Courtesy of Gibbs and Hill, Inc.)

■ 18.6 CAD/CAM

The improvement of productivity is a major preoccupation in industry, and CAD/ CAM is seen as vital to this endeavor. Much of the publicity given to CAD/CAM has focused upon the use of the computer graphics terminal as a drafting board. Such use is important, but is only a small part of CAD/CAM.

Stated simply, CAD/CAM is an integrated documents management system. The system may be viewed schematically, as shown in Figure 18.22. All of the information of the types shown in the figure, needed for the engineering and manufacturing phases of the business, is stored in the main frame of the computer. This information is called the *data base*. The arrowheads point in both directions because information of the kinds shown can be both put into and taken out of the data base. In an idealized system, even sketching and preliminary analyses would be done on the interactive computer terminals of the system, as if paper did not exist. However, in a real system, paper is still a vital part of the operation. For example, one of the advantages of having a central data base is that, at the press of a button, a complete set of documentation can be prepared for a user—with all of the latest changes fully incorporated into it. Furthermore, with modern systems it will be printed with crystal sharpness.

It is worth a pause here to comment upon these two matters: up-to-dateness and printing quality. With regard to the first point, one of the banes of manufacturing is the slowness with which changes arrive at the factory floor. Sometimes, in order to speed the process, companies have resorted to so-called change orders, which can be sketched up quickly on standard-size $8\frac{1}{2} \times 11$ paper and sent to the factory. These are fastened to the factory's blueprints* as authority for the changes to be made. After a

---

* The term *blueprint* is something of an anomaly. At one time, prints were indeed blue, with white lines. Today, the prints may utilize blue lines (sometimes called *blueline prints*), or black ones. In either case, the background color is white, not blue. Nevertheless, the term *blueprint* continues to persist in ordinary conversation.

number of these have accumulated and have to be sorted out, they become a prob-lem in themselves. In a CAD/CAM system there might not be a blueprint at all. The "authority" for the shape of the part may be in the form of a set of data points in the computer to be used for the numerical control (abbreviated NC) of the machines in the factory. Whenever any change in the dimensions of a part is entered into the computer data base, the NC instructions are updated automatically.

Printing quality is also not to be taken lightly. In traditional methods of repro-ducing drawings, the lines on the originals often become faint with time, with the result that prints made from those drawings are hard to read. With prints made auto-matically from computer data, every copy is essentially an original. Some of those who have committed their organizations to CAD/CAM cite the high quality of computer-produced "blueprints" as one of the advantages.

As an example of the potential usefulness of an integrated CAD/CAM system, let us assume we are talking about a firm that designs and constructs power plants. Many power plants share certain basic arrangements, permitting one of the major advantages of CAD/CAM to emerge. Instead of starting from scratch with each new plant, the firm's engineers can call up the layouts of an existing plant design on their terminals and make the necessary modifications for the new design. All the unchanged elements remain intact in the data base. Furthermore, as changes are made, the com-puter programming automatically alters related documentation such as specifications, parts lists, and construction schedules. Up-to-date documentation is vital to the pro-cess of designing power plants, in gaining the necessary approvals, and in inspection. This is especially true in the case of nuclear plants, where documentation containing the very latest changes may have to be available instantly in many locations.

CAD/CAM has also entered the aircraft industry in a major way. Lockheed and Boeing are two companies that have made major commitments to CAD/CAM. Boeing points out that its first model of the 757 had parts that had been made in countries throughout the world, and the entire process was controlled through CAD/CAM.

In using CAD/CAM, it is possible for an engineer to sit at a computer terminal and do all of the basic layout and design on the screen. In fact, it is not possible to design such things as very large integrated circuits any other way. However, not all kinds of design adapt themselves to this style. For example, in order for a particular shape to be drawn on a screen, the software designer must have provided for this possibility in the list of commands available to the user. As a result, some kinds of original design continue to be performed "off-line." Even in such a case, as soon as a certain design is settled upon, it is mandatory that it be put into the data base by entering it through a computer terminal.

Computer-Aided Drafting

A widely used method for creating a drawing on a computer screen proceeds as fol-lows. The screen is considered to be composed of an x-y grid marked off in decimal fractions of an inch, with the origin at the lower left corner of the screen. A pair of cross hairs known as a *cursor* can be moved anywhere on the screen by means of a "mouse." The mouse rests on the table next to the computer and can be rolled from one position to another by the operator as if it were in fact resting directly on the

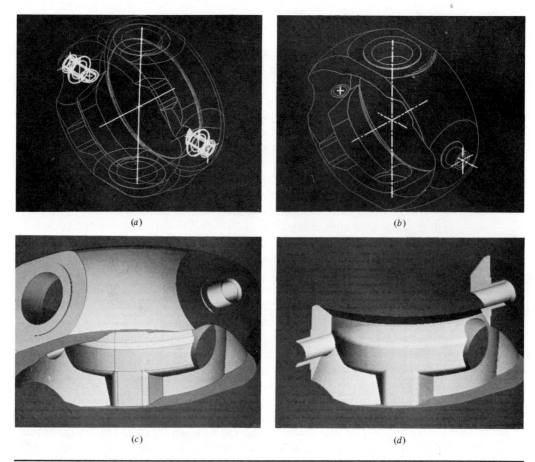

(a)   (b)

(c)   (d)

Figure 18.23

Computer-generated drawings showing successive stages of picturing a gyroscope gimbal. (a) Complete line drawing of the gimbal. (b) All hidden lines not visible to the viewer have been suppressed by the computer, upon command. (c) The drawing has been automatically converted by the computer into a pictorial view, complete with shading. (d) The pictorial view has been cut in half to show a cross section. (Courtesy of Applicon Incorporated)

computer screen itself. As the mouse moves, it transmits its motion electronically to the cursor. Thus, moving the mouse moves the cursor.

As the cursor moves, its location in x-y coordinates (in decimal fractions of an inch) is displayed at the top of the screen. Along one edge of the screen (sometimes also along the top) is a so-called *menu*. The menu is a list of options, such as drawing a line, drawing a circle, dimensioning, drawing centerlines, drawing hidden lines, lettering, and so on. The operator selects an item from the menu by moving the cursor until it is superimposed on the item, and then pressing a button on the mouse.

If, for example, a straight line is to be drawn, a menu item called "LINE" is selected. The cursor is placed at the beginning point of the line, and a button on the mouse is pressed. Subsequently, as the mouse is moved, the cursor moves in unison

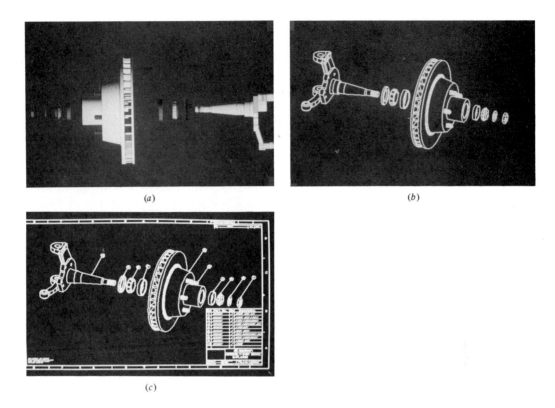

(a)

(b)

(c)

Figure 18.24

Computer-generated pictures showing successive stages of handling diagrams of disk brake assembly drawings. (a) An "exploded" view of the assembly is automatically prepared from the information in the computer data base. (b) A line drawing of the exploded view has been prepared. The computer, upon command, has rotated the assembly so it is viewed from an angle and has also removed all hidden lines. (c) Drawing labels and a bill of materials have been added by the computer, again working from the information in the data base. (Courtesy of Applicon Incorporated)

with the mouse on the screen but appears to be attached to the beginning point of the line by a colored "rubber band," which expands or contracts with the mouse's motion. When the cursor is moved to the desired ending point of the line, the "rubber band" lies exactly in the location of the line that is wanted. At this point, the button on the mouse is pressed again, the colored "rubber band" disappears, and in its place a solid line appears on the screen. In this fashion, line after line is located on the screen until the drawing is finished. (This is a greatly simplified description, of course.)

When it comes time to dimension the drawing, the computer is put into its "dimension" mode by selecting "DIM:" from the menu. Then the cursor is moved to the location where the dimension begins, and the button on the mouse is pressed. Next the cursor is moved to the location where the dimension ends, and the button is pushed again. The computer now knows the magnitude of the dimension because, as the lines were originally placed on the drawing, the computer memorized their coordinates. The only task remaining is to designate where on the drawing the dimension is to be located. The operator does this by placing the cursor on the desired location and pushing the button once more. The computer calculates the dimension and places it in the proper location, complete with leaders and arrowheads.

In many other ways, the computer can save enormous amounts of drafting time. For example, any time repetitive detail is called for, the computer can rapidly repeat the desired arrangement in all the locations specified, rather than the drafter having to laboriously draw each one. This feature is especially useful for drawing standard hardware, such as screws, rivets, nuts, and bolts.

A more elaborate example comes from the airplane business. Each airline that signs a purchase contract for a new aircraft may want to have the interior arrangements customized to its specifications. These may be combinations of standardized modules—seat arrangements, galley design, and the like. The standardized modules are stored in the computer data base and can be called up on the screen instantaneously, saving much laborious drafting work. And, as a remarkable demonstration of the power of the system, if a standard component is changed, it is necessary only to alter the master version of that component, and the changes are then automatically incorporated by the computer into the many places—perhaps thousands—where that component appears in the data base.

One might ask how a drawing of a very large system can be crammed into the tiny confines of a computer terminal screen. The answer is that the complete drawing can be considered a *virtual image* extending in all directions behind the screen. The screen, then, is like a magnifying glass, looking at only the part of the drawing that is of interest. By changing the scale factor, the entire drawing might be condensed onto the screen, but the details would become so fine as to be lost. Conversely, any portion of the drawing can be isolated and enlarged to show any degree of detail. Such a portion is shown in Figure 18.25(a), and then a section of it is enlarged in Figure 18.25(b) to reveal the details of the bolted connections.

The Data Base

A natural outcome of having all the drawings in a data base is that bills of materials can be prepared automatically, with the computer counting up the hundreds or thousands of parts that are needed. Also, the computer can be programmed to examine its internally stored information regarding the shapes and locations of parts to see if any of the pieces have inadvertently been programmed to occupy overlapping sections of space, a condition known as *interference*. In like vein, the computer can be directed to investigate the variations in tolerances of the parts to see if proper fits will occur.

In some kinds of design problems, such as the design of a turbine blade for an aircraft gas turbine, it is possible to have the computer display, for example, the pressure distribution on the blade under planned operating conditions. The engineer, with an interactive terminal, can cause modifications to be made in the shape of the blade to improve the pressure distribution. Then, with a shift in program, the computer can be directed to set up a finite element mesh in the blade shape and analyze it for stress, deflection, and vibration characteristics. These steps might have to be repeated several times until a satisfactory design solution is achieved. Finally, the sets of dimen-

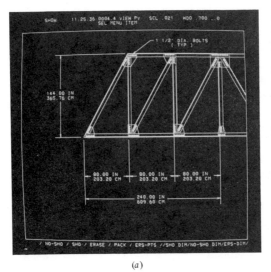

(a)

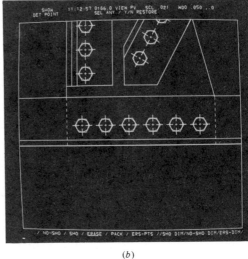

(b)

Figure 18.25

(a) Computer graphics display of a portion of a bridge truss. Upon command from the operator, a small portion of the truss can be shown on the screen in magnified detail, as in (b). (Courtesy of Gibbs and Hill, Inc.)

sional points that are stored in the computer, which describe the shape of the blade in three-dimensional space, can be transferred (perhaps by magnetic tape) to the factory. There, the machine tool that will cut the blade's forging die is controlled directly in accordance with the three-dimensional data points describing the shape. In this case, the "official document" is the data base in the computer, rather than the traditional printed form.

The kind of process just described is also used to delineate complex shapes such as airplane wings. Earlier practice would have called for numerous cross sections of the wing to be drawn with a sharp stylus on metal—full size, and to extremely high

Figure 18.26

Computer results frequently are produced in printed form, rather than graphically. This engineer is carefully checking some intermediate results before proceeding with the next stage of his design analysis.

accuracy—a process known as *lofting*. These full-size shapes would then have been cut out and used to make patterns for the dies, which would stamp out parts with the desired contours. With CAD/CAM, the shapes are computed mathematically and stored as sets of coordinate points in the computer. Then, just as with the turbine blade, a numerically controlled machine cuts the dies, or perhaps machines the part itself, working directly under control of the computer data.

Robotics

A final comment on productivity has to do with *robotics*. Thousands of robots are in use in manufacturing. (A typical unit is shown in Figure 15.9.) Clearly, they do not resemble the popular notion of a robot, i.e., a human-like figure with arms and legs. Frequently they do have arms, because an arm-like arrangement happens to be quite effective for many kinds of tasks. Robots are especially good at boring, repetitive tasks that humans avoid and at hazardous jobs such as welding and heat treatment.

As robots are improved and equipped with the abilities to see, hear, and feel, they will be assigned ever more demanding tasks. They will also have to be made smarter. A robot that is programmed to move its arm from point A to point B and does so without realizing it has smashed a delicate instrument that happened to be in the path of its motion is not the most desirable kind of "employee."

Since robots can work tirelessly around the clock, and sometimes can produce work more accurately than can human beings, they are attractive as a way to increase productivity. As a result, they are likely to be controversial if people come to feel that machines have displaced human employees.

Will Computers Replace Engineers?

To the extent that engineers are used for routine tasks, they will probably be replaced by computers. Jobs requiring a great deal of calculating or detail design tasks that can be systematized are prime candidates for computerization. But no one has yet discovered how to make a computer be creative, although there is plenty of effort in that direction. For example, computers have been used to compose music of a sort and even to write low-grade poems. However, "creativity" of this type depends upon the ability of the computer's designers to specify the creative rules by which the machine operates. Some people believe creativity is the ability to make random connections that turn out to be meaningful. If their view is correct, then computers might eventually be "creative." Nevertheless, the building of a computer with as many random-connection possibilities as the human brain still lies an unforeseeable distance in the future.

In 1957 some experts were predicting that, within ten years, a computer would discover an important new mathematical theorem, write music of value, and reduce theories of psychology to computer routines. Long after the ten-year period was over, these things had not happened, although it might be hasty to declare that they never will.

College engineering students frequently use computers. They typically learn a programming language and are introduced to a variety of prepackaged programs for solving problems. These programs can effectively be treated like "black boxes," meaning that it is necessary to deal only with the inputs and outputs to the black box without concern for the contents.

What this means to the student is the following: It is certainly necessary to learn how to use computers and prepackaged programs, but it is also still necessary to learn the basics of science and mathematics. We should remember that computers are able to do only things for which they have detailed instructions. New situations are continually arising that fall outside the scope of existing computer programs, and there is no reason to suppose that this will change. It is the role of engineers to deal with the new and the unexpected. In doing this, they continually fall back upon their knowledge of fundamentals, and when that knowledge turns out to be insufficient, they even know how to go about gaining new knowledge, because they have learned how to study.

Exercises

18.6.1 Look in your library for books or periodical articles on artificial intelligence. Discuss the pros and cons of providing computers with capabilities approaching human intelligence.

18.6.2 From newspapers, technical magazines, and books, see how many applications of computers you can find that were not mentioned in this chapter.

18.6.3 Collect several examples from newspapers and magazines that describe the use of robotics in industry. Classify these as to type, such as those used for welding or soldering, for transporting parts, for positioning and holding parts, or for handling hazardous substances or working in hazardous environments.

18.6.4 Collect some examples of the use of computers in medicine, beyond those described in this book.

18.6.5 Find some examples, from books or technical magazines, in which computers are used to control industrial processes or transportation networks. Briefly describe the part the computer plays in the process or network. If you can, identify some of the problems that had to be solved. (For a start, review the information on BART in Chapter 4.)

References

J. A. Ball, *Algorithms for RPN Calculators* (New York: John Wiley and Sons, 1978).

G. R. Bertoline, *AutoCAD for Engineering Graphics* (New York: Macmillan, 1990).

H. L. Helms, Jr., *Computer Language Reference Guide* (Indianapolis, Ind.: Howard W. Sams & Co., 1980).

E. B. Koffman and F. L. Friedman, *Problem Solving and Structured Programming in FORTRAN 77* (Reading, Mass.: Addison-Wesley, 1990).

H. Mullish and S. Kochan, *Programmable Pocket Calculators* (Rochelle Park, N.J.: Hayden Book Co., 1980).

Only One Science—Twelfth Annual Report of the National Science Board (Washington, D.C.: National Science Foundation, 1981).

W. J. Tompkins and J. G. Webster (eds.), *Design of Microcomputer-Based Medical Instrumentation* (Englewood Cliffs, N.J.: Prentice-Hall, 1981).

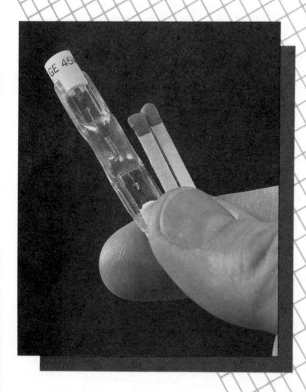

Photographs courtesy of Applicon
Incorporated (top), Deutz-Allis
(center), and General Electric
Company (bottom).

19

Engineering Economics

Most engineers are employed by corporations, and the financial health of their corporate employers is a matter of vital concern to them. Companies are in business to make money, and if they cannot earn a profit, they will soon use up their assets and die. In the case of manufacturing companies, the engineers are counted on to provide a continual flow of new products to keep their companies profitable. If they cannot do this in a financially effective way, their competitors will soon leave them behind. In nonmanufacturing organizations it is no different. An engineering/construction company depends upon its engineers to use its resources efficiently, and so do a utility, a service organization, and a software company.

Engineers are at the focal point of all these business activities. It has been pointed out that engineers often go into management, and the prudent use of money is a central activity of managers. But even if they are not managers, engineers frequently must perform cost analyses. Hence, they need to know the financial meanings of terms such as *assets*, *liabilities*, *net worth*, *cash flow*, *rate of return*, *present worth*, and *depreciation*. This chapter will provide a brief glimpse of these topics. A more extended treatment is provided by courses bearing titles such as *Engineering Economy* or *Engineering Economics*. (The word *economy* means "prudent use of resources," whereas *economics* means "financial considerations.")

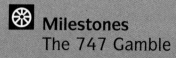

Milestones
The 747 Gamble

In 1965, the Boeing Company and Pan American World Airways joined in one of the greatest engineering development gambles of all time—the 747. Boeing committed $2 billion to the project, in advance, to develop from scratch a radical new airplane that would fly higher, fly faster, and carry more people than any plane ever had. The design depended upon powerful new engines that didn't exist yet. It would have no navigator, but would be guided by an inertial navigation system which was still to be designed. To build the huge new airplane, Boeing would have to construct one of the world's largest buildings. There was little doubt that if not enough planes were sold, Boeing would be bankrupted.

Pan Am, for its part, committed $500 million to the project, half of which would be paid before the delivery of the first plane. For this sharing of the risk, Pan Am would have the privilege of being the first airline to fly the radical new craft. But it was clear also that if the 747 did not succeed, Pan Am, too, would be bankrupted.

According to the initial agreement, the 747 would have a range of 5100 miles, fly at mach 0.9, carry up to 400 passengers, and have a cruising altitude of 35,000 feet and a maximum gross weight of 550,000 pounds. But as soon as the engineers actually went to work on the design, the structural requirements of this huge aircraft began to exert a relentless pressure, and the projected weight began to creep upward. As the weight increased, the range decreased, and so did the projected payload. Thus, by the time a firm contract was signed, the range had dropped to 4460 miles, the speed to mach 0.877, the payload to 370 passengers, and the cruising altitude to 33,000 feet; the maximum gross weight had gone up to 655,000 pounds.

Subsequently, in a process familiar to every aeronautical engineer, the weight continued to creep upward relentlessly, and, as a consequence, the range and payload continued to decline. Pan Am began to put on the heat to get Boeing to adhere to the specifications, and Boeing's engineers found themselves working in a pressure cooker. Pan Am's concerns can be understood by considering a typical disagreement involving an unexpected 13,000-pound increase in the weight of the aircraft. Because of such a weight increase, the plane would be able to carry 58 fewer passengers, with a loss in revenue to Pan Am of $20,000 *on every flight.* Thus, it is understandable that Pan Am applied pressure,

■ 19.1 ASSETS, LIABILITIES, AND NET WORTH

Let us suppose you and four associates are going to start a small manufacturing company, called the ABC Corporation. Each of you puts in $10,000, so you have $50,000 to start with.[1] But you do not think this is enough, so you borrow $50,000

---

[1] Note that commas have been used in this chapter for numbers such as $50,000, contrary to the practice recommended in the chapter on SI metrics. This follows accepted accounting practices, which are not influenced by SI.

and Boeing in turn applied pressure to Pratt and Whitney, which was to come up with the superpowerful new engines that would lift this monster into the air. Since everything depended upon the engineers at both companies to do what had never been done before, it was of course the engineers who felt the heat the most. In the end, the 747 had a maximum gross weight of over 700,000 pounds, and the principal factor that made it possible for this huge machine to fly was the increase in take-off thrust of the Pratt and Whitney engines, from 42,000 pounds to 45,000 pounds, achieved by water injection.

By the time the airplane flew, so many rumors of problems and imminent disasters had circulated that the press had adopted what was referred to as a "death-watch" attitude toward the 747. Each negative development was magnified into a catastrophe, while positive accomplishments were ignored. If disaster was indeed the expectation of reporters, they must have been disappointed, because the 747 slid smoothly into service with as little trouble as had ever been experienced by a new airplane. Furthermore, it began to become apparent that the great Boeing/Pan Am gamble had paid off, as airline after airline fell into line to purchase 747's.

What had the engineers wrought during this prolonged period of tension? Had they, under the enormous pressure of trying to maintain schedules and improve pay-off, secretly cut corners and thus jeopardized safety? Apparently not, judging from the record. At the end of ten years' use, the 747 had compiled one of the most enviable safety records in aviation history. No doubt a careful and protracted period of testing had had its part to play in producing this record, as well as the experienced judgment of those engineers who had controlled the plane's design. For example, some of the test maneuvers imposed on the 747 were to put the airplane repeatedly into steep climbs, deliberately producing full stalls, after which it would fall forward into a dive until coming back into a normal flying position. Most jet aircraft, following a stall, tend to fall off to one side and into a spin, but not the 747. Also, when other models go into a dive, the dive can quickly get out of hand, but not the 747, which tends to be self-correcting. Other test maneuvers called for flying with two engines dead, and take-offs and landings with only three engines. Whatever the pressures, and regardless of the fact that the 747 has been called "The Great Gamble," it was clear that the engineers had not gambled with safety. Instead, they had produced an aviation marvel.

Reference
L. S. Kuter, *The Great Gamble: The Boeing 747* (University, Alabama: University of Alabama Press, 1973).

from a friendly bank, in the form of a long-term loan. You need some furniture, some tools and equipment, and some supplies. Let us say you are ready to start; you have spent $80,000 on your furniture and so on, and still have $20,000 cash in the bank. Your *assets*, then, add up to $100,000, but you owe $50,000 to the bank, and you also "owe" $50,000 to yourselves, the "stockholders." The evidence of your ownership is in the form of *common stock*.[2] (Common stock is the most widely used kind of

[2] The company in this example is very small and probably would not actually issue common stock. The example has been purposely simplified to illustrate the principles involved.

Figure 19.1
The Boeing 747, one of the safest air-craft ever built, was a development project on which Boeing and Pan American Airways together literally "bet their companies." (Courtesy of Boeing Commercial Airplane Company)

stock, although there are other kinds, such as preferred stock.) It is customary to display all of this information on what is called a *balance sheet*:

ABC CORPORATION
Balance Sheet

Assets:		*Liabilities:*	
Cash	$ 20,000	Bank loan	$ 50,000
Furniture, tools,		*Net Worth:*	
equipment, etc.	80,000	Common stock	50,000
	$100,000		$100,000

The two sides of the balance sheet must, by definition, balance. If the business prospers, it will do one of two things, or perhaps both, with its earnings: acquire more assets (i.e., grow) or pay dividends to the stockholders. Let us say that after one year the earnings are as shown in the following *operating statement*:

ABC CORPORATION
Operating Statement

Sales	$200,000
Expenses	175,000
Profit before income taxes	$ 25,000
Estimated income taxes	12,500
Net profit after income taxes	$ 12,500

For simplicity, let us assume that the $50,000 loan has not changed during the year, that no dividends are to be paid to the stockholders at this time, and that all of the $12,500 profit is to be used to expand the business—to purchase more tools, equipment, and supplies. The $12,500 will be shown on the balance sheet as "retained earnings," although the term *earned surplus* is sometimes used. Let us also assume you have maintained the cash in the bank at the $20,000 level so that you can pay your bills promptly. Then, after the one year, your balance sheet will appear as follows:

ABC CORPORATION
Balance Sheet

Assets:		*Liabilities:*	
Cash	$ 20,000	Bank loan	$ 50,000
Furniture, tools,		*Net Worth:*	
equipment, etc.	92,500	Common stock	50,000
		Retained earnings	12,500
	$112,500		$112,500

The owners could have chosen to pay all of the net profit to themselves in the form of dividends, if they had wished. (Presumably, any full-time employees of their company would have paid themselves salaries, as well. The salaries would be included in the operating statement as a part of "expenses.") If they had paid all of the $12,500 profit to themselves, they would have nothing left to expand the business, so the balance sheet would look the same at the end of the year as it did in the beginning. But that would mean the company would not be growing, and most owners choose to strike a balance between the payment of dividends and growth.

The example given for the ABC Corporation is an unrealistically simple one, of course, but it demonstrates the fundamental purposes of balance sheets and operating statements. The left side of the balance sheet shows everything the company owns—the assets—and the right side shows everything it owes—the liabilities. Even the "net worth" is a kind of liability, because this is the part that would be "owed" to the owners if the company should be liquidated, i.e., have all its assets turned into cash.

A Realistic Case

Real operating statements and balance sheets are more complicated than the foregoing, of course, but they can generally be reduced to the simple categories just described. Let us look for a moment at an operating statement (Figure 19.2) and a balance sheet (Figure 19.3) patterned after those of a real company. (Here the company has been named X Corporation.)

In Figure 19.2, the designation "Net sales" should be self-explanatory. The next entry, "Cost of sales," refers to all the expenses that went into the products themselves, such as cost of materials, labor, and the like. The section headed "Operating expenses" contains other costs of doing business, frequently referred to as *overhead*. Costs such as supervision, rent, taxes, insurance, utilities, and fringe benefits (health insurance, vacation, and sick leave) generally are included within the heading "General and administrative."

Some of the entries in the balance sheet of Figure 19.3 may require explanation. "Cash" does not; that is money in the bank. "Accounts receivable" represents the value of products that have been sold to customers but have not yet been paid for. The section headed "Inventories" reflects the fact that many of the company's products have either been finished and are awaiting sale or are in process. A lot of money

"X" CORPORATION
Operating Statement

Net sales	$97,988,000
Cost of sales	51,805,000
Gross margin	$46,183,000
Operating expenses:	
Marketing and selling	15,115,000
Research and development	8,135,000
General and administrative	7,381,000
Interest on long-term debt	986,000
Earnings before income taxes	$14,566,000
Provision for income taxes	6,904,000
Net profit after income taxes	$ 7,662,000

Figure 19.2
The X Corporation operating statement.

has gone into them, and they are valuable because they will soon be turned into cash income. The company lists these as assets. "Land," "Buildings," and "Equipment" should require no explanation.

On the right side of the balance sheet are first the "current liabilities," meaning in effect all the company's liabilities that are not long term. "Accounts payable" is the reverse of accounts receivable. This is the amount owed to the company's suppliers. "Accrued wages" represents the amount owed, but not yet paid, to the company's employees, and "Accrued taxes" represents the amount of taxes owed, but not yet paid, on the date the balance sheet was prepared. The long-term debt is the amount of working capital that has been acquired by the company other than that originally put in by the stockholders. The remaining items are familiar from our earlier example, except this company prefers to use the term *stockholders' equity* instead of *net worth*.

"X" CORPORATION
Balance Sheet

Current Assets:		*Current Liabilities:*	
Cash	$ 1,109,000	Accounts payable	$ 5,262,000
Accounts receivable	21,744,000	Accrued wages	5,029,000
Inventories:		Accrued taxes	1,217,000
Finished goods	3,279,000	Other expenses	2,298,000
Work in process	15,284,000	*Long-term Debt*	12,277,000
Raw materials	10,066,000	*Stockholders' Equity:*	
Fixed Assets:		Common stock	8,234,000
Land	4,020,000	Retained earnings	46,571,000
Buildings	10,406,000		
Equipment	14,980,000		
	$80,888,000		$80,888,000

Figure 19.3
The X Corporation balance sheet.

■ 19.2 PROFIT AND RATE OF RETURN

Using the simple example of the ABC Corporation from the preceding section, we can make a few calculations to see whether the investors handled their money prudently when they put it into the corporation. The bank that made the $50,000 loan earned its standard rate of interest, so presumably it is satisfied with its return. The ABC Corporation charged this interest off as a cost of doing business, and it is included on the operating statement as a part of the $175,000 "expenses" item.

There are several measures of profitability that the proprietors might use to evaluate their success. One of the most popular is to calculate the profit as a percentage of sales, called the *net profit margin*:

$$\frac{\text{net profit after income taxes}}{\text{net sales}} = \frac{\$\ 12,500}{\$200,000} = 6.25\%$$

This is not too bad. Many businesses do worse than this, although such results are not likely to drive investors mad with joy. However, many people believe that the profit as a percentage of sales is not the best guide to profitability but prefer to calculate what is called *return on stockholders' equity*. The stockholders have gotten back $12,500 on their $50,000 investment and could have paid this $12,500 directly to themselves in the form of dividends, if they had wanted to. The fact that they chose to put the money back into the business makes no difference. The effect upon their personal wealth is the same as if they had paid the $12,500 to themselves, because their *equity* ownership in the business increased from $50,000 to $62,500. (We are assuming the ABC Corporation will continue to be successful and is not on the brink of failure.)

The owners, then, might calculate the return on their investment in the following way, called *return on stockholders' equity*:

$$\frac{\text{net profit after income taxes}}{\text{stockholders' equity}} = \frac{\$12,500}{\$50,000} = 25\%$$

This is not a bad rate of return, considering the riskiness of starting new businesses. It is certainly more than can be earned on treasury bonds in normal times, although people who make a business of providing venture capital to new companies generally look for higher rates of return than this, to compensate for the losers. They know that a distressing number of new companies not only fail to make a profit, but also spend all their invested capital with no possibility of recovery.

Even the "return on stockholders' equity" is not the best measure of profitability, according to some. They say, with justification, that the business would not have been able to function at the level it did if not for the bank loan. The loan was used to buy twice as much equipment and hire twice as many people as would have otherwise been possible. They say that without the loan, the magnitude of the company's operation would have been only half as large as it was, and the profit probably would also have been only half as large. Therefore, a rate of return should be calculated that takes *both* the stockholders' equity and the value of long-term loans into account. This calculation, called *return on total capital*, shows the true rate of profitability, they say. In computing this, the net profit after income taxes is combined with the interest paid on the bank loan, and the sum is considered to be the total return. In our example, we will assume an interest rate of 12 percent on the long-term debt, so that the interest paid for one year is $6000. The computation of the *return on total capital* is

$$\frac{\text{net profit after taxes} + \text{interest paid on long-term debt}}{\text{stockholders' equity} + \text{long-term debt}}$$

$$= \frac{\$12,500 + \$6000}{\$50,000 + \$50,000} = 18.5\%$$

This is not bad compared to industry medians, although a new company seeking to prove itself might hope for a higher return. Table 19.1 gives some industry medians for several industries, where average returns on total capital over a five-year period are shown. Returns on total capital for some specific well-known companies are shown in Table 19.2. A number of well-known international oil companies are included in Table 19.2, with returns ranging from 13.2 to 20.0 percent. A lot of public ire has been directed at oil companies for their presumed excess profits, but Boeing and IBM earned more, and so did Coca-Cola. Banks are even lower on the return scale, where two very large banks (Bank America, 12.6 percent, and Citicorp, 10.3 percent) have been included for comparison purposes. Electric utilities, another target for public anger, come even lower on the scale for profitability. Airlines and auto companies are lower yet, raising serious questions regarding their survival. The five-year period in question included both some good years and some recession years,

Table 19.1   Industry Medians, Return on Total Capital

Industry	Return on Total Capital (5-year average)
Electrical equipment	16.6%
Office equipment and services	15.8
Computers	15.6
Oil-field drillers and services	15.4
Soft drinks	15.2
Electronics	14.9
International oil companies	13.9
Industrial equipment and services	13.4
Aerospace and defense	13.3
Heavy construction	13.2
Banks	11.6
Chemicals	10.5
Nonferrous metals	9.4
Forest products	9.0
Steel	7.6
Electric utilities	6.9
Autos and trucks	7.3
Airlines	4.5

Source: *Forbes* Magazine, January 3, 1983.

Table 19.2   Return on Total Capital (5-year average)

Schlumberger	33.4%	Martin Marietta	12.9
Coca-Cola	21.3	E. I. du Pont	12.7
Boeing	21.2	Bank America	12.6
IBM	20.9	Honeywell	12.0
Standard Oil Ohio	20.0	Citicorp	10.3
Hewlett-Packard	19.1	Dow Chemical	10.3
General Electric	18.2	Alcoa	10.2
Fluor	17.4	General Dynamics	9.7
Texas Instruments	16.3	Tenneco	9.6
Northrop	15.5	Lockheed	8.7
Procter & Gamble	15.3	Pacific Gas & Electric	8.1
Standard Oil California	15.2	General Motors	7.9
CBS	15.1	Delta Air Lines	7.8
Xerox	15.0	AT&T	7.6
Exxon	14.4	Consolidated Edison	7.0
Rockwell International	13.6	Trans World	4.0
Westinghouse Electric	13.5	UAL	2.3
Mobil	13.4	Ford Motor	0.4
Shell Oil	13.2		

Source: *Forbes* Magazine, January, 3, 1983.

so it would appear to be reasonably representative. However, the period also was one of heavy inflation, averaging 7 to 8 percent over the five years. As can be seen from Table 19.2, many companies did not manage to keep up with inflation in their returns on invested capital, and so were losing ground.

Since ABC Corporation is a new business that appears to have been successful in its first year, we should mention one more term: *cash flow.* Many businesses get into difficulty because of inadequate cash flow. For example, if our corporation invested all of its cash in equipment and supplies in an attempt to produce more goods, it probably would not have enough left to meet its payroll and pay its suppliers. Thus, even though the operating statement might appear to show a nice profit and our calculations of return on capital investment might look great, there could still not be enough cash on hand to pay the bills. Part of the problem might be that the company expected its customers to pay for goods delivered more expeditiously than turned out to be the case. The company might then be forced to take out a short-term loan—at high interest rates—until it could get its cash flow under control. Cash flow problems sometimes become intractable enough to ruin businesses.

Exercises

19.2.1 Calculate the *net profit margin, return on stockholders' equity,* and *return on total capital* for X Corporation, from the operating statement and balance sheet given in Figures 19.2 and 19.3.

19.2.2 Suppose ABC Corporation had made a $20,000 loss instead of a profit of $12,500. What would its balance sheet look like? (Assume it still has the $80,000 worth of furniture, tools, and equipment it started with; its bank loan is still $50,000; and its common stock is still $50,000.)

Figure 19.4
This movable, watertight floodwall at Washington Harbor, Washington, D.C., includes some of the largest swing gates ever employed. (Courtesy of American Consulting Engineers Council)

■ 19.3 INTEREST AND ANNUITIES

When money is borrowed, there is almost always an agreement to pay an amount of *interest* as well as to repay the original loan. If the interest is i percent per year and the borrowed amount P, called the *principal*, is retained for n years, then the total interest paid (called *simple interest*) is

$$\text{interest} = Pin \tag{19.1}$$

If I lend you $1000 at 8 percent-per-year simple interest for three years, the interest to be paid at the end of three years is

$$\text{interest} = \$1000 \cdot (0.08) \cdot (3) = \$240$$

On the other hand, I might reason as follows. At the end of the first year, you owe me $80 interest. You really should pay me the interest now, rather than waiting until the end of the three-year period. If I let you keep the interest payment instead of giving it to me, then it is the same as if I had made an additional loan to you of $80 (or so I reason), and you should pay me interest on that as well as on the original $1000. This leads to the concept of *compound interest*:

Year	Amount of Principal at Beginning of Year	Interest (at 8%) Earned During Year	Amount of Principal at End of Year
1	$1000	$80	$1080
2	1080	86.40	1166.40
3	1166.40	93.31	1259.71

Compounding, of course, is exactly the method used to calculate our interest when we deposit money in a savings account in a bank. However, banks usually compound the interest daily rather than yearly. This means that the interest rate per day is 1/365 of the annual rate, and for a three-year loan, $n = 3(365) = 1095$. The computations for such a case are too complicated to be done by the step-by-step method of the foregoing example. We need to develop a formula, as follows.

At the end of one period (which could be a year or a day), the interest on the original principal P is Pi, so the amount accumulated at the end of the period is

$$P + Pi = P(1 + i)$$

This is the principal at the beginning of period 2. During period 2, the interest earned is $i \cdot P(1 + i)$. Therefore, at the end of period 2, the principal is

$$P(1 + i) + i \cdot P(1 + i) = P(1 + i + i + i^2) = P(1 + i)^2$$

During period 3, the interest earned is $i \cdot P(1 + i)^2$. Therefore, at the end of period 3, the principal is

$$P(1 + i)^2 + i \cdot P(1 + i)^2 = P[1 + 2i + i^2 + i(1 + 2i + i^2)]$$
$$= P(1 + 3i + 3i^2 + i^3) = P(1 + i)^3$$

Continuation of this process leads us to the general formula for *compound interest*:

$$F = P(1 + i)^n \tag{19.2}$$

where F is the total *future sum* of principal and interest after n periods.

Let us go back to the example where we compound the interest daily on a $1000, three-year, 8 percent loan. As we said before, $n = 3(365) = 1095$ and $i = 0.08/365 = 0.0002191$. Therefore, under these circumstances, at the end of three years F will be

$$F = \$1000(1 + 0.0002191)^{1095} = \$1271.11$$

A comparison of the three methods discussed, for a three-year loan of $1000 at 8 percent, is

Method	Repayment at End of Three Years
Simple interest	$1240.00
Compounded annually	1259.71
Compounded daily	1271.11

Figure 19.5
Fiber-reinforced composites are widely used to produce high-strength, low-cost components. Here a computer-controlled filament winding machine is constructing a rocket motor casing 37 inches in diameter for use on the Peacekeeper missile. (Courtesy of Morton-Thiokol)

Present Worth

In the compound interest formula, $F = P(1 + i)^n$, we invest an amount P at the present time and get back F in the future after n interest periods. Thus, we can think of F as the *future value* of a sum P in the present. Conversely, P can be thought of as the *present worth* of a sum F that we expect to receive in the future. We can solve Equation (19.2) for P and get an equation for *present worth*:

$$P = F\left(\frac{1}{1 + i}\right)^n \tag{19.3}$$

where

P is the present worth
F is the future value of P at interest i
i is the interest rate per interest period
n is the number of interest periods

As an example, suppose we wish to deposit an unknown sum P now, which, after ten years at 8 percent interest compounded annually, will be worth $10,000.

Using Equation (19.3), we substitute the values and get the present worth:

$$P = \$10,000\left(\frac{1}{1 + 0.08}\right)^{10} = \$4632$$

Annuities

An *annuity* is a series of equal payments either from or to a fund. If you put $1000 into a savings account each year to provide for your child's college education, that is an annuity. If you take a loan to purchase a home, agreeing to repay x dollars per month in principal and interest for the next 20 years, that, too, is an annuity.

We can develop a formula to show the relationship between A, the amount placed in a fund at the end of each period, and F, the future amount accumulated at the end of n periods, as follows. The interest rate is i, compounded each period.

The foregoing procedure has been described in such a way that the money A is placed in the fund at the *end* of each of n periods. We can draw a *time line* to portray this process, as follows:

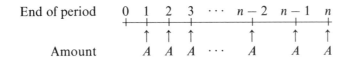

There are indeed n periods. However, there is no money in the fund during period 1, because we are putting our money into the fund at the *end* of each period. (We could just as well have defined our procedure so that A is placed into the account at the *beginning* of each period, but the formula would be slightly different.) Since the first amount is not put in the fund until one period has passed, it earns interest for only $n - 1$ periods, rather than n periods. At the end of the entire time, then, the amount in the fund resulting from the first amount A, using our compound interest formula, Equation (19.2), is

$$A(1 + i)^{n-1}$$

The next amount A is in the fund for $n - 2$ periods, so at the end it contributes the following to the total future sum F:

$$A(1 + i)^{n-2}$$

The progression of succeeding amounts is obvious. The next one is $A(1 + i)^{n-3}$, and so on. But now let us go all the way to the end and work backward. The very last payment A is just the amount needed to bring the total fund up to the desired amount F, but it earns no interest, because the instant it is deposited, we then total up everything to see how much we have aggregated from all the payments. However, the payment we made just before the last one was in the fund for one period, so it contributed the following to the sum:

$$A(1 + i)^{1}$$

The one before that contributed $A(1 + i)^2$, and so on. Adding them all up gives the total future sum F:

$$F = A(1 + i)^{n-1} + A(1 + i)^{n-2} + \cdots + A(1 + i)^2 + A(1 + i) + A$$
$$= A[1 + (1 + i) + (1 + i)^2 + \cdots + (1 + i)^{n-2} + (1 + i)^{n-1}] \qquad (19.4)$$

If we multiply both sides of Equation (19.4) by $(1 + i)$, term by term, we get

$$(1 + i)F = A[(1 + i) + (1 + i)^2 + (1 + i)^3 + \cdots + (1 + i)^{n-1} + (1 + i)^n] \qquad (19.5)$$

If we now subtract Equation (19.4) from Equation (19.5), we get

$$F(1 + i) - F = A[(1 + i) + (1 + i)^2 + (1 + i)^3 + \cdots$$
$$+ (1 + i)^{n-1} + (1 + i)^n - 1 - (1 + i)$$
$$- (1 + i)^2 - \cdots - (1 + i)^{n-2} - (1 + i)^{n-1}]$$

Most of the terms within the brackets can be matched in pairs and eliminate each other, leaving us with only

$$F[1 + i - 1] = A[-1 + (1 + i)^n]$$
$$iF = A[(1 + i)^n - 1]$$
$$F = A\left[\frac{(1 + i)^n - 1}{i}\right] \qquad (19.6)$$

The kind of procedure we have just described leads to what is often called a *sinking fund*. Usually, in a sinking fund, we desire to create a fund of F dollars n periods hence and want to know how large each payment A must be. As an example, suppose we want to accumulate $10,000 at the end of five years, with interest compounded annually at 8 percent, and we want to know how much money we must deposit at the end of each year for the desired result to occur. We use Equation (19.6), solving for A, and get

$$A = F\left[\frac{i}{(1 + i)^n - 1}\right]$$
$$= \$10,000\left[\frac{0.08}{(1 + 0.08)^5 - 1}\right]$$
$$= \$1705 \qquad (19.7)$$

Repayment of a Loan

By using Equation (19.7), we can find out what the amount A of a series of payments would be to repay a loan of P at interest i. This is the familiar case of a loan that is to be repaid with regular installment payments. However, Equation (19.7) is given in terms of a future sum F, whereas we have to put it in terms of the present value P because this is the amount we plan to borrow now, in the present. To do this, we use Equation (19.2), which gives us the relation between F and P, at compound interest:

$$F = P(1 + i)^n$$

If we substitute Equation (19.2) into Equation (19.7), we get

$$A = P(1 + i)^n \left[\frac{i}{(1 + i)^n - 1} \right]$$

$$A = P \left[\frac{i(1 + i)^n}{(1 + i)^n - 1} \right] \tag{19.8}$$

As an example, let $P = \$10,000$, to be repaid monthly in 12 equal payments at 12 percent annual interest. The interest per month is thus $i = 1$ percent, compounded monthly, and $n = 12$. Substituting in Equation (19.8), we get

$$A = \$10,000 \left[\frac{0.01(1 + 0.01)^{12}}{(1 + 0.01)^{12} - 1} \right]$$

$$= \$888.49$$

Exercises

19.3.1 If you lend me $1000 for four years at 9 percent annual interest, compounded annually, how much must I repay you at the end of the four-year period?

19.3.2 If you lend me $1000 for four years at 9 percent annual interest, compounded *monthly*, how much must I repay you at the end of the four-year period?

19.3.3 If you lend me $1000 for four years at 9 percent annual interest, compounded *daily*, how much must I repay you at the end of the four-year period?

19.3.4 What is the future value of $10,000 after ten years, if we assume an annual interest rate of 8 percent compounded daily?

19.3.5 What is the present worth of a sum that will be worth $10,000 in ten years, if we assume an annual interest rate of 8 percent compounded daily?

19.3.6 You have just had a child, and you wish to set up a savings plan that will have accumulated $10,000 18 years hence, to help pay your child's college expenses. You want to pay in a fixed sum each month, which will be compounded monthly at an annual interest rate of 5.75%. How much should you pay in per month?

19.3.7 You start a personal savings plan at age 22, planning to deposit $100 per month in a savings account until age 65, at which time you intend to retire. How much money will you have in your account when you are 65, assuming the interest rate was 6 percent, compounded monthly?

19.3.8 You borrow $5000 in order to buy a car and agree to repay the loan, plus interest, in a series of equal monthly installments over a three-year period. The interest rate is 10 percent annually, compounded monthly. How much is your monthly payment? What will your payments for three years total?

19.3.9 Assume the same conditions as in the preceding problem, except that now you are going to pay the loan back in a series of equal monthly installments over a *one*-year period. How much is your monthly payment? What will your payments total?

19.3.10 Prepare a spreadsheet like the one in Figure 18.21 for the conditions given in Exercise 19.3.9.

■ 19.4 DEPRECIATION

When an asset, such as a truck or a piece of production machinery, is purchased by a company, it is placed on the books at its purchase cost in the "fixed assets" category. However, we know that its value to the company will decline over the years because of wear and tear, or because of obsolescence. Eventually it will have no value to the company, or perhaps only a small amount, which we refer to as the *salvage value*. At such time it must be replaced by new equipment.

Obviously, we must do something about this, because the decline in value is a cost of doing business. Some portion of the original value of the machinery and other equipment owned by a company must be charged off each year as a part of the expense of producing the company's product. Such a charge-off is called *depreciation*, which is the amount we feel the equipment has declined in value each year.

In doing this, we first calculate the *depreciation cost*, which is

depreciation cost = original cost − salvage value

Next we must assume a useful lifetime. Estimation of such a figure is the product of experience. A typical lifetime for machinery and other equipment is often chosen to be ten years. If it turns out that the equipment lasts longer than this without costing too much in repairs, then we will realize—after ten years—that we have been charging off more than we needed to during the past ten-year period. This means our expenses were somewhat overstated during the past, and our profit was actually larger than we thought. But it also means that we cut down on our income tax payments during that period, because the tax we paid was based upon our stated profits. Now, however, if the equipment is completely depreciated, but we continue to use it, we have to stop charging off depreciation expense for that item. Our computed profits will rise as a result, and so will our taxes.

It should be apparent from the foregoing that the method of computing and charging depreciation will be subjected to close scrutiny by tax examiners. Of the several recognized methods for computing depreciation, we will describe only one, called the *straight-line method*. In this method, the depreciation cost is simply divided

Figure 19.6
A three-axis, numerically controlled profiler simultaneously machines three F-14 nacelle bulkheads on spindles 40 inches apart. Such methods greatly increase productivity, provided the production runs are large enough to justify the costly equipment that is needed. (Courtesy of Grumman Aerospace Corporation)

by the number of years of estimated life:

$$\text{annual depreciation} = \frac{P - S}{n} \tag{19.9}$$

where

> P is the original cost
> S is the salvage value
> n is the estimated life in years

Each year, the value of that particular asset, as shown on the balance sheet, is reduced by the annual depreciation charge. At the end of n years, the value of the asset is S dollars, the salvage value. Thus, if a machine cost \$10,000 new and is assumed to have a life of ten years with a salvage value of \$600, then the annual depreciation is

$$\frac{\$10,000 - \$600}{10 \text{ years}} = \$940 \text{ per year}$$

At the end of the first year of operation, this particular asset will be listed at a value of \$10,000 − \$940 = \$9060. At the end of the second year it will be listed at \$8120, and so on. The value at which an asset is listed on the books is called its *book value*. This value may bear only an approximate relation to the amount the equipment would actually bring if sold on the open market.

◼ 19.5 ECONOMIC DECISIONS

Decisions are usually based upon an analysis of anticipated costs, as between two or more possible courses of action. One typical method is to compare the estimated *annual costs* of the alternatives. The method is simple and straightforward and is popular for that reason.

   For example, let us suppose we want to make a decision between two different kinds of production machinery that will be used to produce a new product. The production rate is 100,000 units per day. In alternative A, each production machine costs \$10,000 and can produce 20,000 units daily. Therefore, five such machines will be needed, at a "first cost" of \$50,000, and five operators will be necessary, each with an annual salary of \$20,000. The life of each machine is assumed to be ten years, and the annual maintenance on each unit is assumed to be \$750 per year.

   In alternative B, each production machine costs \$150,000, but each one is capable of a higher production rate than in A—namely, 50,000 units daily. Thus, even though each machine costs a great deal more than in A, only two machines will be required to achieve the desired production rate, at a "first cost" of \$300,000. This also means we will only need two operators, again at \$20,000 each per year. We again assume a machine life of ten years, but, because the machines are more complex than in A, we estimate annual maintenance costs per machine of \$5000.

   We assume an interest rate for capital of 9 percent and assume that taxes and insurance costs will be 5 percent of the purchase cost of the equipment. Depreciation

Table 19.3 Cost Comparison for New Production Equipment
(Daily Production Rate, 100,000 Units)

	Alternative A	Alternative B
First cost	$50,000	$300,000
Production rate per machine, per day	20,000 units	50,000 units
Machine life, years	10	10
Number of machines needed	5	2
Number of operators needed @ $20,000 each	5	2
Interest rate	9 percent	9 percent
Taxes and insurance	5 percent	5 percent
Depreciation	Straight-line	Straight-line
Annual maintenance, per machine	$750	$5000
Annual costs:		
Depreciation	$ 5000	$ 30,000
Interest on first cost	4500	27,000
Taxes and insurance	2500	15,000
Maintenance	3750	10,000
Labor	100,000	40,000
	$115,750	$122,000

will be figured on a straight-line basis. We can tabulate the cost comparison as shown in Table 19.3. The annual estimated cost for alternative *A* is $6250 less than that for alternative *B*. Unless we have overlooked something, alternative *A* appears more attractive, even though we must hire five people instead of two. Obviously, the machines of alternative *B* cost too much to justify buying them, in spite of their improvement in production rate.

However, if the manufacturer of the machine in alternative *B* should come up with a new model costing, say, $200,000, but with a production rate of 100,000 units per day, matters would be different, because only one machine and one operator would be needed. Let us call this alternative *C* and estimate the annual costs for just this one machine, assuming an annual maintenance cost of $10,000.

Depreciation	$20,000
Interest on first cost	18,000
Taxes and insurance	10,000
Maintenance	10,000
Labor	20,000
Total	$78,000

Under these circumstances the new model machine would be very attractive.

We will give one more example of cost analysis, of a type that often arises in major public works projects. This is based upon calculation of a *benefit/cost ratio*, wherein both the future benefits and future costs are estimated and compared.

A hypothetical water resources project involves a large dam and a system of canals to deliver irrigation water. We estimate that the project will cost $100 million and have a life of 100 years. The irrigation benefits, in terms of the value of the crops to be produced from the new land that is put into production, is estimated to be $2.5 million annually. A portion of the acreage to which water from the dam is to be delivered is already in production, but is irrigated with water pumped from deep wells. The pumped water will be replaced with water that flows by gravity from the dam; no new agricultural production will occur under these conditions on this portion of the land, but the annual pumping costs, which come to $1.5 million, will no longer be incurred. This is obviously a $1.5 million benefit. In our example, we will assume the impounded waters will be used for recreational purposes, so the recreation and "wildlife" benefits are estimated at $1 million per year. (It should be realized that all of the foregoing estimated benefits may be hotly disputed by opponents of the dam, but we will use the numbers given for our analysis.)

Now we must consider the costs. First, we estimate the annual operation and maintenance costs at $800,000. Next, we must do something about the $100 million construction cost. To simply divide the cost by the life of 100 years, which gives $1 million per year, ignores the fact that the $100 million could be placed somewhere else and earn interest. Since the $100 million presumably will have come from the taxpayers, the taxpayers may think they could put the money into bonds, for example, with greater benefit than they will get from the dam. Therefore, we need to treat the cost of the dam as though it were borrowed from the bank, so to speak, with annual repayments of principal and interest for 100 years. This is simply the installment payment process we discussed earlier, which produced Equation (19.8):

$$A = P\left[\frac{i(1 + i)^n}{(1 + i)^n - 1}\right]$$

Let us assume the dam is to be built at a time when interest rates are low and use $i = 3.5$ percent, compounded annually. Then, with $n = 100$, Equation (19.8) gives us

$$A = \$100,000,000\left[\frac{0.035(1 + 0.035)^{100}}{(1 + 0.035)^{100} - 1}\right]$$

$$= \$3,616,000$$

We can now make our benefit/cost analysis, as has been done in Table 19.4. The benefit/cost ratio is computed at 1.13, so the dam is assumed to be a good investment because the benefits are greater than the costs. On the other hand, if the dam's opponents successfully argue that the benefits have been overestimated, the benefit/cost ratio could quickly become less than unity, making it appear unattractive.

The benefit/cost ratio is also very sensitive to interest rates. If interest rates go up, say, to 7 percent, then Equation (19.8) gives us

$$A = \$100,000,000\left[\frac{0.07(1 + 0.07)^{100}}{(1 + 0.07)^{100} - 1}\right]$$

$$= \$7,008,000$$

Table 19.4   Benefit/Cost Analysis for New Water
Resources Project

Cost of project	$100,000,000
Annual estimated benefits:	
Irrigation benefits	$2,500,000
Pumping savings	1,500,000
Recreation and wildlife benefits	1,000,000
	$5,000,000
Annual estimated costs:	
Annual capital cost at 3.5% for 100 years	$3,616,000
Annual operation and maintenance	800,000
	$4,416,000
Benefit/cost ratio = 1.13	

If the $800,000 annual operation and maintenance cost is added to the foregoing, the total annual cost is then $7,808,000. This gives us a benefit/cost ratio of only 0.64, which makes the dam decidedly unattractive.

Exercises

19.5.1 A certain assembly operation takes 2 minutes to perform in a manufacturing firm that produces 24,000 such assemblies per year. An engineer has designed an assembly fixture that will reduce the assembly time to $1\frac{1}{2}$ minutes. Hourly cost for an assembler is $10 per hour. The new assembly fixture is estimated to cost $10,000. How many years of using the new fixture would it take to save enough money to equal its cost? Would the investment be worthwhile, in your view?

19.5.2 You, as a new engineer with the manufacturing firm of Exercise 19.5.1, redesign the assembly fixture so it allows the operation to be performed in 20 seconds. Furthermore, it is now possible to use less skilled personnel costing $7.50 per hour. But the fixture will now cost $20,000 to construct. What is the payback period (number of years for the savings to equal the cost of the fixture) under these new circumstances?

19.5.3 A manufacturing firm has an old factory that is fully depreciated, but can be kept operating at a satisfactory rate by spending $300,000 per year in maintenance costs. A new factory is proposed that would cost $2,000,000, but would need only 15 people to operate it instead of the current 30 people. The lifetime over which the new factory is to be depreciated is 20 years. However, it is believed the old factory will last at least as long as 20 years, because of the willingness to spend $300,000 annually in maintenance. The new factory will require $150,000 in maintenance per year because, even though the equipment will be new, it will be more automated and complex. Production volume will not change, because the market will not allow more to be sold successfully. Make an annual cost comparison of the two alternatives, and decide which would be the better course.

Item	Present Case	Alternative
Capital investment	—	$2,000,000
Factory life	At least 20 yr	20 yr
No. of personnel @ $15,000 each	30	15
Interest rate on capital	—	9%
Annual taxes and insurance	$50,000	$150,000
Depreciation	—	Straight-line
Annual maintenance	$300,000	$150,000

19.5.4 Suppose, in the case of Exercise 19.5.3, that the new factory would allow the production rate to be doubled, from 200,000 units per year to 400,000 per year. The costs of operating the new and old factories are as given in Exercise 19.5.3, and the sales force believe they could sell 400,000 units per year if the sales price could be dropped from its present level of $10 per unit to $7. The selling cost is to be taken as 30 percent of the sales price, and other costs (except for actual manufacturing cost) at 20 percent of the sales price. Under these circumstances, does the $2,000,000 investment in the new plant look like a good idea? Calculate the annual return on capital investment. (Assume the manufacturing cost is $4 per unit when production is 200,000 per year, and $2 per unit if the new factory is built and the production rate is doubled.)

19.5.5 In the *benefit/cost ratio* example of the text, make a plot showing how the annual estimated costs vary as a function of assumed interest rate. Estimate the interest rate at which the benefit/cost ratio equals unity, i.e., the "break-even" point.

References

D. G. Newnan, *Engineering Economic Analysis*, 2nd ed. (San Jose, Calif.: Engineering Press, 1983).

J. A. White, M. H. Agee, and K. E. Case, *Principles of Engineering Economic Analysis* (New York: Wiley, 1977).

Appendix A

CODE OF ETHICS,
NATIONAL
SOCIETY OF
PROFESSIONAL
ENGINEERS

Preamble

Engineering is an important and learned profession. The members of the profession recognize that their work has a direct and vital impact on the quality of life for all people. Accordingly, the services provided by engineers require honesty, impartiality, fairness and equity, and must be dedicated to the protection of the public health, safety and welfare. In the practice of their profession, engineers must perform under a standard of professional behavior which requires adherence to the highest principles of ethical conduct on behalf of the public, clients, employers and the profession.

I. Fundamental Canons

Engineers, in the fulfillment of their professional duties, shall:

1. Hold paramount the safety, health and welfare of the public in the performance of their professional duties.
2. Perform services only in areas of their competence.
3. Issue public statements only in an objective and truthful manner.
4. Act in professional matters for each employer or client as faithful agents or trustees.
5. Avoid deceptive acts in the solicitation of professional employment.

II. Rules of Practice

1. Engineers shall hold paramount the safety, health and welfare of the public in the performance of their professional duties.
    a. Engineers shall at all times recognize that their primary obligation is to protect the safety, health, property and welfare of the public. If their professional judgment is overruled under circumstances where the safety, health,

property or welfare of the public are endangered, they shall notify their employer or client and such other authority as may be appropriate.

   b. Engineers shall approve only those engineering documents which are safe for public health, property and welfare in conformity with accepted standards.

   c. Engineers shall not reveal facts, data or information obtained in a professional capacity without the prior consent of the client or employer except as authorized or required by law or this Code.

   d. Engineers shall not permit the use of their name or firm name nor associate in business ventures with any person or firm which they have reason to believe is engaging in fraudulent or dishonest business or professional practices.

   e. Engineers having knowledge of any alleged violation of this Code shall cooperate with the proper authorities in furnishing such information or assistance as may be required.

2. Engineers shall perform services only in the areas of their competence.

   a. Engineers shall undertake assignments only when qualified by education or experience in the specific fields involved.

   b. Engineers shall not affix their signatures to any plans or documents dealing with subject matter in which they lack competence, nor to any plan or document not prepared under their direction and control.

   c. Engineers may accept assignments and assume responsibility for coordination of an entire project and sign and seal the engineering documents for the entire project, provided that each technical segment is signed and sealed only by the qualified engineers who prepared the segment.

3. Engineers shall issue public statements only in an objective and truthful manner.

   a. Engineers shall be objective and truthful in professional reports, statements or testimony. They shall include all relevant and pertinent information in such reports, statements or testimony.

   b. Engineers may express publicly a professional opinion on technical subjects only when that opinion is founded upon adequate knowledge of the facts and competence in the subject matter.

   c. Engineers shall issue no statements, criticisms or arguments on technical matters which are inspired or paid for by interested parties, unless they have prefaced their comments by explicitly identifying the interested parties on whose behalf they are speaking, and by revealing the existence of any interest the engineers may have in the matters.

4. Engineers shall act in professional matters for each employer or client as faithful agents or trustees.

   a. Engineers shall disclose all known or potential conflicts of interest to their employers or clients by promptly informing them of any business association, interest, or other circumstances which could influence or appear to influence their judgment or the quality of their services.

   b. Engineers shall not accept compensation, financial or otherwise, from more than one party for services on the same project, or for services pertaining to the same project, unless the circumstances are fully disclosed to, and agreed to by, all interested parties.

    c. Engineers shall not solicit or accept financial or other valuable consideration, directly or indirectly, from contractors, their agents, or other parties in connection with work for employers or clients for which they are responsible.

    d. Engineers in public service as members, advisors or employees of a governmental body or department shall not participate in decisions with respect to professional services solicited or provided by them or their organizations in private or public engineering practice.

    e. Engineers shall not solicit or accept a professional contract from a governmental body on which a principal or officer of their organization serves as a member.

5. Engineers shall avoid deceptive acts in the solicitation of professional employment.

    a. Engineers shall not falsify or permit misrepresentation of their, or their associates', academic or professional qualifications. They shall not misrepresent or exaggerate their degree of responsibility in or for the subject matter of prior assignments. Brochures or other presentations incident to the solicitation of employment shall not misrepresent pertinent facts concerning employers, employees, associates, joint ventures or past accomplishments with the intent and purpose of enhancing their qualifications and their work.

    b. Engineers shall not offer, give, solicit or receive, either directly or indirectly, any political contribution in an amount intended to influence the award of a contract by public authority, or which may be reasonably construed by the public of having the effect or intent to influence the award of a contract. They shall not offer any gift or other valuable consideration in order to secure work. They shall not pay a commission, percentage or brokerage fee in order to secure work except to a bona fide employee or bona fide established commercial or marketing agencies retained by them.

III. Professional Obligations

1. Engineers shall be guided in all their professional relations by the highest standards of integrity.

    a. Engineers shall admit and accept their own errors when proven wrong and refrain from distorting or altering the facts in an attempt to justify their decisions.

    b. Engineers shall advise their clients or employers when they believe a project will not be successful.

    c. Engineers shall not accept outside employment to the detriment of their regular work or interest. Before accepting any outside employment, they will notify their employers.

    d. Engineers shall not attempt to attract an engineer from another employer by false or misleading pretenses.

    e. Engineers shall not actively participate in strikes, picket lines, or other collective coercive action.

    f. Engineers shall avoid any act tending to promote their own interest at the expense of the dignity and integrity of the profession.

2. Engineers shall at all times strive to serve the public interest.

    a. Engineers shall seek opportunities to be of constructive service in civic affairs and work for the advancement of the safety, health and well-being of their community.

    b. Engineers shall not complete, sign, or seal plans and/or specifications that are not of a design safe to the public health and welfare and in conformity with accepted engineering standards. If the client or employer insists on such unprofessional conduct, they shall notify the proper authorities and withdraw from further service on the project.

    c. Engineers shall endeavor to extend public knowledge and appreciation of engineering and its achievements and to protect the engineering profession from misrepresentation and misunderstanding.

3. Engineers shall avoid all conduct or practice which is likely to discredit the profession or deceive the public.

    a. Engineers shall avoid the use of statements containing a material misrepresentation of fact or omitting a material fact necessary to keep statements from being misleading or intended or likely to create an unjustified expectation; statements containing prediction of future success; statements containing an opinion as to the quality of the engineers' services; or statements intended or likely to attract clients by the use of showmanship, puffery, or self-laudation, including the use of slogans, jingles, or sensational language or format.

    b. Consistent with the foregoing, engineers may advertise for recruitment of personnel.

    c. Consistent with the foregoing, engineers may prepare articles for the lay or technical press, but such articles shall not imply credit to the author for work performed by others.

4. Engineers shall not disclose confidential information concerning the business affairs or technical processes of any present or former client or employer without his consent.

Appendix B

RADIAN MEASURE

Frequently, angles are given in *radians* instead of in degrees. A radian is an angle that subtends an arc equal in length to the radius of the circle of which the arc is a part. A full circle has 360° in it, and the circle has a circumference of $2\pi r$. From the definition of a radian, one radian subtends an arc of length r, so $r/2\pi r$ is the proportion of 360° that corresponds to one radian. We can write this

$$1 \text{ radian} = \frac{r}{2\pi r} \, 360°$$

$$= \frac{360°}{2\pi} = 57.29578°$$

From the foregoing, we see that $360° = 2\pi$ radians. It follows that

$$180° = \pi \quad \text{radians}$$

$$90° = \pi/2 \text{ radians}$$

$$60° = \pi/3 \text{ radians}$$

$$45° = \pi/4 \text{ radians}$$

$$30° = \pi/6 \text{ radians}$$

and so on.

Appendix C

EXPONENTIAL AND LOGARITHMIC FUNCTIONS

An important relationship involving logarithms frequently pops up in mathematical manipulations and sometimes leaves students mystified. That relationship is the following:

if $\quad \ln x = y$ (C.1)

then $\quad x = e^y$ (C.2)

In these expressions, $\ln x$ means logarithms to the base e—so-called *natural* or *Napierian* logarithms—where $e = 2.7182. \ldots$ Sometimes the expression $\log_e x$ is used when natural logarithms are meant, and sometimes $\ln x$ is used. In this book, $\ln x$ is used for natural logarithms, and $\log x$ is used for logarithms to the base 10.

Now, the important thing to realize is that Equations (C.1) and (C.2) are two equivalent ways of writing the same function. If the two are plotted on the x-y plane, they trace the same curve. (See Figure C.1.)

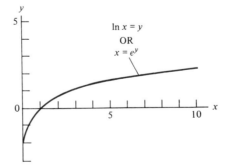

Figure C.1
An exponential curve, whether plotted as $\ln x = y$ or $x = e^y$.

567

Equations such as (C.1) and (C.2) can be written for any base, not just the base e. Thus,

$$\log_a x = y \tag{C.3}$$

$$x = a^y \tag{C.4}$$

This means, of course, that for so-called *common* logarithms to the base 10, we can write

$$\log x = \log_{10} x = y \tag{C.5}$$

$$x = 10^y \tag{C.6}$$

In mathematical manipulations, it is the base e that will show up most often.

No proofs are given for the foregoing expressions. Remember when you use them that x must always be a positive number.

Appendix D

DETERMINANTS

Very often, the mathematical model of an engineering problem produces a set of simultaneous linear algebraic equations. This is the case, for example, for the Wheatstone bridge in Chapter 17. In fact, such systems of equations arise frequently in all branches of engineering. Structural engineers, for example, often find themselves confronted by systems of simultaneous equations, sometimes of a high order.

In Chapter 17, we have the following third-order system to solve:

$$3i_1 - i_2 - 2i_3 = 4$$
$$i_1 - 4i_2 + i_3 = 0 \tag{D.1}$$
$$2i_1 + i_2 - 7i_3 = 0$$

The solutions for the unknowns i_1, i_2, and i_3 give us the values of the electric currents flowing in the three loop circuits of the Wheatstone bridge.

Let us set up a third-order system of equations in the following symbology:

$$a_{11}x_1 + a_{12}x_2 + a_{13}x_3 = k_1$$
$$a_{21}x_1 + a_{22}x_2 + a_{23}x_3 = k_2 \tag{D.2}$$
$$a_{31}x_1 + a_{32}x_2 + a_{33}x_3 = k_3$$

The system of Equations (D.2) is structurally the same as that of Equations (D.1) but has been generalized. The x values are the unknowns, and the a's and k's are constants. We now arrange the coefficients $a_{11}, a_{12}, \ldots$ in what is called the *determinant* of the coefficients, D:

$$D = \begin{vmatrix} a_{11} & a_{12} & a_{13} \\ a_{21} & a_{22} & a_{23} \\ a_{31} & a_{32} & a_{33} \end{vmatrix} \tag{D.3}$$

The horizontal lines of coefficients are referred to as *rows*. The vertical lines are *columns*. Note that the first subscript of each coefficient gives the row number, and the second gives the column number. Thus, the element a_{32} is the one in the third row and second column.

To evaluate a determinant, which means to reduce it to a single number, D, we proceed as follows. We start with the first element and multiply it by the determinant remaining after we have eliminated all the elements belonging to the same row and column as does the first element. We represent this with the following product:

$$a_{11}\begin{vmatrix} a_{22} & a_{23} \\ a_{32} & a_{33} \end{vmatrix} \tag{D.4}$$

We then select the second element of the top row, and again write its product with the determinant that remains after we have stricken out the elements belonging to the same row and column as does the second element:

$$a_{12}\begin{vmatrix} a_{21} & a_{23} \\ a_{31} & a_{33} \end{vmatrix} \tag{D.5}$$

And we do the same with the third element in the top row:

$$a_{13}\begin{vmatrix} a_{21} & a_{22} \\ a_{31} & a_{32} \end{vmatrix} \tag{D.6}$$

We combine Equations (D.4), (D.5), and (D.6) to get the value of D, but when we do so, we alternate signs, as follows:

$$D = a_{11}\begin{vmatrix} a_{22} & a_{23} \\ a_{32} & a_{33} \end{vmatrix} - a_{12}\begin{vmatrix} a_{21} & a_{23} \\ a_{31} & a_{33} \end{vmatrix} + a_{13}\begin{vmatrix} a_{21} & a_{22} \\ a_{31} & a_{32} \end{vmatrix} \tag{D.7}$$

We now have three second-order determinants in Equation (D.7). Any second-order determinant can be evaluated by the following rule: We take the product of the two elements along the *principal diagonal* (going from upper left to lower right) and subtract the product of the two elements along the other diagonal. That is, the value of the first determinant in Equation (D.7) is

$$a_{22}a_{33} - a_{23}a_{32}$$

and so on. Therefore, the full evaluation of Equation (D.7) is

$$D = a_{11}(a_{22}a_{33} - a_{23}a_{32}) - a_{12}(a_{21}a_{33} - a_{23}a_{31}) + a_{13}(a_{21}a_{32} - a_{22}a_{31})$$

which, upon rearrangement, becomes

$$D = a_{11}a_{22}a_{33} + a_{12}a_{23}a_{31} + a_{13}a_{21}a_{32} - a_{13}a_{22}a_{31}$$
$$- a_{11}a_{23}a_{32} - a_{12}a_{21}a_{33} \tag{D.8}$$

Equation (D.8) may, in fact, be taken as the definition of the evaluation of a third-order determinant.

The foregoing has been specialized for a third-order determinant, but the process can be extended to the case of an nth-order determinant. Such a determinant is written

$$D = \begin{vmatrix} a_{11} & a_{12} & \cdots & a_{1n} \\ a_{21} & a_{22} & \cdots & a_{2n} \\ \vdots & \vdots & & \vdots \\ a_{n1} & a_{n2} & \cdots & a_{nn} \end{vmatrix} \tag{D.9}$$

To evaluate D in Equation (D.9), we proceed as before, starting with the first term and multiplying it by the determinant produced by eliminating the row and column to which the first term belongs. Then we form the next product by taking the second term and multiplying it by the determinant produced by eliminating the row and column to which the second term belongs. We do the same with the third term, and so on, until we come to the end with the term a_{1n}. Then we combine all these products, affixing a plus sign to the first, a minus sign to the second, a plus to the third, a minus to the fourth, and so on, alternating signs. When we have done so, we will have a sum of n terms, each one consisting of the product of an element and a determinant; but each determinant is now one order smaller than the one we started with. Now we must evaluate each of the n determinants, using the process just described. Each time we do this, we obtain a group of determinants one order less than we had in the preceding step. Eventually, we will get down to a series of second-order determinants, which are readily evaluated as before by taking the difference of the products along the two diagonals.

Obviously, if n is very large, this process becomes incredibly cumbersome, and there are better ways to go about evaluating determinants than the one given here. However, such methods belong to the more advanced topic of *matrix theory* and will not be included here.

Returning to our system of three simultaneous equations in Equations (D.2), we can now form three new determinants. First, we write what is called a *column vector* of the constants, k_1, k_2, and k_3:

$$\begin{Bmatrix} k_1 \\ k_2 \\ k_3 \end{Bmatrix} \tag{D.10}$$

The three new determinants are formed by successively replacing the first column in the determinant of Equation (D.3), then the second column, and finally the third column with the column vector (D.10). We will designate these three new determinants as D_1, D_2, and D_3:

$$D_1 = \begin{vmatrix} k_1 & a_{12} & a_{13} \\ k_2 & a_{22} & a_{23} \\ k_3 & a_{32} & a_{33} \end{vmatrix} \tag{D.11}$$

$$D_2 = \begin{vmatrix} a_{11} & k_1 & a_{13} \\ a_{21} & k_2 & a_{23} \\ a_{31} & k_3 & a_{33} \end{vmatrix} \tag{D.12}$$

$$D_3 = \begin{vmatrix} a_{11} & a_{12} & k_1 \\ a_{21} & a_{22} & k_2 \\ a_{31} & a_{32} & k_3 \end{vmatrix} \tag{D.13}$$

By repeating the process we used to evaluate D, we will get

$$D_1 = k_1(a_{22}a_{33} - a_{23}a_{32}) - a_{12}(k_2a_{33} - k_3a_{23}) + a_{13}(k_2a_{32} - k_3a_{22}) \tag{D.14}$$

$$D_2 = a_{11}(k_2a_{33} - k_3a_{23}) - k_1(a_{21}a_{33} - a_{23}a_{31}) + a_{13}(k_3a_{21} - k_2a_{31}) \tag{D.15}$$

$$D_3 = a_{11}(k_3a_{22} - k_2a_{32}) - a_{12}(k_3a_{21} - k_2a_{31}) + k_1(a_{21}a_{32} - a_{22}a_{31}) \tag{D.16}$$

The solutions to Equations (D.2) can then be written

$$x_1 = \frac{D_1}{D} \qquad x_2 = \frac{D_2}{D} \qquad x_3 = \frac{D_3}{D} \tag{D.17}$$

which is known as *Cramer's rule*. Note that it is necessary for $D \neq 0$, or else Equations (D.17) cannot produce a solution, because we would be attempting to divide by zero.

We will now use Cramer's rule to evaluate the unknowns in Equation (D.1).

$$D = \begin{vmatrix} 3 & -1 & -2 \\ 1 & -4 & 1 \\ 2 & 1 & -7 \end{vmatrix} = 3 \begin{vmatrix} -4 & 1 \\ 1 & -7 \end{vmatrix} - (-1) \begin{vmatrix} 1 & 1 \\ 2 & -7 \end{vmatrix} + (-2) \begin{vmatrix} 1 & -4 \\ 2 & 1 \end{vmatrix}$$

$$= 3(28 - 1) + (-7 - 2) - 2(1 + 8) = 81 - 9 - 18 = 54$$

$$D_1 = \begin{vmatrix} 4 & -1 & -2 \\ 0 & -4 & 1 \\ 0 & 1 & -7 \end{vmatrix} = 4 \begin{vmatrix} -4 & 1 \\ 1 & -7 \end{vmatrix} - (-1) \begin{vmatrix} 0 & 1 \\ 0 & -7 \end{vmatrix} + (-2) \begin{vmatrix} 0 & -4 \\ 0 & 1 \end{vmatrix}$$

$$= 4(28 - 1) + (0 - 0) - 2(0 - 0) = 108$$

$$D_2 = \begin{vmatrix} 3 & 4 & -2 \\ 1 & 0 & 1 \\ 2 & 0 & -7 \end{vmatrix} = 3 \begin{vmatrix} 0 & 1 \\ 0 & -7 \end{vmatrix} - 4 \begin{vmatrix} 1 & 1 \\ 2 & -7 \end{vmatrix} + (-2) \begin{vmatrix} 1 & 0 \\ 2 & 0 \end{vmatrix}$$

$$= 3(0 - 0) - 4(-7 - 2) - 2(0 - 0) = 36$$

$$D_3 = \begin{vmatrix} 3 & -1 & 4 \\ 1 & -4 & 0 \\ 2 & 1 & 0 \end{vmatrix} = 3 \begin{vmatrix} -4 & 0 \\ 1 & 0 \end{vmatrix} - (-1) \begin{vmatrix} 1 & 0 \\ 2 & 0 \end{vmatrix} + 4 \begin{vmatrix} 1 & -4 \\ 2 & 1 \end{vmatrix}$$

$$= 3(0 - 0) + (0 - 0) + 4(1 + 8) = 36$$

$$i_1 = \frac{D_1}{D} = \frac{108}{54} = 2 \text{ A}$$

$$i_2 = \frac{D_2}{D} = \frac{36}{54} = \frac{2}{3} \text{ A}$$

$$i_3 = \frac{D_3}{D} = \frac{36}{54} = \frac{2}{3} \text{ A}$$

Cramer's rule can be generalized to n equations by forming a column vector of n k's and replacing each successive column of determinant D with the column vector to produce n determinants $D_1, D_2, D_3, \ldots, D_n$ and get the n solutions

$$x_1 = \frac{D_1}{D} \qquad x_2 = \frac{D_2}{D} \quad \cdots \quad x_n = \frac{D_n}{D} \tag{D.18}$$

Appendix E
SUMMATION

The symbol $\sum_{i=1}^{n} x_i$ is understood to mean

$$\sum_{i=1}^{n} x_i = x_1 + x_2 + x_3 + \cdots + x_i + \cdots + x_n$$

where i is called the *summation index* and takes on all the values from 1 through the final value n. The symbol Σ is the capital Greek letter *sigma* and signifies a summing process. Thus, if we have five values ($n = 5$) of the variable x, such as

$$x_1 = 3 \qquad x_2 = 2 \qquad x_3 = 4 \qquad x_4 = 3 \qquad x_5 = 2$$

then

$$\sum_{i=1}^{5} x_i = x_1 + x_2 + x_3 + x_4 + x_5$$
$$= 3 + 2 + 4 + 3 + 2$$
$$= 14$$

Sometimes, when the meaning is clear, the summation index may be omitted and the summation written simply $\Sigma\, x$. Alternative forms, also used only when no ambiguity can occur, are $\Sigma\, x_i$ and $\Sigma_i\, x_i$. They all mean the same thing. In addition to the letter i, the letters j and k are also often used for index notation, although it should be clear that *any* letter could be used.

Some examples are

$$\sum_{i=1}^{n} x_i y_i = x_1 y_1 + x_2 y_2 + x_3 y_3 + \cdots + x_n y_n$$

$$\sum_{i=1}^{n} x_i^2 = x_1^2 + x_2^2 + x_3^2 + \cdots + x_n^2$$

$$\sum_{j=1}^{n} (x_j + y_j) = (x_1 + y_1) + (x_2 + y_2) + \cdots + (x_n + y_n)$$

Appendix F

CALCULUS

Following is a highly useful geometrical interpretation of calculus. No proofs will be given; such matters belong to the formal courses in calculus that come in the first two years of an engineering education.

We begin with the *derivative*, a topic that arises in differential calculus. If we have a relationship involving the variables y and x, where x is the independent variable, we can say that y is a *function of* x and write this in shorthand notation as

$$y = f(x) \tag{F.1}$$

This does not yet tell us exactly what the function is. For example,

$$y = x$$
$$y = 2x$$
$$y = x^2$$
$$y = 225x^3 + 26x^2 + 3x + 4$$

are all functions of x. The symbol $f(x)$ is a way of indicating *any* function of x.

The symbol we usually use for the *derivative* is

$$\frac{dy}{dx}$$

Sometimes we use the symbol $f'(x)$ in place of dy/dx. It means the same thing as dy/dx. Therefore,

$$f'(x) = \frac{dy}{dx}$$

575

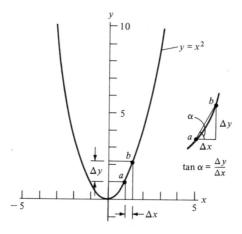

Figure F.1

Graph of $y = x^2$, showing the first step in defining the derivative.

The meaning to be attached to the symbol dy/dx is that it is the rate at which y changes, caused by a change in x. This will be made clearer by reference to Figure F.1. In this figure we have shown a particular function of x, $y = x^2$. If we look at a particular point on this curve, such as a (where $x = 1$, $y = 1$), and move a small distance along it to point b, then x will have changed by the small amount Δx (pronounced "delta x"), and y will have changed by the small amount Δy. The ratio $\Delta y/\Delta x$ is the *average* rate at which a change in y occurs, caused by a change in x, in the small interval under consideration. In calculus, if we let the interval Δx become smaller and smaller until it is infinitesimal, then Δx has almost shrunk down into the point a, so that the ratio $\Delta y/\Delta x$ no longer is to be thought of as an *average* rate of change, but is the *exact* rate of change of y with respect to x at point a. Under these circumstances, we drop the delta notation and write dy/dx, the derivative. In the precise development of calculus, this process is carried out by determining the so-called *limit* of $\Delta y/\Delta x$ as Δx approaches zero. However, we will not deal with that topic here, because it constitutes one of the central themes of courses in calculus.

The derivative can also be thought of as the *slope* of the curve at the point where it is calculated. This can be seen by considering the small right triangle in Figure F.1, with base Δx, altitude Δy, and angle α. This right triangle represents an enlargement of the section of the curve from point a to point b. It can be seen that the line segment ab cuts across the curve. It is almost, but not quite, coincident with a line drawn tangent to the curve at point a. From trigonometry, we know that

$$\tan \alpha = \frac{\Delta y}{\Delta x} \tag{F.2}$$

When Δx shrinks until it is infinitesimally small, then the line segment ab becomes exactly coincident with a line drawn tangent to the curve at a. We define the *slope* of this tangent by its angle α. At this point, as we saw before, $\Delta y/\Delta x$ becomes exactly

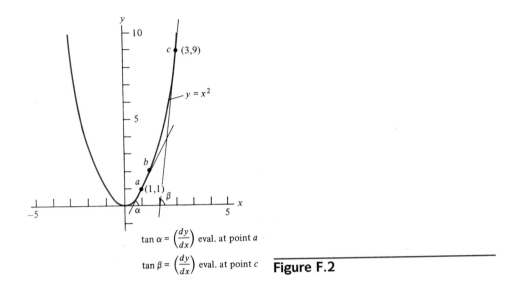

$$\tan \alpha = \left(\frac{dy}{dx}\right) \text{ eval. at point } a$$

$$\tan \beta = \left(\frac{dy}{dx}\right) \text{ eval. at point } c$$ **Figure F.2**

dy/dx, and we write

$$\tan \alpha = \frac{dy}{dx} \qquad\qquad\qquad\qquad (\text{F.3})$$

For example, in Figure F.2, where point a on the curve $y = x^2$ has coordinates $x = 1$ and $y = 1$, we can calculate the slope of the curve at a, as follows. First we must calculate the derivative of the function $y = x^2$. We have not yet said anything about how derivatives are calculated. Following is a formula, without proof, for calculating the derivative of a simple function like the one we are dealing with:

if $y = ax^n$

then $\dfrac{dy}{dx} = nax^{n-1}$ \qquad\qquad\qquad\qquad (F.4)

In tables of derivatives, this would be written

$$\frac{d}{dx}(ax^n) = nax^{n-1} \qquad\qquad\qquad (\text{F.5})$$

Thus, we can see that, if $y = x^2$, then $a = 1$ and $n = 2$, so

$$\frac{dy}{dx} = 2x \qquad\qquad\qquad\qquad (\text{F.6})$$

At the point $x = 1$, $y = 1$, by substitution in Equation (F.6) we have

$$\tan \alpha = \left(\frac{dy}{dx}\right)_{x=1} = 2(1) = 2$$

Therefore, $\alpha = 63.43°$.

At point c in Figure F.2, where $x = 3$ and $y = 9$, by substitution again in Equation (F.6) we calculate

$$\tan \beta = \left(\frac{dy}{dx}\right)_{x=3} = 2(3) = 6$$

and β thus is equal to 80.54°.

Some other simple formulas for calculating a derivative (or, as an alternative way of saying it, differentiating a function) follow without proof. In your calculus courses, much time will be spent on proving that these formulas are correct and in providing you with methods for calculating the derivatives of more complicated functions.

$$\frac{d}{dx}(C) = 0 \qquad \text{where } C \text{ is a constant}$$

$$\frac{d}{dx}(x) = 1$$

$$\frac{d}{dx}(ax^n) = nax^{n-1}$$

$$\frac{d}{dx}(\ln x) = \frac{1}{x}$$

$$\frac{d}{dx}[\ln (x + a)] = \frac{1}{x + a}$$

$$\frac{d}{dx}(\sin ax) = a \cos ax$$

$$\frac{d}{dx}(\cos ax) = -a \sin ax \qquad\qquad\qquad\qquad\qquad (F.7)$$

Now for a little integral calculus. The process we call *integration* is the inverse of differentiation, and it is symbolized by the sign $\int$. Looking at the formulas for differentiation just listed, we can write a corresponding table of integrals as follows. Note that each case in the table below is just the inverse relationship of the corresponding formula in the list of derivatives. In other words, if you start with a particular function and differentiate it, and then integrate the result, you get back the original function. However, note that a so-called arbitrary constant C is included in each integration formula. This is because whenever you differentiate a constant, you get zero. Thus, in integrating, you cannot ever be sure whether a constant was there originally. You must arbitrarily assume the existence of such a constant when you integrate. In an actual physical problem, methods are available to evaluate such a constant and fix a numerical value on it. The details on how to do this will be left for more advanced courses.

$$\int 0 \cdot dx = C \qquad \text{where } C \text{ is a constant}$$

$$\int 1 \cdot dx = x + C$$

$$\int nax^{n-1}\, dx = ax^n + C$$

$$\int \frac{dx}{x} = \ln x + C$$

$$\int \frac{dx}{x+a} = \ln(x+a) + C$$

$$\int a \cos ax\, dx = \sin ax + C$$

$$\int a \sin ax\, dx = -\cos ax + C \qquad\qquad (F.8)$$

A more customary way of writing the last two expressions is

$$\int \cos ax\, dx = \frac{1}{a} \sin ax + C$$

$$\int \sin ax\, dx = -\frac{1}{a} \cos ax + C$$

This is because a constant can always be taken "outside" the integral sign without altering the expression. That is,

$$\int a \cos ax\, dx = a \int \cos ax\, dx$$

and so on.

At this point, the student who is unfamiliar with calculus may be troubled by the presence of the term dx in each integration formula. This comes about because it is possible to treat the term dx as if it were itself an algebraic quantity and move it about in an equation in the same way as for any algebraic variable. Thus, we could rewrite all the formulas in (F.7) as follows:

$$d(C) = 0 \cdot dx \qquad \text{where } C \text{ is a constant}$$

$$d(x) = 1 \cdot dx$$

$$d(ax^n) = nax^{n-1}\, dx$$

$$d(\ln x) = \frac{dx}{x}$$

$$d[\ln(x+a)] = \frac{dx}{x+a}$$

$$d(\sin ax) = a \cos ax\, dx$$

$$d(\cos ax) = -a \sin ax\, dx \qquad\qquad (F.9)$$

Now, a comparison of each formula from Equation (F.9) with the corresponding one in Equation (F.8) shows more clearly that they indeed are inverses of each other. The term dx that we moved about in the preceding expressions is called the *differential* of x. The term dy, then, is the differential of y. Their ratio, dy/dx, is, of course, the derivative of y with respect to x, as we saw before.

Appendix G

SOLUTIONS TO DIFFERENTIAL EQUATIONS OF CHAPTER 17

Solution to Equation (17.9), the Differential Equation for a Capacitive Circuit

Equation (17.9) is

$$\frac{dq}{q} = -\frac{1}{RC}\, dt \tag{G.1}$$

We integrate both sides, getting

$$\int \frac{dq}{q} = -\frac{1}{RC} \int dt \tag{G.2}$$

From the integral formulas given as Equation (F.8), we obtain the integrals of both sides of Equation (G.2) and write

$$\ln q + C_1 = -\frac{1}{RC}\, t + C_2 \tag{G.3}$$

where C_1 and C_2 are arbitrary constants of integration. We can combine these into a single constant C_3, where $C_3 = -C_1 + C_2$, and write

$$\ln q = -\frac{1}{RC}\, t + C_3 \tag{G.4}$$

From the properties of logarithms we know that

if $\quad\ln q = y$

then $\quad q = e^y$

Therefore, Equation (G.4) can be changed to

$$q = e^{(-t/RC + C_3)}$$
$$= e^{C_3} e^{-t/RC}$$
$$= B e^{-t/RC} \tag{G.5}$$

where, because e^{C_3} is a constant, we have designated it with $B = e^{C_3}$. This is the solution given as Equation (17.11).

Solution of Equation (17.17), the Differential Equation for an Inductive Circuit

Equation (17.17) is

$$\frac{di}{V/R - i} = \frac{R}{L} dt \tag{G.6}$$

We note that, because of Ohm's law, V/R will be the final value of the current, I_f, in the circuit after the current stops changing and di/dt has gone to zero. So, substituting $I_f = V/R$, and making a further slight rearrangement, we get

$$\frac{di}{i - I_f} = -\frac{R}{L} dt \tag{G.7}$$

We integrate both sides:

$$\int \frac{di}{i - I_f} = -\frac{R}{L} \int dt \tag{G.8}$$

From the integral formulas given as Equation (F.8), we obtain the integrals of both sides of Equation (G.8) and write

$$\ln (i - I_f) + C_1 = -\frac{R}{L} t + C_2 \tag{G.9}$$

We can combine constants C_1 and C_2 by setting $C_3 = -C_1 + C_2$ and writing

$$\ln (i - I_f) = -\frac{R}{L} t + C_3 \tag{G.10}$$

if $\ln q = y$

then $q = e^y$

so

$$i - I_f = e^{(-Rt/L + C_3)}$$
$$= e^{C_3} e^{-Rt/L}$$
$$= A e^{-Rt/L} \tag{G.11}$$

The arbitrary constant A can now be evaluated. When $t = 0$, then $i = 0$, so from Equation (G.11),

$$0 - I_f = Ae^{-R \cdot 0/L} = Ae^0 = A$$

so $A = -I_f$. Substituting this value of A in Equation (G.11), we get

$$i = I_f - I_f e^{-Rt/L}$$
$$= I_f(1 - e^{-Rt/L}) \tag{G.12}$$

which is the complete solution to Equation (17.17) and is given as Equation (17.18) in Chapter 17.

Solution to Equation (17.20), the Differential Equation for the Waveform Example of Chapter 17

Equation (17.20) is

$$V = \frac{dq}{dt}R + \frac{q}{C} \tag{G.13}$$

If we rearrange this equation (you should carry out the algebra), we get

$$\frac{dq}{q - CV} = -\frac{1}{RC}dt \tag{G.14}$$

This equation is structured just like Equation (G.7) in the foregoing section. The solution is

$$\ln(q - CV) = -\frac{1}{RC}t + C_3 \tag{G.15}$$

where C_3 is an arbitrary constant.

We next use the properties of logarithms as we did in the foregoing section, and get

$$q - CV = e^{(-t/RC + C_3)}$$
$$= e^{C_3}e^{-t/RC} \tag{G.16}$$

$$q - CV = Ae^{-t/RC} \tag{G.17}$$

We evaluate the arbitrary constant A by noting that $q = 0$ when $t = 0$. If we substitute these values into Equation (G.17), we get

$$0 - CV = Ae^0$$
$$A = -CV$$

Substituting this value of A in Equation (G.17), we get our final solution:

$$q - CV = -CVe^{-t/RC}$$
$$q = CV(1 - e^{-t/RC}) \tag{G.18}$$

which is the solution given as Equation (17.21) in Chapter 17.

Appendix H
ENGINEERING GRAPHICS

In Chapter 11, some of the basic concepts of design layouts, working sketches, and working drawings were presented. Much was left unsaid regarding acceptable practices in dimensioning, making sectional views, and the use of common conventions. Additional material on these and related topics are given in this appendix. The material is selected from American National Standards ANSI Y14.3-1975, ANSI Y14.5-1973, and ANSI 14.6-1978 and is reproduced through the courtesy of the American Society of Mechanical Engineers.

Principal Views

The kinds of working drawings we customarily use are prepared in accordance with the rules of *orthographic projection*. In this technique, the lines and surfaces belonging to an object are projected along lines of sight that are perpendicular to imaginary planes surrounding the object. If the lines of sight are not taken perpendicular to the imaginary planes, types of projections other than orthographic are obtained. If the lines of sight are at an angle, we get *oblique* projections, and if they emanate from a point, we get *perspective* projections.

Customarily, in orthographic projection, we consider the object of interest to be located inside an imaginary glass box, as shown in Figure H.1. In that figure, the three surfaces of the imaginary glass box located closest to the observer are labeled *front view*, *top view*, and *side view*. The reader should understand that there are three more surfaces to this imaginary glass box that have been left out of the figure. These might have been labeled *bottom view*, *back view*, and *left-side view*. (It should be apparent that the side view in Figure H.1 is actually a *right-side view*.)

In Figure H.1, if parallel "lines of sight," shown as dashed lines, are projected perpendicular to the imaginary surfaces of the glass box, then points of intersection between the lines of sight and the imaginary surfaces can be found. If we connect these intersection points on one of the imaginary surfaces, we get a "view" of the

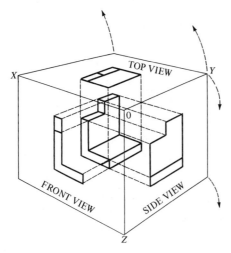

Figure H.1

An imaginary glass box surrounding an object, showing the three principal views.

object from that direction. Thus, if the observer stands directly in front of the imaginary surface marked "front view," the intersections traced out by the lines of sight perpendicular to that surface show what the object looks like when viewed from that direction. Similarly, if the observer stands facing the surface marked "side view," then a projection onto that surface shows what the object looks like when viewed from that side. Finally, if one could be suspended in mid-air and look directly down on the object, the "top view" shows what the object looks like when viewed from that direction.

Considering Figure H.1 again, let us imagine that lines O-X and O-Z are hinged and that any connection along the line O-Y is severed. Then the plane containing the top view is rotated upward about hinge O-X, as shown by the dashed lines with arrows. The plane containing the side view is rotated outward about hinge O-Z, also shown by the dashed lines with arrows. When each plane has been rotated 90 degrees, it will lie in the same plane as the front view, and we will have the situation shown in Figure H.2(a). This figure shows three views—front, top, and side—which are sufficient to show everything that is needed to describe this object. There are three other views not shown in Figure H.2(a): the bottom view, the rear view, and the other side view. The six together are known as *principal views*.

Note, in Figure H.2(a), that the side view is essential in order to describe the object completely. As an illustration, all three of the side views shown in Figure H.2(b) would be compatible with the front and top views of Figure H.2(a), which shows why the side view is necessary. In some instances, however, only two views are sufficient. This is true for cylindrical objects, such as in some of the working sketches in the example of Chapter 11.

It should be noted in the three views of Figure H.2(a) that none of the lines or edges is hidden from view—all are visible. However, if we had taken a side view from the left side instead of the right, one of the edges would be invisible. We would show

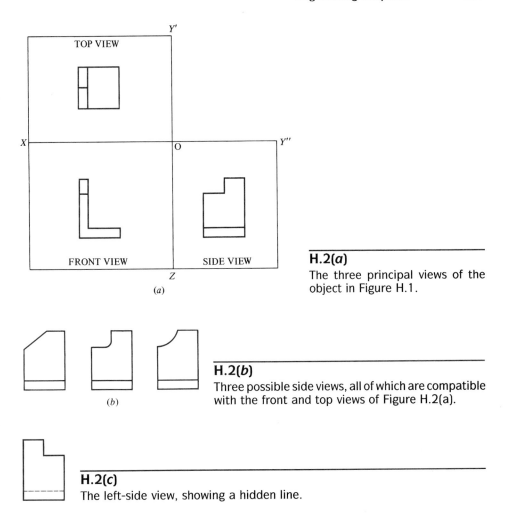

H.2(a)
The three principal views of the object in Figure H.1.

(a)

H.2(b)
Three possible side views, all of which are compatible with the front and top views of Figure H.2(a).

(b)

H.2(c)
The left-side view, showing a hidden line.

such a "hidden line" by a dashed line instead of a solid one, as has been done in Figure H.2(c).

Line Conventions

Certain commonly used line representations are shown in Figure H.3. Two line weights are shown: one thick (approximately 0.032 inch), the other thin (approximately 0.016 inch). Visible lines and cutting-plane lines are thick; all others are thin. Visible lines should connect at their intersections and corners without leaving gaps. Hidden lines should begin and end with a dash in contact with the line to which they connect. Hidden lines should be omitted when not needed for clarity.

Dimension lines should terminate with neat, filled-in arrowheads, as shown, and should connect to their extension lines without leaving gaps.

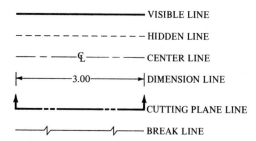

Figure H.3
Some common line representations.

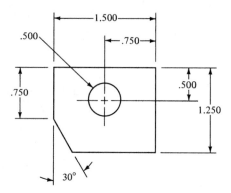

Figure H.4
Placement of dimensions.

Dimensioning

Dimensions should be placed on drawings neatly and with ample space to prevent crowding. Preferably, they should all be horizontal so they can be read without rotating the drawing. They should generally be placed outside the outline of the part being dimensioned. See Figure H.4 for typical placement of dimensions. Dimensions should also be aligned and grouped for uniform appearance, as in Figure H.5. Di-

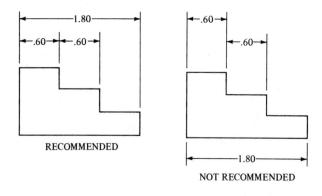

Figure H.5
Grouping of dimensions.

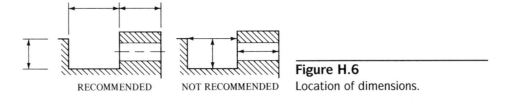

Figure H.6
Location of dimensions.

RECOMMENDED   NOT RECOMMENDED

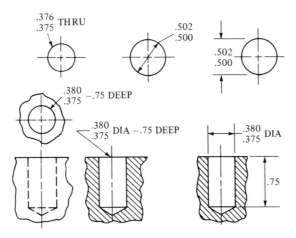

Figure H.7
Representation of round holes.

mension lines in general are not placed so that they connect directly to the outline of the object, but instead connect to extension lines, as shown in Figure H.6.

Round holes can be dimensioned in a number of ways, as in Figure H.7.

Counterbored and countersunk holes are typically dimensioned as in Figure H.8. Note the abbreviations CBORE for "counterbore" and CSK for "countersink."

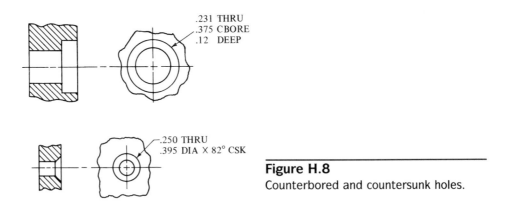

Figure H.8
Counterbored and countersunk holes.

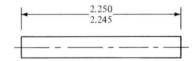

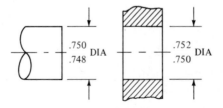

Figure H.9
Limit dimensioning.

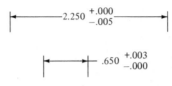

(*a*) Unilateral Tolerancing

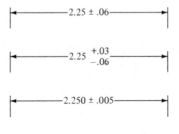

(*b*) Bilateral Tolerancing

Figure H.10
Plus and minus tolerancing.

Tolerances are generally shown by one of two methods: *limit dimensioning*, as in Figure H.9, or *plus and minus tolerancing*, as in Figure H.10. Plus and minus tolerancing may be either unilateral or bilateral, as shown. Frequently, tolerances may be indicated in a general note, such as "ALL TOLERANCES ARE ±.03, UNLESS OTHERWISE INDICATED."

Threads and Fasteners

There are three general methods for showing screw threads on drawings (see Figure H.11). The simplified method is preferred unless misunderstanding could occur.

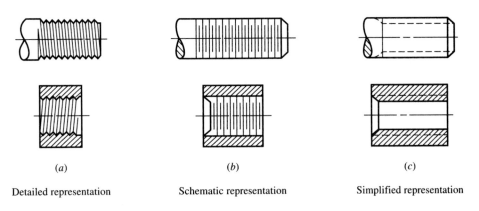

(a)   (b)   (c)

Detailed representation   Schematic representation   Simplified representation

Figure H.11
Methods of representing screw threads. Method (c) is usually preferred.

Threads are usually designated by a note such as that in Figure H.12. For the nominal diameter, three decimals are usually shown; numbered sizes may be used. If a numbered size is used, the decimal equivalent should also be given, in parentheses, as

NO. 10(.190)-32 UNF-2A

The number of threads per inch is frequently chosen to conform with a standard, as listed in handbooks. The most common thread series are UNC (Unified Coarse), UNF (Unified Fine), and UNEF (Unified Extra Fine). The "class symbol" (Figure H.12) represents the precision with which the dimensions are controlled. Class 1 provides the loosest tolerances; Class 2 represents the bulk of commercial usage; Class 3 calls for close fits and extra precision and is more expensive than Class 1 or 2. The "internal or external symbol" is "A" (for external threads) or "B" (for internal threads). Threads are considered to be right-hand unless otherwise specified.

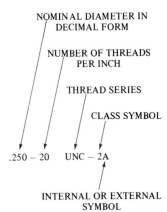

NOMINAL DIAMETER IN
DECIMAL FORM

NUMBER OF THREADS
PER INCH

THREAD SERIES

CLASS SYMBOL

.250 – 20   UNC – 2A

INTERNAL OR EXTERNAL
SYMBOL

Figure H.12
Thread specification.

There are other kinds of threads, such as square threads, Acme threads, buttress threads, and taper threads, that are not covered here. The information herein relates to the common V-threads found on commercial fasteners.

Sectional Views

Sectional views are used to clarify internal structure and are obtained by passing imaginary cutting planes through the object. A *full section* is shown in Figure H.13. The cutting plane line is identified with letters such as A-A, as shown. If the cutting plane line is obvious, it may be omitted. Sometimes a *half section* is useful, as in Figure H.14, because it shows both interior and exterior structure.

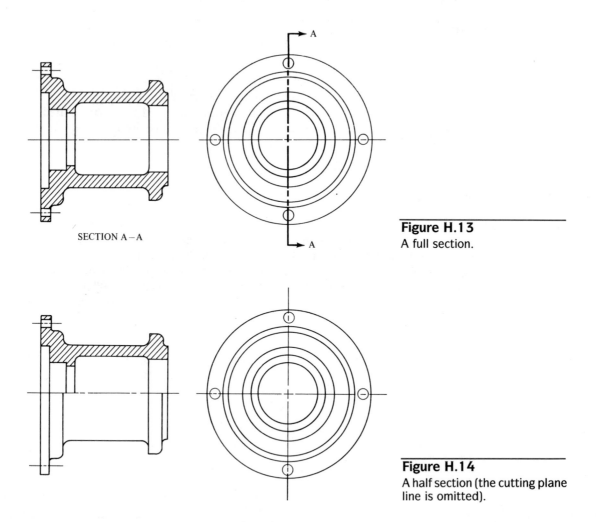

SECTION A–A

Figure H.13
A full section.

Figure H.14
A half section (the cutting plane line is omitted).

Appendix I

UNIT SYMBOLS

Unit	Symbol	Comments
acceleration	m/s^2	
ampere	A	Electric current
angstrom	Å	10^{-10} m
angular acceleration	rad/s^2	
angular velocity	rad/s	
angular velocity	r/min; rpm	Revolutions per minute
angular velocity	r/sec; rps	Revolutions per second
area	m^2	
bar	bar	100 kPa
barrel (petroleum)	bbl	42 gal
becquerel	Bq	Activity (of ionizing radiation source)
British thermal unit	Btu	
calorie	cal	4.1868 J
Calorie ("kilocalorie")	Cal	4186.8 J
candela	cd	Luminous intensity
centimeter	cm	
coulomb	$C = A \cdot s$	Quantity of electricity
curie	Ci	3.7×10^{10} Bq
day	d	
decibel	dB	Sound intensity level
degree	°	Plane angle
degree Celsius	°C	Temperature
degree Fahrenheit	°F	Temperature

Sources: *Standard Practice for Use of the International System of Units* (Philadelphia, Pa.: American Society for Testing and Materials, 1989); *Metric Guide for Educational Materials* (American National Metric Council, Washington, D.C. 20036, 1977).

(Continued)

(Continued)

Unit	Symbol	Comments
degree Kelvin	K	Thermodynamic temperature
degree Rankine	°R	"Absolute" temperature
density	kg/m^3	
dyne	dyn	10^{-5} N
electric field strength	V/m	
electron volt	eV	$1.602\ 19 \times 10^{-19}$ J
entropy	J/K	
erg	erg	10^{-7} J
farad	F = C/V	Electric capacitance
foot	ft	
foot-pound (force)	ft-lbf	Energy; torque
gallon	gal	
gauss	G	10^{-4} T
gram	g	
gray	Gy	Absorbed dose
hectare	ha	$10^4\ m^2$
henry	H = Wb/A	Electric inductance
hertz	Hz = 1/s	Frequency
horsepower	hp	
hour	h	
inch	in.	
joule	J = N·m	Work; energy; quantity of heat
kelvin	K	Thermodynamic temperature
kilogram	kg	Mass
kilogram-force	kgf	
kilowatt-hour	kWh	3.6×10^6 J
lambert	L	$3.183 \times 10^3\ cd/m^2$
liter	l	$1000\ cm^3$; use L in U.S.
lumen	lm = cd·sr	Luminous flux
luminance	cd/m^2	
lux	lx = lm/m^2	Illuminance
magnetic field strength	A/m	
maxwell	Mx	10^{-8} Wb
megahertz	MHz	
megawatt	MW	
meter	m	Length
metric ton	t	1000 kg; called "tonne"
mho	mho	Electrical conductance; now called siemens
micron	$\mu;\ \mu m$	$1\ \mu m$
mile	mi	
minute	min	Time
minute	′	1/60 of 1°
mole	mol	Amount of substance

(*Continued*)

Unit	Symbol	Comments
newton	$N = kg \cdot m/s^2$	Force
oersted	Oe	79.577 A/m
ohm	$\Omega = V/A$	Electric resistance
pascal	$Pa = N/m^2$	Pressure; stress
poise	P	$0.1 \; Pa \cdot s$
pound	lb	
pound-force	lbf	
pound-mass	lb	
pressure	$lbf/in.^2$; psi	Pounds per square inch
pressure	$Pa = N/m^2$	
rad	rd	0.01 Gy
radian	rad	Plane angle
radiant intensity	W/sr	
roentgen	R	$2.58 \times 10^{-4} \; C/kg$
second	s	Time
second	$''$	1/60 of $1'$
siemens	$S = A/V$	Electrical conductance
slug	slug	$lb\text{-}sec^2/ft$
specific heat	$J/kg \cdot K$	
steradian	sr	Solid angle
stokes	St	$10^{-4} \; m^2/s$
stress	$lbf/in.^2$; psi	Pounds per square inch
stress	$Pa = N/m^2$	
tesla	$T = Wb/m^2$	Magnetic flux density
thermal conductivity	$W/(m \cdot K)$	
ton (short)	ton	2000 lb
tonne	t	Metric ton = 1000 kg
torr	torr	133.3 Pa
velocity	mi/hr; mph	Miles per hour
velocity	ft/sec	
velocity	m/s	
viscosity, dynamic	$Pa \cdot s$	
viscosity, kinematic	m^2/s	
volt	$V = W/A$	Electric potential difference
volt-ampere	VA	
volume	m^3	
volume	cm^3	Cubic centimeter
volume	ft^3	Cubic foot
volume	$in.^3$	Cubic inch
watt	$W = J/s$	Power
watt-hour	$W \cdot h$	
weber	$Wb = V \cdot s$	Magnetic flux
yard	yd	

Appendix J
SI UNITS

J.1 Units Accepted for Use with SI

Unit	Symbol or Abbreviation	Value in SI Units	Comments
day	d	8.64×10^4 s	
degree Celsius	°C	1°C = 1 K	Formerly degree centigrade
degree (of arc)	°	$\pi/180$ rad	
hectare	ha	10^4 m^2	Land or water area only
hour	h	3600 s	
liter	L	10^{-3} m^3	
metric ton	t	1000 kg	Use of the word *tonne* is discouraged
minute	min	60 s	
minute (of arc)	'	$\pi/10\,800$ rad	Cartography only; decimalized degree preferred
second (of arc)	"	$\pi/648\,000$ rad	Cartography only; decimalized degree preferred
tonne (*see* metric ton)			

Source: *Standard Practice for Use of the International System of Units* (E380-89a) (Philadelphia, Pa.: American Society for Testing and Materials, 1989).

J.2 Units Temporarily in Use with SI

Unit	Symbol	Definition	Comments
kilowatt-hour	kWh	3.6 MJ	Electric energy
barn	b	10^{-28} m^2	Cross section
bar	bar	100 kPa	Pressure (meteorology)
curie	Ci	3.7×10^{10} Bq	Activity (of a radionuclide)
roentgen	R	2.58×10^{-4} C/kg	Exposure (x- and gamma rays)
rad	rd	0.01 Gy	Absorbed dose
rem	rem	0.01 Sv	Dose equivalent

Source: *Standard Practice for Use of the International System of Units* (E380-89a) (Philadelphia, Pa.: American Society for Testing and Materials, 1989).

J.3 Nonapproved Units for SI

The following is a list of units that were used formerly but are now considered undesirable. Abbreviations or symbols are given in parentheses.

angstrom (Å); equals 10^{-10} m
are (a); a unit of area = 100 m^2
standard atmosphere (atm); equals 101.325 kPa
calorie (cal); equals 4.1868 J
Calorie (Cal); frequently called "kilocalorie"; equals 4186.8 J
candlepower (candlepower); a unit of luminous intensity = 1 cd
degree centigrade (°C); replaced by degree Celsius
degree Kelvin (designated now simply as "kelvin")
dyne (dyn); equals 10^{-5} N
erg (erg); equals 10^{-7} J
gal (Gal); a unit of acceleration = 0.01 m/s^2
gamma (γ); a unit of magnetic flux density = 10^{-9} T
gamma (γ); a unit of mass = 10^{-9} kg
gauss (Gs or G); a unit of magnetic flux density = 10^{-4} T
kilogram force (kgf); equals 9.806 65 N
maxwell (Mx); a unit of magnetic flux = 10^{-8} Wb
metric carat; a unit of mass = 200 mg
mho (mho); a unit of electric conductance = 1 S
micron (μ); equals 1 μm
millimeter of mercury (mm Hg); a unit of pressure $\cong$ 133.3 Pa, at 0°C
oersted (Oe); a unit of magnetic field strength $\cong$ 79.577 A/m
poise (P); a unit of dynamic viscosity = 0.1 Pa·s
stere (st); a unit of volume = 1 m^3
stokes (St); a unit of kinematic viscosity = 10^{-4} m^2/s
torr; a unit of pressure $\cong$ 133.3 Pa

Source: *Standard Practice for Use of the International System of Units* (E380-89a) (Philadelphia, Pa.: American Society for Testing and Materials, 1989).

Appendix K

DENSITIES OF
SELECTED
MATERIALS
(APPROXIMATE)

Material	Average Density kg/m³	Average Density lb/ft³
Acid, sulfuric (87%)	1795	112
Air (15°C, 760 mm)	1.225	0.0765
Alcohol (ethyl)	802	50.1
Alcohol (methyl)	809	50.5
Aluminum (cast-hammered)	2643	165
Ammonia (20°C, 760 mm)	0.712	0.044
Asbestos	2451	153
Asphaltum	1298	81.0
Benzene	737	46.0
Brass (cast-rolled)	8553	534
Brick (hard)	2051	128
Bronze, phosphor	8874	554
Carbon	1800–2100	112–131
Carbon dioxide (0°C, 760 mm)	1.997	0.125
Carbon monoxide (0°C, 760 mm)	1.250	0.0780
Cement, Portland	1505	94.0
Clay (damp)	1761	110
Coal (bituminous)	1346	84.0
Concrete	2309	144
Copper (cast-rolled)	8906	556
Earth (moist, packed)	1538	96.0

Source: R. C. Weast and M. J. Astle (eds.), *CRC Handbook of Chemistry and Physics*, 62nd ed. (Boca Raton, Fla.: CRC Press, 1981); T. Baumeister, E. A. Avalone, and T. Baumeister III (eds.), *Marks' Standard Handbook for Mechanical Engineers*, 8th ed. (New York: McGraw-Hill, 1978).

(Continued)

(Continued)

Material	Average Density	
	kg/m³	lb/ft³
Gasoline	705–731	44.0–45.6
Glass, common	2595	162
Gold (cast-hammered)	19 300	1205
Granite	2644	165
Graphite	2163	135
Gravel (wet)	2019	126
Helium (0°C, 760 mm)	0.1785	0.0111
Hydrogen (0°C, 760 mm)	0.0899	0.005 61
Ice	897	56.0
Iron (pure)	7860	491
Iron (gray cast)	7079	442
Kerosene	801	50.0
Lead	11 370	710
Limestone	2484	155
Magnesium	1740	109
Manganese	7608	475
Marble	2725	170
Masonry	2160–2640	135–165
Methane (20°C, 760 mm)	0.667	0.0416
Mercury	13 570	847
Nickel	8602	537
Nitrogen (0°C, 760 mm)	1.251	0.0781
Oil, crude	801–977	50.0–61.0
Oil, lubricating	914	57.1
Oxygen (0°C, 760 mm)	1.429	0.0892
Paper	929	58.0
Platinum (cast-hammered)	21 300	1330
Potassium	860	53.7
Quartz	2645	165
Rubber	946	59.1
Salt	2165	135
Sand (dry, packed)	1602–1922	100–120
Sandstone	2290	143
Shale	2758	172
Silicon	2320–2340	145–146
Silver (cast-hammered)	10 510	656
Snow (fresh-fallen)	128	8

(*Continued*)

Material	Average Density	
	kg/m³	lb/ft³
Sodium	970	60.6
Steel (cold-drawn)	7832	489
Sulfur	2001	125
Sulfur dioxide (0°C, 760 mm)	2.927	0.183
Tin (cast-hammered)	7352	459
Titanium	4500	281
Tungsten	19 220	1200
Uranium	18 740	1170
Water (4°C)	1000	62.4
Water, sea	1025	64.0
Wood (Douglas fir, dry)	513	32
Wood (mahogany, dry)	705	44
Wood (sugar maple, dry)	689	43
Wood (white oak, dry)	770	48
Wood (southern pine, dry)	610–673	38–42
Wood (redwood, dry)	417	26
Zinc (cast-rolled)	7049	440

Appendix L

UNIT CONVERSIONS

Conversion values are given to four significant digits. Pounds, ounces, and tons are given in avoirdupois units unless otherwise noted. T = temperature in kelvin. T_R = degrees Rankine. t_C = degrees Celsius. t_F = degrees Fahrenheit.

Multiply	By	To Obtain
abamperes	10	amperes
abcoulombs	10	coulombs
abfarads	10^9	farads
abhenries	10^{-9}	henries
abohms	10^{-9}	ohms
abvolts	10^{-8}	volts
acres	43 560	square feet
"	4047	square meters
"	4840	square yards
acre-feet	43 560	cubic feet
" "	3.259×10^5	gallons (U.S. liquid)
amperes	1	coulombs per second
angstroms	10^{-10}	meters
ares	0.024 71	acres
"	100	square meters
astronomical units	1.496×10^{11}	meters
atmospheres	76	centimeters of mercury
"	29.92	inches of mercury
"	33.90	feet of water
"	14.70	pounds per square inch (force)
"	1.013×10^5	pascals

Sources: *Standard Practice Manual for Use of the International System of Units* (E380-89a) (Philadelphia, Pa.: American Society for Testing and Materials, 1989); R. G. Hudson, *The Engineers' Manual*, 2nd ed. (New York: Wiley, 1947); *Metric Guide for Educational Materials* (Washington: American National Metric Council, 1977); E. A. Mechtly, *The International System of Units, Fundamental Constants and Conversion Factors* (Champaign, Ill: Stipes Publishing Co., 1977); *Reference Manual for SI (Metric)* (Inland Steel Company, 1976).

(Continued)

(*Continued*)

Multiply	By	To Obtain
bars	0.9872	atmospheres
"	10^6	dynes per square centimeter
"	14.51	pounds per square inch (force)
"	10^5	pascals
barrels (petroleum)	42	gallons (U.S. liquid)
becquerels (radioactivity)	1	disintegrations per second
British thermal units	778.2	foot-pounds-force
" " "	1055	joules
" " "	2.931×10^{-4}	kilowatt-hours
Btu per minute	17.58	watts
Btu per pound	2326	joules per kilogram
bushels	1.244	cubic feet
"	0.035 24	cubic meters
"	4	pecks
calories	4.187	joules
"	10^{-3}	kilocalories
calories per gram	4187	joules per kilogram
Calories (kilocalorie)	4187	joules
candelas per square meter	3.141×10^{-4}	lamberts
carats (metric)	2×10^{-4}	kilograms (mass)
centimeters	0.3937	inches
centimeters of mercury	0.013 16	atmospheres
" " "	0.4461	feet of water
" " "	0.1935	pounds per square inch (force)
" " "	1333	pascals
circular mils	5.067×10^{-6}	square centimeters
" "	7.854×10^{-7}	square inches
cords	8 ft. $\times$ 4 ft. $\times$ 4 ft.	cubic feet
coulombs (quantity of electricity)	1	ampere-seconds
cubic centimeters	3.531×10^{-5}	cubic feet
" "	6.102×10^{-2}	cubic inches
" "	10^{-6}	cubic meters
" "	10^{-3}	liters
" "	2.642×10^{-4}	gallons (U.S. liquid)
cubic feet	2.832×10^4	cubic centimeters
" "	1728	cubic inches
" "	0.028 32	cubic meters
" "	0.037 04	cubic yards
" "	7.481	gallons (U.S. liquid)
" "	28.32	liters
cubic inches	16.39	cubic centimeters
" "	5.787×10^{-4}	cubic feet
" "	1.639×10^{-5}	cubic meters
" "	2.143×10^{-5}	cubic yards
" "	4.329×10^{-3}	gallons (U.S. liquid)
cubic meters	35.31	cubic feet
" "	61 024	cubic inches
" "	1.308	cubic yards
" "	264.2	gallons (U.S. liquid)

(*Continued*)

Multiply	By	To Obtain
cubic yards	27	cubic feet
" "	46 656	cubic inches
" "	0.7646	cubic meters
" "	202.0	gallons (U.S. liquid)
curies	3.7×10^{10}	becquerels
days	24	hours
"	1440	minutes (time)
"	8.640×10^{4}	seconds (time)
degrees (angle)	60	minutes (angle)
" "	0.017 45	radians
degrees Fahrenheit	—	degrees Celsius: $t_C = (t_F - 32)/1.8$
degrees Celsius	—	kelvin: $T = t_C + 273.15$ K
degrees Fahrenheit	—	kelvin: $T = (t_F + 459.67°R)/1.8$
degrees Rankine	—	kelvin: $T = T_R/1.8$
degrees per second (angle)	0.1667	revolutions per minute
degrees Kelvin (*see* kelvin)		
density: pounds-mass/in.^{3}	27 680	kilograms per cubic meter (mass)
drams	1.772	grams-force
"	0.0625	ounces-force
dynes	1.020×10^{-3}	grams-force
"	7.233×10^{-5}	poundals
"	2.248×10^{-6}	pounds-force
"	1	gram-centimeters/s^2 (mass)
"	10^{-5}	newtons
electron volts	1.602×10^{-19}	joules
ergs	9.479×10^{-11}	British thermal units
"	7.378×10^{-8}	foot-pounds-force
"	10^{-7}	joules
"	1	dyne-centimeters
ergs per second	1.341×10^{-10}	horsepower
" " "	10^{-7}	watts
farads (electric capacitance)	1	coulombs per volt
fathoms	6	feet
feet	0.3048	meters
feet per second	0.3048	meters per second
feet of water	0.029 50	atmospheres
" " "	0.8827	inches of mercury
" " "	0.4336	pounds per square inch (force)
feet of water (39.2°F)	2989	pascals
foot-candles	10.76	lumens per square meter (lux)
" "	10.76	lux
" "	1	lumens per square foot
foot-pounds-force	1.285×10^{-3}	British thermal units
" " "	1.356×10^{7}	ergs
" " "	1.356	joules

(Continued)

Multiply	By	To Obtain
force: lbf	4.448	newtons
force: kgf ("kilopond")	9.807	newtons
force (1 kg·m/s²)	1	newtons
frequency (1/s)	1	hertz
furlongs	40	rods
gallons (U.S. liquid)	3.785×10^{-3}	cubic meters
" " " "	0.1337	cubic feet
" " " "	231	cubic inches
" " " "	4	quarts (U.S. liquid)
gallons (U.S. dry)	4.405×10^{-3}	cubic meters
gallons (U.K. liquid)	4.546×10^{-3}	cubic meters
gals (unit of acceleration)	10^{-2}	meters per second per second
gammas (mass)	10^{-9}	kilograms (mass)
gammas (magnetic flux density)	10^{-9}	teslas
gausses	10^{-4}	teslas
gills	0.25	pints (U.S. liquid)
grads	1.571×10^{-2}	radians
grains	1.429×10^{-4}	pounds
grams	10^{-3}	kilograms
grams-force	0.035 27	ounces-force
" "	0.032 15	ounces-force (troy)
" "	2.205×10^{-3}	pounds-force
hectares	2.471	acres
"	10^4	square meters
henries (inductance)	1	webers per ampere
horsepower	42.41	British thermal units per minute
" "	33 000	foot-pounds per minute (force)
" "	550	foot-pounds per second (force)
" "	745.7	watts
horsepower-hour	2.684×10^6	joules
inches	2.540	centimeters
"	2.540×10^{-2}	meters
inches of mercury (32°F)	0.033 42	atmospheres
" " " "	0.4912	pounds per square inch (force)
" " " "	3.386×10^3	pascals
inches of mercury (60°F)	3.377×10^3	pascals
joules (energy, work, heat)	1	newton-meters
"	9.478×10^{-4}	British thermal units
"	0.7376	foot-pounds-force
"	2.778×10^{-4}	watt-hours
"	0.2388	calories
"	2.388×10^{-4}	kilocalories
joules per kilogram	4.300×10^{-4}	Btu per pound
kelvin	—	degrees Celsius: $t_C = T - 273.15$ K
"	—	degrees Fahrenheit: $t_F = 1.8T - 459.67°$ R
"	—	degrees Rankine: $T_R = 1.8T$

(*Continued*)

Multiply	By	To Obtain
kilocalories	4.187×10^3	joules
"	10^3	calories
kilograms-force (kgf)	70.93	poundals
" " "	2.205	pounds-force
" " "	9.807	newtons
kilograms-mass (kg)	1	kilograms
" " "	0.068 54	slugs (mass)
" " "	2.205	pounds-mass
kilograms per cubic meter	0.062 43	pounds per cubic foot
kilograms per square meter (force)	1.422×10^{-3}	pounds per square inch (force)
kilometers	3281	feet
"	0.6214	miles
"	10^3	meters
kiloponds (kgf)	9.807	newtons
kilowatts	10^3	watts
kilowatt-hours	3.600×10^6	joules
kips (1000 lbf)	4.448×10^3	newtons
kips per square inch	6.895×10^6	pascals
knots (international)	1.151	miles per hour
" "	0.5144	meters per second
lamberts	3183	candelas per square meter
leagues (nautical)	5556	meters
leagues (U.S. survey)	4828	meters
light years	9.461×10^{15}	meters
liters	10^{-3}	cubic meters
"	0.035 31	cubic feet
"	0.2642	gallons (U.S. liquid)
"	10^3	cubic centimeters
lumens (luminous flux)	1	candela-steradians
lumens per square foot	1	foot-candles
lumens per square meter	1	lux
lux (illuminance)	1	lumens per square meter
lux (lm/m²)	0.0929	foot-candles
mass: lb	0.4536	kilograms (mass)
maxwells	10^{-8}	webers
meters	1.094	yards
"	3.281	feet
"	39.37	inches
"	6.214×10^{-4}	miles (U.S. survey)
meters per second	3.281	feet per second
metric carats	2×10^{-4}	kilograms
metric tons (tonnes)	10^3	kilograms
mhos	1	siemens
microns	10^{-6}	meters
miles (nautical)	1852	meters
miles (U.S. survey)	1609	meters
" " "	5280	feet
" " "	1.609	kilometers
" " "	1760	yards

(*Continued*)

Multiply	By	To Obtain
miles per hour	88	feet per minute
" " "	0.8688	knots (international)
milliamperes	10^{-3}	amperes
millibars	10^2	pascals
millimeters	0.039 37	inches
"	10^{-3}	meters
millimeters of mercury (0°C)	133.3	pascals
millivolts	10^{-3}	volts
mils	10^{-3}	inches
miner's inches	1.5	cubic feet per minute
minutes (angle)	2.909×10^{-4}	radians
newtons	1	kilograms-meters per second per second ($\text{kg}\cdot\text{m/s}^2$)
"	0.2248	pounds-force
"	10^5	dynes
"	0.1020	kilograms-force
"	7.233	poundals
newton-meters	0.7376	pound-feet (force)
oersteds	79.58	amperes per meter
ohms (electric resistance)	1	volts per ampere
ounces (troy)	0.083 33	pounds (troy)
" "	1.097	ounces (avoirdupois)
ounces-force	0.2780	newtons
" "	28.35	grams-force
" "	0.0625	pounds-force
ounces-force (troy)	31.10	grams-force
parsecs	3.086×10^{16}	meters
pascals (pressure, stress)	1	newtons per square meter
"	0.9872×10^{-5}	atmospheres
"	2.953×10^{-4}	inches of mercury (32°F)
"	7.501×10^{-3}	millimeters of mercury (0°C) (torr)
"	1.450×10^{-4}	pounds per square inch (force)
pecks (U.S.)	8.810×10^{-3}	cubic meters
pennyweights	1.555×10^{-3}	kilograms (mass)
picas (printer's)	4.218×10^{-3}	meters
pints (U.S. liquid)	4.732×10^{-4}	cubic meters
" (U.S. dry)	5.506×10^{-4}	cubic meters
points (printer's)	3.515×10^{-4}	meters
poises (absolute viscosity)	10^{-1}	pascal-seconds
poundals	0.1383	newtons
"	1.383×10^4	dynes
"	0.031 08	pounds-force
pounds (avoirdupois)	7000	grains
pounds (troy)	0.8229	pounds (avoirdupois)
" "	5760	grains
pound-feet (force)	1.356	newton-meters

(Continued)

Multiply	By	To Obtain
pounds-force (lbf)	453.7	grams-force
" " "	16	ounces-force
" " "	32.18	poundals
" " "	4.448	newtons
pounds-mass (lb)	0.4536	kilograms (mass)
pounds per cubic foot	16.02	kilograms per cubic meter
pounds per square inch (force): psi	0.068 03	atmospheres
" " " " " "	2.036	inches of mercury
" " " " " "	6895	pascals
" " " " " "	6.895×10^{-3}	megapascals
pressure: psi	6895	pascals
pressure: atmospheres	1.013×10^5	pascals
quarts (U.S. liquid)	9.464×10^{-4}	cubic meters
" " "	0.2500	gallons (U.S. liquid)
radians	57.30	degrees (angle)
"	63.65	grads
"	0.1592	revolutions
rads (radiation dose absorbed)	10^{-2}	joules per kilogram (grays)
rods (U.S. survey)	16.5	feet
roentgens	2.580×10^{-4}	coulombs per kilogram
revolutions	2π	radians
sections (U.S. survey)	640	acres
" " "	2.590×10^6	square meters
siemens (electric conductance)	1	amperes per volt
"	1	mhos
slugs (mass)	14.59	kilograms (mass)
square centimeters	10^{-4}	square meters
statamperes	3.336×10^{-10}	amperes
statcoulombs	3.336×10^{-10}	coulombs
statfarads	1.113×10^{-12}	farads
stathenries	8.988×10^{11}	henries
statohms	8.988×10^{11}	ohms
statvolts	299.8	volts
steres	1	cubic meters
stokes (kinematic viscosity)	10^{-4}	square meters per second
tablespoons	1.479×10^{-5}	cubic meters
teaspoons	4.929×10^{-6}	cubic meters
temp. (degrees Celsius) + 273.15	1	absolute temp. (kelvin)
" " " + 17.78	1.8	temp. (degrees Fahr.)
temp. (degrees Fahr.) + 459.67	1	absolute temp. (deg. Rankine)
" " " − 32	5/9	temp. (degrees Celsius)
teslas (magnetic flux density)	1	webers per square meter
teslas	10^4	gausses
therms	10^5	British thermal units (Btu)
tonnes (metric tons)	10^3	kilograms

(Continued)

Multiply	By	To Obtain
tons (long)	1016	kilograms
" "	2240	pounds
tons (metric)	10^3	kilograms
" "	2205	pounds
tons (short)	907.2	kilograms
" "	2000	pounds
tons (of refrigeration)	1.2×10^4	British thermal units per hour
tons (nuclear equivalent of TNT)	4.184×10^9	joules
torr (mm Hg, 0°C)	133.3	pascals
volts (electric potential)	1	watts per ampere
watts (power)	0.056 88	British thermal units per minute
" "	10^7	ergs per second
" "	1.341×10^{-3}	horsepower
" "	1	joules per second
watt-hours	3600	joules
webers (magnetic flux)	1	volt-seconds
"	10^8	maxwells
yards	0.9144	meters
"	3	feet

Answers to Problems

■ CHAPTER 3

3.5.5 **(c)** $x = 5.657$ **(d)** $x = 0.7155$ **(e)** $x = 5.687$
3.5.10 **(a)** 9.982 m   **(b)** 96.54 km/h   **(c)** 3856 kg   **(d)** 3.781×10^4 N
      (e) 19.82 m^3 **(f)** 1.055×10^6 J   **(g)** 2.069×10^8 Pa   **(h)** 21.1°C
      (i) 393 K   **(j)** 3.729×10^5 W   **(k)** 4085 kg

■ CHAPTER 7

7.1.1 **(a)** 2.9×10^7 **(b)** 4.37×10^3 **(c)** 1.39×10^{-1}
      (d) 1.376×10^0 or just 1.376   **(e)** 1.32×10^{-4} **(f)** 10^{-3} **(g)** 10^6
      (h) 6.93×10^{-7}
7.1.2 **(b)** 2259 mi/hr   **(d)** 1.613×10^4 mm^2 **(e)** 40.39 mi/hr
      (f) 285.3 m/d
7.2.1 **(a)** 1084   **(b)** 109.71   **(c)** 0.396   **(d)** 9.12×10^3 **(e)** 1.443
      (f) 0.0056   **(g)** 137   **(h)** 1.86×10^{-5}
7.3.6 slope $= 1.303$ **7.3.7** slope $= 1.303$
7.4.1 $m = 2.44$; $b = -5.33$; thus, $y = 2.44x - 5.33$ **7.4.2** $y = 2.2x^{2.8}$
7.4.3 $y = 1.8e^{3.5x}$

■ CHAPTER 8

8.1.2 $\bar{X} = 9.86$; median $= 9.75$; mode $= 9$; best measure is probably the mode because
      it occurs four times and represents an actual shoe size.
8.1.3 mean $= 44.25$; median $= 40$

Note: Answers are given to those problems that have numerical solutions.

8.1.4 $\bar{X} = 17.6$; median $= 18$; mean is *less* than the median because the group is left-skewed.

8.1.5 $\bar{X} = \$25,800$; median $= \$23,850$; median represents the data better because 10 of the 12 people earn less than the mean.

8.1.6 weighted arithmetic mean $= 2,000$ sec

8.2.2 range $= 3.5$; mean dev. $= 1.0$; $\sigma = 1.11$

8.2.3 range $= 60$; mean dev. $= 15.6$; $\sigma = 18.3$

8.2.4 range $= 10$; mean dev. $= 1.49$; $\sigma = 2.33$

8.2.5 Group A: $\bar{X} = 11.29$; median $= 10$; $\sigma = 4.62$
Group B: $\bar{X} = 8.57$; median $= 10$; $\sigma = 4.03$
Group A is skewed to right; Group B is skewed to left.

8.2.6 range $= 0.0040$ in.; weighted arith. mean $= 2.4991$ in.; $\sigma = 0.0011$ in.

8.2.7 $\bar{X} = 972.5$ kWh; $\sigma = 331$ kWh

8.5.1 .95 confidence interval $= 200\,900\,\Omega \pm 243\,\Omega$
.99 confidence interval $= 200\,900\,\Omega \pm 320\,\Omega$

8.6.1 For $\bar{X}$-chart: UCL $= 21.6\,\Omega$; CL $= 19.9\,\Omega$; LCL $= 18.2\,\Omega$
For R-chart: UCL $= 5.25\,\Omega$; CL $= 2.3\,\Omega$; LCL $= 0$

8.7.1 $y_e = 2.8 + 5.21x$ **8.7.2** $y_e = 0.127 + 0.83x$ **8.8.1** $r = 0.996$

8.8.2 $r = 0.56$

■ CHAPTER 9

9.1 **(b)** 3.785 L/gal   **(c)** 0.4047 ha/acre   **(d)** 1.316×10^{-3} atm/mm Hg
(e) 12.97 ft-lbf/sec $= 1$ Btu/min   **(f)** 0.3048 m/s$^2 = 1$ ft/sec^2
(g) 175.08 kg $= 1$ lb-sec^2/in.   **(h)** 2.178×10^{-7} kg/m$^3 = 1$ ton/(mi)^3
(i) 1.102 ton/tonne   **(j)** 3.600×10^9 J/MW·h   **(k)** 9.552 rpm $= 1$ rad/s
(l) 1.341 hp/kW   **(m)** 6.895×10^{-6} GPa/psi

9.2 **(b)** 311 K   **(c)** 3.89 K   **(d)** $-269.3°C$ **(e)** 1922 K

9.4 **(a)** 0.031 slugs; 0.452 kg   **(b)** 6211 slugs; 90 620 kg
(c) 0.0485 slugs; 0.708 kg   **(d)** 0.259 slugs; 3.78 kg

9.5 **(b)** 413.2 nm   **(c)** 7.065 kPa   **(d)** 7.065 kPa   **(e)** 139.3 A/m
(f) 186.3×10^{-8} Wb   **(g)** 0.173 S   **(h)** 5.736×10^{-2} N
(i) 5.736×10^{-4} J

9.6 **(b)** 17.4 kJ   **9.8** 0.2 C; 0.2 A   **9.9** 10 Gy; 10 J/kg

9.10 **(a)** 981 N   **(b)** 445 N   **(c)** 9810 N

9.11 **(a)** 3760×10^{21} nW   **(b)** 92×10^6 km   **(c)** 8377×10^{-9} kN
(d) 0.376×10^{-3} MJ   **(e)** 76.4 kW   **(f)** 4400×10^{-6} m
(g) 33×10^{-6} m^2 **(h)** 1.010 L   **(i)** 0.530 mA

9.12 $\tau = 2.01$ s   **9.14** K.E. $= 13.5$ kg·m^2/s$^2 = 13.5$ N·m $= 13.5$ J

9.15 K.E. $= 9.95$ ft-lbf   **9.16** K.E. $= 3.88$ ft-lbf   **9.17** K.E. $= 5.27$ J

9.18 $Q = 49.3 \times 10^3$ J   **9.19** $Q = 46.8$ Btu

Note: Answers are given to those problems that have numerical solutions.

9.20 power $= 34.3$ hp $= 25\,600$ J/s $= 25\,600$ W   **9.21** 11 250 psi

9.22 $\sigma = 77.52$ MPa

■ CHAPTER 11

11.5.2 0.585″   **11.5.3** 2.500; 2.750; 0.125″ clearance on radius

11.5.4 0.041″; 0.059″

11.6.1 2.749/2.746; 2.753/2.750; min. looseness $= 0.001″$; max. looseness $= 0.007″$

11.6.2 1.028″ min.

■ CHAPTER 14

14.2.3 **(b)** $\mathbf{B} = (19.4\mathbf{i} + 72.4\mathbf{j})$ N   **(c)** $\mathbf{C} = (-70.7\mathbf{i} - 70.7\mathbf{j})$ lb

  (d) $\mathbf{D} = -60\mathbf{j}$ N

14.2.4 **(b)** $\mathbf{C} + \mathbf{D} = (14\mathbf{i} - 14\mathbf{j})$ N; 19.8 N; 45°   **(c)** $\mathbf{R}_1 + \mathbf{R}_2 + \mathbf{R}_3 = 0$

14.2.5 $F_x = 5.80$ N; $F_y = 1.55$ N   **14.2.6** $R = 8.63$ N; angle with 6-N force $= 11.5°$

14.2.7 $F_{O-A} = 5.96$ N; $F_{O-B} = 4.51$ N   **14.2.8** $F_{O-A} = 11.78$ N; $F_{O-B} = 5.49$ N

14.2.9 $F_x = 34.3$ lb; $F_y = 54.1$ lb   **14.2.10** $F_x = 64.3$ N; $F_y = 76.6$ N

14.2.11 $F_{\text{tension}} = 65.3$ N; $F_{\text{shear}} = 44.6$ N   **14.2.12** $F_x = 65.5$ lb; $F_y = -45.9$ lb

14.2.13 perpendicular $= 0.940W$; parallel $= 0.342W$

14.2.14 perpendicular $= 97.6$ N; parallel $= 21.6$ N compression

14.2.15 $\mathbf{R} = (-58.9\mathbf{i} - 33.5\mathbf{j})$ lb; 67.8 lb; 29.6°   **14.3.5** $M_A = 0.866FL$ clockwise

14.3.6 $M_A = M_B = 0$ **14.3.7** $F = 8.31$ N   **14.5.4** $R_A = 0$; $R_B = 24$ N

14.5.5 horizontal $= 0.866W$; vertical $= 1.5W$ **14.5.6** $W = 177.8$ lb

14.5.7 $T = 0.78W$ **14.5.8** $\theta = 16.7°$

14.5.9 front $= 1236$ lb; rear $= 2118$ lb; rear axle limit exceeded by 38 lb

14.5.10 $F = W \sin \theta$

14.6.1 **(a)** At A: horiz. $= 0$; vert. $= 1333$ lb upward

  At F: horiz. $= 0$; vert. $= 667$ upward

  (b) AB: 1666 lb (comp.); AC: 1000 lb (tens.); BD: 500 lb (comp.);

  BE: 834 lb (comp.); CE: 1000 lb (tens.); DF: 834 lb (comp.);

  EF: 500 lb (tens.); BC: 2000 lb (tens.); DE: 667 lb (tens.)

14.6.2 AB: 5000 lb (tens.); CB: 7072 lb (comp.); CD: 0; AC: 5000 lb (tens.);

  BD: 5000 lb (tens.)

14.6.3 At A: horiz. (to the right) $= 10\,000$ N; vert. (upward) $= 12\,500$ N

  At D: horiz. $= 0$; vert. (downward) $= 12\,500$ N

  AB: 12 500 N (comp.); BC: 16 000 N (tens.); AC: 10 000 N (comp.);

  CD: 12 500 N (tens.)

14.6.4 At A: horiz. $= 0$; vert. $= 15\,710$ lb (upward)

  At N: horiz. $= 0$; vert. $= 14\,290$ lb (upward)

  DE: 24 750 lb (comp.); JK: 28 560 lb (tens.)

14.7.3 in first 45 sec, $\Delta s = 1945$ ft; next 45 sec, $\Delta s = 5945$ ft; $v = 176$ ft/sec

Note: Answers are given to those problems that have numerical solutions.

14.7.4 500 sec   **14.7.5** 1.55 sec   **14.7.7** 262.8 mph; 12.4° west of north
14.7.8 **(a)** 1320 ft/min^2 **(b)** 50 160 ft   **14.8.1** $F = 39.3$ N; $a = 9.807$ m/s^2
14.8.2 $W = 8.82$ lbf; $m = 0.274$ slugs; $a = 32.2$ ft/sec^2 **14.8.3** $F = 54.3$ lb
14.8.4 2.43 m
14.8.5 **(a)** $T = W$; net force $= 0$ **(b)** $T = 0$; net force $= W$, downward
(c) $T_1 = 2W$; net force $= W$, upward
14.8.6 **(a)** $P = 200$ lb   **(b)** $P = 200$ lb   **(c)** $P = 200$ lb   **(d)** $P = 0$
(e) $P = 300$ lb

14.8.7 **(a)** 5311 lb   **(b)** 5000 lb   **14.8.8** $a = \dfrac{m_2 - m_1 \sin \alpha}{m_1 + m_2} g$

14.8.9 $S = m_2 \mathbf{a}$ **14.8.10** 25.2 m/s

■ CHAPTER 15

15.1.3 $\Delta L = 7 \times 10^{-6}$ m   **15.1.4** $\Delta L = 2.76 \times 10^{-4}$ in.   **15.1.5** $d = 0.357$ in.
15.1.6 $d = 0.00675$ m   **15.2.1** 8997 psi at root of beam; 0.477 in.
15.2.2 107.9 MPa; 32.4×10^{-3} m   **15.2.3** 4265 psi at center of beam; 0.167 in.
15.2.4 37.27 MPa; 8.09×10^{-3} m   **15.2.5** 1799 psi; 0.037 in.
15.2.6 40.00 MPa; 6.67×10^{-3} m   **15.3.1** 2.67 in.   **15.3.2** 5.66 in.
15.3.3 0.1137 m
15.3.4 **(a)** 14 690 psi; 0.417 in.   **(b)** stress does not change, but $\Delta = 1.209$ in.
15.3.5 12″ × 5″ beam weighing 31.8 lb/ft is smallest I-beam that will work
15.3.6 24″ × 7″ beam weighing 79.9 lb/ft will work
15.3.7 5″ × 3″ beam weighing 10 lb/ft will work

■ CHAPTER 16

16.1.4 $\Delta T = 0.37$ K $= 0.66$°F   **16.1.5** 64.7 m   **16.1.6** $v = 80.2$ ft/sec
16.1.7 1.77×10^8 J; efficiency $= 0.78$ **16.2.3** $V_2 = 40.6$ ft^3
16.2.4 36 200 Btu   **16.2.5** 246°F $= 392.3$ K   **16.2.6** no change
16.2.7 1.46 J/h; coeff. of performance $= 2.62$ **16.3.2** 22.76×10^3 J
16.3.3 90 000 ft-lb

■ CHAPTER 17

17.1.4 $C = 2.73 \times 10^{-5}$ F   **17.1.5** $C = 2.48 \times 10^{-3}$ F   **17.1.6** $L = 156.3$ H
17.1.7 $i = 1.1$ A; $p = 121$ W   **17.1.8** no more than 17 bulbs
17.1.9 no more than 2 bulbs   **17.1.10** $q = 0.5$ C   **17.1.13** $RC = 0.20$ s
17.1.15 $t = 6$ s   **17.1.16** v(at $t = 0.005$ s) $= 0.84$ V; v(at $t = 0.01$ s) $= 1.48$ V
17.1.17 v (at $t = 0.005$ s) $= 0.79$ V; v (at $t = 0.01$ s) $= 1.50$ V

Note: Answers are given to those problems that have numerical solutions.

17.1.18 98 W; 123.5 Ω **17.2.2** 1815 Ω **17.2.3** 14.15 Ω
17.2.4 $R_{eq} = 10\ \Omega$; i (each resistor) = 1 A; i (total) = 10 A   **17.2.5** 7.19 Ω
17.2.6 $i_1 = 0.572$ A; $i_2 = 0.176$ A; $i_3 = 0.230$ A;
current in 6-Ω and 7-Ω resistors = i_1;
current in upper 3-Ω resistor and 5-Ω resistor = i_2;
current in right-hand 3-Ω resistor and 2-Ω resistor = i_3;
current in left-hand 3-Ω resistor = $i_1 - i_2$;
current in middle 4-Ω resistor = $i_2 - i_3$;
current in left-hand 4-Ω resistor = $i_1 - i_3$
17.2.7 $R_{eq} = 1357\ \Omega$ **17.2.8** $R_1 = 392\ \Omega$
17.2.9 $R/2$; it decreases   **17.2.10** $R_{eq} = 41.8\ \Omega$
17.2.11 (a) $R_{eq} = 20\ \Omega$ (b), (c) same as for (a)   (d) $R = 10\ \Omega$
17.2.12 $R = 4.36\ \Omega$ **17.3.2** inductance dominates; current lags voltage
17.3.3 inductance dominates; sketch is exactly like that in Figure 17.17
17.4.1 $V_L/V_S = 40$ **17.4.2** $R_L = 1250\ \Omega$ **17.4.3** $V_L = 3$V; $i_C = 3$ mA

■ CHAPTER 18

18.4.1 (a) Y = A*X**2 + B*X + C
(b) R = (3*X**2 + 2)/(X + 1)
(c) T = (A*X**3 + B*X**2)/(X + C)
(d) A = A + 100
(e) C = (A**2 + B**2)**0.5
(f) S = X/((SIN (X))**0.5 + 1)
18.4.2 BA, AB, YY, Y, Y4, X45
18.4.3 (a) IF (R.GT. 0.0) THEN
GOTO 2
ENDIF
(b) IF (R.GE. 0.0) THEN
GOTO 250
ENDIF
(c) IF ((B**2 − 4*A*C) .LT. 0.0) THEN
GOTO 300
ENDIF
(d) IF (ABS(X**2 − A) .LT. 0.001) THEN
GOTO 800
ENDIF

■ CHAPTER 19

19.2.1 net prof. margin = 7.82%; return on stockholders' equity = 13.98%; return on
total capital = 12.89%

Note: Answers are given to those problems that have numerical solutions.

19.2.2 *Assets:*   *Liabilities:*

Cash	$ 0	Loan	$50,000
Furn., etc.	80,000	*Net Worth:*	
		Stock	50,000
		Loss	(20,000)
	$80,000		$80,000

19.3.1 $1411.58   **19.3.2** $1431.41   **19.3.3** $1433.26   **19.3.4** $22,253.46

19.3.5 $4493.68   **19.3.6** $26.50   **19.3.7** $242,251.43

19.3.8 $161.33 monthly; $5807.88 total   **19.3.9** $439.58 monthly; $5274.96 total

19.5.1 5 years; marginally worthwhile   **19.5.2** 2.89 years

19.5.3 alternatives essentially even   **19.5.4** annual ret. on investment $= 20\%$

19.5.5 break-even point: $i = 4.1\%$

Note: Answers are given to those problems that have numerical solutions.

Index

Greek Alphabet

A	α	alpha	N	ν	nu
B	β	beta	Ξ	ξ	xi
Γ	γ	gamma	O	o	omicron
Δ	δ	delta	Π	π	pi
E	ε	epsilon	P	ρ	rho
Z	ζ	zeta	Σ	σ	sigma
H	η	eta	T	τ	tau
Θ	θ	theta	Y	υ	upsilon
I	ι	iota	Φ	ϕ	phi
K	κ	kappa	X	χ	chi
Λ	λ	lambda	Ψ	ψ	psi
M	μ	mu	Ω	ω	omega

Formulas:

$$a^x a^y = a^{x+y}$$

$$(a^x)^y = a^{xy}$$

$$a^{1/y} = \sqrt[y]{a}$$

$$a^{x/y} = \sqrt[y]{a^x}$$

$$\frac{a^x}{a^y} = a^{x-y} \quad (\text{if } a \neq 0)$$

$$a^{-x} = \frac{1}{a^x} \quad (\text{if } a \neq 0)$$

$$a^0 = 1 \quad (\text{if } a \neq 0)$$

if $\sqrt[x]{a} = b$, then $a = b^x$

$$(ab)^x = a^x b^x$$

$$\left(\frac{a}{b}\right)^x = \frac{a^x}{b^x} \quad (\text{if } b \neq 0)$$

$$\log_{10} x = \log x$$

$$\log_e x = \ln x$$

$$\log_a xy = \log_a x + \log_a y$$

$$\log_a \frac{x}{y} = \log_a x - \log_a y$$

$$\log_a x^k = k \log_a x$$

if $\ln x = y$, then $x = e^y$ (where $\ln x = \log_e x$)

if $\log_a x = y$, then $x = a^y$

Binomial Theorem

$$(a \pm b)^n = a^n \pm na^{n-1}b + \frac{n(n-1)}{2!} a^{n-2}b^2$$
$$\pm \frac{n(n-1)(n-2)}{3!} a^{n-3}b^3 + \ldots$$

Arithmetic Progression

$$a, \quad a+d, \quad a+2d, \quad a+3d, \ldots$$

Geometric Progression

$$a, \quad ar, \quad ar^2, \quad ar^3, \ldots$$

Quadratic Equation

If $ax^2 + bx + c = 0, \quad a \neq 0$

then $x = \dfrac{-b \pm \sqrt{b^2 - 4ac}}{2a}$

Any Triangle

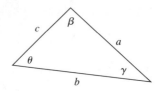

Law of Sines:

$$\frac{a}{\sin \theta} = \frac{b}{\sin \beta} = \frac{c}{\sin \gamma}$$

Law of Cosines:

$$a^2 = b^2 + c^2 - 2bc \cos \theta$$

Sum of Angles:

$$\theta + \beta + \gamma = 180° \quad \text{(for any triangle)}$$